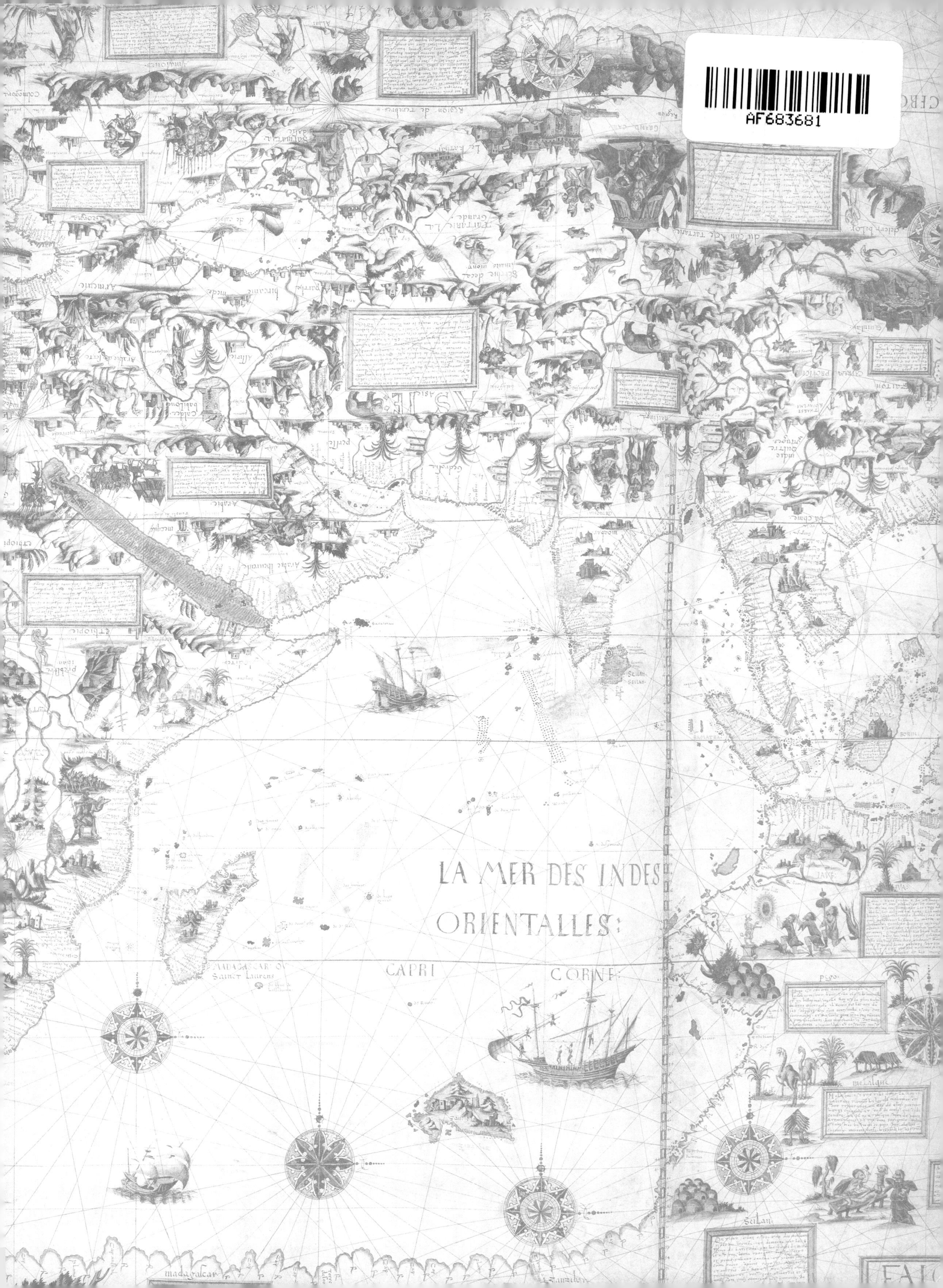
LA MER DES INDES
ORIENTALLES:
CAPRI
CORNE:
MADAGASCAR OV
Saint Laurens

IN SEARCH OF UTOPIA

ART AND SCIENCE IN THE ERA OF THOMAS MORE

Jan Van der Stock

FOR VERONIQUE VANDEKERCHOVE (1965–2012)
Chief Curator of M-Museum Leuven and initiator
of this exhibition

FOR JAN ROEGIERS (1944–2013)
Emeritus Professor of History at the KU Leuven and
a source of inspiration for this exhibition

IN SEARCH OF UTOPIA

ART AND SCIENCE IN THE ERA OF THOMAS MORE

Jan Van der Stock

DAVIDSFONDS UITGEVERIJ

Amsterdam University Press

AUTHORS

[CVD] Chet Van Duzer
[CK] Cecile Kruyfhooft
[DVH] Daan van Heesch
[EDP] Els De Palmenaer
[EM] Elizabeth Morrison
[EV] Emmanuelle Vagnon
[HI] Hannah Iterbeke
[JH] Jan Herman
[JL] Jeroen Luyckx
[JLB] Jens Ludwig Burk
[JP] Jan Papy
[JS] Jochen Sander
[JVG] Joris Van Grieken
[KB] Koenraad Brosens
[KS] Katharina Smeyers
[KVC] Koenraad Van Cleempoel
[LC] Lorne Campbell
[LMM] Linda M. Müller
[LW] Lieve Watteeuw
[MAC] Miguel Ayres de Campos
[MB] Maarten Bassens
[MDS] Marcus de Schepper
[MF] Malve Anna Falk
[MU] Matthias Ubl
[PVDB Paul Vandenbroeck
[RJ] Robrecht Janssen
[SH] Stephanie Heremans
[SMK] Scot McKendrick
[SVDE] Stéphanie Van den Eynde
[WW] Wendy Wauters

Contents

Catalogue

UTOPIA OF THOMAS MORE (1516)

A Golden Book from Leuven Conquers the World

HOLDING THE UNIVERSE

Dreaming of Space and Time

Foreword

On 20 October 2016, the *In Search of Utopia* exhibition opens in M-Museum Leuven. It brings together an extraordinary number of exceptional works of art, most of them by masters who were active in Flanders and Brabant in the late fifteenth and early sixteenth century. For both the City of Leuven and University of Leuven the exhibition commemorates the fact that December 2016 is the 500th anniversary of the publication of *Utopia*, written by the English humanist and statesman Thomas More. Dirk Martens, the master printer from Aalst and Antwerp who, at the time, had a printing press in the centre of Leuven, printed the book and marketed it. The renowned humanist Desiderius Erasmus came up with the title and acted as intermediary between his good friends More and Martens.

In his soon-to-be-famous book More brilliantly analysed the lamentable social and economic conditions prevailing in England at that time before describing a 'utopian' alternative that was to be found on an island where everything was 'perfectly' regulated. *Utopia* was not intended as a viable political programme; it was an intellectual exercise – one, moreover, that touched on many subjects that are still highly topical today. The celebration of *500 Years of Utopia* also sees the launch of a citywide project in which a range of themes are explored from both international and local perspectives.

Preparations for the ambitious exhibition project began in 2009 under the inspirational leadership of Professor Jan Van der Stock and the late Veronique Vandekerchove (1965–2012). They were joined by Lien De Keukelaere, Annelies Vogels, Hanna van Zutphen, Professor Barbara Baert and Professor Lieve Watteeuw. Together they coaxed loans from national and international lenders, mobilized funders and sponsors, and enthused a team of highly motivated colleagues and associates.

The exhibition was of course preceded by considerable academic research. This was carried out under the aegis of Illuminare – Centre for the Study of Medieval Art (KU Leuven) and focused on the works of art that would come to Leuven. The extensive examination of Mechelen's *Besloten Hofjes*, the systematic study of compositions from the tradition of Hieronymus Bosch, the analysis and conservation of the unique, large-scale city portrait of Leuven and various other objects, are the subjects of doctoral dissertations, master's theses and bachelor's papers. The results are published in the present exhibition catalogue and will be presented to the international academic community at the *Imaging Utopia: New Perspectives on Northern Renaissance Art* symposium. The aim of the exhibition is to make recent academic and scientific research accessible to the general public and at the same time to highlight Leuven as a dynamic and vibrant city.

The organization of *In Search of Utopia* is in the joint hands of KU[N]ST Leuven vzw, in which the City of Leuven and the University of Leuven collaborate, and the museum team of M-Museum Leuven. We thank everyone who has contributed to making the complex organization of the exhibition a success.

LOUIS TOBBACK

Mayor of the City of Leuven

RIK TORFS

Rector of KU Leuven

KATLIJN MALFLIET

Vice-rector of Culture, KU Leuven

Co-chair KU[N]ST Leuven vzw

DENISE VANDEVOORT

Alderman for Culture of the City of Leuven

Chair of M-Museum, Leuven

Co-chair of KU[N]ST Leuven vzw

In Search of Utopia — The Exhibition

JAN VAN DER STOCK

Utopia. A word whose meaning has, down the centuries, caromed between naive dream and powerful action. Half a millennium ago the isle of Utopia was conceived by the European humanists Thomas More[1] and Desiderius Erasmus,[2] to give their ideal but entirely fictional place a name and a tangible shape. Utopia is located 'nowhere', far beyond the horizon.[3] And though the perfection of that island is obviously an illusion, the dream of an ideal world is certainly not. It's a very real idea. The desire for it acts like a powerful engine, setting people in motion with the promise of happiness but also with the risk of disappointment and failure. This is the essence of the story: the journey to Utopia is one of repeated attempts; the isle itself recedes and is always out of reach. Either the dream remains intact or it becomes a nightmare.

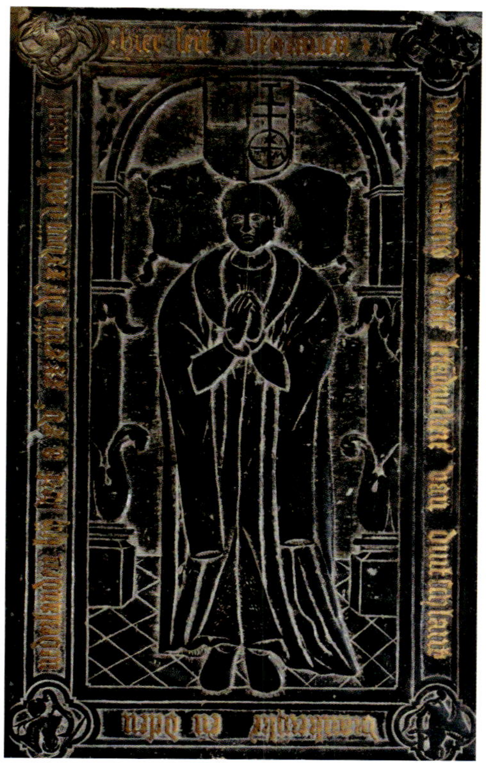

[fig. 1] Grave-slab of Dirk Martens with his only known portrait, 1534. The slab is now incorporated in an eighteenth-century monument, Aalst, Church of St Martin.

Five hundred years ago, in mid-December 1516, Dirk Martens, Leuven's and indeed Flanders's foremost printer, published a book called *Utopia*. That slim volume symbolically marked the start of a new era in European intellectual thought.[4] The booklet's full title is *On the Best State of a Commonwealth and on the New Island of Utopia. A Truly Golden Handbook, No Less Beneficial than Entertaining.* Martens had published his first book in Aalst in 1473 (fig. 1). Later he founded a printing house in Antwerp, before finally moving to Leuven.[5] His premises were on the corner of the present Naamsestraat and Standonck-

1
Thomas More, *Thomas Morus* in Latin (1475–1535), is the author of *Utopia*. More was an English humanist, lawyer and statesman. In 1499 he met Erasmus (cat. 13A and 14A), who often stayed with him. It was during one such visit that Erasmus wrote *In Praise of Folly* (*Moriae Encomium* in Latin, explicitly referring to More). The two men wrote to each other frequently. According to his own words, More was in Leuven for a short time in 1508; but his presence cannot be confirmed by primary sources. In 1515 More was in Bruges as an envoy of Henry VIII. He also visited a number of other places, including Antwerp, where he called on his friend Pieter Gillis. It was there, in Antwerp, that he wrote *Utopia*. In 1529 More became Lord Chancellor of England, a post he held until 1532. In 1535 he was executed, having fatally clashed with Henry on the principle of papal authority. Four hundred years later he was canonized by the Catholic Church. For a good biography of More, see Ackroyd 1998.

2
Desiderius Erasmus of Rotterdam (1466/1469–1536) was the most famous Dutch humanist of his day. He was also a writer, philosopher, priest and Augustinian canon. In 1502 he was offered and declined the position of professor at Leuven's university. His best known works are *In Praise of Folly*, dedicated to his friend Thomas More, and the *Enchiridion*, in which he examines the meaning of Christianity for man. For a good biography, see Schoeck 1990–1993 and Jardine 1991.

3
The original title of More's text was *Nusquama* or 'Nowhere'. The eventual title, *Utopia*, was suggested by Erasmus.

4
See the essay on this topic by Jan Papy in this catalogue (pp. 31–39).

5
In 1473/74 Martens worked with Joannes de Westfalia. He died in Aalst on 28 May 1534. Rouzet 1975: 140–143 and Heireman 1973.

straat, just across the road from the building now occupied by the University Halls.[6] And there he gradually became the printer-publisher of choice for several of the celebrated humanists who taught and conducted their research in Leuven in the early sixteenth century – a process in which his good friend Erasmus had a large hand. The cream of Europe's intelligentsia ran into one another on the banks of the Dyle.

The name Utopia is a play on words, typical of the humour of the time ('utopia' and 'utopian' are derived from the title of the book, not vice versa); a neologism, coined by fusing the Greek *eu-topos* or 'good place' and *ou-topos* or 'not place'. More introduces a mysterious narrator to tell the tale of the island of Utopia (fig. 2), one Raphael Hythlodaeus or 'nonsense merchant', 'peddler of nonsense' (more wordplay), who claims to have spent no less than five years there. This device enabled More to dissociate himself from the story

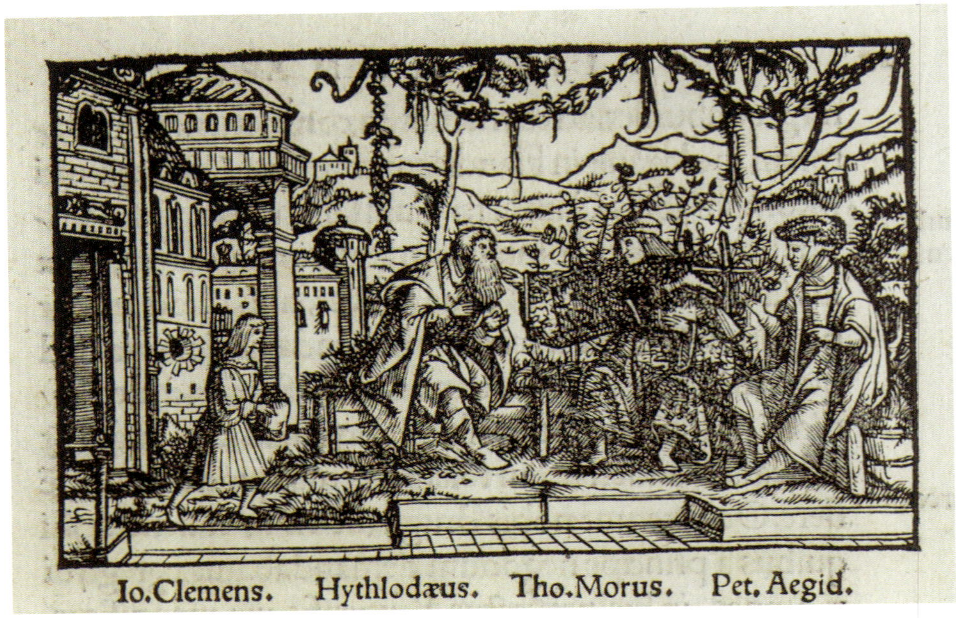

[fig. 2] Ambrosius Holbein, *Raphael Hythlodaeus tells Thomas More and Pieter Gillis the story of Utopia. The humanist John Clement, then More's servant, is also present*, in: Thomas More, *De optimo reip. statu, deque noua insula Utopia, libellus uere aureus*, fol. 25, Basel, 1518, KU Leuven, University Library (cat. 7.2).

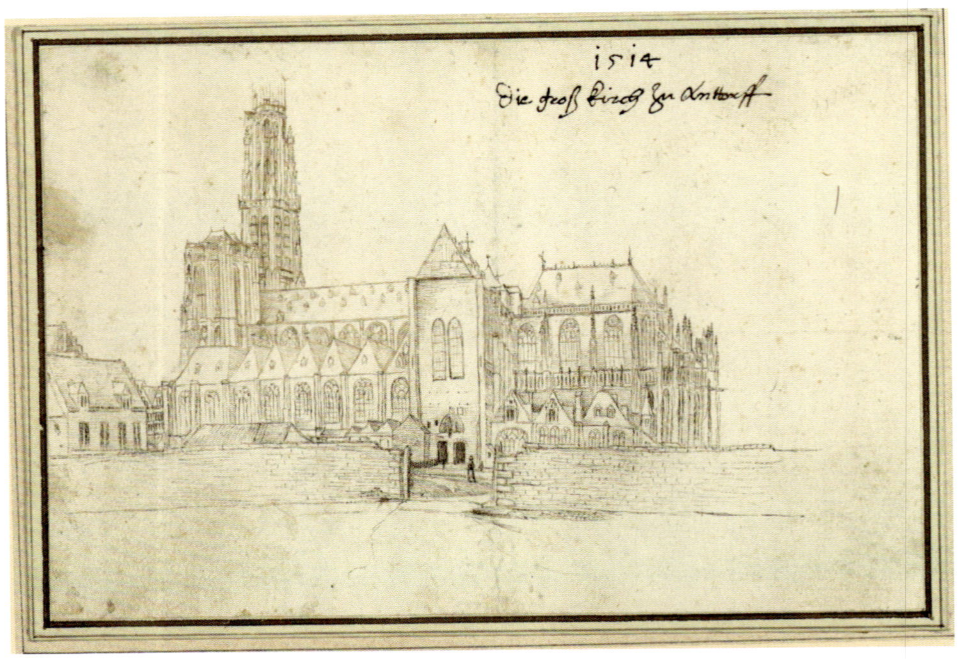

[fig. 3] *The Church of Our Lady at Antwerp* (with inscription *Die gross kirch zu Anttorff*, and date 1514, by Albrecht Dürer), 1507-1514, Vienna, Graphische Sammlung Albertina. It was outside this church that Thomas More set the fictitious meeting with Raphael Hythlodaeus.

6
In 1516 this building housed the Faculty of Law. Martens's Leuven printer's address (from 1512 to 1529) was *E regione Scolae juris civilis*.

7
Pieter Gillis, or Petrus Aegidius in Latin (1486–1533), was a humanist and Antwerp's town clerk. He was not only a good friend of Erasmus and More but also of other humanists such as Rodolphus Agricola and painters like Quinten Metsys. In *Utopia* it is he who introduces More to the fictitious traveller Raphael Hythlodaeus, who tells More all about the island.

in the book; after all, he is simply repeating hearsay. But Hythlodaeus seems very like a literary alter ego of Desiderius Erasmus. For in *Utopia* Erasmus is conspicuous by his absence even though much of what is said was directly inspired by him.

A central theme in humanist thought and writing was the question of whether man and the world are makeable. They found a positive answer, primarily in education and training. More needs only a few dozen pages to produce a sharply drawn picture of a society that now, five hundred years later, seems surprisingly modern in remarkably many respects. He describes the redistribution of wealth, childcare, free education for all, equal opportunities based on aptitude and ability, and democratic elections. He cites the importance of elegant and pleasant gardens in cities, advocates getting rid of red tape, and explains in some detail how working only six hours a day – three before noon and three after – is highly beneficial for Utopia. And, he adds neatly, Utopians invariably spend their spare time in meaningful intellectual activity, in which men and women are equal. In Utopia there is freedom of religion and euthanasia is a common practice. The Utopians have an unrivalled command of astronomy and have no truck with astrology. That this book came off the press in 1516 is astonishing. At other moments More is a man of his time, referring without a qualm to unlimited colonization and merciless capital punishment for those who don't agree with it. Then again, More can hardly be blamed for knowing nothing of the twentieth century's wretched history. In any case, *Utopia* is not a party manifesto. It's an intellectual experiment, a successful attempt to articulate a coherent utopia without taking into account all its practical consequences.

The starting point for the project in 2016 is a piece of Leuven history, namely the printing, in 1516, of the first edition of *Utopia*. But the exhibition, *In Search of Utopia*, ranges far beyond that straightforward fact. It explores the European fascination with man as an individual, man's dream of a better world and his curiosity about the boundless mystery of the universe, time, and the planets. It's a quest for knowledge, wealth and power, and – yes – for happiness as well. Around 1516 that dream took shape in the most diverse ways, illustrated in the exhibition by some ninety outstanding fifteenth- and sixteenth-century works of art.

In Search of Utopia is arranged in four chapters. The first, *Thomas More's 'Utopia'– A Golden Book from Leuven Conquers the World*, focuses on *Utopia* the book and the intellectual and cultural context in which it was produced. It is constructed around Thomas More's really rather unremarkable-looking 'golden' book (cat. 1), which yet acquires a sacred aura by being placed on a golden pedestal. A number of later editions and translations provide symbolic evidence of the speed with which More's ideas about the ideal state spread to a wide international audience (cat. 7.1, 7.2, 8, and 9). The story of *Utopia* the book is also the story of a close friendship, and its principals are naturally well represented: More himself, of course (fig. 4, cat. 3, 5 and 6), Erasmus, and the Antwerp town clerk Pieter Gillis, three friends and three extraordinarily incisive minds (cat. 13A–B and 14A–B).[7] Also occupying a place of honour is the earliest and unique portrait of the city of Leuven, which dates from around 1540 and in which the town unblushingly casts itself in the role of cultural and intellectual centre (cat. 2).

The second chapter, *Beyond Utopia – Images of Paradise and Hell*, follows the rhythm of passion, devotion, and the fear of death. All three are fundamental drivers of human enterprise that are seamlessly woven together. The passionate *Garden of Earthly Delights*, the crypto-erotic *Roman de la Rose* and the enigmatic *Festival of the Archers* by an anonymous Antwerp artist provisionally known as the Master of Frankfurt are the quintessence of this section (cat. 22, 24). The *Festival of the Archers*, which dates from 1493, is a key piece: a group

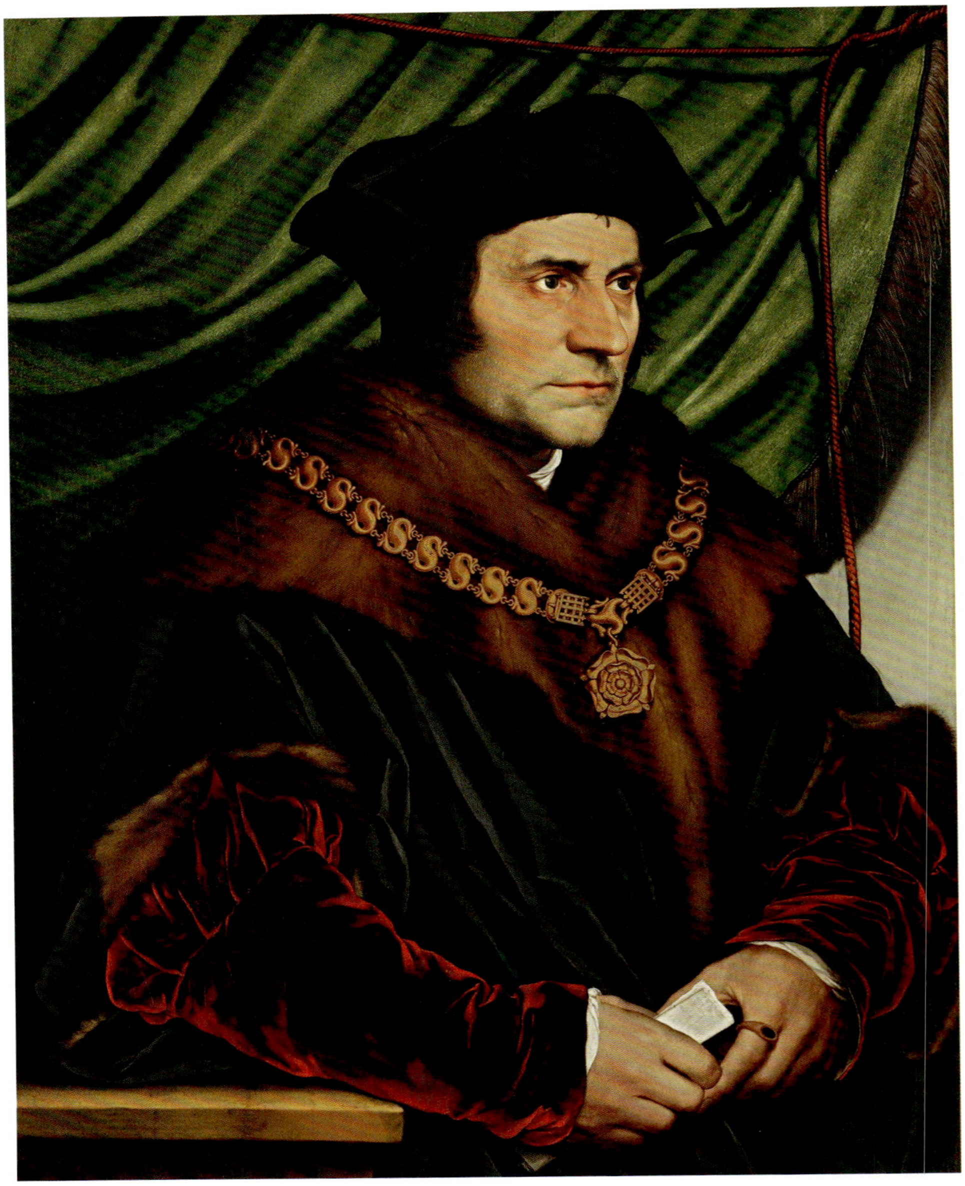

[fig. 4] Hans Holbein the Younger, *Portrait of Thomas More*, 1527. New York, Frick Collection.

of people are enjoying themselves in an enclosed garden, the *locus amoenus*, a 'pleasant place'. Outsiders try in vain to scale the fence. Armed guards man the gate, keeping out uninvited guests (fig. 5). The work that was produced more than half a millennium ago in Antwerp is almost uncannily meaningful today.

The courtly erotic *Festival of the Archers* and the *Roman de la Rose* have their religious counterparts in the three mysterious and mystical *Besloten Hofjes* or 'enclosed gardens' from Mechelen, made in the first decades of the sixteenth century (cat. 32–34). These highly detailed mixed-media devotional objects are *the* revelation of the entire exhibition. Artists – nuns, in this case – could hardly get any closer to the utopian ideal of paradise. These fascinating creations epitomize the perfect convergence of profound contemplation of the paradisal garden and unbridled imagination. The *Hofjes* were specially conserved for this exhibition by a team of eight specialists who have worked on them for more than three years. Besides Paradise on earth, sublimated in the *Besloten Hofjes*, a blood-red room houses the most feared anti-utopia: burning Hell, the symbolic representation of utter dehumanization and mortal fear of evil (cat. 36–41). There is no escape. Or is there?

The third chapter, *Beyond the Horizon – Imaging the Unknown*, is a search for *Terra Incognita*, the 'unknown land', the seductive and mysterious utopian world beyond the horizon. Creating an image of the unknown, the strange, gives solid form to the desire to go in search of it. Mythical unicorns, exotic animals and costly dyestuffs guaranteed to bring a profit are hoisted aboard European ships (cat. 46 and 54). All kinds of extraordinary monsters and bizarre wild men inhabit still unknown areas (cat. 49–51). When the French cartographer Pierre Desceliers in 1550 draws his *Mappa Mundi*, his map of the world, on parchment for the king of France the entire earth is literally within grasp (cat. 55). By the mid-sixteenth century the outlines of the continents and oceans are known fairly accurately: only the large land mass that is assumed to lie to the south, *Terra Australis* or Australia, is still largely guesswork. The interiors of America, Africa and Asia are crammed with utopian fantasies, loosely based on travellers' tall tales and vivid imagination (cat. 52, 53 and 55). The numerous scenes on the world map give tangible shape to the anticipated riches. They were undoubtedly intended to persuade the king to support the expeditions of those who dared to venture far beyond the horizon.

The fourth and final chapter is *The Universe in Your Hand – Dreams of Space and Time*. By around 1516, space and time have become rationalized and measurable – as if in an ultimate utopia man seeks to comprehend and grasp the universe and eternity. Science and art merge in grandiose form. *The Movements of the Universe*, a colossal Brabant tapestry woven in 1500, perfectly encapsulates the mystery (cat. 69). An angel sets the universe in motion by turning a crank. The tapestry is further populated by all kinds of personifications, mythological figures and constellations. Time, traditionally seen as a perpetual cycle, is represented in the *Labours of the Months* by Simon Bening a full generation before Pieter Bruegel the Elder (cat. 73). And in the portrait of a girl by Jan Gossaert (cat. 64), perhaps the most beautiful child portrait to come from the sixteenth-century Netherlands, the young princess holds the universe in one hand, as only a child can, in search of Utopia.

Utopia, first published in Leuven, was more than just the title of a book; it became a new word and a new literary and artistic genre. In 2016 Leuven celebrates the quincentenary of *Utopia* with a citywide project and this exhibition. With *500 Years of Utopia* Leuven has decided on engagement and action. Artists depict desires and fears, driven by utopia and its obverse, dystopia. It is a nuanced story of those who dream of paradise, who are lured

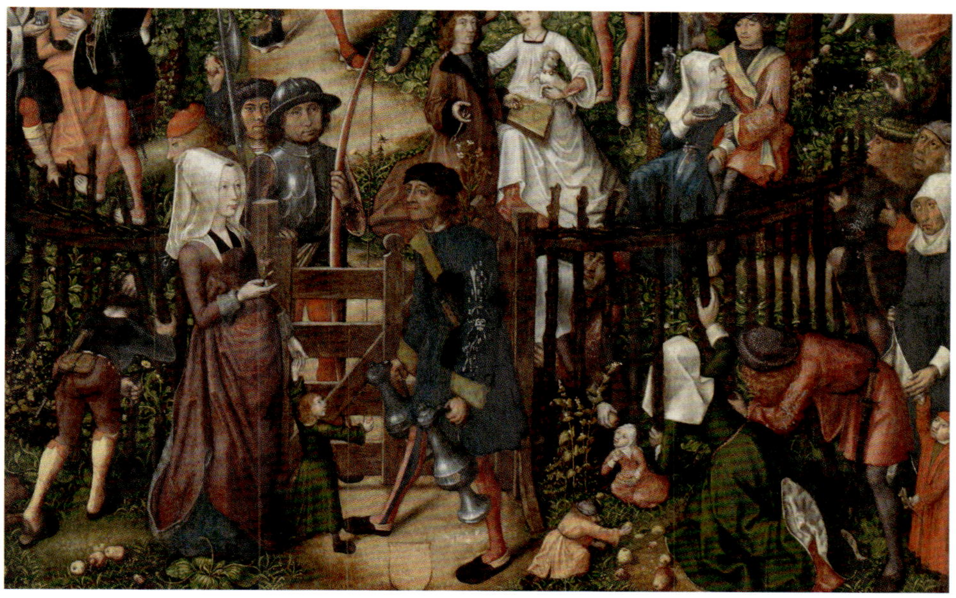

[fig. 5] Master of Frankfurt, *Festival of the Archers* (detail: in vain the crowd attempts to access the closed garden), 1493. Antwerp, Koninklijk Museum voor Schone Kunsten (cat. 24).

by the horizon, who long for the universe yet have a mortal fear of hell. That was the case five hundred years ago, when More's *Utopia* was published in Leuven, and it's no different today. More's little handbook trenchantly analyses the society of his time, corrupted by greed and envy, and sets out a utopian alternative. That alternative lies beyond the horizon. You will never set foot on its shore, but the island inspires travel, the making of journeys. Today, the scenario More and his friends dreamt up is more relevant than ever. People are still looking for their Utopia. The story takes commitment.

Europe-America-Utopia: Visions of an Ideal World in the Sixteenth Century

HANS COOLS

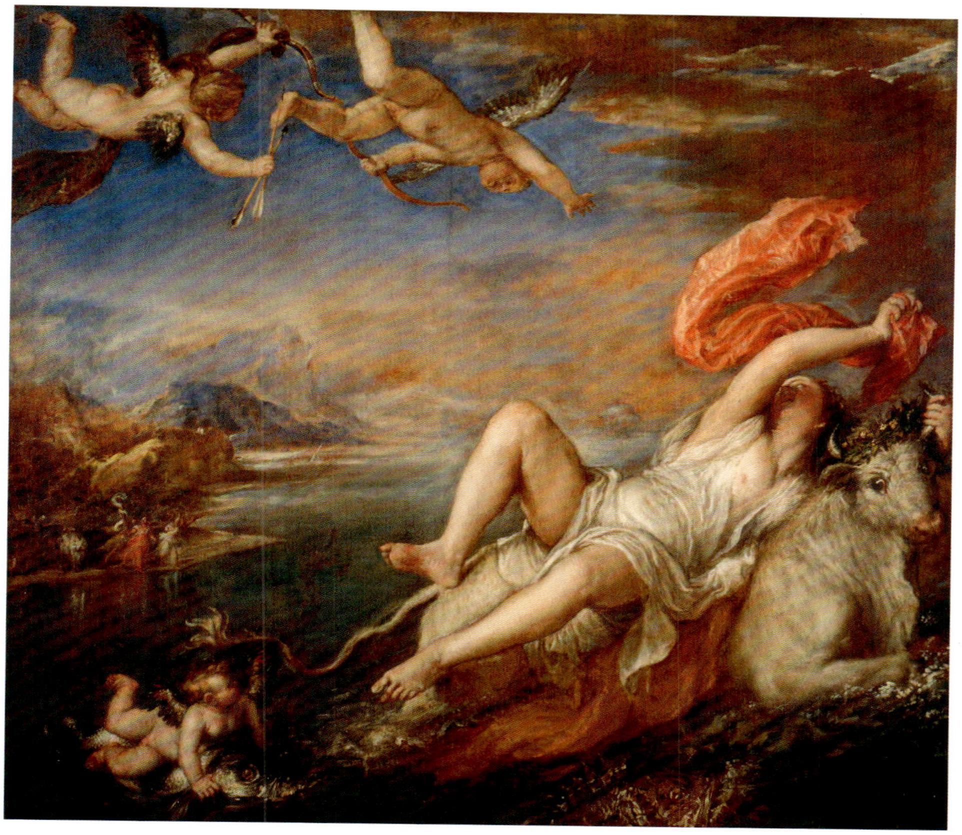

[fig. 6] Titian, *Europa*, 1560-1562, Boston, Isabella Stewart Gardner Museum.

In Greek mythology, Europa was a Phoenician princess. One day, as she and her friends were playing on the beach at Sidon, a magnificent white bull caught their attention. The huge creature seemed remarkably tame and friendly and in no time at all had allowed Europa to scramble up onto his back. Thus mounted, the princess paraded along the beach. But bit by bit the bull drew nearer to the water then gently waded in, deeper and deeper, and before long he was far out to sea with Europa still clinging to his back, to the great dismay of her astonished friends. The great bull swam all the way to Crete, where at last he resumed his true shape as Zeus, supreme deity and ruler of the gods. Not until Europa had borne him several sons – the exact number varies with the different versions of the myth – did Zeus return to Olympus, abandoning his Phoenician princess. Eventually she married Asterion, the king of Crete, who adopted her sons and brought them up as his own. This unlikely tale, which was told by several authors including Ovid (43 BCE–17 CE), who added it to his *Metamorphoses* (II, 836–875), quickly became highly popular in the sixteenth century.

There is a reason for this. Europa as a name may have been old by then, but it was not until the sixteenth century that the abducted princess gave her name to a continent. Up to that point the peoples who lived on this neck of land would have described themselves

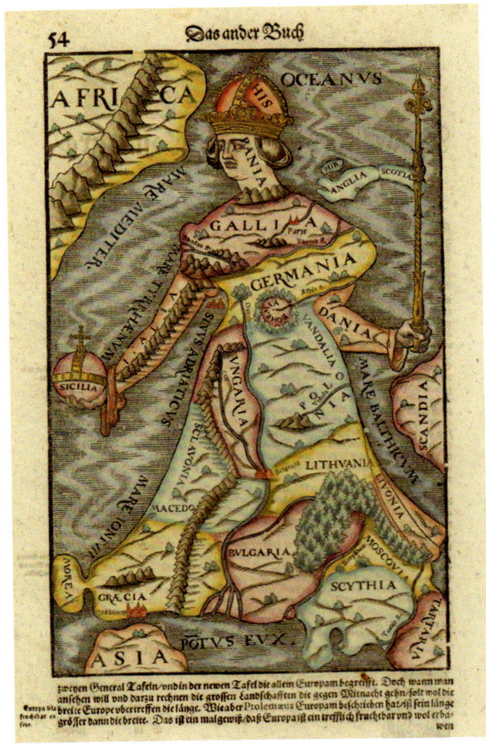

[fig. 7] *Europa Regina*, coloured woodcut in: Sebastian Münster, *Cosmographia*, Basel, 1570: p. 54.

first and foremost as Christians. Their part of the world was that of *christianitas*. That (Latin) Christendom stemmed from Roman culture, had grafted itself onto the structures of the Roman Empire, and had survived the decline and fall of that empire in North Africa and the West.

The *Méditerranée* was still regarded as the 'mid-point' of the known world. Along its coasts Asians, Africans and Europeans had long rubbed shoulders and for ages Europeans had exploited that interaction, especially in their pursuit of African and Asian riches and knowledge. But all that was about to change. For at the very end of the fifteenth century a new continent was 'discovered' on the other side of the Atlantic. In the following decades the Mediterranean waned in importance as political dynamics, economic growth and cultural flourishing gradually shifted westwards to the Atlantic coasts. Explorers sailed out of Europe's Atlantic seaports and a few years later fleets loaded with plundered American gold and particularly silver sailed back in (cat. 53 and 54). Much of that silver was used to help pay for the growing wars between European states, but some also went to Asia, where merchants exchanged it for spices and porcelain (cat. 46). The seventeenth and eighteenth centuries saw no diminution of this prosperous state of affairs along the Atlantic coasts, as shipowners fitted out their slave ships there and received mountains of sugar in return.

In short, from the second half of the fifteenth century Europeans enlarged their world. They made further explorations of Africa and Asia and conquered (large parts of) America (cat. 55.). The economies of those continents integrated and Europeans steered that process. They also capitalized on it, reaping huge rewards, to the great and often fatal detriment of some of the earth's other inhabitants – the decimation of native Americans and the massive transhipment of African slaves being the most obvious examples of the price paid by outsiders for European supremacy.[1]

Thus, in the sixteenth century Europa put off the guise of ingénue, all too easily enticed and ravished, and stepped forth as a queen. Artists have portrayed this process countless times. Around 1560 for instance, Titian still depicted Europa as a distraught

1 Pagden 2001: 50–87.

nymph (fig. 6), helplessly carried away by the bull as ominous copper-coloured clouds cover the sky. The painting is fraught with calamity.[2] But the Europa depicted on an anthropomorphic map that first appeared in a 1588 reprint of Sebastian Münster's (1498–1552) famous *Cosmographia* displays a very different character (fig. 7). In this woodcut she advances as would a haughty empress, crowned and bearing orb and sceptre. The world belongs to her, and in particular to her head, Spain. This image soon became extremely popular: not one of the countless editions and translations of the *Cosmographia* – which continued to appear until 1628 – was without it.[3]

Unknown America often functioned as Europe's opposite. For unlike Africa and Asia, to which 'explorers' also sailed in those years, America had never been written about before in the Old World. So the new continent fuelled Europeans' imagination. They projected their dreams onto it, fashioning the New World according to their own taste and understanding. At virtually the same time as Europa made her imperious appearance in the *Cosmographia*, the Dutch artists Johannes Stradanus (1523–1605) and Philips Galle (1537–1612) depicted the landing of Amerigo Vespucci (1454–1512) on America's virgin shores (fig. 8). Their engraving portrays Vespucci as a modern crusader. While cannibals stoke a fire in the background the Italian explorer gazes at the naked princess personifying America as she lounges in her hammock, then wakes her with a kiss.

Vespucci's (partly fictional) account of his travels soon appeared in print. His often somewhat imaginative ethnological observations were eagerly seized upon and had a profound influence on the way Europeans saw the original inhabitants of the American continent.[4] Thomas More (cat. 3) had also read Vespucci's travels. More to the point, he makes specific reference to them in *Utopia* (cat. 1), in which he introduces Raphael Hythlodaeus, the discoverer of Utopia, as a veteran of Vespucci's expeditions.[5] But unlike Vespucci, who saw the native Americans as uncouth and uninhibited savages, More represented the Utopians as a highly disciplined folk who despite holding property in common maintained a monogamous lifestyle.[6]

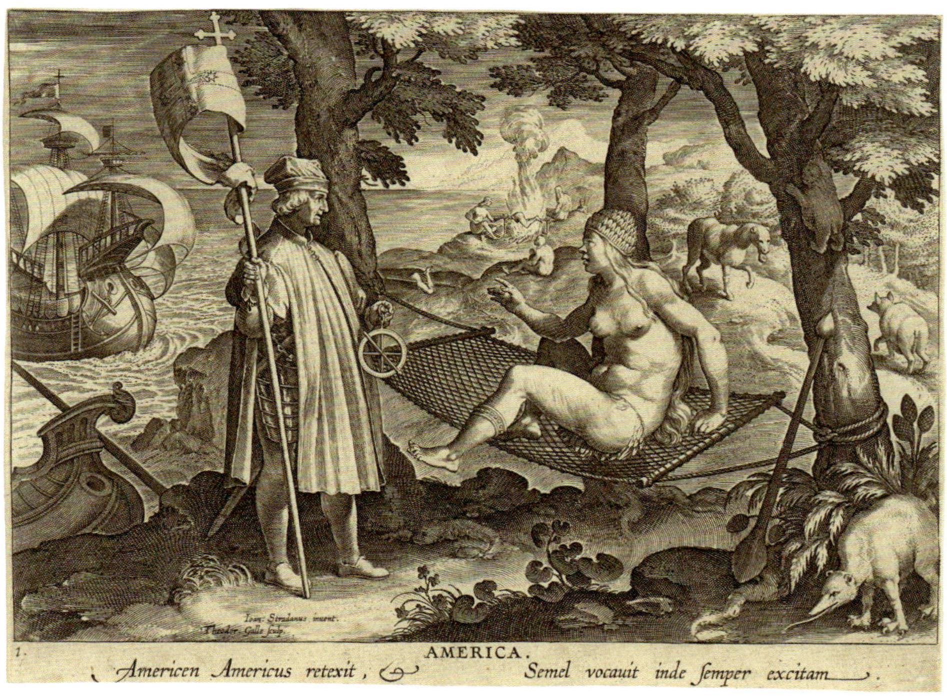

[fig. 8] Johannes Stradanus and Philips Galle, *Amerigo Vespucci discovers America*, engraving in the series: *Nova Reperta*, c.1590, Brussels, Royal Library of Belgium, Print Room.

2
Greengrass 2014: 29–30; Moormann & Uitterhoeve 1997: 106–107; FitzRoy 2015.

3
Meurer 2008: 355–370.

4
Fernández-Armesto 2006: 98–135.

5
See the essay by Jan Papy in this catalogue (pp. 31–39).

6
Cave 1991: 211–216.

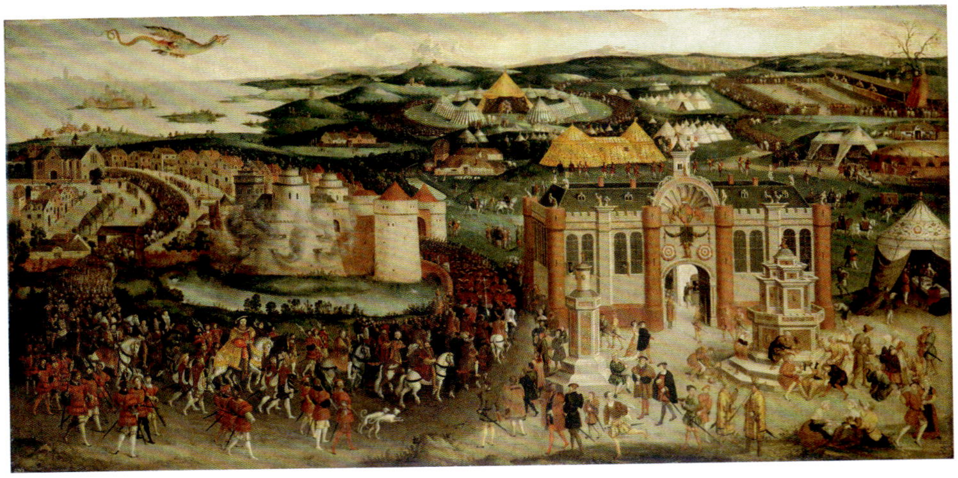

[fig. 9] *The Field of the Cloth of Gold*, c.1545. Royal Collection Trust, Hampton Court.

More's Utopians were not completely unspoiled. They derived their name from a benevolent conqueror who nearly eighteen hundred years before the arrival of Hythlodaeus had acquainted the people with the wonders of classical Greek civilization. Utopians thus held up a mirror to Europeans. They were their better selves. Unlike Europeans, Utopians shared their possessions, were concerned for the common weal, and abided by the law; they were peace-loving and their prince did not indulge in dynastic wars.

Thomas More did indeed have a keen eye for the power politics of European monarchs. As the English monarch's counsellor and envoy he was himself a player in their game. More's patron, Henry VIII (1491–1547) was a distrustful man very prone to feeling let down by those he should have been able to rely on (fig. 10). That lack of trust was partly attributable to England's recent past. In 1485 his father, Henry VII (1457–1509), had unexpectedly emerged as the winner of the Wars of the Roses and founded a new dynasty, the Tudors. After decades of civil war Henry VII brought peace and concentrated on restoring his virtually bankrupt exchequer. Nevertheless, in the years that followed his claim to the throne was contested by a succession of pretenders. Backed by foreign rulers, the more ostensibly credible among them presented an all too real threat.

The uncertainty that threatened to undermine the Tudor reign continued under Henry VIII. Henry had a short answer for doubters: Edmund de la Pole (c.1472–1513), for instance, a Yorkist claimant who had been imprisoned in the Tower of London for years, was sent to the block. He also agonized obsessively over the succession, more so than other monarchs. His marriage to Catherine of Aragon (1485–1536) had resulted in six pregnancies. Two boys and a girl survived their birth but only the latter, Mary (1516–1558), known to history as Bloody Mary, lived into adulthood.

Henry was prepared to take drastic steps to ensure the continuation of the Tudor line. When he failed to wrest an annulment of his marriage to Catherine from the pope he successively married Anne Boleyn (c.1501–1536), had a special ecclesiastical court declare his earlier union with Catherine invalid, and finally broke with Rome. None of which produced the hoped-for result. On 29 January 1536, the very day of Catherine's funeral, Anne gave birth to a stillborn son. Less than six months later, Anne herself went to the scaffold, accused of witchcraft and adultery. A like fate had already befallen More and many other perceived opponents of the king.[7]

Henry VIII was not the only early modern monarch to take vigorous action in order to ensure a male heir and buttress a still fragile claim to the throne. In France, Louis XII (1462–1515) lost no time in ridding himself of his wife Jeanne once he had become king in 1498. His new bride was Anne (1477–1514), Duchess of Brittany and widow of Louis's rival

7 Guy 1988: 116–153.

and predecessor Charles VIII (1470–1498). Between them, Anne's two spouses sired a dozen children, but only two daughters from her second union reached marriageable age. The eldest, Claude (1499–1524), married France's new king, Francis I (1494–1547; fig. 11), in 1514.[8]

Even though they had lost the Hundred Years' War, England's monarchs had not given up their claim to the French throne. At Calais a strip of mainland French soil still remained in English hands. Moreover, in 1513 Henry VIII had taken Tournai, then a French enclave in Burgundian territory.[9] Francis I and Henry VIII were therefore not only contemporaries, in many ways they were also rivals. Both were physically well-favoured athletic princes, each with a reputation to uphold. Both were fascinated by the culture of chivalry, with its emphasis on individual heroism and physical exertion. Yet both knew the other primarily from hearsay. It was not until the spring of 1520 that they came face to face for the first time, at a meeting arranged to enhance the friendship between England and France. Two dazzling encampments were built within the Pale of Calais on a site that came to be known as the Field of the Cloth of Gold (fig. 9). The meeting also included a tournament that lasted several days in which the flower of the two nations competed against each other in all kinds of events. Even the two kings themselves had a wrestling match that ended abruptly when Francis floored Henry.[10]

The outcome of that contest was something of a handicap in the negotiations that French and English diplomats were meanwhile conducting behind the scenes. They anticipated a good deal from an alliance between their two princes; it was for that purpose, after all, that the meeting had been organized. For a year earlier, Charles V (1500–1558; fig. 12; cat. 19) had beaten Francis I in the election of a new Holy Roman Emperor. As a result the shy young Habsburg prince had become the most eminent ruler of Christendom. From his parents and grandparents Charles had already inherited the various principalities of the Low Countries and the Iberian kingdoms along with their territories in Italy and the New World. What the title of emperor conferred was not so much actual power as prestige: he could now claim to be the successor to the Roman emperors of Antiquity. Charles's chief advisor, Mercurino di Gattinara (1465–1530), aimed to turn that claim into real power too. As far as he was concerned, Charles should aspire to nothing less than world dominance: *monarchia universalis*.[11]

Not surprisingly that aspiration evoked considerable hostility, not least from Francis I. He hoped with English support to put a damper on Habsburg ambitions. Up until then, France had been the strongest power on the European playing field. The country had more inhabitants and therefore more tax revenue than the Netherlands, the Iberian kingdoms and England put together.[12] Comparatively speaking, moreover, its government was highly centralized. But now that Francis had lost the imperial election to Charles, and France was almost completely encircled by Habsburg possessions, its dominance was under threat.

In the following decades, the Habsburg and Valois French monarchies would contend for European supremacy in a practically unbroken series of wars. The battle was largely fought out in Italy. That area was rich. The Renaissance was flourishing there in cities and royal courts. Moreover, it lay midway along the Mediterranean – whoever controlled Italy had Europe in his power, so contemporary policymakers believed.

Like Francis I, Henry VIII feared Habsburg supremacy. Yet only an alliance with Charles V offered any prospect of gaining or retaining at least part of those French lands to which he laid claim by right of descent. Thus Henry was constantly changing sides. In the end that vacillation brought only loss to his dynasty. In 1558 Henry's daughter Mary was forced to give up Calais, England's last possession on the French mainland.[13]

Ultimately, the Habsburgs proved stronger than the Valois. By the terms of the Peace of Cateau-Cambrésis, signed in 1559, Francis I's son Henry II (1519–1559) resigned

8 Quilliet 1986: passim.

9 Davies 1998: 1–14.

10 Knecht 1994: 170–175.

11 Blockmans 2000: 52–68.

12 Bonney 1991: 365–366.

13 Greengrass 2014: 270.

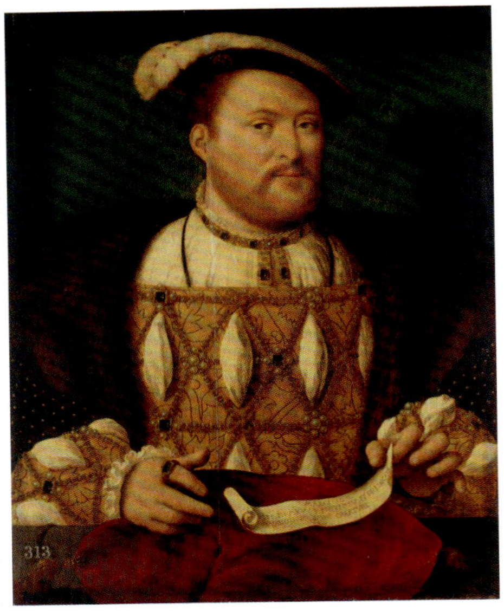

[fig. 10] Joos van Cleve, *Portrait of Henry VIII*, c.1530-1535. Royal Collection Trust, Hampton Court Palace.

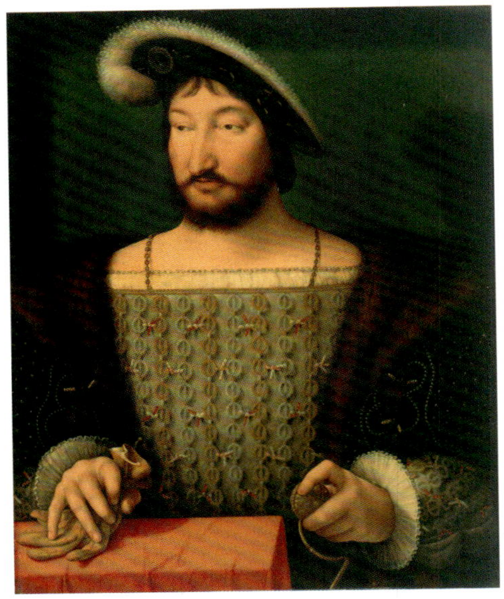

[fig. 11] Joos van Cleve, *Portrait of Francis I*, c.1532-1533. Philadelphia Museum of Art, John G. Johnson Collection.

his claims in Italy. Henceforth, the Habsburg dynasty was the European superpower par excellence. But the victory was a Pyrrhic one. Philip II (1527–1598), who succeeded Charles V in 1555, was immediately confronted with an empty treasury. For the rest of his long reign he was almost constantly short of the financial wherewithal to fulfil the obligations of his status. He also suffered a series of humiliating defeats. In 1588 the English fleet destroyed his Armada and in France he failed to achieve the outcome he wanted from the civil wars that raged in the second half of the sixteenth century. Rather the contrary in fact, as the French crown finally went not to his daughter Isabella (1566–1633) but to Henry IV of Bourbon (1553–1610). Admittedly, in 1594 that former Protestant leader did convert to Catholicism, but he also granted civil rights to the reformers. Last but not least, a rebellion broke out in the Low Countries that would last well beyond Philip II's own lifetime. Not until 1648, after eighty years of war, would Philip II's grandson Philip IV (1605–1665) acknowledge that the northern part of the Netherlands was permanently lost to the Habsburg dynasty.

One of duties that Charles V had also undertaken as Holy Roman Emperor was to preserve the unity of the *christianitas*. But neither he nor his son Philip II succeeded, though they both waged a number of wars. As early as 1517 – almost the same time as the first edition of *Utopia* was published – the German Augustinian friar and professor of theology Martin Luther (1483–1546) had begun to question the pope's ability to provide faithful Catholics with entry to paradise in the form of indulgences. By doing so the pope detracted from God's omnipotence, said Luther. His message spread like wildfire. In no time at all Luther's theses were being printed and discussed in any number of places. Reformers rose up everywhere. The precise content of Christian faith and the intermediary role of the Church thus became the topic of heated public debate. Town councils and rulers of all kinds determined the local results of these debates, if only to maintain public order. They thus acquired an influence on the content of faith and brought the various communities that had arisen here and there under their control.[14]

The Catholic Church responded slowly to the challenges the reformers presented. The initial assumption was that condemning Luther's theses and burning his books – proceedings in which Leuven theologians took the lead – would be enough to see off the problem.[15] But the convocation of a council was, for the time being, ruled out. For that kind of general ecclesiastical gathering, at which the Church's teachings were to be redefined,

14 Pettegree 2005: 17–39.

15 Gielis 2009: 57, 75.

implicitly encroached on papal authority. Eventually, in 1545 and with powerful backing from Charles V, Paul III (Alessandro Farnese, 1468–1549) made it happen but not before he had strengthened the Roman institutions – including the establishment of the Inquisition as the central authority for combating heresy. For almost twenty years, a changing group of Council fathers met periodically in Trento and Bologna. By the time that process ended they had sharpened the definition of Catholic teaching, worked out a reform and training programme for the clergy, and decreed guidelines for liturgy and religious art. As a result of those proceedings the distance that separated the various Protestant denominations, which had gone through similar heart-searching in the same years, became clear.[16]

While Amerigo Vespucci and his fellow explorers so confidently explored the globe, in Europe the Christian civilization they thought they represented was torn asunder. Moreover, it appeared that the *monarchia universalis* for which the Habsburg dynasty strove was nothing but a chimera.

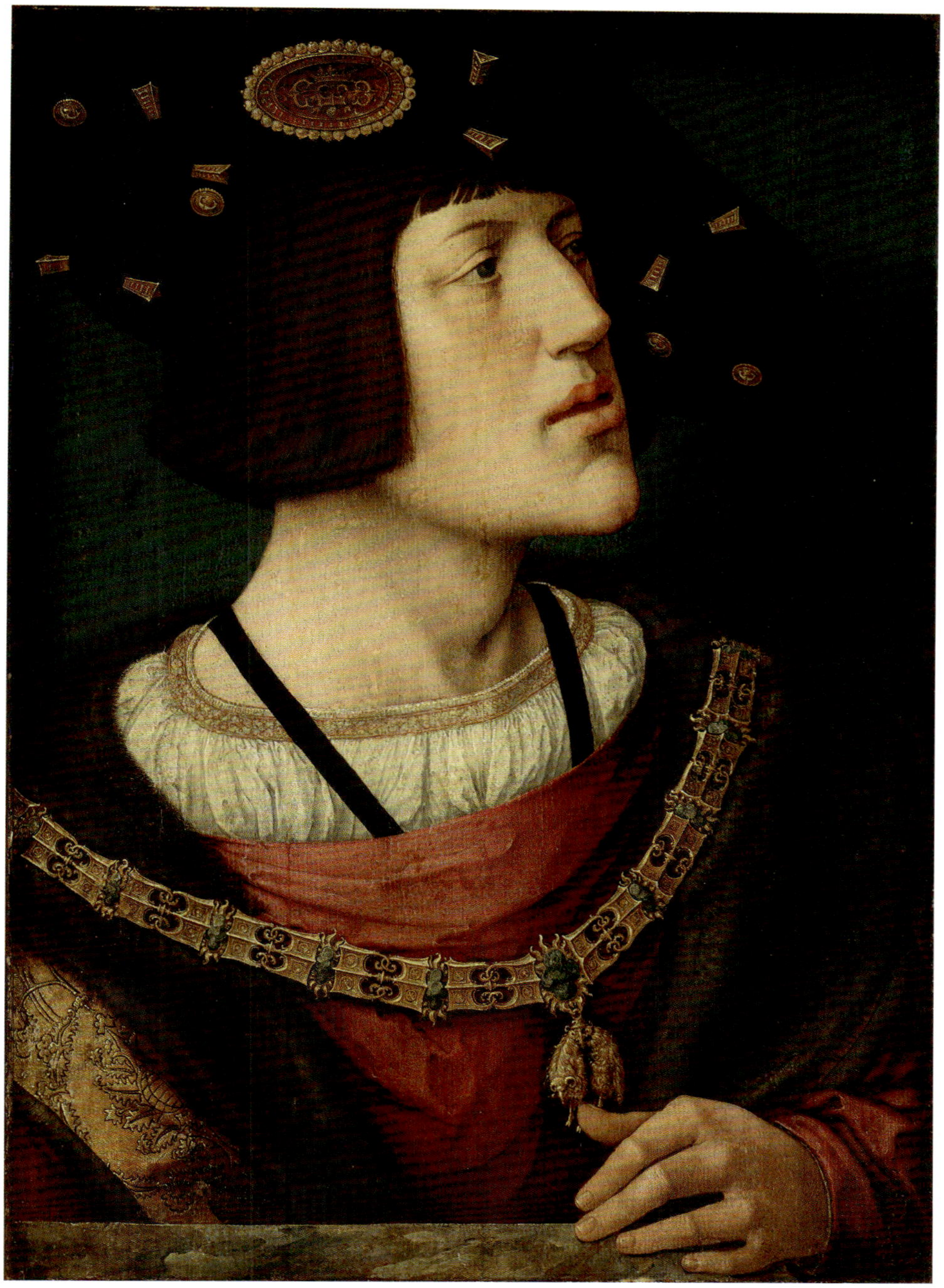

[fig. 12] Bernard van Orley, *Portrait of Charles* V, 1519-1520, Budapest, Szépmüvészeti Muzeum.

16 O'Malley 2013: 49–76.

In that bleak indeterminate climate, in which numerous Europeans dreamed of a better world, Thomas More's *Utopia* found a ready market. In the late 1540s and early 1550s – not coincidentally the decades in which the religious division in Europe finally took place – the book appeared successively in Italian (1548), French (1550; cat. 8), English (1551) and Dutch (1553; cat. 9) translations.[17] More himself, meanwhile, was dead, beheaded in 1535 for his unwavering refusal to acknowledge Henry VIII's takeover of the Church to serve his own dynastic interests. His memory, however, lived on. In the late nineteenth and early twentieth century, a time of renewed Catholic militancy, it was still vivid enough for consecutive beatification and canonization (1886 and 1935). Despite his dislike for utopias, More's steadfastness was also admired by John Paul II (Karol Wojtyła, 1920–2005). In 1980, he commended Thomas More to professional politicians as their patron saint. Whether they will be guided in their actions by More's ideals of altruism, care for the common weal and respect for the law remains to be seen. But perhaps dreams do not always have to be delusions?

17 Greengrass 2014: 178–179. See the contributions by Jan Papy (pp. 31–39, 73, 129) and Marcus de Schepper (p. 103) in this catalogue.

Thomas More, *Utopia* and Leuven. Tracing the Intellectual and Cultural Context

JAN PAPY

From *Nusquama* to *Utopia*

Ever since its publication in December 1516, *Utopia* has constantly fascinated and also constantly confused (cat. 1). To begin with, the title itself, which was suggested to Thomas More by Erasmus, is a Greek neologism made up of *ou* and *topos*. A nod to the existing Greek *atopia* – 'placelessness' or 'uniqueness', 'unusualness'.[1] That 'Noplace' is subsequently revealed to be a newly discovered island, somewhere in the New World. The explanatory subtitle of the work, printed by Dirk Martens in his premises known as *De Gulden Toirtse* on the corner of Naamsestraat and Standonckstraat in Leuven, runs as follows: *On the Best State of a Commonwealth and on the New Island of Utopia. A Truly Golden Handbook, No Less Beneficial than Entertaining*. That perfect 'best' society has at least one definite place: in a story. Though even there it is embedded in a dialogue in which a sunburnt voyager, called Raphael Hythlodaeus, aims from the very first pages to transport the reader to an imaginary world (fig. 13; cat. 7.1 and 7.2). But just how reliable is his account of the five years he spent on Utopia? For his name also gives us pause. Not his first name: Raphael is, after all, familiar as the name of the angel of healing, Tobias's faithful companion and guide, and accordingly of the patron saint of pilgrims and travellers. Hythlodaeus, on the other hand, is once again compounded of highly suggestive Greek elements: *hythlos* ('nonsense' or 'idle talk') and *hodaios* ('merchant', 'peddler') as used in Homer's Odyssey (8, 163 and 15,

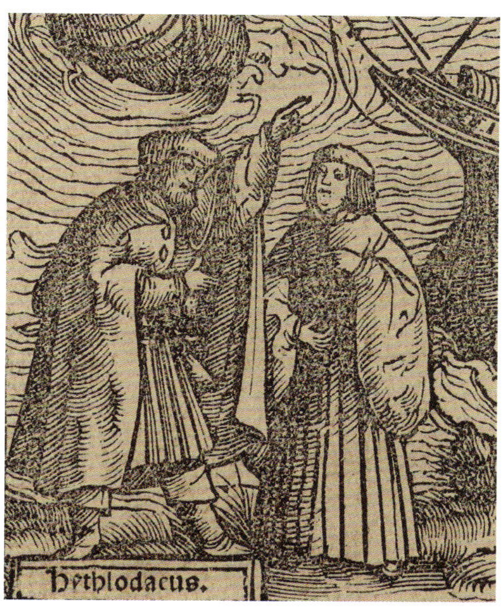

[fig. 13] Ambrosius Holbein, *Map of Utopia* (detail: *Raphael Hythlodaeus tells the story of Utopia to Thomas More*) in: Thomas More, *De Optimo reip. statu, deque noua insula Utopia, libellus uere aureus*, Basel, 1518, fol. 12v, KU Leuven, University Library (cat. 7.1).

1 See inter alia Plato, *Phaedrus*, 229e and 251d.

2 Since the interpretation given by G.J. Vossius in his letter to Samuel Sorbière, who published a French translation of *Utopia* in Amsterdam in 1643 (cf. G.J. Vossius, *Opera*, Amsterdam 1695–1701, IV: 240–341), Hythlodaeus's name is also often explained as a compound of *hythlos* ('idle talk') and *daïos* ('well-versed in').

445).[2] More's imaginary traveller, witness of that wonderful world of Utopia, is a purveyor of tall tales. More, the man of letters, sets his readers off with a twirl.

Even then the reader must wait for that tale. For what follows is not the story of the island of Utopia but a lengthy debate about whether Hythlodaeus would make a useful member of a king's council. Humanists, who since Petrarch had been driven to search for a new intellectual stance between the poles of *vita activa* and *vita contemplativa*, invariably defined themselves as 'political and social counsellors'.[3] But within that debate is yet another, again narrated by Hythlodaeus, on the rising poverty and associated crime seen in both rural and urban England, escalating tyranny and corruption, and the lamentable social conditions that leave those dispossessed of home or work through no fault of their own with no option but to steal or starve.[4] And still the reader has not arrived in Hythlodaeus's Utopia but is borne along on a reflection by More, the humanist-counsellor, on political and social developments in England (cat. 3). There the Tudor dynasty was installed, central authority was fortified by a generously filled treasury, a loyal bureaucracy and a standing army, and Henry VIII inherited a land at peace. Yet he dragged his people into a spiral of senseless and costly wars.[5] Thomas More, the humanist, embarks on his pacifist credo.

Only at the start of Book Two does Hythlodaeus's tale of Utopia finally begin. As it is undoubtedly to this second part of the work that *Utopia* as a whole owes its fame it is worth knowing that More actually wrote it first, as a separate work, while he was in Antwerp, working at it in his spare moments during his diplomatic mission in May to October 1515.[6] What we know as Book One was written last, and quite some time after his return from Antwerp to England in late October 1515.[7] According to Erasmus, spare time was now a thing of the past for More.[8] He was once again sucked into the hectic political life of a turbulent England: Cardinal Wolsey asked him to join the king's council.[9] And so for More the broad discourse about the nature of a true Christian state had suddenly become an internal debate.

This genesis has its implications for the interpretation of *Utopia* as a whole. Moreover, not only did More merge his 'two utopias' into a single work, he also left it to Erasmus to see the book through the press (cat. 13A and 14A). If Book Two – what one might call a proto-*Utopia* – was once an autonomous and coherent work in itself, the final assemblage that More sent to Erasmus in early September 1516 in the form of a manuscript entitled *Nusquama* ('Nowhere') gave a new direction to what would appear in Leuven as *Utopia*.[10] Not only had More added Book One, he had also written a new, completely revised preamble and afterword to the dialogue and, in consultation with Erasmus, incorporated his *Letter to Pieter Gillis* as a kind of 'note to the reader'. Book One was not inserted at the beginning without good reason, however. More obviously saw this long dialogue as an introduction to the story of the ideal state of Utopia, while the *Letter to Pieter Gillis*, written in a style typical of humanist Latin openings or dedicatory letters and rhetorically laden with philosophical and literary platitudes, immediately hallmarked *Utopia* as a work belonging to that tradition. It goes without saying that it was intended for erudite friends who shared those cultural and politico-philosophical backgrounds.[11]

3 Skinner 1978.

4 Schulte Herbrüggen 1997a: 3–22.

5 Hoskins 1976. See the essay by Hans Cools in this catalogue (pp. 21–28).

6 On this English trade mission to strengthen the ties and trade relations with Flanders following the accession of Francis I, see Rogers 1947: 16–20.

7 Surtz & Hexter 1965: xv–xxiii.

8 In a letter to Ulrich von Hutten dated 1519; see Allen 1906: no. 21; CWE 7: 24 and Hexter 1952: 99–102.

9 Elton 1972: 87–91; Guy 1980: 7.

10 Allen 1906: no. 339 and 461; CWE 4: 66 (September 1516, *Nusquama*), 79, 93, 98, 125 (first reference to *Utopia*, letter from Gerard Geldenhouwer to Erasmus), 131, 163–164. See also Rogers 1947: 75.

11 Kristeller 1980.

Erasmian Humanism: Language, Humour and (In)Accessibility

Just as Erasmus, in his *Moriae Encomium* (*Praise of Folly*; fig. 15), had played on the Greek *moros* ('fool') and More's own name, simultaneously making his work a 'Praise of More', Thomas More likewise made it clear from the start, with his book's very title, *Utopia*, that Greek neologisms were meant for those able to understand them.[12] For example, running

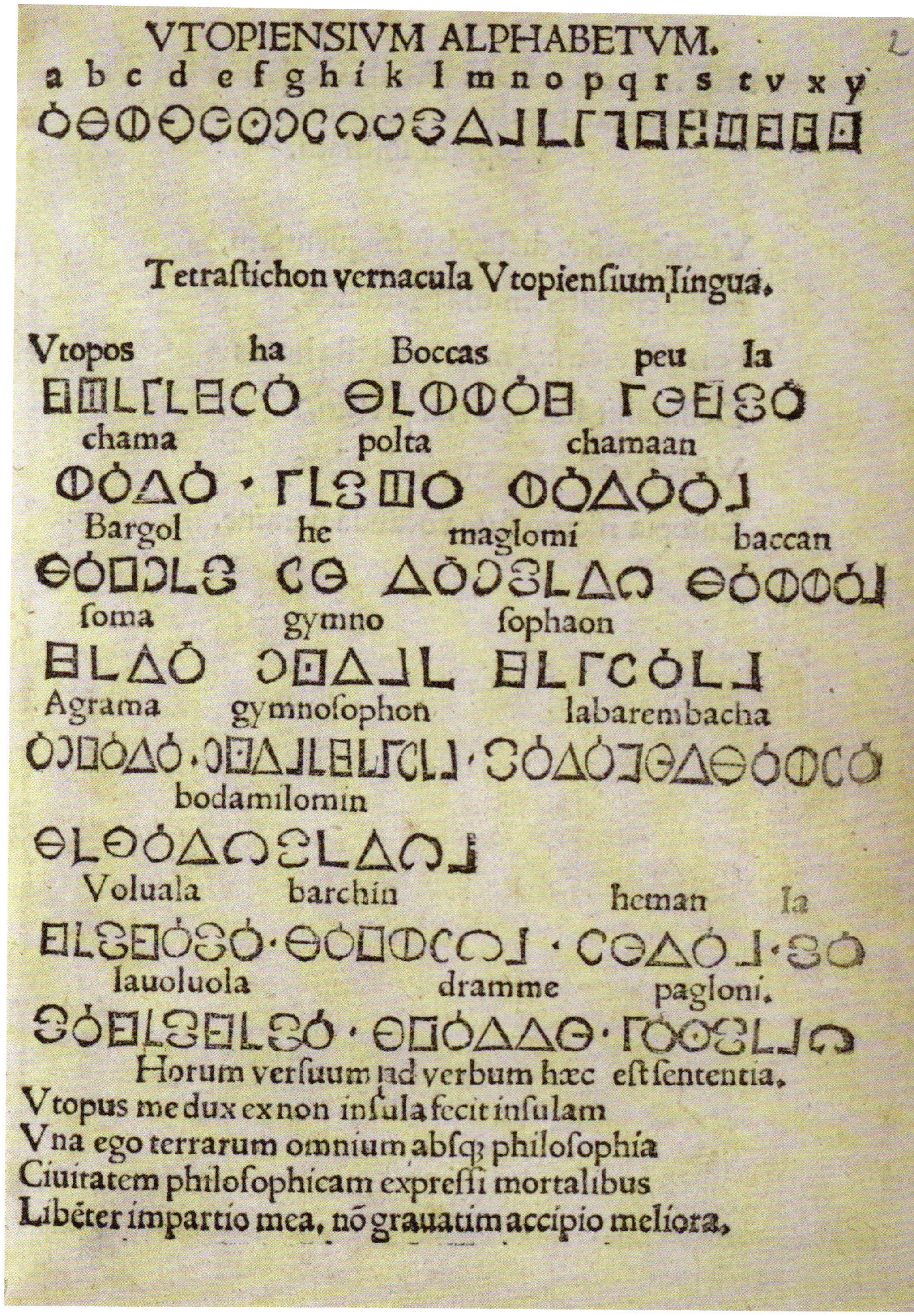

VTOPIENSIVM ALPHABETVM.

a b c d e f g h i k l m n o p q r s t v x y

Tetraſtichon vernacula Vtopienſium lingua.

Vtopos ha Boccas peu la
chama polta chamaan
Bargol he maglomi baccan
ſoma gymno ſophaon
Agrama gymnoſophon labarembacha
bodamilomin
Voluala barchin heman la
lauoluola dramme pagloni.

Horum verſuum ad verbum hæc eſt ſententia.

Vtopus me dux ex non inſula fecit inſulam
Vna ego terrarum omnium abſq; philoſophia
Ciuitatem philoſophicam expreſſi mortalibus
Libēter impartio mea, nō grauatim accipio meliora.

[fig. 14] *The Utopian Alphabet, designed by Pieter Gillis*, in: Thomas More, *Libellus vere aureus…de noua insula Utopia*, Leuven, 1516, Brussels, Royal Library of Belgium (cat. 1).

through the 'invisible' city state of Utopia, shown in a map of the island contributed by his humanist friend Gerard Geldenhouwer (cat. 1 and 7.1),[13] is a river, the Anydrus – 'not water'. That waterless river irrigates the capital Amaurotum, a name that is in turn derived from the Greek *amauros*, 'obscure' or 'unknown'. The seriocomic use of Greek can hardly be misunderstood: *Utopia* is a work for humanists who deplore the old order and advocate a new society – the Christian-humanist society of reason and faith portrayed by Erasmus. The enthusiastic fictional narrator Hythlodaeus, who prefers Greek literature to Latin, gives their common idealism a classical semblance. In him, More's humanist friends find their almost heroic stature. In his grievances and project they perceived the reformer

12 Goldhill 2002; Nelson 2004: 19–48.

13 On the map of Utopia, which in the later edition published in 1518 by Froben in Basel was the work of Ambrosius Holbein, but in the first edition of 1516 was still attributed by Gerard Geldenhouwer to 'quidam egregius pictor' (cf. Allen 1906: no. 380), see Schulte Herbrüggen 1997b.

who carried their principles to a logical – or illogical – conclusion: the image of the unachievable but always desirable Utopia. Social satire sought and found an antithesis: an ideal, imaginary island.

Language plays a very particular part in Utopia. In 1515 Pieter Gillis (cat. 13B and 14B), first in a line of humanist town clerks in Antwerp and personally introduced to More by Erasmus, devised a twenty-two letter alphabet for the Utopians, a people of 'Greek origin' (cat. 1). His Utopian language was indeed somewhat related to Greek, though it 'resembles Persian in most respects',[14] and in fact was more akin to Zend.[15] The Utopian quatrain that Gillis interpolated at the front of the book was accompanied by a word-for-word Latin translation:

Vtopos ha Boccas peu la chama polta chamaan
Bargol he maglomi baccan soma gymno sophaon.
Agrama gymnosophon labarembacha bodamilomin.
Voluala barchin heman la lauoluola dramme pagloni.

Vtopus me dux ex non insula fecit insulam.
Vna ego terrarum omnium absque philosophia
Ciuitatem philosophicam expressi mortalibus.
Libenter impartio mea, non grauatim accipio meliora.[16]

Utopus, my ruler, converted me, formerly not an island, into an island.
Alone of all lands, without the aid of abstract philosophy,
I have represented for mortals the philosophical city.
Ungrudgingly do I share my benefits with others; undemurringly do I adopt whatever is better from others.

The philosophical city of Utopia, as Pieter Gillis lets it be understood from the opening lines of *Utopia*, is built on the wisdom of the fabled 'gymnosophists' of the East, the 'naked philosophers', the Brahmans of India. They perform no manual labour; they have a deep aversion to vanity and laziness. Here the exotic touches the far-flung boundaries of Gillis's and More's dazzling erudition. For they could only have learned about the gymnosophists in any detail from Philostratus's *Apollonius*, published, with a Latin translation by Alemannus Rhinucinus, just a few years earlier in Venice; or from *De optimo cive* by the Italian humanist Bartolomeo Platina (1421–1481), printed in Lyons three years before *Utopia* appeared; circulating in the Middle Ages was St Ambrose's *De moribus Brachmanorum*, and

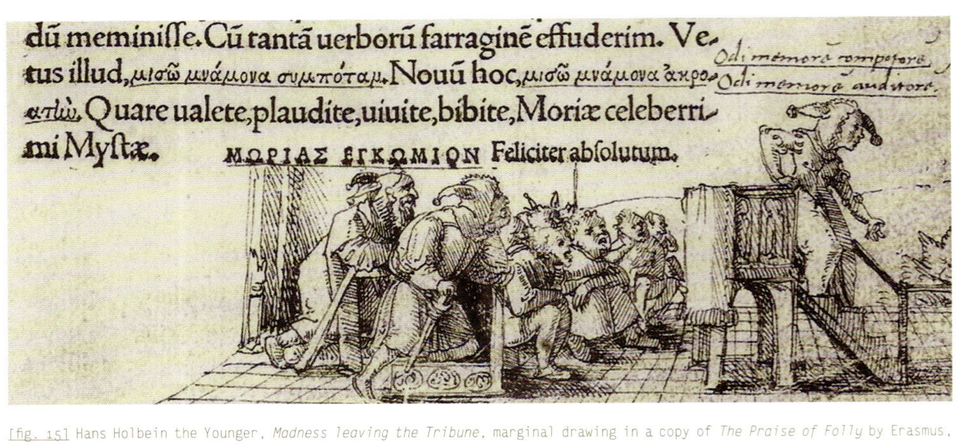

[fig. 15] Hans Holbein the Younger, *Madness leaving the Tribune*, marginal drawing in a copy of *The Praise of Folly* by Erasmus, 1515, Basel, Kupferstichkabinett.

14
Surtz & Hexter 1965: 180, l. 23–24.

15
Pons 1930: 589–607. A new interpretation of the languages that Gillis had incorporated into his Utopian alphabet is found in Vielle 2009 and Vielle 2013.

16
'Zend' is the earlier name of the Old Persian language of the Avesta, the writings that contain the teachings of Zoroaster. Since *zend* possibly means 'commentary', and the commentary on the Avesta is written in Middle Persian, the language of the original text is now called Avestan. Surtz & Hexter 1965: 18.

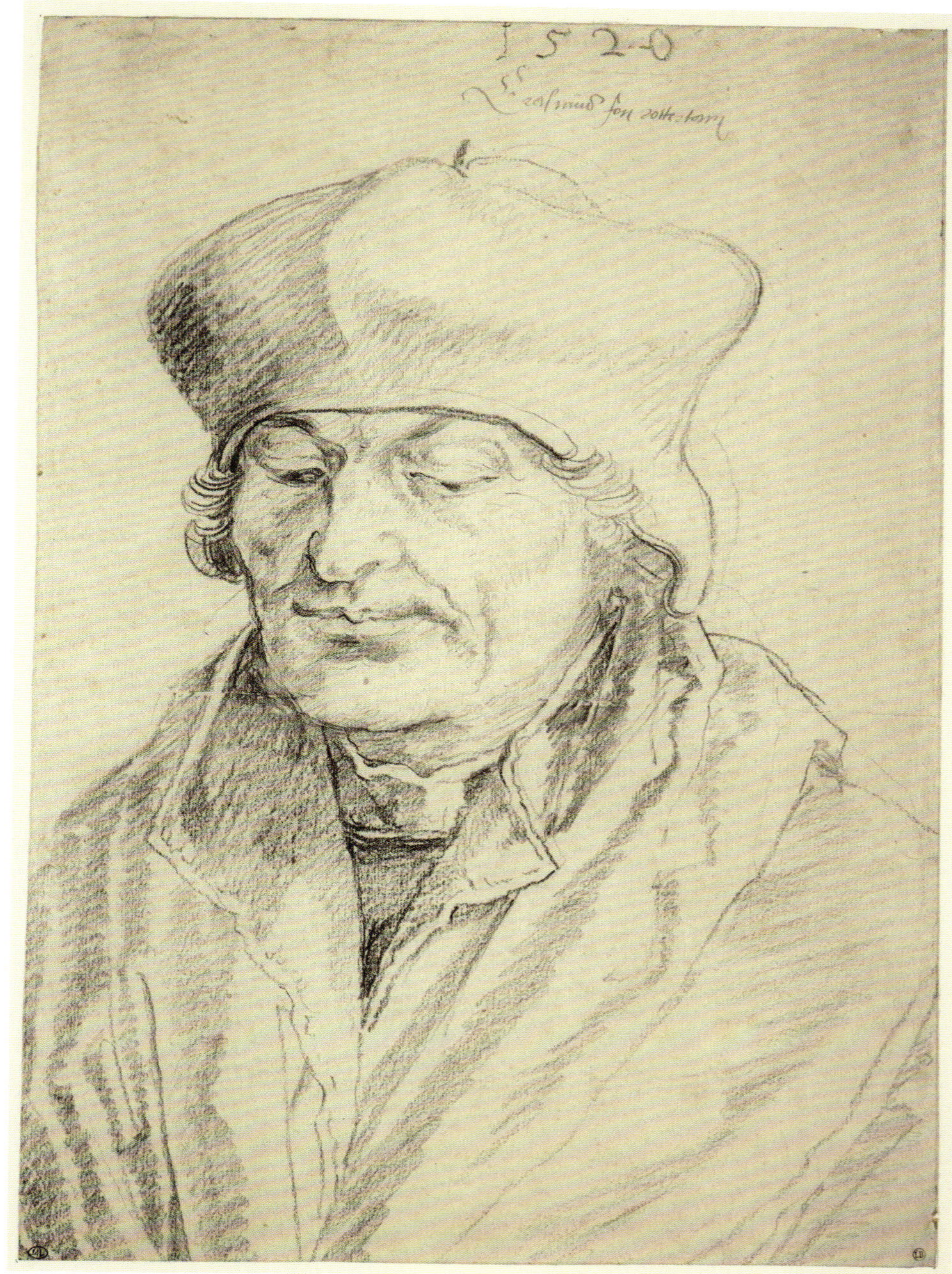

[fig. 16] Albrecht Dürer, *Portrait of Erasmus*, drawing, 1520, Paris, Musée du Louvre, Département des Arts graphiques.

those Indian sages were also mentioned by St Jerome and classical authors such as Cicero, Plutarch and Lucian.[17]

More and Erasmus had already collaborated on the work of Lucian of Samosata (*c.*125–*c.*180) in fact, having started translating the Greek satirist together in the years before the writing of *Utopia*. In 1505–1506 Erasmus stayed with More at his London home, the Old Barge in Bucklersbury, and there they competed in making Latin translations of Lucian's Greek dialogues. The result of that translation work – a typical humanist preoccupation and to be understood in the context of the circle of London humanists surrounding the theologian John Colet (1467–1519), in which the study of Greek for the purpose of Biblical exegesis was key – was published in late 1506 by Josse Bade (Judocus Badius) in Paris.[18] But their 'Latin Lucian' was not the only reason why More asked his friends Erasmus and Gillis to oversee the publication of Utopia. Erasmus had written his serio-satirical *Moriae Encomium* at More's home in 1509 and published it in Paris in 1511, with a dedication to More.[19] In 1515 he introduced More to Pieter Gillis. Gillis, to whom

17
Derrett 1962: 18–34; Derrett 1966.
18
Thompson 1939: 855–881; Schulte Herbrüggen 1997a: 51.
19
Allen 1906: no. 222; CWE 2: 161–164.

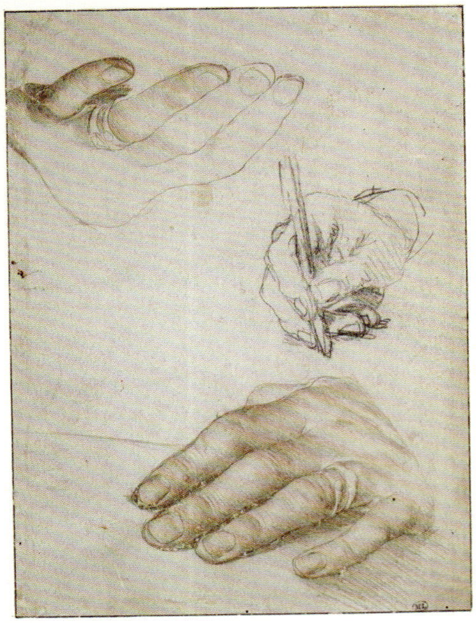

[fig. 17] Hans Holbein the Younger, *Three Studies of Hands for the Portrait of Erasmus*, drawing, Paris. Musée du Louvre, Département des Arts graphiques.

More dedicated Utopia, would act with Erasmus in getting the book printed by Dirk Martens, for whom, he had worked as a corrector as a young man.[20] Erasmus's hand is also evident in these developments. He had used Martens (fig. 1), the only printer in Leuven just then, to print many of his works. Since his own *Moriae Encomium* had been reprinted in Leuven in 1512, Martens – the prototype of the printer-humanist – would also have been the obvious choice when it came to printing More's *Utopia*.[21]

Without Erasmus's intervention (cat. 13A and 14A; fig. 16, 17) More's *Utopia* would have looked quite different. For in September 1516, before the manuscript was handed over to Martens for printing, Erasmus added the so-called *parerga* – a handful of epigrams and the introductory letter to Pieter Gilles.[22] Thus More's utopian flight of imagination was more than just a gratuitous, literary form of escapism. His *Utopia*, printed less than a year before Luther nailed up his ninety-five theses on 31 October 1517, was written as a reflection on the political and social developments in his own England. It was the response of the erudite humanist and man of letters, of the statesman and thinker, to an extraordinarily eventful and nervous century in which European powers evolved from medieval states and kingdoms, the Church was reformed and religious barriers were erected, and civilians were dragged into bloody strife. Erasmus's prefatory matter made it clear that the work was embedded in the humanist intellectual milieu. Consequently, More's pithy dialogue could only be read 'humanistically'. If the reader was struck by the many notable echoes of Erasmus's own *Praise of Folly*, he also recognized many classical and contemporary thinkers on politics and society. For instance, if Utopia's constitution is 1760 years old at the time that Hythlodaeus visits the island, the educated reader picks up the allusion to the Greek philosopher Plutarch's *Life of Agis*. Agis (265–241 BCE) was a king of Sparta who sought to restore the laws and institutions initiated by the legendary Lycurgus, to redistribute the land in Sparta, increase the number of citizens and also allow non-citizens to share the advantages of his reforms, idealistically setting the ball rolling by giving up his own lands and other assets.[23] Then again, More himself also displayed striking parallels with the social, political and religious views encountered in *Utopia* in more than one place in his other works. In his letter to the monk John Batmanson, dated 1519, More reverts to man's

20 Heireman 1973: 117–125.

21 Kronenberg 1967: 134–136; Heireman 1973: 159–163; Adam & Vanautgaerden 2009. For a general study on More and his printers, see Devereux 1975.

22 Allen 1963; Cave 2008.

23 Schoeck 1977: 277–278.

selfish greed in terms that echo Hythlodaeus's words in *Utopia*.[24] And in a letter to William Gonell, perhaps written a year earlier, More's scale of moral values is that of *Utopia*: first virtue, then study that leads to love of God and true charity.[25]

Lucian, Plato, Augustine: Biblical Humanism or Dystopia?

In *Utopia* More exposes human folly with a generous helping of irony – an irony also evident in his Latin epigrams. It is the laughter, invariably and unvaryingly serious, of Erasmus's *Moriae Encomium*. *Utopia* – that voyage to an imaginary world – should positively be read in a 'Lucian' spirit: poised between the present reality of apparent objectivity and gripping fantasy.[26] Like Amerigo Vespucci's imagination-stirring account of the New World, the fictional elements in More's utopian world have an extra dimension. Like Lucian, More gives his dialogue and characters imaginary names that are usually rooted in Greek and are in themselves a commentary on their function in the dialogue. But before everything else, More regarded Lucian's *True History* (Vera Historia), written in 165 CE, as the exemplar of an antique 'science-fiction' story in which the traveller arrives at the Island of the Blessed to discover that of all the great men of the past, Plato alone is missing. It was said that he was 'living in his imaginary city under the constitution and the laws that he himself wrote'.[27] Plato, model for the entire dialogue from *Utopia*'s beginning, is lost in mental space.

Yet for all its humorous irony, *Utopia* cannot be dismissed simply as a witty satire or piece of comic fiction.[28] For it is just this passage from Lucian's *True History*, after all, that More refers to as a key moment in the great Platonic dialogue, *The Republic*: the moment when Socrates argues how philosophers should engage in society and politics. Glauco, to whom Socrates is talking, sees the state of the philosopher as something that exists only in 'the city whose home is in the ideal,' which 'can be found nowhere on earth.' To which Socrates replies, 'Well, perhaps there is a pattern of it laid up in Heaven for him who wishes to contemplate it and so beholding to constitute himself its citizen.'[29] This ideal state is also the Utopia of Raphael Hythlodaeus, the Portuguese voyager who sailed with Vespucci to the New World and had seen the wondrous isle of the Utopians.

Twining through Hythlodaeus's detailed descriptions is a skein of humanist nostalgia, a longing for a primal state of social innocence, for a Pythagorean life of community, collective ownership and mutual charity. It is the peaceful world of Ovid's *Metamorphoses* and Augustine's *City of God* (cat. 35), where 'the righteous men in primitive times were made shepherds of cattle rather than kings of men'.[30] If More saw a mirror of his own time in late antiquity's crisis of conscience, his *Utopia* – one of the many possible outcomes of the eternally recurring in the history of Providence – had to be the counter-image of what Europe had become and had left us as a legacy.[31] The ideal commonwealth of Utopia lay in between Augustine's divine and human cities. It is a building under construction in which a utopian catharsis releases a Christian-humanist Erasmus-inspired society.[32] For if More's *Utopia* was often interpreted as the tantalizing revelation and successful literary portrayal of an ideal society, a moneyless state with few laws and with property held in common, his real concern lies elsewhere. The question that exercises More is what process should be followed in order to put such a society into practice. One in which private property is not protected and privileged, in which one must learn to resist the 'idols of society'. It is with that end in view that gold is kept not in a well-guarded treasury but in the form of chamber pots, slave chains and other ignominious and publicly embarrassing objects – a nod to Plato's *Republic* and *Laws*,[33] and also to Vespucci's account

24 Rogers 1947: 165–206; Rogers 1961: 114–144.

25 Rogers 1947: 120–123; Rogers 1961: 103–107.

26 Margolin 1989: 307: 'L'habileté de More, avec ses interventions personnelles, soit dans sa lettre-préface à son ami Gilles, soit dans les premières pages du récit lui-même, où il évoque sa présence physique à Anvers, ses rencontres, dont celle d'Hythlodée, a pour résultat que l'île d'Utopie se présente à nous à la fois comme une réalité qui s'intègre dans les récits de voyages et de découvertes de l'époque (le nom de Vespucci est cité dès le début), une île lointaine, encore très mystérieuse, isolée au milieu de l'Atlantique, dans l'hémisphère austral, bien au-delà de l'équateur et de la zone torride des tropiques, à mi-distance de la Flandre et de l'Equateur, et comme une création littéraire ou philosophique. Elle se situe donc dans un espace à la fois réel et imaginaire.'

27 Lucian, *Verae historiae*, II, 18.

28 Branham 1985: 23–43.

29 Plato, *Republic*, 592 A–B.

30 Augustin, *City of God*, 19, 15.

31 Prévost 1978: xvii. See also the essay by Hans Cools in this catalogue (pp. 21–28).

32 Prévost 1978: xviii.

33 Socrates restricts the possession of gold and precious metals to the category of 'guardians'(see Plato, *The Republic*, 416e–417a); in the *Laws* a law forbids any private person to possess, trade in or exchange gold or silver (see Plato, *Laws*, 742a). These ideas were inspired by Lycurgus's legislation in Sparta.

of the New World, where the indigenous inhabitants prefer feathers to gold and pearls.[34] The town planning and political, social and religious endeavours of Utopus, the conqueror from whom the isle of Utopia takes its name, have enabled the Utopians to rise from a level of relatively crude barbarism to a level of civilization that is, in fact, matched nowhere in the world.[35] Learning is a monastic rhythm of manual work and time for development of both men and women, the giving up of senseless pleasure and the assumption of a Pythagorean life of community, collective ownership and mutual charity. The reward is security and support in sickness and in old age. It made Guillaume Budé (1468–1540), the eminent scholar of ancient Greek and a friend of Erasmus, rename Utopia as *Hagiapolis* ('Holy City').[36] It caused William Tyndale (*c.*1494–1536), the exiled English theologian and Bible translator who would be executed in Vilvoorde, to dub More's *Utopia* fantastic 'poetry'.[37]

'Infection' from the mainland – with, as was well-known, the moral, social and cultural decline that resulted from private property – was avoided and resisted by massive excavations and therapeutic and appropriate education. If More's *Utopia* is a pointed analysis of that moral infection and if, in his ironic story of prevailing political practice, he offers a contrasting ideal imaginary counter-narrative then 'Morus' – More's fictional self in the dialogue – is realistic and sober in his final words: 'there are very many features in the Utopian commonwealth which it is easier for me to wish for in our countries than to have any hope of seeing realized.'[38]

If the 'placeless' *Utopia* by the humanist man of letters Thomas More seems like of one of his most 'Erasmian' literary productions, embedded in a specific intellectual and cultural context,[39] then his literary and philosophical dream of a distant world, distant in time, distant in space, still found a 'place' on the island of Utopia. His remarkable neologism and his new genre continue to inspire. They brought about a new way of thinking about the gap between what should be and what is.

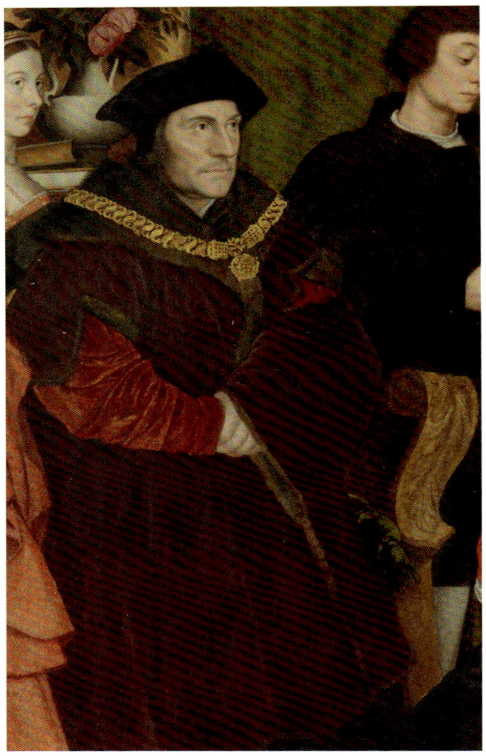

[fig. 18] Rowland Lockey, follower of Hans Holbein the Younger, *Portrait of Thomas More and his Family* (detail: *Thomas More*), 1593, London, National Portrait Gallery (cat. 5).

34 See the *Quatuor Americi Vespucij navigationes*, St. Dié 1507, I, sig. c1v, cited in Surtz & Hexter 1965: 428. See also cat. 48.

35 Surtz & Hexter 1965: 236–237.

36 Schoeck 2002: 97, with reference to Surtz & Hexter 1965: 12. On Budéas as a reader of *Utopia*, see De la Garanderie 1989: 327–338.

37 Cited thus in Logan 1983: 8, who in turn refers to Chambers 1935: 143.

38 Surtz & Hexter 1965: 246–247.

39 Papy 2000: 118–128; Mout 2015: 60, 65–76; Papy 2015: 84–88, 90–97.

Erotic Utopia: the 'Garden of Earthly Delights' in Context

PAUL VANDENBROECK

Utopia: Thomas More created the term in the sense we know today. But long before that learned humanist devised his ideal society, many had dreamed of 'a different life' with forms of society and types of experience that bestowed only pleasure. It didn't have to lead to theoretical musings. Popular cultures produced fantasies of concupiscent paradises located somewhere on this earth but with a supernatural timeless dimension, ruled by a fairy or other eldritch being, where one wallowed in physical pleasure till the Day of Judgement arrived. Ecclesiastical and bourgeois culture had no idea what to do about these dreams, shunned them as diabolically inspired, but still wondered whether they might not actually be true. This half-earthly, half-supernatural pseudo-paradise was known as the Grail.

In the fourteenth and fifteenth centuries, however, what appeared in the aforementioned myth as a fantasy was put into practice by heretical movements such as the Brethren of the Free Spirit. These free-thinking groups recruited their members from the fringes of society: folk who had nothing to lose. They sought to revive the state of blissful perfection that they attributed to 'Adamic' man. In a flurry of 'communism', eschatological mysticism and end-time expectation (chiliasm or millenarianism: the Millennium based on natural law and the 'Spirit') they practiced the cult of Adam in order to reawaken the lost paradise right here and now. Their utopia encompassed nudity, uninhibited sexual practices (including homosexuality, group sex and other carnal pursuits), the rejection of every religious and social institution, the renunciation of personal property, and so forth. The Brethren of the Free Spirit, the Picards of Bohemia, the heretics of Durango in the Basque country, the Ortlibarii of the Upper Rhine, the Turlupins in France and the Brussels chiliasts led by Heilwige Bloemardinne (1265–1335) were just a few of such movements,

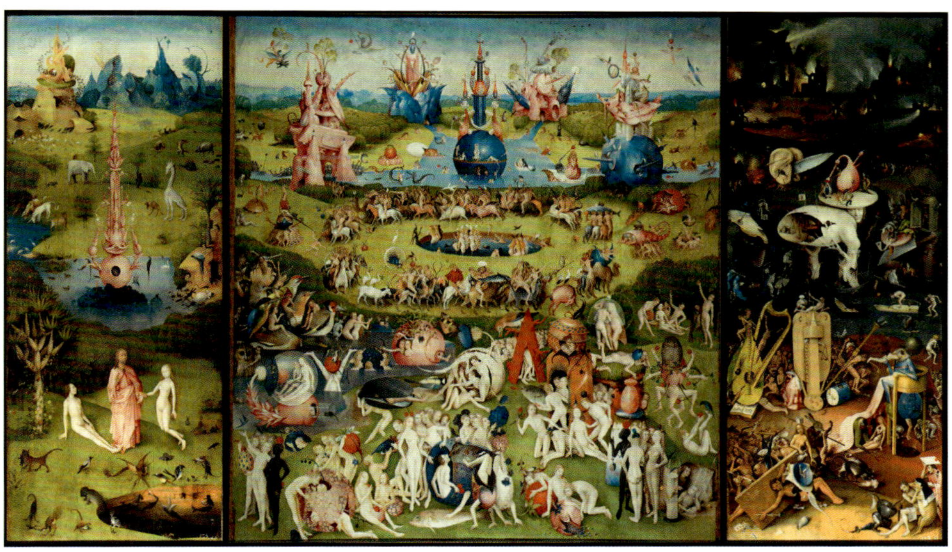

[fig. 19] Hieronymus Bosch, *Triptych with the Garden of Earthly Delights*, interior, c.1495-1505, Madrid, Museo del Prado.

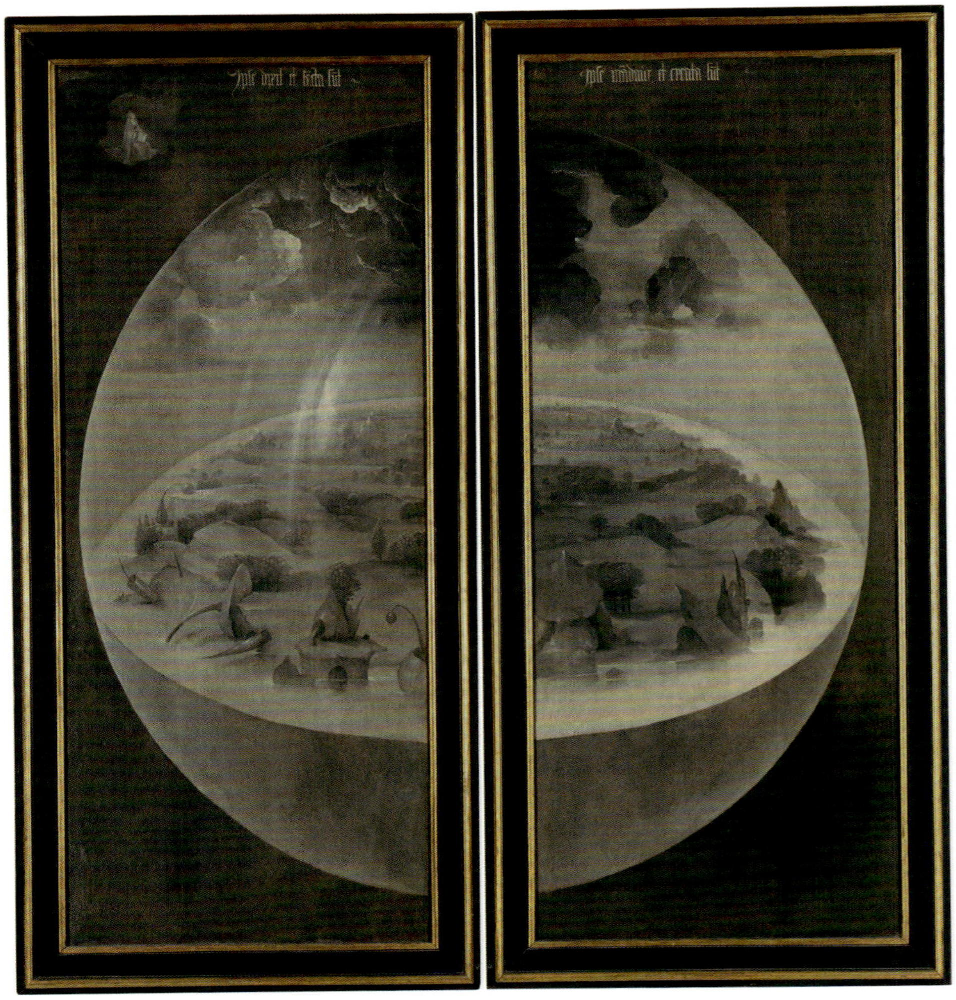

[fig. 20] Hieronymus Bosch, *Triptych with the Garden of Earthly Delights*, exterior, c.1495-1505, Madrid, Museo del Prado.

which took root more of less throughout Europe. In 'establishment' eyes of the time they were the ultimate threat to society's social and religious foundations. To the powers that be, these utopias represented not just an intellectual exercise but a pernicious belief that must at all costs be converted – by fire and sword if need be.

The burgher class, which from the twelfth to the fifteenth century had established and hardened its economic and political power, created its own concept of culture, its own image of man. Its members were, nonetheless, equally fascinated by the idea of free love, boundless pleasure, and the absence of any of the social strictures that as a class they deemed essential to civilization. From that convergence of wild longing for pleasure and moral prohibition emerged phantasmagoria like Hieronymus Bosch's triptych, known by its modern title of *The Garden of Earthly Delights*.

The original title – if there ever was one – of Bosch's most famous work is unknown (fig. 19).[1] The triptych is a 'history of the world', beginning on the outside with the Creation then, inside, pausing in the Garden of Eden for the joining of Adam and Eve, lingering long over the 'course of the world' on the centre panel, and ending in Hell on the right wing.

On the triptych's exterior (fig. 20) – though not copied in the tapestry (cat. 22) – is an enormous transparent sphere that fill the panel's surface entirely: encapsulated within it is a circular flat earth. Above and outside the sphere sits the tiny figure of God the Father, and at the very top of the panels are lines from the Psalms: 'Ipse dixit et facta sunt, ipse mandavit et creata sunt' (Ps. 33, 9 & 148, 5: 'For he spake and it was done; he commanded and it stood fast'). Conspicuous in the landscape are the outlandish artificial-looking organic and inorganic forms that appear so often in Bosch's works. They're generally

1
Hieronymus Bosch (c.1450–1516), the so-called *Garden of Earthly Delights*, c.1480–1485, oil on panel, 220 x 389 cm, Madrid, Museo Nacional del Prado, inv. P02823. See Vandenbroeck 1989: 9–210; Vandenbroeck 1990: 9–192. These studies include all the requisite bibliographic references, which are not repeated here. The content of that very detailed analysis of this triptych is still largely valid, even though new interpretations have appeared. The only point that I would now reformulate is the issue of the patron. On this topic, see Vandenbroeck 2016.

associated with temptations of saints or scenes of hell,[2] but they occur in the Garden of Eden as well. They allude to the rampant sexual power and 'charging' of Nature at the beginning of time. They were created by God but are no less dangerous because of that. It borders on the heterodoxy that Bosch saw the Earthly Paradise – which was thought of as a 'lower' paradise for the reasonably good souls in the Hereafter – as implicitly 'natural' and sexually charged.

Depicted on the left wing of the open triptych is the First Marriage in the Earthly Paradise. God, in the visible person of the Son, unites the first human couple (fig. 21). Countless animals and the same sort of dream-like organic and inorganic forms as those on the exterior panels demonstrate the sexual drive in nature. At the point where the diagonals intersect – the very centre of the panel – the phantasmagoric paradisal fountain is home to a little owl surrounded by birds. In the fifteenth and sixteenth centuries the owl was a symbol of seduction and deceit by seemingly innocent Evil. A refrain from the collection of poems by Jan van Stijevoort (*c.*1524) brings all these ideas together and explains the Boschian motif: marriage, as it was instituted by God, has been corrupted by those who have twisted God's instruction to 'Be fruitful and multiply' into an endorsement of unfettered lust and sexuality for its own sake.[3] In connection with the subject of marriage it alludes to the perversion of marriage instituted by God. Thus we already have an indication of what the contents of the centre panel are going to be. The role of Nature is clear from the semi-natural, semi-artificial shapes of the fountain: around 1500, structures of this kind symbolized the *opus naturae*, the 'work of Nature'. Moreover, the paradisal fountain is a kind of tree-fount, traditionally an image of fertility and life force (see also cat. 29).

The centre panel is one of the most famous paintings in the history of Western art. In the mid-ground is a circular pool in which both dark- and fair-skinned young women seductively display their naked selves. A horde of men, highly aroused as judged by their attitudes and mounted on all kinds of real and fabulous creatures, one of which has a distinctly phallic shape, rides anticlockwise around them (fig. 22). Here too, both literally

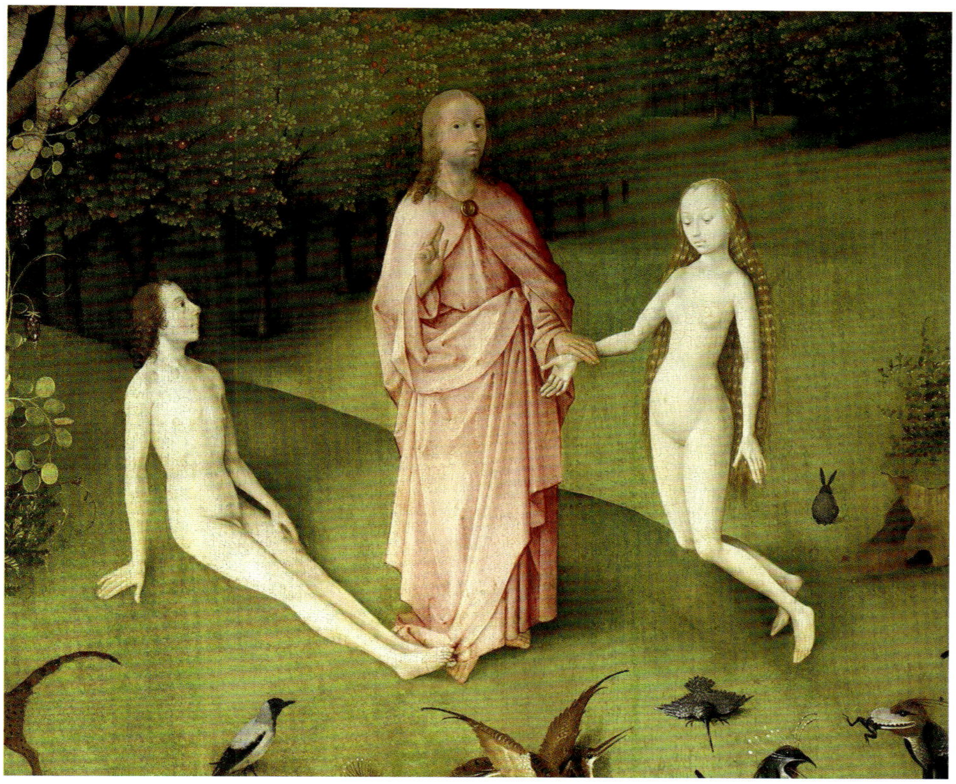

[fig. 21] Hieronymus Bosch, *Triptych with the Garden of Earthly Delights* (detail), c.1495-1505. Madrid, Museo del Prado.

2
On the *Hell* panel of the so-called *Garden of Earthly Delights* such sphere-peel and enormous fruits exist in the underdrawing, but were ommitted at the final painting stage. See: Vandenbroeck 1981: 107–119; Madrid 2000.

3
Lyna & Van Eeghem 1930: 148, CCV, vv. 19–22; Vandenbroeck 1989: 110–112.

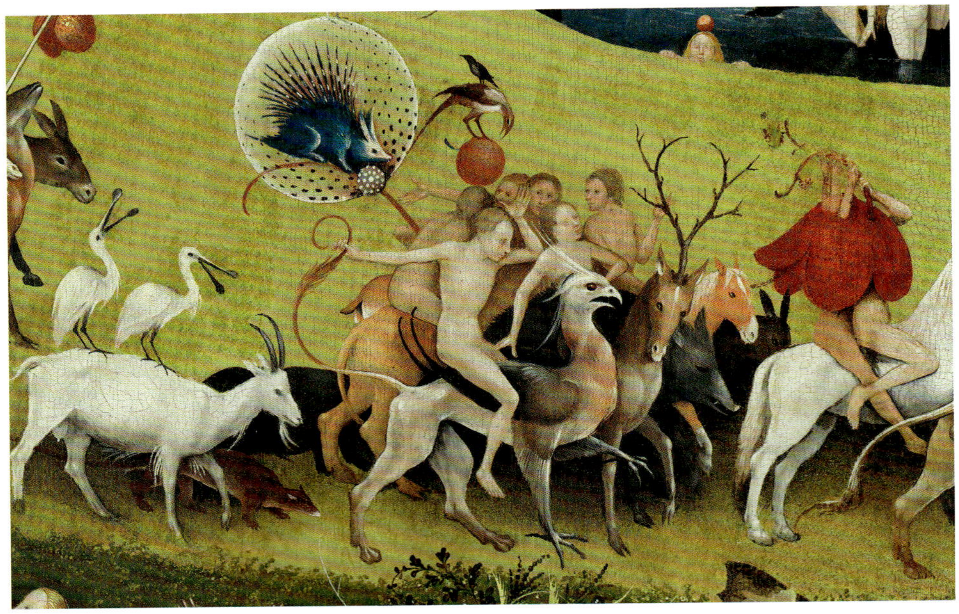

[fig. 22] Hieronymus Bosch, *Triptych with the Garden of Earthly Delights* (detail), c.1495-1505. Madrid, Museo del Prado.

and figuratively, everything revolves around woman's sexually enticing power and the effect it has on men. It's a central theme in the composition. Riding in a circle has long been synonymous with 'profane' and sinful – and in particular lustful – man. Giving in to the urge was seen as a relapse into an animal pattern that eternally repeats itself as a ride around a closed circle. Riding wild animals was an image of sinfulness, and/or 'savagery': as such it was a topical motif in late medieval images of wild men. In the foreground and background of Bosch's painting teeming crowds of people indulge in all sorts of pleasure. There is not a single piece of 'cultural' information in the scene. Apparently mankind dwells in an entirely natural state, as it was imagined in the fifteenth century: *quemque trahit voluptas sua* – 'each to his own pleasure' – seems to be the motto.

A Paradisal Nature and her Art …

Rising in the background are a colossal fountain and four fantastical buildings in pink, red and blue; four rivers branch off in four directions. Composition, constructions and choice of colour allude to the architecture of Heavenly Paradise as medieval authors imagined it. Certain components (planes, spheres, discs, spines…) occur in many of Bosch's works as elements of Nature. They have erotic connotations. The sexualization of Nature was an old holistic view of the *creatio continua*. *Natura* and *ars* are directly related, Nature herself is an artist who creates endless new *mirabilia naturae* from amorphous primal matter (fig. 23). But the nature of the matter itself, and the impulsiveness in nature and man, pervert this creative power. Evil and the sexual urge are thus inherent characteristics of Nature, which is seen as 'feminine'. Creative power, Evil, love, the (false) paradise and the feminine are all strung together in a single chain of association.

… in Earliest Times and End Times

The whole picture, moreover, corresponds to what people in Bosch's day imagined life was like in earliest times. That period, between Adam and Noah, was the first of the Seven Ages of world history. *Luxuria* or 'lust' predominated and was a prefiguration of the life

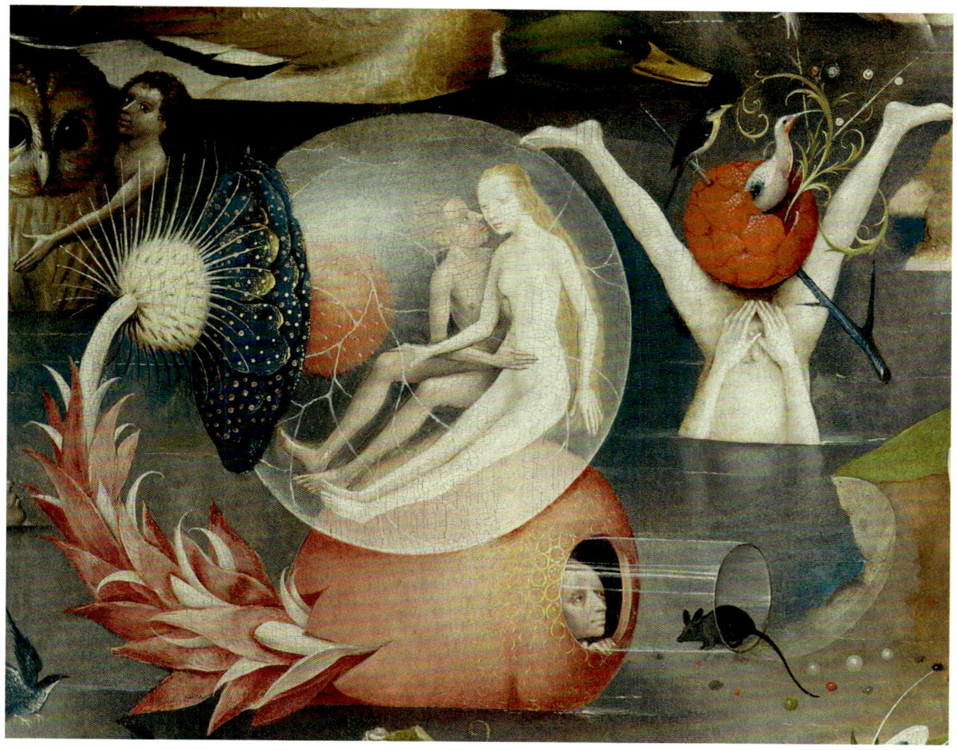

[fig. 23] Hieronymus Bosch, *Triptych with the Garden of Earthly Delights* (detail), c.1495-1505. Madrid, Museo del Prado.

of mankind before the End Times, as recounted in the New Testament. The outcome, therefore, is predictable: Hell, painted on the right panel.

Paradoxically, despite its lustful character this earliest of times was seen as a Golden Age (*aetas aurea*). Mankind lived a life of uncultivated pleasure, sensuality and freedom from care. Oddly enough, that vision of ancestral enjoyment took no account of the rather cogent notion of original sin, from which every man and woman since Adam and Eve suffers (cat. 26, 28–31). Apparently, medieval man felt an irrepressible urge to extend that paradisal existence to a long historical period (cat. 57).

The False Love-Paradise

But Bosch does not offer a purely 'historical' painting of primordial time. The outside of the triptych emphasizes the idea of natural sexuality and fertility/ procreation. Inside, on the left panel the First Marriage is the principal subject with 'Be fruitful and multiply' as a motto. But the theme of the 'days of Noah' also refers to marriage. At the bottom left of the centre panel we see a group of people pointing to the events taking place on the left wing (fig. 24). It's the visual equivalent of a topos in medieval literature: those who live wanton lives point to the biblical injunction to be fruitful as the justification for their promiscuity. In his profligate carnality man imagines himself to be living in a paradise, but it's a false and dangerous delusion… Medieval literature contains countless references to false love-paradises. Worldly love is frequently and moralistically interpreted as a false paradise. Though the authors of the extraordinarily successful *Roman de la Rose* (cat. 23), which was still being widely read in the sixteenth century, saw it rather differently. Their allegorical poem pushes sensual mysticism to unprecedented heights by promising true Heavenly Paradise to those who follow Nature's commandments and take sexual passion as a guiding principle. The false paradise is reserved for those who disregard that 'law'. This

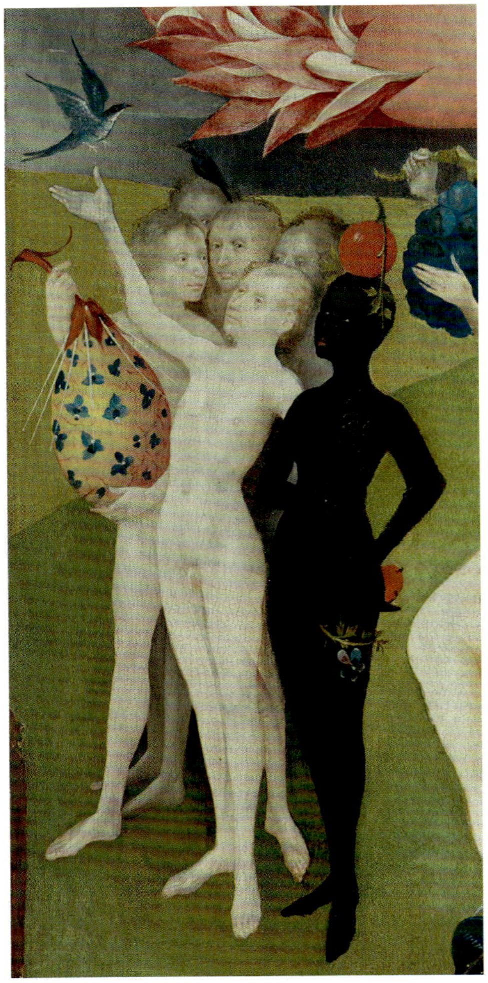

[fig. 24] Hieronymus Bosch, *Triptych with the Garden of Earthly Delights* (detail), c.1495-1505, Madrid, Museo del Prado.

theory was, of course, at direct odds with the views on marriage held by the bourgeoisie and the Church, which, while accepting the role of sex solely for purposes of procreation, feared sexuality as a cause of chaos and corruption. Proceeding from the idea of the false (love-)paradise, Bosch also makes use of the topical elements from paradisal iconography. It might be mentioned that in his oeuvre Bosch gives credence to *two* paradises in the Hereafter: the 'old' Earthly Paradise as a place for the 'reasonably good' who must undergo further purification, and the Heavenly Jerusalem which, logically, he does *not* represent because no mortal can begin to picture it. Bosch has painted only the upward flight to it.

The middle panel merges the iconography of both paradises: the 'sensuous' Earthly one (which is painted after the fashion of the stereotypical *locus amoenus*) and the 'abstract' Heavenly one (the quadrangle of fanciful constructions around the central fountain in the background). Bosch leaves the border between the two very vague; moreover, various elements make it clear that this can be either the 'lower' or the 'higher' paradise.

Where has Bosch looked for his inspiration? When he painted this triptych the worldly garden of love had already been a hugely popular subject for three centuries, first in courtly literature and later in bourgeois literature too (fig. 25). But Bosch has brought together different traditions: the garden of love, the love-paradise and the pseudo-paradise, the *locus amoenus*-nature painting, mankind in the earliest age, the doctrine of nature, fancies about paradises in the hereafter, theories of marriage and love, and religious warnings ('as in the days of Noah'). He has fused all this into one umbrella theory of the history of mankind, from Adam to Hell.

Bosch must have been aware of a popular belief, held quite generally in the Low Countries and German regions (and elsewhere), in a paradise on earth where every physical desire, especially the erotic ones, was smoothly and effortlessly satisfied. It was located between man's worldly existence and the hereafter: you did not have to be dead to enter it, but if you stayed there for more than a year you could never leave it again. Those who did stay would linger there until the Last Judgment, after which they would be damned forever. The Church and the erudite branded this place a false paradise, a sinful and diabolic 'utopia'. Hence the warning that come the Day of Judgment, all who had stayed there were doomed. It is precisely this that Bosch has portrayed in his *Garden of Earthly Delights*: his basic material persisted in popular belief but the view he presented of it was characteristic of an anti-popular culture. In his eyes, the Grail is a false, deceptive and diabolic utopia that can only lead to Hell – which is therefore also depicted in all its fearful menace on the right wing.

Bosch's triptych deals with world history and the dramatic tension between man's basic instinct and morality. The work is a mirror of marriage – see the paradisal marriage on the left wing – and a warning against the sexual perversion of it. A religious connection (from Creation to Hell), a popular myth (the Grail) and perhaps a historical theme (Adamic prehistory to the Flood) facilitate a reflection about sexual passion and control, wildness and primitivism, and the links between nature, ethics and patterns of behaviour.

Bosch was the first to depict an 'erotic utopia'. Yet the triptych remains unique in European art history. Whatever utopia was imagined to be has almost never led to a pictorial representation of it.

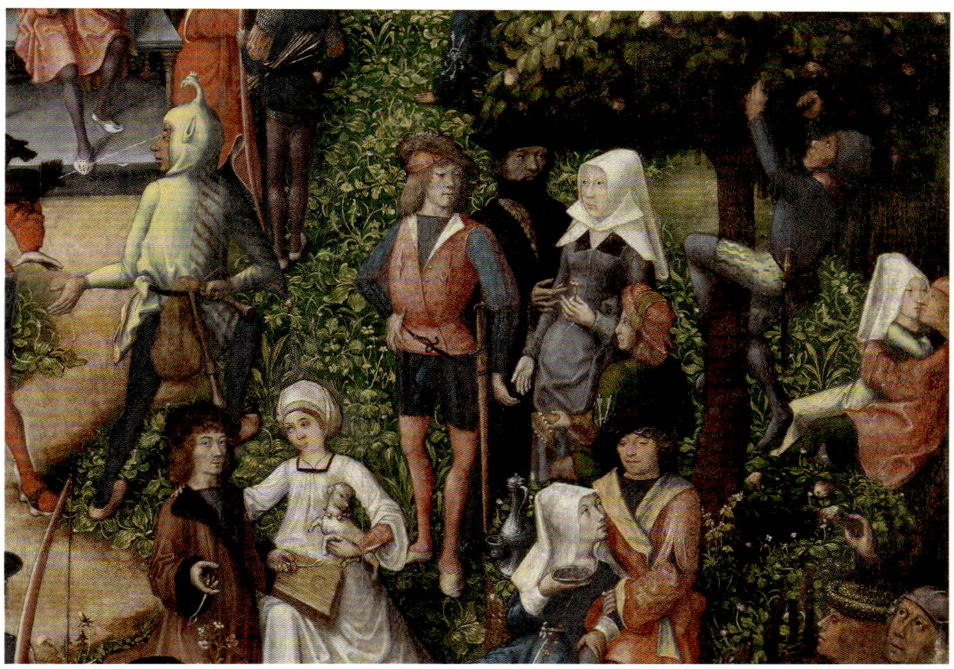

[fig. 25] Master of Frankfurt, *Festival of the Archers* (detail), 1493, Antwerp, Koninklijk Museum voor Schone Kunsten (cat. 24).

The Enclosed Garden: a Utopian and Mystical Sanctuary

BARBARA BAERT

'Besloten Hofjes' or 'Enclosed Gardens' are the collective terms for a very special group of altarpieces that were produced in the Low Countries and the Rhineland, very often in convents and beguinages, in the late Middle Ages and Early Modern period (cat. 26, 32–34).[1] These simulated gardens reflect the devotee's desire to recreate the lost, utopian paradise. The predominant motif in Enclosed Gardens is the *locus amoenus* or 'pleasant place' in which the mystical marriage between the devout soul and the all-highest will take place.[2]

An Enclosed Garden is basically an upright shallow box, sometimes with painted panels that can be closed like shutters, that is filled not with the traditional narrative carving but with material remains from holy places such as Jerusalem: stones, relics and little packets of earth are combined with *papier mâché* and wax medallions, silver-mounted gems, souvenirs of pilgrimages and parchment scrolls. The precious remains are reverently wrapped in pieces of silk or embroidery then surrounded with a profusion of decorative components in the form of flowers made of textiles and silk thread, semi-precious stones, pearls and *paperolles* (delicate scrolls of paper made by a technique known as quilling). The final result resembles an intoxicating garden that is indeed enclosed by a hedge and filled with ornaments in a whole range of media and qualities. The material remains such as stones and bones are hidden amongst an abundance of artificial vegetation and artfully imitated flowers.[3] Sometimes little sculptures made of pipeclay or wood are also added – perhaps a Calvary group (cat. 32), which also references the topography of Jerusalem, or a scene from the Old Testament such as Daniel in the lion's den (fig. 26).The tiny figures were bought from the so-called Mechelen *cleynstekers*.[4] But the fabricated flora and ornamented

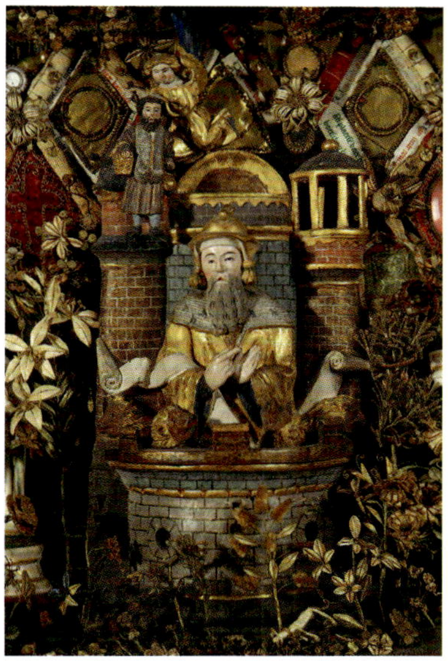

[fig. 26] *Enclosed Garden with Daniel in the Lion's Den* (detail), Musea en Erfgoed Mechelen, Collection Gasthuiszusters.

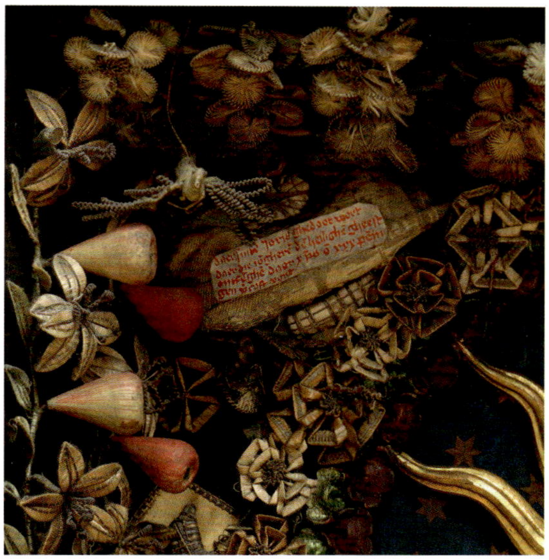

[fig. 27] *Enclosed Garden with the Garden of Eden* (detail), c.1500, Antwerp, Koninklijk Museum voor Schone Kunsten (cat. 26).

1 Van der Stock 1991: 447; Scholten 1996: 13–15; De Nijn & Eeman 1998; Triest 1998; Wolfson 2001: 64–65.

2 On this topic see inter alia Baert 2016d.

3 According to some sources the genre and technique date back to the fourteenth century. At that time the Enclosed Gardens had more modest contents with paper flowers behind glass, sometimes with a religious statuette in the middle. Gradually the paper was replaced by silk, silver or gilt and increased in size, while the statuette made way for larger groups of figures; De Nijn & Eeman 1998: 49.

4 This sculpture production is dealt with at length in Kruip 2006a.

garden was predominantly in female hands, as it had been for centuries. Thus the Enclosed Gardens evidence a division between the masculine canonical world of the figurative (the carved groups and the painting on the shutters) and the feminine decorative world of domestic art (the ornamental garden elements, the wrapped souvenirs).[5]

Enclosed Gardens are ambiguous objects that occupy a place somewhere between altarpieces and domestic furnishings.[6] They may have been used in the context of both private devotion and the *vita communis* of the convent.[7] And because they contain relics (though not exclusive ones) and all manner of material remains of a 'lesser' sort, such as stones and bones and even baglets of earth, they are also (*ditto*) shrines (fig. 27). Perhaps the Enclosed Garden can best be thought of as a kind of cabinet of curiosities whose contents are an intimate translation of the aspirations and spirituality of the religious woman.[8] Briefly put, in the late medieval convents of the Low Countries a unique form of 'mixed media' recycled and found art developed that articulated feminine devotional identity in a quite extraordinary way and maintained a stratal relationship with sacred topography.[9] Wrapping these fragments of stone and bone in fine handiwork divested them of their material mundaneness and elevated them to its antithesis – fecund life itself. In this essay I want to look at the Enclosed Gardens as the sanctuary of interiorization, as the sublimation of the sensorium (especially smell) and as a unique form of creative process.

The Sanctuary of Interiorization[10]

The Dutch word 'hof', as in Besloten Hofje – enclosed garden – is an archetype of paradise whose semantic roots are in the Persian language and culture. The Old Persian for 'paradise' is *paridœza*, rendered in Greek as *peri* (around) *teichos* (wall).[11] The garden alludes to the seclusion of the convent itself and more specifically to the garth – not coincidentally referred to as the paradise garden – where water plashes from the fountain and flowers bloom.[12] The Enclosed Gardens are also a material expression of the Old Testament Song of Solomon, in which the love between man and woman is lauded in the garden of love.[13] The bridegroom invites the bride, saying 'I come to my garden, my sister, my bride; I gather my myrrh with my spice, I eat my honeycomb with my honey, I drink my wine with my milk. Eat, friends, drink, and be drunk with love' (Song 5, 1).

The garden is a material synecdoche of paradise and also a medium through which to interiorize that 'utopian' metaphor for purity. The Enclosed Gardens are a means of seeking, finding and entering the unspotted 'spot' in the soul, and as such are consistent with many late medieval spiritual exercises. Effectively they are vehicles by which to make a mental journey to Jerusalem (or, by extension, to any holy place) without the need to move physically.[14] These 'virtual pilgrimages' are known from a genre of manuscript in which the holy places were visualized and described and thus could be 'visited' mentally. In much the same way Enclosed Gardens facilitated the spiritual journey to a physically unreachable utopian place where one can saunter in the garden or search for the natural objects hidden among the bushes and experience the delights of discovering a treasure (a relic, a souvenir).[15]

The conspicuous energy expended on the detailed description of holy places in Jerusalem can also be found in the accounts and journals penned by pilgrims. The texts are exemplars of a narrative *horror vacui*.[16] The genre is a literary counterpart to the material, repetitive and accumulative character of the Besloten Hof.[17] Jerusalem is more likely to be inaccessible than not, but it returns to the viewer via its material fragment, its synecdoche, its relics and its dead remains: hidden, wrapped and decorated and thereby sublimated. These material traces are accumulatively present in the Enclosed Gardens.

5
Silk flowers were made by both female religious and by the craft guild of the passementerie makers. The latter had a monopoly on the manufacture of ribbons, edgings and all kinds of textiles that served as decoration on furniture and clothing. Baert 2016b and Iterbeke 2015.

6
Hamburger 1990; Vandenbroeck 1994; Hamburger 1997; Hamburger 2004: 13–145; Schmitt 2004; Hamburger & Suckale 2005: 21–39; Schmitt 2004.

7
For nuanced points of view within the often methodologically rigorous split between a so-called secular and so-called sacred space, see Webb 2005: 27–49.

8
For the ambiguous typology of the Enclosed Gardens within the phenomenon of the relic, the rarity, the fetish and by extension the investigation of what constitutes 'remains' and what a 'thing' is, see the studies by Laube 2011: 31 ('Urgent questions arise: were the objects regarded as sacred or as profane? Can the idiosyncratic or strange object really be out of the ordinary or were biblical or dogmatic meanings nonetheless attributed to it?'); Weddigen 2008: 3–6; Stielau 2011: 5–16, 5 ('As an object meant to embrace another object, the etui sits in an undefined interstice between received categories. It is not comfortable a work of art. Nor, when ornamented and carefully crafted, can it be placed in the vast category of functional objects known as "material culture".') and Rudy 2015: 26–29; 235–236.

9
It appears from fifteenth- and sixteenth-century wills, inventories and accounts that these shrines were found not only in convents but also in lay collections. A will made in 1490 by one Catherine Le Quien states: 'Item, audit Colard Leurion je donne ung petit gardinet où est la Nativité de Nostre Seigneur, enclose de gourdines de soye verde, et le custode de bos.' That men also owned such shrines appears from the will of a certain Phillip d'Escamaing, dated 1455: 'Item, je donne à demiselle d'Escamaing, femme à Robert, ung gardinet ordonné de fleurs de soye, auquel est la pourtraiture de Nostre Seigneur.' Société historique et archéologique de Tournai 1897: 271, 334. See also Kaspersen 2004; Bynum 2011; Baert 2016b.

10
Baert 2014; Baert 2016a: 27–44; Baert 2016b; Baert 2016c.

11
Stammler 1962: 106–116; Grimm 1977.

12
It may be that the groups of figures in the Enclosed Gardens were an analogous reminder of the monumental Calvaries and / or Holy Sepulchres that were sometimes encountered in convents and béguinages.

13
Beumer 1954: 411–439; Reinitzer 1982; Asiedu 2001: 299–317.

14
Rudy 2011: 110–118.

The sublimated smell of Enclosed Gardens can be seen as a natural form of metonymy for the idealized spot, particularly Paradise, Jerusalem or by extension the utopian place. The Early Modern period marked a turning point in the mystical tradition of the 'love in the garden'. Although the earliest treatises in which the allegory of the spiritual garden forms the central theme date from as early as the thirteenth century, they became especially popular in the fifteenth and early sixteenth centuries.[18] They were read by the laity as well as the cloistered and were promoted by the Devotio Moderna. In *Die geestlicke boomgaert der vruchten*, printed around 1500, and in Gerard Leeu's slightly earlier *Thoofkijn van devotien* (1487), for example, the spiritual relationship between the soul and God is expressed in the drinking (hence tasting) of the water of life and the scents of the Garden of Eden (fig. 28).[19] If texts and prayers use taste and smell to channel devotional experiences and spiritual insights, why should looking at motifs in the shape of flowers and fruits not fulfil exactly the same spiritual function?[20] This reasoning certainly applies to the Enclosed

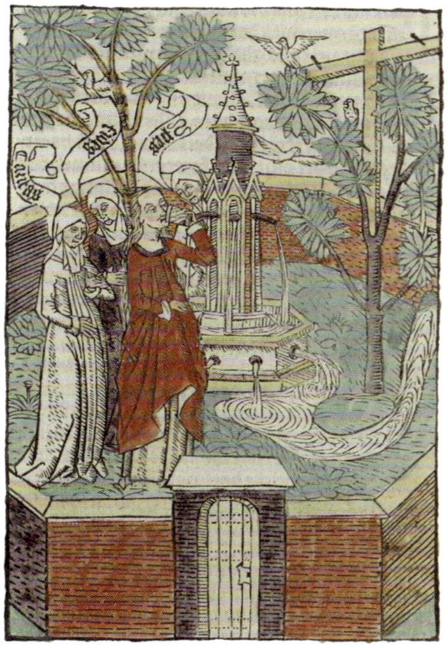

[fig. 28] Illustration from Gerard Leeu's *Thoofkijn van devotien*, Antwerp, 1487, Ghent, University Library.

Garden, which by its very nature materializes the mystical delights of the garden and is thus a garden that 'smells' in all its exuberance. The garden emits its perfume because of the nearness of God and the mystical union that takes place in it. 'Awake, O north wind, and come, O south wind! Blow upon my garden that its fragrance may be wafted abroad. Let my beloved come to his garden, and eat its choicest fruits' (Song 4, 16). The garden smells like the lost Paradise or the Heavenly Jerusalem with its natural fragments and relics and remains enhanced by its flowery perfume. The eleventh-century poem by Archbishop Guibinus of Lyons sings of the glory of Paradise and the Heavenly Jerusalem in very olfactory terms: 'Paradisi amena regio / Quam possedit quondam primus homo, / Quam pulcra es sanctis animabus et requies. // In te spirant odora gramina, / Rubet rosa, albescunt lilia, / Et arbusto profundunt balsama / Quam pulcra es sanctis animabus et requies. // Pulcher hortus, mellita flumina, / Sonat aura lenis per nemora, / Ibi flores et mala punica, / Quam pulcra es sanctis animabus et requies.'[21]

15
Koldeweij 2006; Rudy 2011: 114 ('To enter them visually is to be mesmerized by their dizzying array of flowers. To move in their first layer of interpretation is to penetrate the garden fence using prayer as a vehicle. Entering the box, the viewer enters the Holy Land on the scale of a dollhouse, an idealized microcosm of female enclosure.')

16
Rudy 2011: 117.

17
Rudy 2011: 118 ('In wrapping miles of thread around tiny armatures, were the nuns pacing off equal distances toward Rome, Jerusalem, or any of the local shrines? Not only do these hofjes provide an alternative to travelling to Jerusalem, but they provide an alternative to the bookish pilgrimage. They anticipate a body of images that let go of the tangible traces and move one step closer to the purely imagined.')

18
Falkenburg 1994: 20.

19
Thoofkijn van devotien is a Middle Dutch translation of Pierre d'Ailly's *Le jardin amoureux de l'*âme; Falkenburg 1994: 36–37.

20
Falkenburg 1994: 83 ('In the perspective of "mirrored piety" it is possible to look to the garden tracts for greater insight into the nature of devotional attitudes associated with the consumption of fruit and flowers in the "Andachtsbilder".')

21
Lohmeyer 1919: 45.

The manual work in the Enclosed Garden (fig. 29, 30) involves a twofold conjuration of time: time that envelops and advances like a viscid jelly, and time that encapsulates, cocoon-like, that enfolds fate, that entices the instantaneous into precise detailing. Once the 'time' factor is introduced we also realize that the Hofje represents a scenography of supreme meditation. Labour-intensive work, particularly when it's repetitive, like embroidery or weaving, is known to induce a flow of concentration: after a while the hands appear to continue their work independently, as if separate from the body, autonomous beings on automatic pilot. In turn the artistic 'thinking' hand – now an automaton with its own particular memory – creates a new space, a space in the spirit for meditation and prayer. The garden is never finished and demands loving care.

The 'making' of an Enclosed Garden is an action that goes in tandem with intense concentration – for which, read 'pray'. Moreover, if the meaning of the Enclosed Garden coincides with its own production process, the Hofje itself *is* the prayer space. Consequently, another sense (in addition to smell) presents itself on the periphery of this particular artistic process – the sense of hearing. It is not inconceivable that as they worked the women would have sung, prayed, muttered, or uttered words of some kind or other, perhaps a rhythmic humming sound? This rhythmic translation of hands to song, voice and speech is said to have helped to memorize or correctly execute intricate actions in textile working.[22]

As a cabinet full of curiosities, baubles and relics all twined and woven together into a garden, might one not also call an Enclosed Garden a 'nest'? Etymologically, 'nest' is related to the Hofje in the sense of the latter as a 'niche'. The French *niche* probably derives from the verb *nicher* ('to build a nest'), which comes from the Latin *nidicare* or *nidificare*, from *nīdus*: nest.[23] Thus the spatial connotation of niche emerged through a formal resemblance to the most intimate possible enclosure for something that is extraordinarily precious and fragile. Is it not tempting, therefore, to think that in a manner of speaking, scent and song envelop the nest in intimacy?

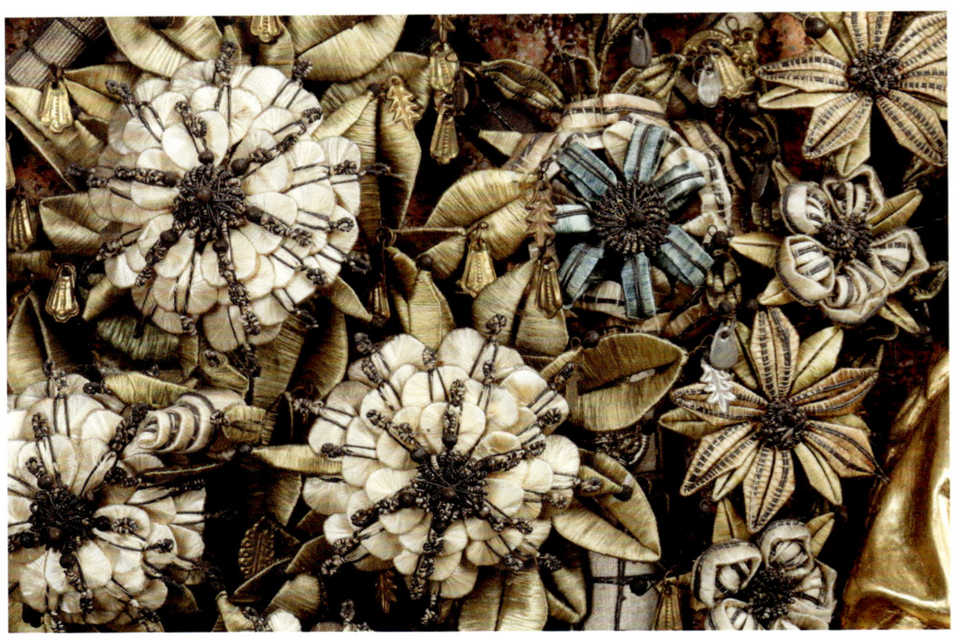

[fig. 29] *Enclosed Garden with SS Elisabeth, Ursula and Catherine* (detail), 1520-1530, Musea en Erfgoed Mechelen, Collection Gasthuiszusters (cat. 33).

22 Harlizius-Klück 2004 is a challenging study that starts from Greek semantic roots (Odysseus mythography) for the linguistic idiom relating to weaving. In the process she comes upon the track of the origin of mathematics as a cosmological model.

23 Philippa, Debrabandere, Quak et al. 2007: 416.

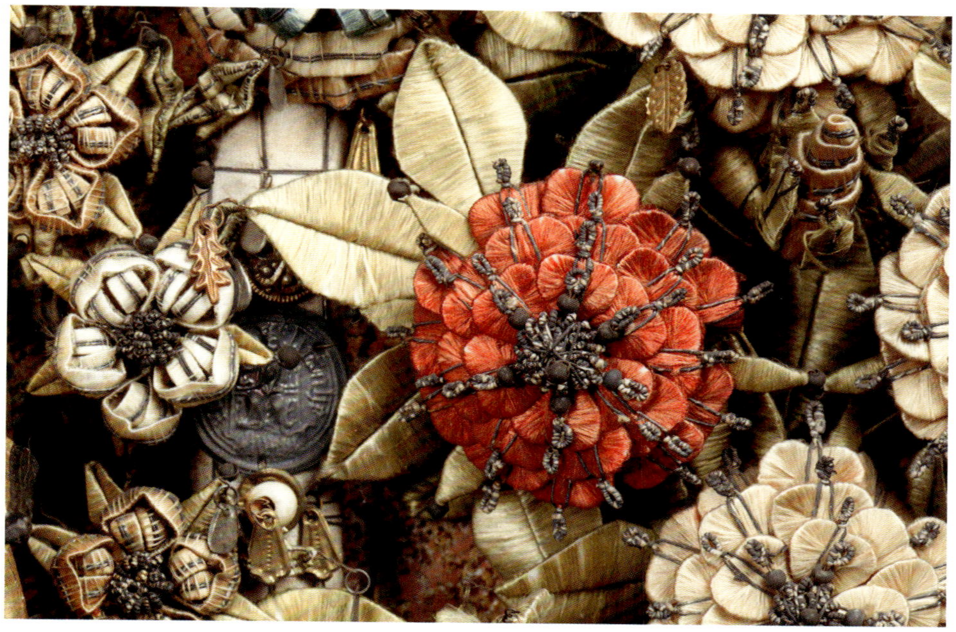

[fig. 30] *Enclosed Garden with SS Elisabeth, Ursula and Catherine* (detail), 1520-1530, Musea en Erfgoed Mechelen, Collection Gasthuiszusters (cat. 33).

Beyond the Horizon: Terra Incognita[1]

EMMANUELLE VAGNON

'The island of the Utopians is two hundred miles across in the middle part, where it is widest'

Thomas More gave his work the title of *Utopia*, which is Greek for 'not place'. In so doing he evoked a paradox that allowed him to describe in detail a real, specific and rationally organized area that is simultaneously untraceable, abstract and purely imaginary, nothing but language and imagery. More plays with the pliability of the human mind: his utopian island hovers between actual, probable and desirable, rather in the way that equivocal new islands appeared on world maps: whether they were real or merely speculative was unclear. The imaginary aspect of the places he describes is entirely in keeping with what at the time was more or less routine on maps and in travellers' tales and stories about the discovery of unknown worlds.

Utopia begins with a fictive encounter in Antwerp in which the author strikes up a conversation with the 'Utopian', Raphael Hythlodaeus, who had sailed with Amerigo Vespucci.[2] More implores him to describe the isle of Utopia in the greatest detail, which Hythlodaeus proceeds to do. Utopia, he tells More, was a peninsula until its inhabitants cut a wide channel to sever their territory from the mainland, thus turning their land into an island. That isolation, and the treacherous rocks and reefs that made entering Utopia's harbour too hazardous for strange ships, allowed their island to become completely autonomous, a place where an ideal society could be created that no longer suffered from the mistakes of the old world.

In describing his fictional island More exploited the conventions of the travel literature and map collections that were so popular during the Renaissance. The empty spaces between the islands, the blank places on the maps and the uncertainty about the world's contours left plenty of room for his ideal isle. It was neither more nor less real than the unknown areas that had been described or depicted on western maps since Antiquity – at any rate it was every bit as plausible as they were.

The French historian Jacques Le Goff once called the Indian Ocean the 'oneiric horizon' of medieval Europe: a distant, almost inaccessible place of legends and wonders. In fact that could be said of every region that was far from the Mediterranean and the Roman Empire: sub-Saharan Africa, the North, China. Whether the medieval tales were pedigree extensions of antique legends or mongrelized with dashes of popular fables and travellers' geographic descriptions, the weird and wonderful always began where actual knowledge stopped or became confused and testimonies no longer afforded any kind of certainty. Contrary to the stock cliché, people did not see the world as a closed and finite space demarcated by God's will.

As heirs of Antiquity medieval Christian scholars began to take an interest in describing the world at a very early date. In their view the world was a sphere, most of which was not yet known to man, had not yet been revealed by God. The earth was not,

1 For suggested further reading in relation to this essay, see Le Goff 1977; Schaer & Sargent 2000; Lestringant 2005; Hiatt 2008; Hofmann, Richard & Vagnon 2012; Gauthier Dalché 2013.

2 See also the essay by Jan Papy in this catalogue (pp. 31–39).

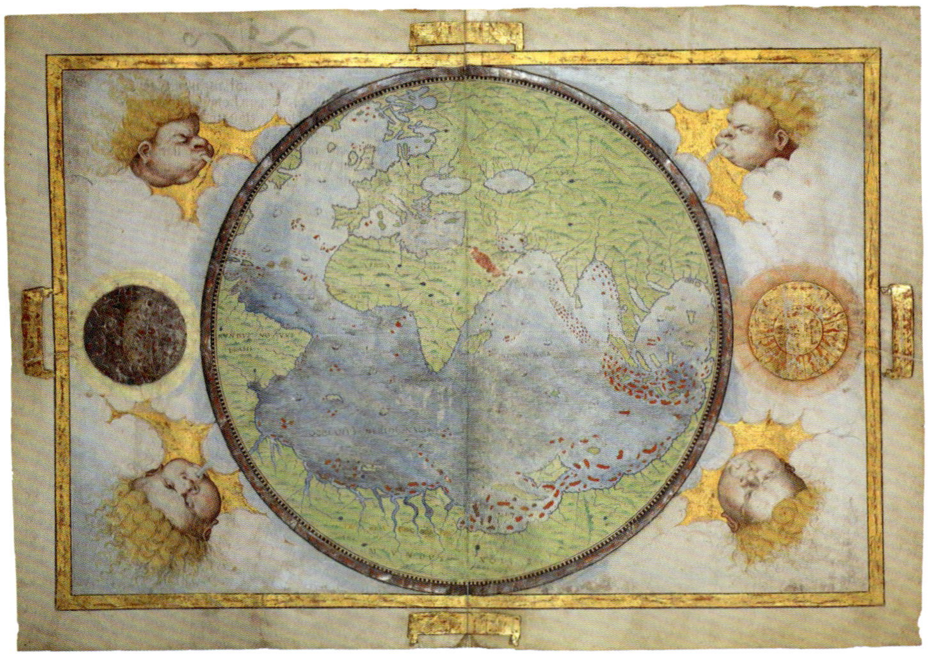

[fig. 31] Pedro and Jorge Reinel, Lopo Homem and Antonio da Hollanda, *Circular Mappamundi*, in: *Miller Atlas*, fol. 1, c.1519. Paris, Bibliotèque nationale de France, BnF Cartes et Plans (cat. 53).

therefore, a flat disk but a solid sphere in the centre of the universe (fig. 31). The habitable area, which historians called the *oecumene*, covered only a small part. Elsewhere on the globe there might be other worlds – on maps, therefore, man gave his imagination free rein.

In their simplest version, the diagrams derived from the *Etymologiae*, an etymological encyclopaedia compiled by the learned bishop and church father Isidore of Seville in the seventh century, only represent the *oecumene*. Often – but not always – they have a symbolically circular shape. East is at the top and the surface of the *oecumene* is divided between the three continents, with Asia above, Europe below left, and Africa below right. Sometimes the Earthly Paradise is also depicted; sometimes Jerusalem is shown in the centre. The three continents are portioned out to the three sons of Noah (the ancestors of post-diluvian humanity, it was believed), Asia going to Shem, Europe to Japheth and Africa to Ham (fig. 32; cat. 42). This simple and distinctly symbolic diagram remained in use in diverse forms until the sixteenth century (cat. 43), along with other diagrams and more detailed world maps on which regions and peoples and flora and fauna were indicated.

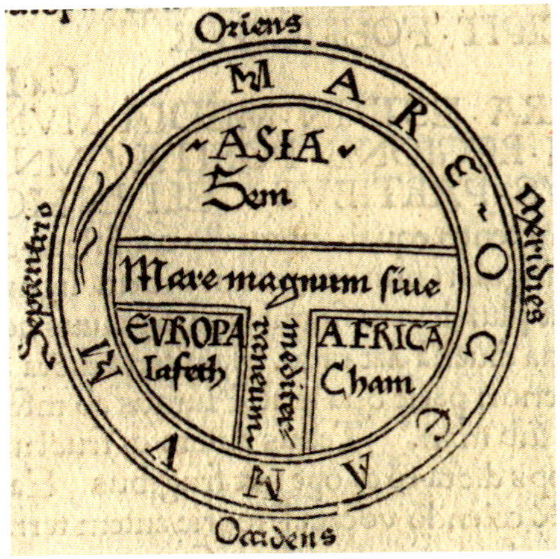

[fig. 32] *Earliest example of a printed T and O map*, in: Isidore of Seville, *Etymologiae*, published by Günter Zainer, Augsburg, 1472.

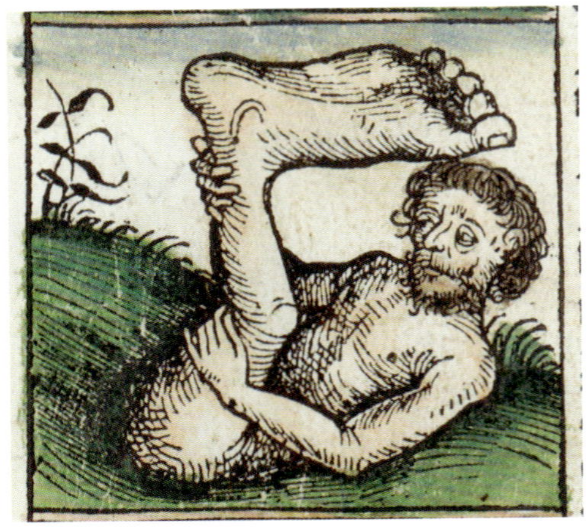

[fig. 33] Michael Wolgemut and Wilhelm Pleydenwurff, *Antipode*, In: Hartmann Schedel, *Liber chronicarum*, published by Anton Koberger, Nuremberg, 1493, fol. 12r, Private Collection (cat. 50).

The other model from Antiquity, Ptolemy's *Geography*, was hugely influential during the Renaissance (cat. 44–45). In this case the *oecumene* occupies half the northern hemisphere, spread out across approximately 180 degrees of longitude, from the Isles of the Blessed in the west to Seres (China) in the east, and in longitude reaching from slightly south of the equator to just north of the British Isles. What lies beyond Meroe, a city on the east bank of the Nile, or south of Taprobana, as Sri Lanka was once called, was *terra incognita*.

From the fifteenth century onwards new maps were made by mixing both models of the habitable world with the stories that travellers brought back from their journeys. In the handsome manuscript of *Geographia* in the Royal Library of Belgium, for instance, the Ptolemaic model is combined with a wonderfully varied staffage of peoples and animals from world maps and elements from illustrated sea charts (cat. 44). The world map in the so-called *Nürnberger Chronik* (1493) by Hartmann Schedel (cat. 50) combines several cartographic traditions: Ptolemy's trapezoidal world whose shape represents the projection of a part of the globe on a map; the antique winds that blow from all sides; and the three Biblical figures, Ham, Shem, and Japheth, peeping in from outside the frame to see their respective continents. The part-human, part-animal creatures in a column on the left of the page are reminiscent of the classical and medieval tradition of the 'monstrous races' (cat. 49–50).

Some medieval world maps suggest that there is unknown territory in other parts of the earth. In the fifth century, in line with Greek theories, Macrobius and Martianus Capella wrote that the known world covered only that part of the globe which lay between the North Pole and the equator, which were respectively considered too cold and too hot for habitation. Theoretically it was possible that other lands lay on the other side of the oceans. Those zones and their inhabitants were given all sorts of names, always related to their position on the globe. The most well known are the 'antipodes' – etymologically: 'with the feet opposite' (fig. 33). The term was first used for the people and eventually for their assumed habitat 'down under' as well. Alfred Hiatt has shown that even in Antiquity the antipodes were seen as the place of the possible and the opposite, as the place where the other held up a critical mirror to society as understood in our world. The Middle Ages saw those classical ideas reshaped into a Christian concept in which two forms of universalism went hand in hand: that of the Empire that sought to be a world-imperium and that of Christian Revelation, which was meant for all men, to the farthest corners of the globe.

'All men'. In which case you have to know what a man is. Augustine was already pondering this question in the fourth century. According to earlier Greek and Latin treatises on *paradoxa* and *mirabilia* all kinds of monstrous races lived in far-flung regions on the edge of the world, especially in Africa and India: creatures with deformed bodies – sciapods, blemmyae, cynocephali (cat. 49) – or wild men completely covered with hair (cat. 51). In *De Civitate Dei* (*The City of God*, or in French *La Cité de Dieu*; see cat. 35) Augustine subjects these pagan legends to a rigorous critique: on the one hand, it was theoretically possible that other zones did exist beyond the *oecumene*, zones that were inaccessible to men of the habitable world because of their terrible heat or the boundless tracts of ocean that intervened; on the other, however, the message of the Bible was meant for all men. If these antipodal regions existed, argued Augustine, they were either uninhabited or inhabited by creatures that were not men, the monstrous races of which everyone spoke without ever having seen them. World maps (*mappa mundi*) from the Middle Ages and Renaissance – up to which time Augustine's views were generally accepted – testify to the debate that was carried on in pagan Antiquity and by Augustine thereafter. On these maps, moreover, these monsters also occur outside the antipodes

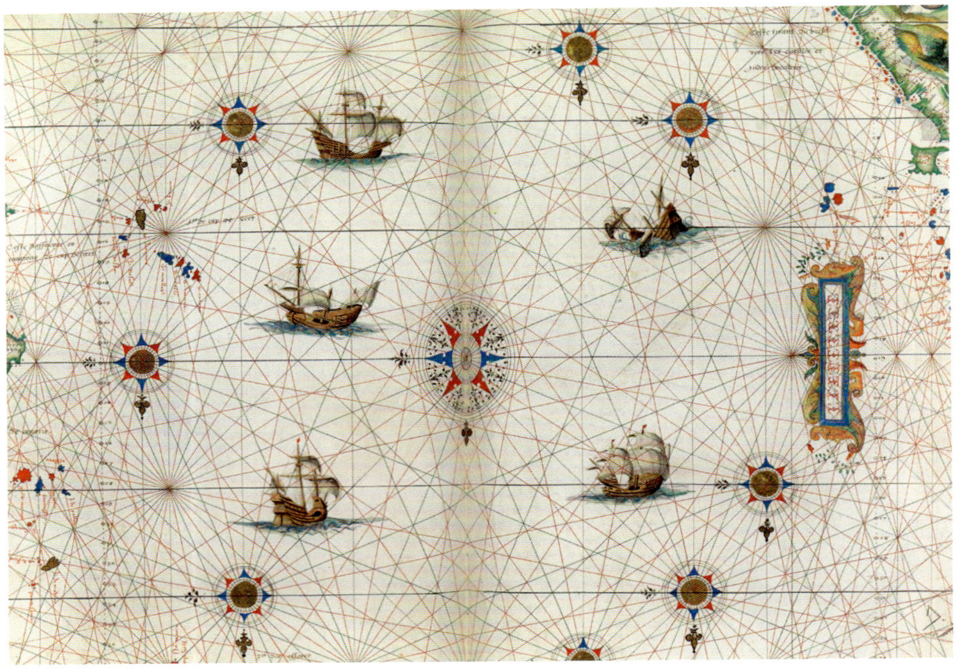

[fig. 34] *Sailing Ships on the Atlantic*, in: *Atlas de Dauphin*, Dieppe, *c.*1538, fol. 19v-20, The Hague, Royal Library (cat. 52).

(cat. 55). The cultural history of the West from Antiquity to the Renaissance is thus the background of the exploration of *terrae incognitae* and the encounter with new peoples.

It is against that background that the great navigators sailed out from Europe. When Christopher Columbus headed west across the Atlantic he expected to arrive in China, which was already shown on Ptolemy's maps and was described in the thirteenth century by Marco Polo. Bartolomeo Diaz (1488) and Vasco da Gama (1498) sailed around Africa in the wake of Alexander the Great's generals and Marco Polo to reach the Indian Ocean. Stories about heroes who discovered new worlds so that they could be conquered and converted to Christianity turned the voyages into the ideological 'great discoveries' that are etched in the Western memory (fig. 34). Through their encounters with other

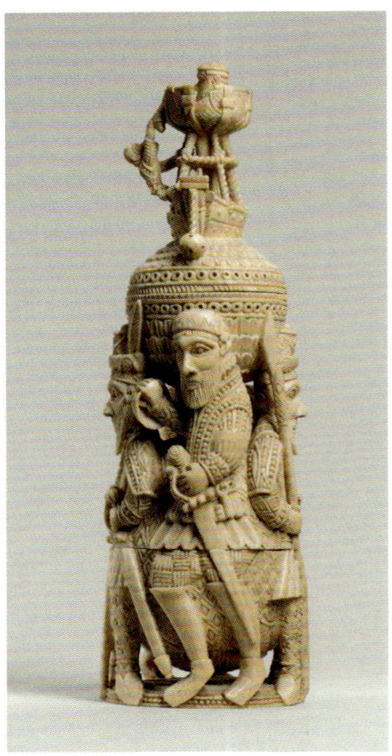

[fig. 35] Edo/Bini People, Kingdom of Benin (in present-day Nigeria), *Afro-Portuguese salt cellar*, first half of the sixteenth century, Paris, Musée du Quai Branly (see also cat. 47).

peoples Columbus and da Gama were seen as successors to Jason and Odysseus. Paintings and tapestries glorified the exploits of men of their stamp (cat. 46).

European expansion began during the Renaissance with Portuguese voyages of discovery along the coast of Africa. Following the conquest of Ceuta in 1415, ships sailed down the west coast and their crews seized or founded factories and towns where gold, salt, ivory and slaves were traded (fig. 35). The exploration of Africa and the Indian Ocean only confirmed and reinforced what was already known thanks to classical geography and Arab cartography. America, however, was another matter altogether: the discovery of that continent was a revelation. Explorers saw these new territories as a tangled string of islands and archipelagos. They had never been described, not even when the Roman Empire was at its widest. Whereas Columbus continued to think that he had arrived in Asia until, on his last voyage, he became convinced that he had found the site of the Garden of Eden, Amerigo Vespucci boldly claimed to have identified the *mundus novus*. In 1507 a group of scholars and cartographers from the Gymnasium Vosagense in Saint-Dié, amongst whom was Martin Waldseemüller, linked Vespucci's name to the new continent. Nevertheless, the Europeans' first impressions of what were, to them, the surprising animal and human inhabitants of South America were expressed in the accustomed words and concepts of classical and medieval culture (cat. 54).

On sixteenth-century maps these discoveries were represented as an improvement on the scientific knowledge of Antiquity and as a political, economic and Christian conquest of the world. Clearly marked on these maps was the division of the newly discovered lands between Portugal and Spain, as sanctioned by papal authority (in the Treaty of Tordesillas, signed in 1494, for example). Known places invariably lay next to *terrae incognitae*, unknown parts whose very existence was far from proven, though that did not prevent all sorts of ideas being formed about them. The world maps represent these new worlds ever more accurately but at the same are heavily laden with symbolism – every bit as weightily as the medieval *mappa mundi* were after their own fashion (cat. 55). The sixteenth-century maps and atlases were based on mathematical principles and mariners' observations, but no less did they embody the contemporary view of the world and of man's desires (cat. 52–53).

The classical theory of the antipodes resurfaced, now in the form of an immense land to the south of America, Africa and Asia. It appears on Pierre Descelier's *mappa mundi* of 1550 (cat. 55), for instance. This new-style *terra incognita* was the cartographic response to the hypothesis that there ought to be balance and symmetry between the regions of the southern and northern hemispheres as well as to what Magellan and his successors saw of Tierra del Fuego: to them that island was the beginning of an unknown continent, which the map-maker nonetheless populated with all kinds of animals and characters (fig. 36).

Unknown, still inaccessible territory: the new worlds seemed to be asking to be conquered, exploited and colonized. In the part of South America to the south of the tropic of Cancer (*Circulum Cancri*) or the equator are wild lands with endless forests, exotic animals and colourful parrots. The magnificent maps from the *Atlas du Dauphin* (cat. 52) and the *Miller Atlas* (cat. 53) are not useful sea charts but illustrated books about the world over which princes could pore to their heart's content while being instructed about the dangers and the riches of *Terra Incognita* (fig. 37) There they could see how large numbers of naked savages armed with spears and sticks are captured by a mere handful of conquistadors dressed in European clothes and armed with swords and guns (cat. 55) – to 'civilize' them or enslave them? The map of Brazil from the *Miller Atlas* shows 'Indians' in feather headdresses cutting down and collecting exotic wood so it can be shipped to Europe in the majestic galleons bobbing about in the Atlantic (cat. 53). Painted images of

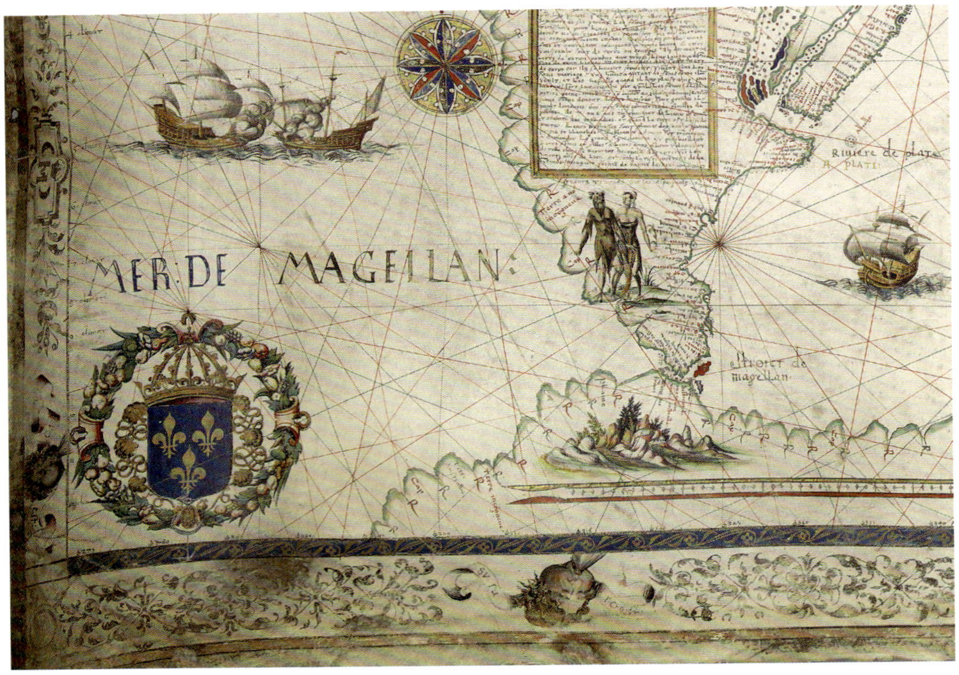

[fig. 36] Pierre Desceliers, *Mappamundi* (detail: *Tierra del Fuego*), 1550, London, British Library (cat. 55).

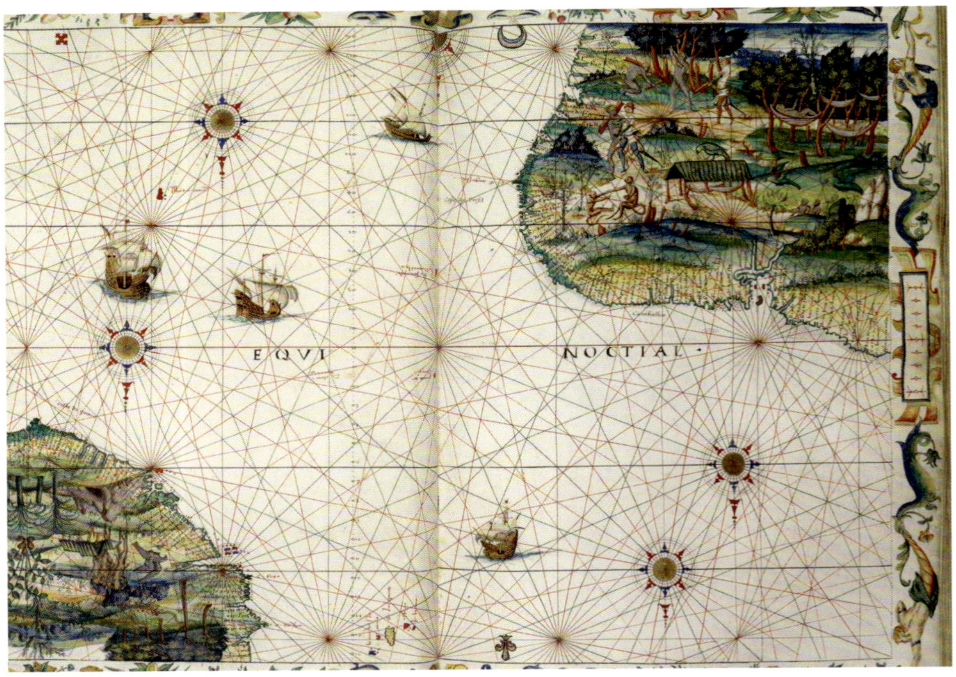

[fig. 37] *Sailing Ships on the Atlantic*, In: *Atlas de Dauphin*, fol. 11v-12r, Dieppe, *c*.1538, The Hague, Royal Library (cat. 52).

naked natives evoke the utopia of a human nature that has preserved intact the perfection of the very beginning (cat. 56–57); the sculptural nude was based on models from classical Antiquity (cat. 54).

The success of Thomas More's book made the 'not place' of Utopia a commonplace of literature and political thought: now the word 'utopia' tends to be used for projects that sprout from the imagination or ideal societies whose practicability is moot. In the Renaissance it was different: the previously undescribed worlds that appeared on the maps led many to hope that there they would encounter that ideal society, unsullied by our civilization, or be able to achieve it in one of the still virgin territories, as an inverted mirror of the old world.

The antique and medieval maps evoke an Earthly Paradise of inaccessible antipodes, those of the Renaissance *terrae nondum cognitae*, 'lands not yet known. Sometimes the desire for wealth and fame was projected onto these areas, sometimes the longing for an origin untarnished by history – undamaged nature and innocent humanity. It was not until the nineteenth century that these unexplored, vaguely sketched or merely suggested territories – the famous blank areas on the map – disappeared, as lands and seas were described with ever increasing accuracy. Utopia, though, does not appear. Tongue in cheek, More confesses that a little problem has cropped up – he has neglected to ask his 'Utopian' a vital question: 'For it didn't occur to us to ask, nor him to say, in what part of the New World Utopia is to be found. I would give a sizeable sum of money to remedy this oversight, for I'm rather ashamed not to know the ocean where this island lies about which I've written so much.'[3]

3 More 2008: 16.

The Ultimate Utopia: a Universe in the Hand

KOENRAAD VAN CLEEMPOEL

Classical Antiquity and the Renaissance: Measuring the Sublime Beauty of the Cosmos

Ever since Antiquity man has desired to reduce the universe to a human scale, to better its understanding.[1] That could be seen as the beginnings of a scientific approach, but at the same time it seems distinctly at odds with what Thomas More set out to achieve in his *Utopia*: here fiction, there reality; there an attempt to grasp reality, here an attempt to escape it by imagining an ideal world. And yet the two things are connected: the sixteenth century was an age of continuous discovery, both terrestrial and celestial. In that respect, therefore, it was not so different from the classical period when Greek astronomers and mathematicians mapped and measured the cosmos and formulated models to explain the apparent motion of planets and stars.

Despite a difference of almost 1900 years, there is a bridge between classical cosmography and sixteenth-century cartography: the porous border between objective observation and the imagination that fills the gaps. Thomas More deals with the tension between reality and fiction in much the same way.

The fictional *Utopia* appropriates the factual voyages of Amerigo Vespucci to the New World as a historical and psychological context. The great Italian explorer, sailor and

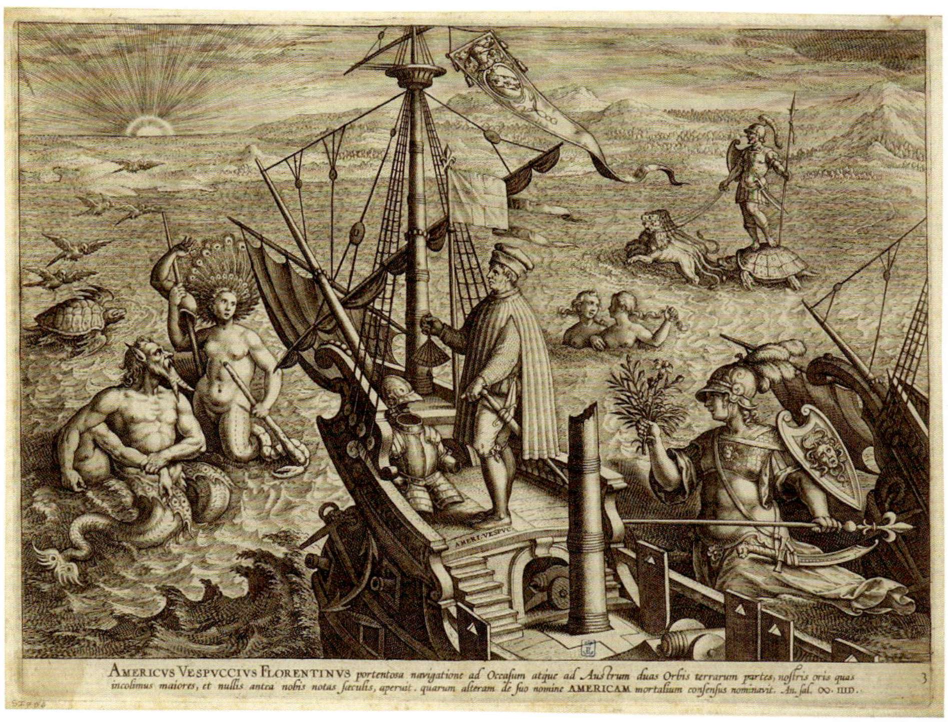

[fig. 38] Johannes Stradanus and Philips Galle, *Amerigo Vespucci aboard his Ship*, engraving in the series: *Nova Reperta*, c. 1590. Brussels, Royal Library of Belgium, Print Room.

1 On this topic, see inter alia: King 1978: 197–198.

cartographer was the first to demonstrate that Brazil and the West Indies constituted a separate and hitherto uncharted landmass, unconnected to any of the continents already known. In 1507 he published an account of his journeys between 1499–1503 in a literary genre of nautical tales (fig. 38). He narrates how, in 1504, he left behind a garrison of twenty-four men on Cape Frio in Brazil. More latched on to this particular and cast his fictitious character, Raphael Hythlodaeus, an avid traveller, as one of the twenty-four. More's fictive alter ego learns how Hythlodaeus and five companions from the garrison set out to explore the new country: 'After many days' journey … they found towns and cities, and commonwealths that were both very populous and not badly governed.' An insatiable explorer, Hythlodaeus went on to travel through many countries, until 'at last, by strange good fortune, he got via Ceylon to Calicut, where he opportunely found some Portuguese ships; and so … he finally returned to his own country.'[2]

The fascination of newly discovered lands and islands with unknown peoples, languages, customs, plants and animals is recognizable in the numerous travelogues and images produced in the sixteenth century. They evidence man's incessant curiosity combined with amazement and imagination. In that respect, the travels genre comes close to More's premise of a newly discovered island. Fiction and reality start to resemble a diptych that can hinge in either direction.[3]

The wonderment at the motion of the heavenly bodies, the observation of them and reflection on those observations, was a recurring theme in Greek philosophy. Around 150 BCE they were synthesized in the *Almagest*, a book by the Roman astronomer Claudius Ptolemy of Alexandria (*c.*90–168). His work was the conduit for the great revival of interest in cosmography in the Renaissance. It relies largely on the geocentric model devised by Plato (*c.*424–*c.*347/348 BCE) and Aristotle (384–322 BCE), which consists of ten concentric spheres that are in constant rhythmic motion, like a huge clock. In the centre is the stationary earth and around it seven planets orbit at different speeds, each on its own sphere: in outward order these are the Moon, the Sun, Venus, Mercury, Mars, Jupiter and Saturn. The eighth sphere carries the fixed stars – the *firmamentum*; the ninth sphere is the crystalline *caestallinum* and the tenth sphere is the *primum mobile*. Sometimes, the ninth and the tenth spheres are combined. The entire universe is borne on Atlas's shoulders.

The earliest reference to a mechanical model of this system appears in Cicero's *De re publica* (60 BCE) where he cites the words of Sulpicius Gallus who alludes to an invention of Archimedes (*c.*287–212) of around two hundred years earlier. In fact two spheres are mentioned: one, a very ancient invention, was a solid compact sphere with the stars that appear in the sky traced upon it, which stood in the temple of Vesta;[4] the second was openwork and showed the movement of the sun, moon and planets. Gallus explains that the second sphere, created by Archimedes and displaying the motions of the sun and moon, and the five planets, or wandering stars, could not be represented by the ancient solid globe, and that 'in this, the invention of Archimedes was admirable … In fact, when Gallus moved this sphere or planetarium, we observed the moon distanced the sun as many degrees by a turn of the wheel in the machine, as she does in so many days in the heavens. From whence it resulted, that the progress of the sun was marked as in the heavens, and that the moon touched the point where she is obscured by the earth's shadow at the instant the sun appears above the horizon.'[5]

Through his study of hydraulics and mechanics Archimedes was familiar with the way that gearwheels worked and knew that by combining them things could be made to revolve at very precise speeds. The passage from Cicero was known during the Renaissance and the mechanical sphere constructed by Pierre de Fobis (cat. 75) is obviously an interpretation of the Archimedean model. The three Greek columns – albeit of the

2 More 2008: 25 and 26.

3 See the essay by Emmanuelle Vagnon in this catalogue (pp. 55–61).

4 Ovid, *Fasti*, VI: 277.

5 Cicero, *De re publica*, I, 22, transl. by Francis Barham (http://oll.libertyfund.org/titles/cicero-treatise-on-the-commonwealth--5#toc_list)

Corinthian order – that surround the central cylinder, clearly reference de Fobis's inspiration from the antique. Moreover, the fact that the columns stand on uncultivated ground may be a metaphorical allusion to the revival of classical culture after a period of darkness.

Another reference to a Greek model of the heavens comes from the poet Claudian (*c.*370–*c.*404). He speaks of a glass sphere with an annual revolution of the sun and a monthly movement of the moon. Nearly 1900 years later, during the Renaissance, this model would also be reconstructed. In 1552 Mercator (1512–1594) made several scientific instruments for Charles V, including a small globe encased in a glass sphere on which the constellations had been engraved with a diamond and inlaid with gold. Sadly, this jewel-like instrument has not as yet been rediscovered, although Mercator's description of it survives.[6]

Spheres and Astrolabes

How closely all this relates to More's world appears from a passage in the correspondence between two of his close friends. On 18 January 1517 Pieter Gillis (cat. 13B and 14B) – to whom More dedicated *Utopia* – wrote to Erasmus (cat. 13A and 14A) as follows: I felt it my duty to write to you, since Nikolaus of Bavaria, who is an expert in astronomy, was leaving for your part of the world. He has with him several astrolabes and spheres to be sold there….'[7] These two instruments represent yet another link between classical Antiquity and the Renaissance. In each case the underlying astronomical and geometrical knowledge is Greek, but the finest examples come from the sixteenth century.

No spheres survive from classical Antiquity itself but there are still several descriptions from that period. A distinction is made between mechanical spheres – such as the one made by Archimedes referred to above – and static spheres. In the latter category there are two variants, the best known being the armillary sphere (fig. 39; cat. 65–68): a globe nestles at the centre of a framework of rings and hoops on which stars and planets are marked and which demonstrates the apparent effect of celestial bodies turning around the earth. An armillary sphere provides a schematic representation of geometric

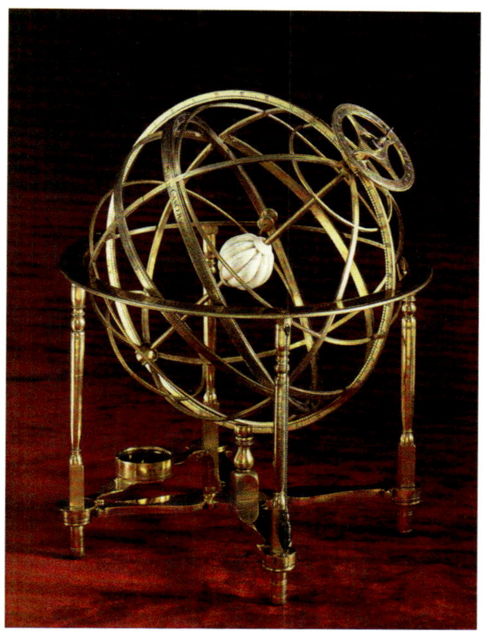

[fig. 39] *Ptolemaic Armillary Sphere*, Leuven, *c.*1540, Private Collection.

6
Watelet 1994: 368. Jan Van Raemdonck discovered this manuscript at the end of the nineteenth century and made a reconstruction drawing on the basis of it.

7
Erasmus 2006: 156, letter 515. The letter was sent from Antwerp to Leuven.

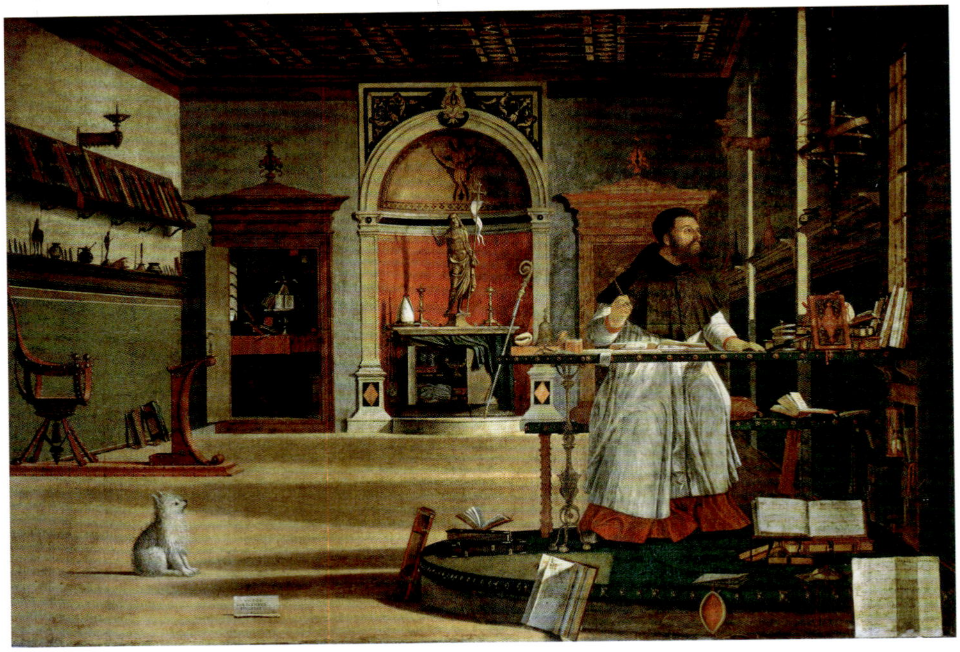

[fig. 40] Vittore Carpaccio, *St Augustine in his Study*, 1503, Venice, Scuola di San Giorgio degli Schiavoni.

relationships as perceived from the earth. It is the most evocative of astronomical instruments and was often depicted as an attribute for astronomers, mathematicians and scholars (fig. 40; cat. 77). The armillary sphere in Jan Gossaert's painting (cat. 64) of the young Princess of Denmark pointing to the meridian to indicate the latitude of her country, is exceptional. And by holding the delicate instrument in close proximity to the jewel on her bodice she lays even stronger emphasis on its precious and sublime nature. With her right forefinger she not only points to the latitude; the obliquely angled ring – which represents the horizon – leads the viewer's eye to her magnificent jewel. This sphere has a handle at the bottom, thus the little Danish princess literally holds the universe in one small hand. The type of armillary sphere that rests in a horizontal ring on one foot is more common (cat. 65–68).

The second type of instrument that Pieter Gillis mentions in his letter to Erasmus is the astrolabe (cat. 71 and 72). Like the armillary sphere it connects Greek knowledge to

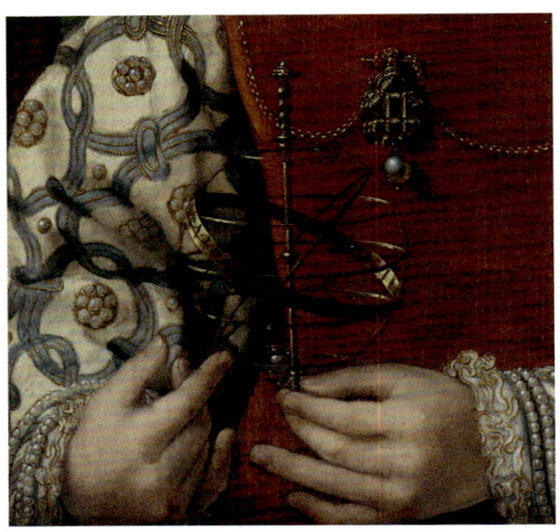

[fig. 41] Jan Gossaert, *Portrait of a Girl Holding an Armillary Sphere (Princess Dorothea of Denmark)* (detail), c.1530, London, The National Gallery (cat. 64).

the Renaissance. The underlying geometrical and astronomical principles were devised by Hipparchus (190–120 BCE), and later recorded once more in Ptolemy's *Almagest*. No astrolabes are known from Antiquity. Some do survive from the Middle Ages, but it was particularly in the Renaissance that the instrument was popular. Around 1650 it fell out of use. In the Islamic world there was also a rich tradition of astrolabe schools. Possibly they also made the connection – Toledo being one possible conduit – between the knowledge of Antiquity and the Middle Ages.

An astrolabe is more mathematically complex than an armillary sphere as it is created by means of a stereographic projection in which the terrestrial and celestial spheres coincide in a single plane. Unlike a sphere, an astrolabe is a two-dimensional instrument: an openwork disc – the 'rete' – represents Heaven (the stars and the sun) and moves above a disk that represents a particular latitude.

The magnificent tapestry from Toledo (fig. 42; cat. 69) uses the rete of an astrolabe as its central motif to represent the sky; on the left an angel sets the stars in motion by turning a crank. But the astrolabe was more than an iconographic motif: it was relatively simple way of making countless calculations – drawing up horoscopes, determining the position of a number of stars and the sun above the horizon, the exact hour (to about three minutes), and so forth. Calculations always start by measuring the angle of a celestial body (the sun during the day, the stars at night) above the horizon.

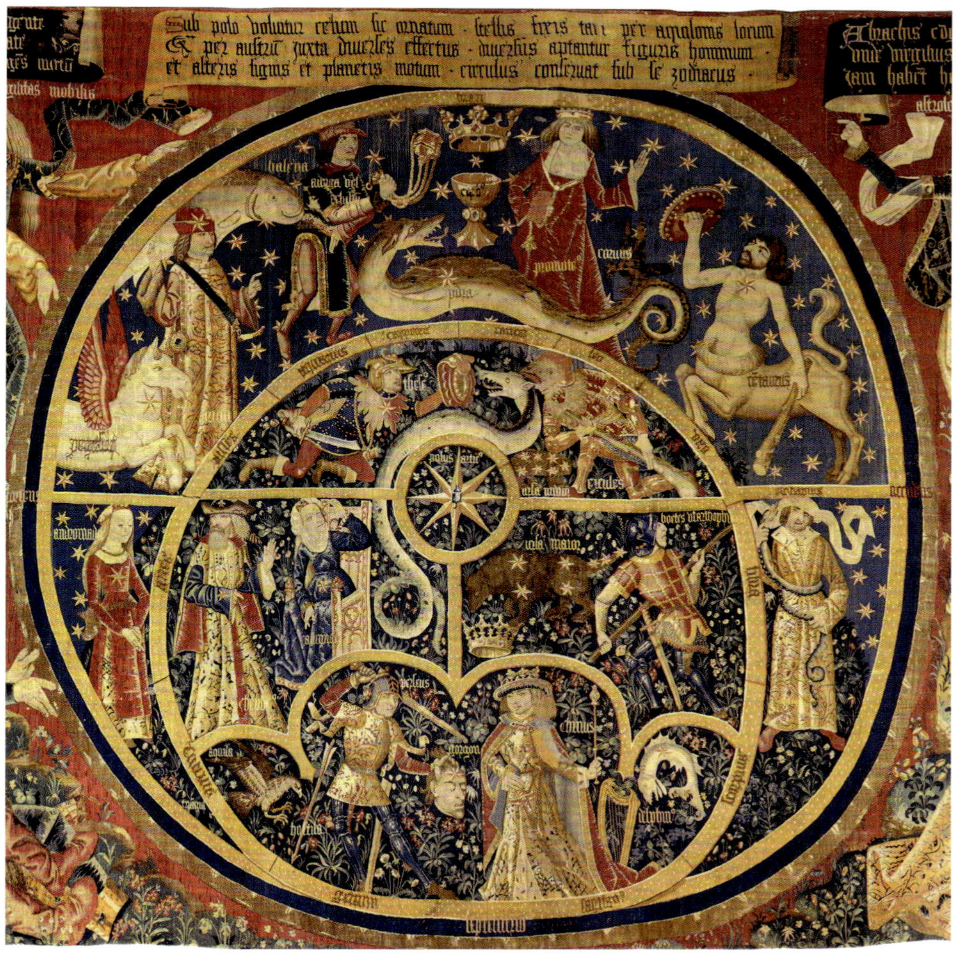

[fig. 42] Flanders/Tournai, *The Movements of the Universe* (detail), *c.*1490-1510, Toledo, Museo de tapices y textiles de Toledo (Cathedral of Toledo: Colegio de Infantes) (cat. 69).

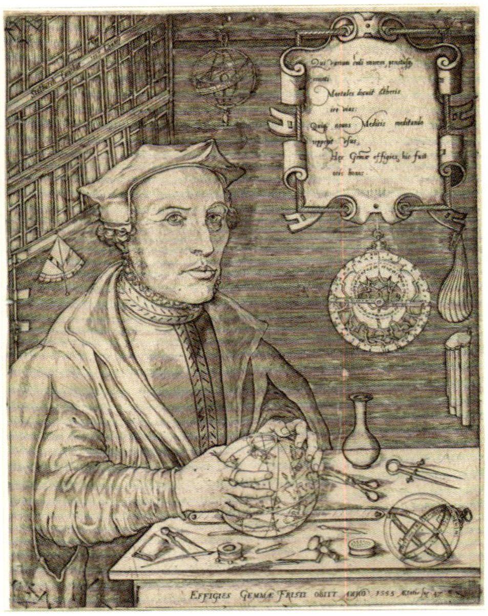

[fig. 43] Jan van Stalburch, *Posthumous Portrait of Gemma Frisius*, 1557, Brussels, Royal Library of Belgium, Print Room.

Leuven: Synthesis of Knowledge and Beauty

In 1548, thirteen years after Thomas More was executed, his countryman John Dee visited Leuven's university and was highly impressed by the meetings and conversations he had enjoyed. Back in Cambridge the English astronomer and mathematician more than once stated his admiration for his colleagues in the university town.[8] He brought back with him several scientific instruments that had been invented and made there. That combination of research into new instruments and the skill needed to actually produce them was unique. So high were the quality and accuracy of the Leuven-made instruments, moreover, that they permitted much greater precision in calculation than had ever been possible before.

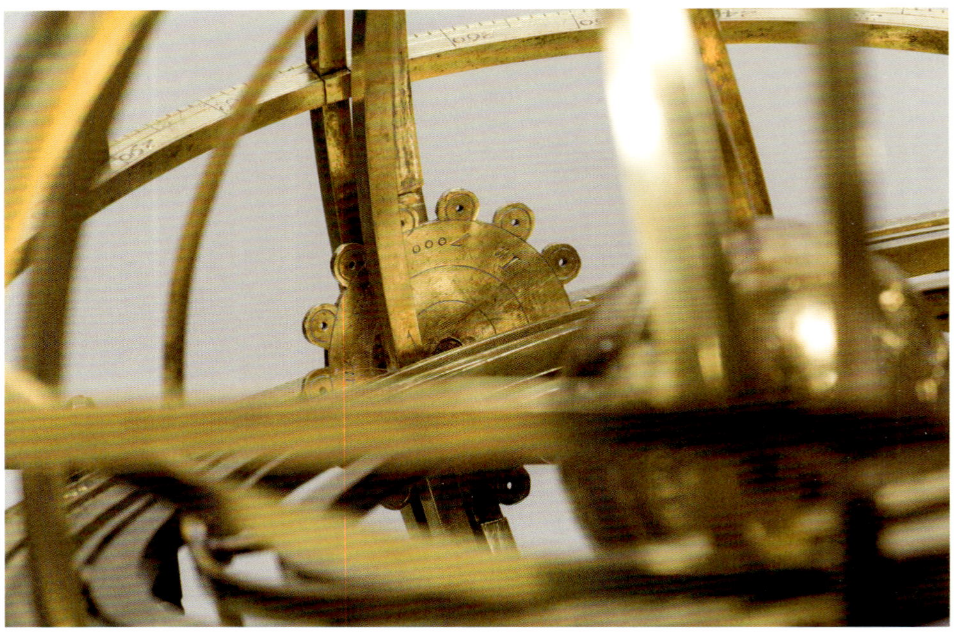

[fig. 44] *Ptolemeïsche armillaire sfeer* (detail), Leuven, 1573, Munchen, Bayerisches Nationalmuseum (cat. 66).

8 Fauvel, Flood & Wilson 2013: 78–79.

The excellent reputation of the Leuven instruments was justified. At its root was the collaboration between Gemma Frisius (1508–1555) (cat. 59) and Gerard Mercator (1512–1594) (cat. 63 and 71). Frisius held the chair of medicine and was Charles V's personal physician. In 1530 he began to devise scientific instruments. The relationship between medicine and the study of the heavenly bodies was not uncommon in the Middle Ages and Renaissance – but Frisius's exceptional application and the number of new instruments he created or improved was. He would quickly have recognized the brilliant intuition of his young pupil Mercator. Together they laid the foundation for the so-called Leuven School, which between around 1540 and 1580 produced Europe's most sought-after globes and instruments. Their status was raised even higher by Frisius's treatises on the instruments.[9] A posthumous portrait by Jan van Stalburch from 1557 (fig. 43) shows him seated at a table whose surface is covered with all kinds of tools and equipment.[10] His hands rest on a celestial globe and on the edge of the table is an instrument known as the astronomical rings, one of his own inventions. Behind him hang an astrolabe and an armillary sphere. It is not clear whether Frisius actually made instruments with his own hands; a signature on an early astrolabe suggests that he was its designer: *Authore Gemma Frisio et exaratu a Gualtero Arsenio Lovanij 1554.*[11]

The high level of finish on the Leuven instruments was coupled with new astronomical additions. When Mercator was working out the positioning of his stars on the rete of his astrolabe, for instance, he took into account Copernicus's theories of heliocentricity. Nicolaus Copernicus's *De revolutionibus orbium coelestium* had only just been published (1543), but Frisius and Mercator lost no time in embracing its ideas. A second example is the theory of trepidation or precession of the equinoxes. Four surviving armillary spheres made by Arsenius (cat. 65 and 66)[12] – after a model by Mercator – show the aforementioned Ptolemaic spheres but between the eighth and ninth sphere is a wheel that allows the eight inner spheres to tilt between the outer one: *De motu octavae sphaerae.* According to the theory of the ninth-century Arab astronomer Thābit ibn Qurra that was one revolution every 7000 years, and this is also engraved on the wheel of Arsenius's armillary sphere (fig. 44). It shows an astronomical model that seeks an explanation for the motions of those planets that were inconsistent with Ptolemy's concept. The Leuven practice of converting several astronomical theories into models on a human scale was unique in Europe. It was yet another reason why the Leuven instruments were in such great demand; why, via the offices of Plantin in Antwerp, they were dispatched to destinations all over Europe.

9
The best intellectual biography of Gemma Frisius is Hallyn 2008, in which the various treatises are discussed at length.

10
23.4 x 30 cm. There are two known copies, in Amsterdam (Rijksprentenkabinet) and Brussels (Royal Library of Belgium, Print Room). Van Sommeren 1890: 376, no. 1951.

11
This may have been one of the astrolabes that were intended as gifts on the marriage of Philip II and Mary Tudor in 1554. They display two different types of universal projection. Van Cleempoel 2002: 87–89, no. 9.

12
Now in Brussels, Greenwich, Chicago and Munich, see Van Cleempoel 2002: no. 37, 39–41.

UTOPIA OF THOMAS MORE (1516)

A Golden Book from Leuven Conquers the World

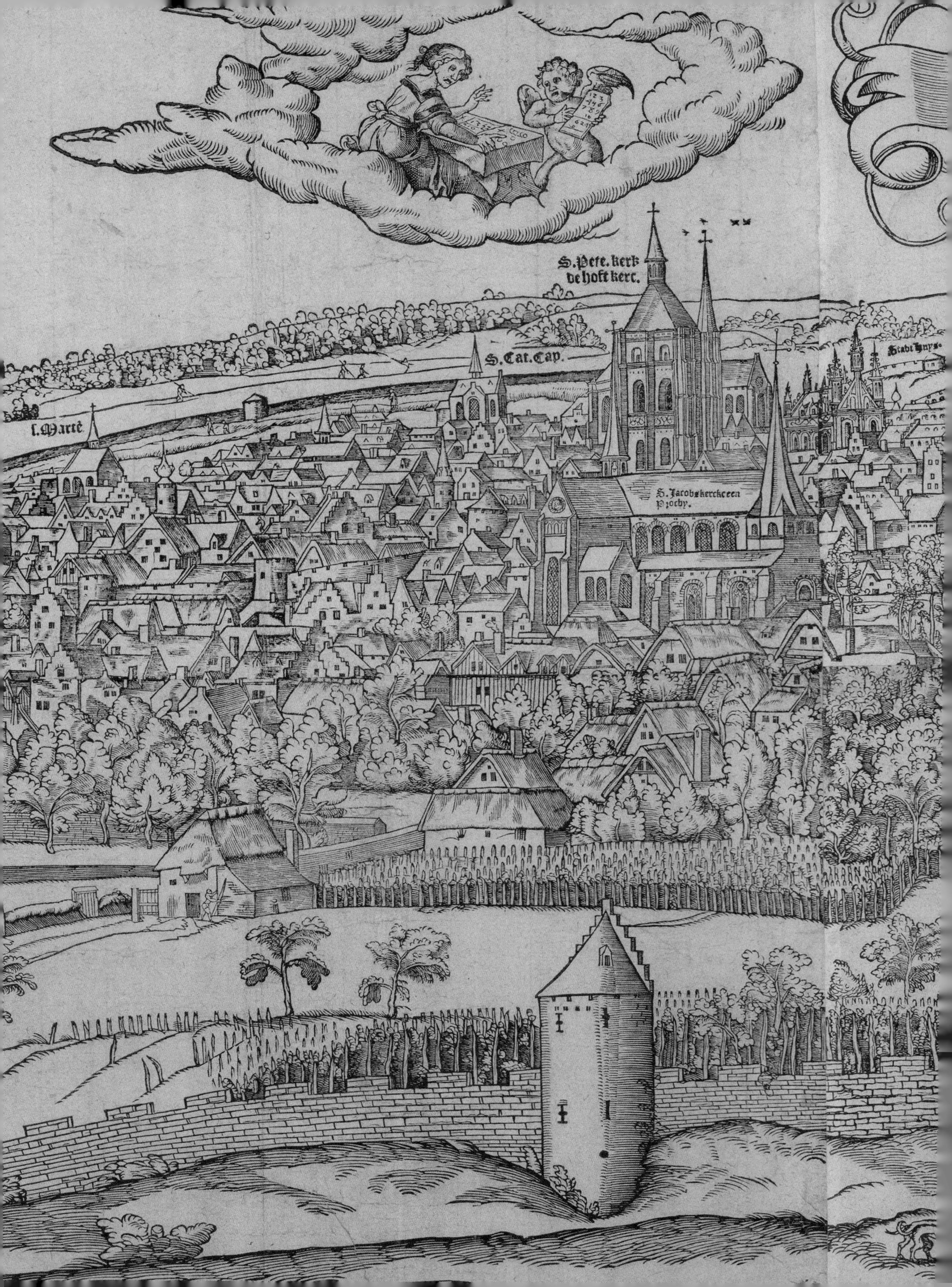
S.Pete. kerk
de hoft kerc.
S.Cat.Cap.
Stadt huys.
s.Marte
S.Jacobskerckeen
prochy.

Utopia the Book, in Leuven and the Low Countries

JAN PAPY

[CAT. 1] DETAIL

Utopia: a book whose every page is informed by Low Countries humanism. It opens and closes with Erasmus's circle of friends. Thomas More and the thirty-two-year-old Erasmus met for the first time in London in 1498. They saw each other again at Eltham the following year. A pivotal moment in their relationship came in 1509 when Erasmus, who had just returned from Italy and was staying with More, dedicated *Praise of Folly* to him. Six years later More was in the Netherlands as a royal envoy. He was lodged in the Prinsenhof in Bruges and while there he found time to visit Brussels and Mechelen and Antwerp. It was in the last of these that he met up with his friends Desiderius Erasmus and Pieter Gillis, Antwerp's town clerk, as well as Hieronymus Busleyden, a member of the Great Council of Mechelen, a powerful patron, and friend of Erasmus. While Erasmus had been mulling over the idea of writing *Praise of Folly* during his trip to Rome, More had been contemplating a political study: *Utopia.* During his days in Antwerp he also discussed with Erasmus, newly appointed as councillor to Prince Charles (the later Charles V), the latter's *Institutio Principis Christiani* (*Education of a Christian Prince*). The two humanist friends exchanged thoughts about government and administration, a new, pure Christianity, and the role of the humanist-counsellor. It was at this time that Erasmus's *Praise of Folly* and his new Latin translation of the New Testament came under sharp attack by the Leuven theologian Martinus Dorpius: More composed a long and elegant letter in defence of Erasmus and humanist scholarship in general. That letter, approximately two-thirds the length of Utopia, was written in the Netherlands. Moreover, in Antwerp, one of northern Europe's busiest trading ports, new perspectives presented themselves to the literary More: wonderful tales of new discoveries by Amerigo Vespucci were stirring people's imagination.

In 1516 More sent the final draft of *Utopia* to Pieter Gillis. Erasmus happened to be staying with Gillis at the time and he immediately saw the unique merit of More's literary work. At Erasmus's prompting, More had dedicated his satirical dialogue to their common friend Gillis. Like Erasmus, More deplored the destructive political ambitions of the 'heads of Christendom', Henry VIII, Francis I, and Popes Julius II and Leo X. Like Erasmus (and Plato) he sought a better world and with the Greek satirist Lucian, whose writings he and Erasmus had translated, he gave the high and mighty a brisk dressing down.

Between them, Erasmus and Gillis, who were to oversee the printing, turned *Utopia* into a humanist monument. They sought letters of commendation to buttress More's text. First of all, Gillis provided an introductory letter to Hieronymus Busleyden in which he presented More's island and fiction and received in return a fitting letter from Busleyden to More to introduce the utopian state. At More's request the Leuven professor of *poetica* and rhetoric, Johannes Paludanus, sent a letter of commendation and a brief panegyric to bestow on the new publication the requisite grandeur. Meanwhile, Erasmus persuaded Dirk Martens, the Leuven university printer whose press had also produced his own work, to print the daring work in Leuven. While Gillis himself provided the Utopian alphabet and a quatrain in that language, Erasmus's friend the historian and satirist Gerard Geldenhouwer, who had often worked with Martens, produced the famous map of the Island of Utopia and penned a prefatory poem to the effect that More, taking to heart Horace's maxim, had brought forth a work both pleasant and profitable. The Aalst poet Cornelius Grapheus, who would later fill Gillis's shoes as town clerk of Antwerp, brought the list of luminaries to an end. He leads the reader to More's wonderful island, a new-found world where people lead new ways of life in harmony with each other and nature, without greed, and with human virtue. Erasmus's orchestration has not failed in its objective: his humanist programme adorns the gates of *Utopia*.

1

Thomas More
Libellus vere aureus … de optimo reip. statu, deq(ue) noua Insula Utopia
(The first edition of *Utopia*)

4°: [54] ff.
Leuven, Dirk Martens, 1516
Brussels, Royal Library of Belgium, inv. LP A 1945 (= V.H. 4802)

BINDING

Violet morocco (early nineteenth century, commissioned by Van Hulthem), boards with gilt double-fillet borders, tight back gold tooled (decoration and lettering), gilt edge, board edges gilded at corners, dentelle on inside of boards.

PROVENANCE

Three ex libris (at least one of which later transferred to more recent pastedown): (1) Hendrik Jozef Rega (1690-1754), Leuven professor of medicine; (2) P.J. Baudewijns (1757-1817), Brussels Latin teacher; (3) Charles van Hulthem (1764-1832), Ghent collector, ex libris designed by the neoclassicist architect Ti(e)l(e)man Franciscus Suys (1783-1861) and engraved by Adolphe Jouvenel (1798-1867); French bibliographical note by Van Hulthem on front endpaper.

BIBLIOGRAPHY

Allen 1906: nos. 461, 467, 474, 477, 481, 484, 487, 491, 502, 508, 530, 534; Nijhoff & Kronenberg 1923-1971: 555, no. 1550; Gee 1933: 87-88; Rogers 1947: nos. 25, 28; Gibson 1961: 5, no. 1; Allen 1963: 91-107; More 1965: CLXXXIII-CXCII; Derrett 1966: 61-66; More 1966; Kronenberg 1967: 134-136; Heireman 1973: 192-194, nos. A273-274, 278, no. M136; Prévost 1978: 215-240; Smith 1971: 261-262; Smith 1981: 20-29; Nink 1993: 28-43; Schulte Herbrüggen 1997b: 215-230; Cave 2008a; Roggen 2008; Adam & Vanautgaerden 2009: 86-87, 221 no. 154; Vielle 2009; Pettegree & Walsby 2011: 941, no. 21636; Adam 2011: 27, 30; Scafi 2013; Vielle 2013: 203-222.

The creation of *Utopia* is well documented. In September 1515 Thomas More stays for a time with Pieter Gillis (1486–1533) in Antwerp and 'meets' Hythlodaeus, who tells him about the unknown island and its remarkable inhabitants. By the end of March 1516 Book II is finished, likewise the introduction to Book I. In mid-1516 More writes the rest of Book I and the conclusion to Book II. On 3 September he sends the manuscript of what he is still calling *Nusquama* ('Nowhere') to Erasmus, relying on him to keep an eye on the printing process. He also asks him to see about getting some recommendations from scholars and prominent personalities. In October Erasmus reports to More that Gillis has written a dedicatory epistle to the influential Hieronymus Busleyden (1470–1517) and that Dirk Martens (1446/47–1534), who was very pleased to have been asked to print the work, is already getting illustrations made. Between 15 and 31 December, *Utopia* comes off the press. In early January 1517 the first copies arrive in England.

VTOPIAE INSVLAE FIGVRA

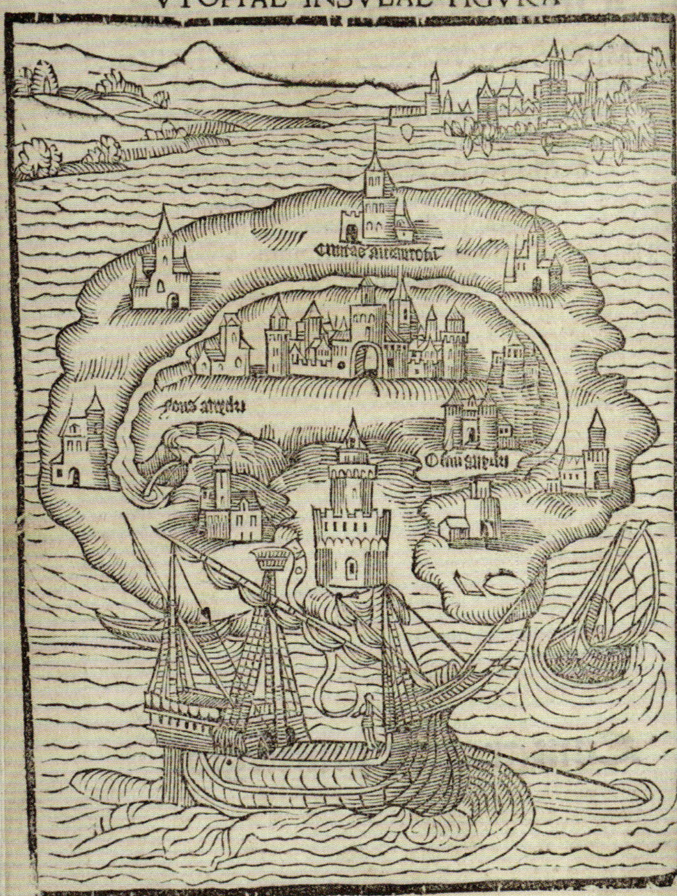

VTOPIENSIVM ALPHABETVM.

a b c d e f g h i k l m n o p q r s t v x y

Tetrastichon vernacula Vtopiensium lingua.

Vtopos ha Boccas peu la

chama polta chamaan

Bargol he maglomi baccan

soma gymno sophaon

Agrama gymnosophon labarembacha

bodamilomin

Voluala barchin heman la

lauoluola dramme pagloni.

Horum versuum ad verbum hæc est sententia.

Vtopus me dux ex non insula fecit insulam
Vna ego terrarum omnium absq; philosophia
Ciuitatem philosophicam expressi mortalibus
Libēter impartio mea, nō grauatim accipio meliora.

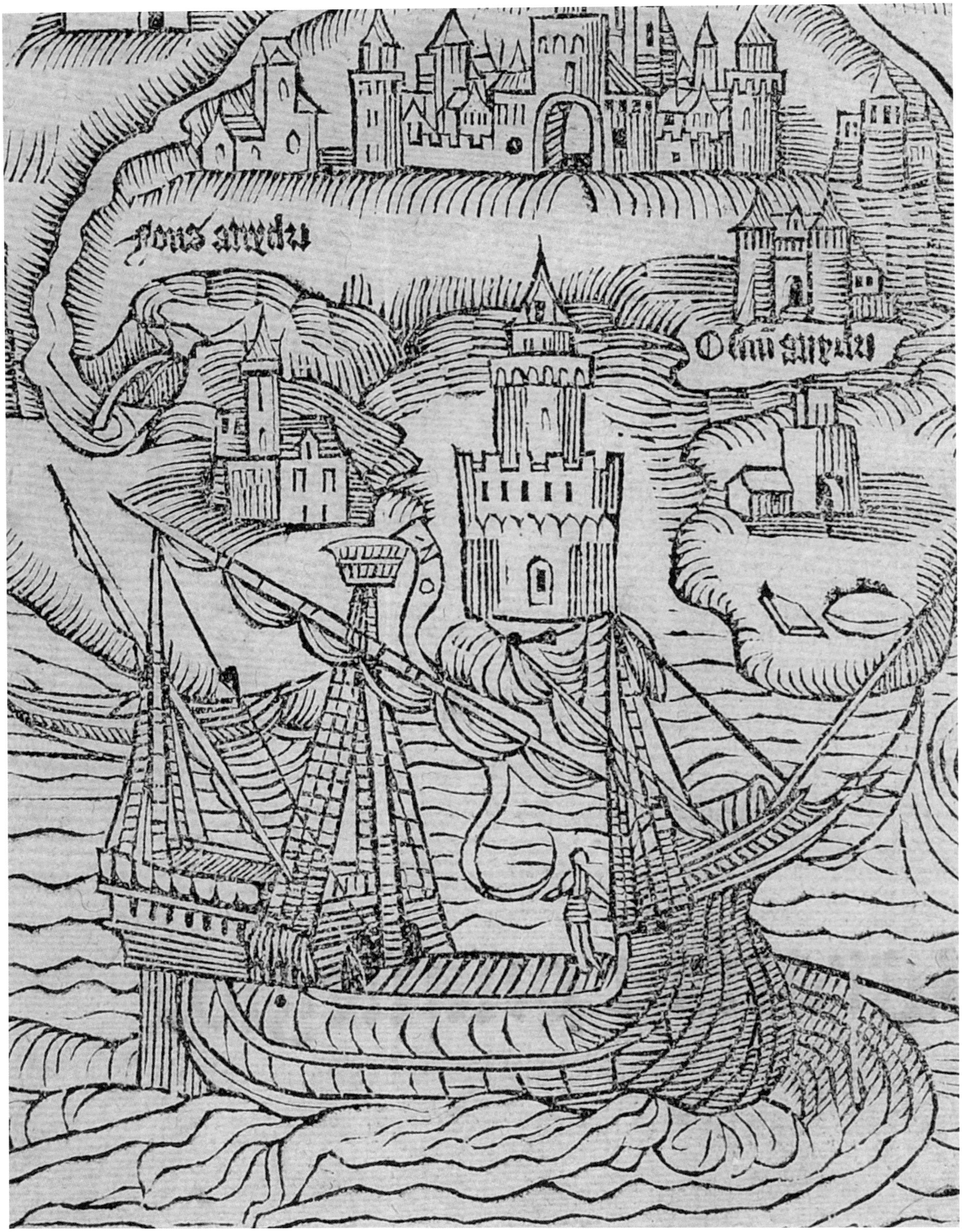

[cat. 1] detail

More and Dirk Martens were no strangers to each other. By 1516 they had already been in touch through business for twelve years, and they would continue to work together for a further seven. As for Erasmus, he had had no fewer than sixty-one editions brought out by Martens. In 1512 Martens had already reprinted Erasmus's *Praise of Folly*, which was dedicated to More. After *Utopia*, in 1519, he would publish another two editions of More's translations of Lucian. *Utopia* may have been intended as a *strena* (New Year's gift – étrenne) and the first printing was probably on the low side (300 copies?).

The text was surprising and new to its readers, but they may have been even more startled by the paratext, particularly the map of Utopia and the Utopian alphabet. The map, designed to make the text more credible, depicts a horseshoe-shaped island seen in semi-bird's-eye view, with the river Anydrus ('not water') in the form of a Greek omega. In the background is a hilly coastline with a many-towered city, in the foreground a cog with a crow's nest and a pennant at the masthead that bears the letters 'N' and 'O', as does a smaller craft. The rather roughly cut map is anonymous and the significance of the letters is also unclear. Martens believed that illustration was important, but the book also had to be affordable for a general readership.

The Utopian alphabet, whose letters form a set of elementary geometric shapes (circle, arc, triangle, right angle, square) facing in different directions and possibly modified with diacritical marks, is pretty strange as well. It appears to be based on oriental languages and travels in contemporary publications such as Bernhard von Breydenbach's *Peregrinatio in terram sanctam* (1486–1502), *Calcoen* (Antwerp 1504: a narrative of Vasco da Gama's second voyage to Calicut, 1502–1503), or Balthasar Springer's *Die reyse van Lissebone … na … Naguaria … voorbi Callicuten* (Antwerp 1508). The reference to 'gymnosophists' links it to India: while there is no suggestion of Sanskrit in the 'Utopian', there are traces of Malayālam, the Dravidian language of Kerala (India).

From the moment it appeared, *Utopia* was received as a wonderful and extraordinary book that would quickly establish More's reputation as a humanist and was to influence profoundly the Western view of society and the world. For Dirk Martens the book was his masterpiece, for Leuven humanism it was a great contribution to world heritage. Worldwide, some thirty copies still survive.

MDS

2

Monogrammist AP (?)
View of Leuven, City of Art, Culture and Science

Antwerp (?)
c.1540
Woodcut, 516 x 3660 mm
Brussels, Royal Library of Belgium, Print Room, inv. S. 23172 (unique)

PROVENANCE

Rudolph Weigel in Leipzig; before 1861 for 225 fr. acquired by the Royal Library of Belgium in Brussels.
The woodcut has been specially conserved for the exhibition by Julie Swennen. The conservation was funded by the Pastoor Manoël de la Serna Fund, King Baudouin Foundation.

BIBLIOGRAPHY

Alvin 1862: 15; Passavant 1862: 152-153; Merlo 1864: 139; Petit 1877: 47-53; Merlo 1895: 1088-1089; Van Even 1895: 181; Van 't Hoff 1955: 92; Utrecht/Leuven 1959: no. 98; Van Hasselt 1961: 138; Van Uytven 1964: no. 111; Rotterdam 1969: no. 252; Roegiers 1969: no. 1; Heireman 1973: no. A206; Van Buyten 1975: no. 7; Roegiers 1976: no. 2; Appuhn 1976: 50-53; Van der Heijden & Roegiers 1994: no. BB-26; Bessemans, Honoré, Smeyers et al. 1998: no. 9; Derez 1999: 8; Padmos & Vanpaemel 2000: no. 1; Derez & Verbrugge 2005: 10; Luyckx 2013: 33-65.

OTHER VERSIONS

Reduced-scale nineteenth-century reproduction by Jules Petit (530 x 2680 mm), in M-Museum Leuven (inv. LP/920) and other places. Except in Heireman 1973 this reproduction was always shown.

This monumental print is the earliest known view of the university town of Leuven. It portrays the city's notable buildings and landmarks as seen a couple of decades after 1516.[1] It consists of eighteen individual woodcuts glued together to form a single large frieze. Its dimensions – a good fifty by three hundred and fifty centimetres – make it early modern Europe's widest surviving city view.[2] The only known impression of this town portrait is held in the Royal Library of Belgium.

A bird's-eye view from the west and a town's profile are combined to produce an extraordinarily detailed portrait of a city, recognizable by its topographical position, its streets and squares, and its characteristic buildings.[3] Over fifty architectural and geographical landmarks are also identified by means of discreet letterpress texts. At top centre of the frieze is the coat of arms of the Holy Roman Empire, set in the middle of a banderole inscribed with 'CIVITAS LOVANIENSIS'. There are also coats of arms in the top left and right corners, respectively those of the Duchy of Brabant and the City of Leuven. Hovering on clouds above the city are eight allegorical groups that allude to the seven liberal arts and philosophy, the fundamental principles of Leuven's university. Down below, in the middle of the

1 On this subject see inter alia Roegiers 2008.

2 On monumental woodcuts, see Apphun & Von Heusinger 1976 and Silver & Wyckoff 2008.

3 Huvenne 1991: 46; Luyckx 2013: 33–65.

4 That students were among the target buyers has already been suggested in Van Even 1895: 181.

5 Passavant 1862: 152–153.

6 This attribution is discussed in depth in a forthcoming publication by the present author. See also Hollstein, Dutch, XIII: 12–16.

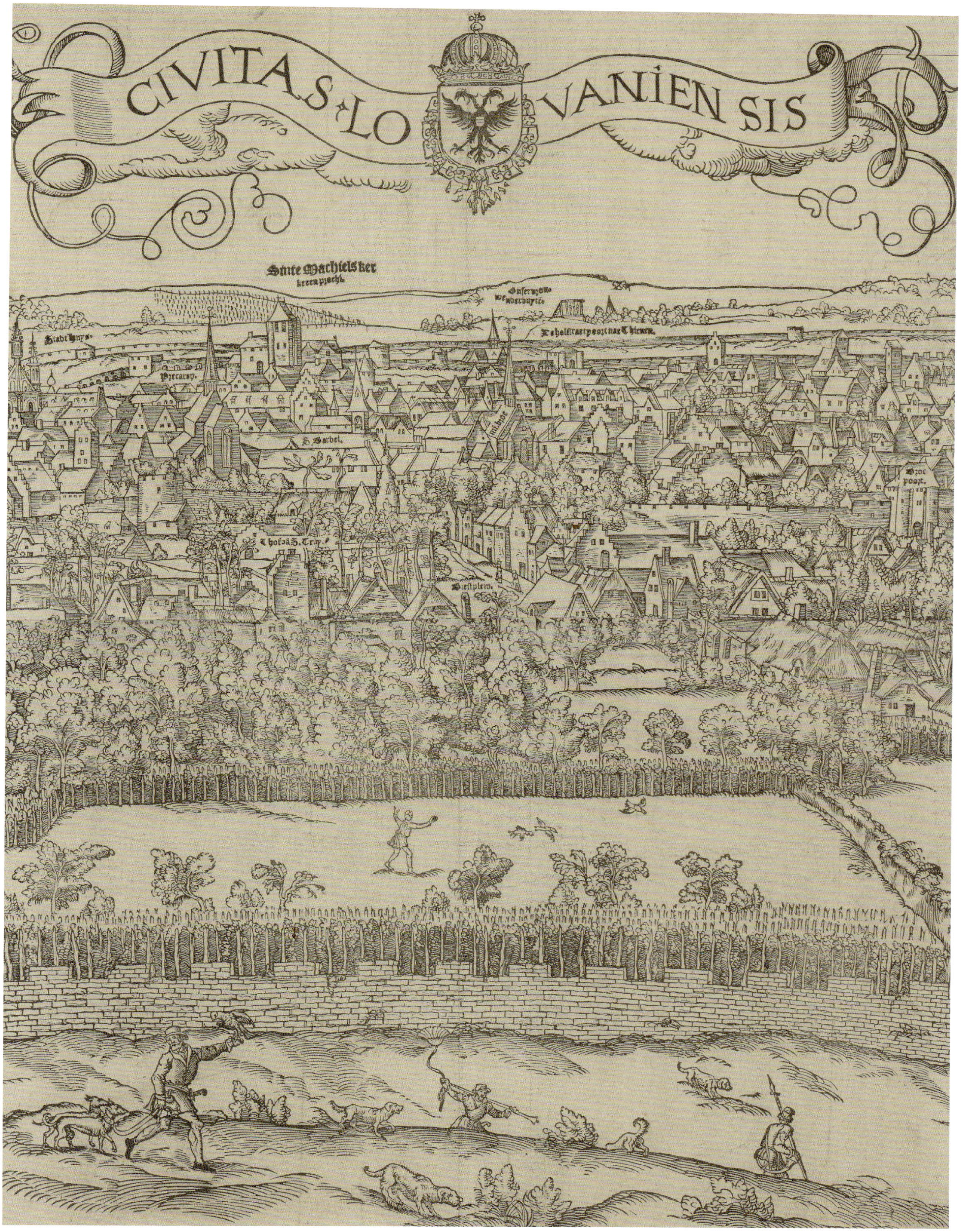

[cat. 2] detail

frieze, is the densely built centre of the town, including St Peter's Church and the Town Hall, surrounded by the old twelfth-century ramparts. A short distance away are the castle on the Keizersberg and various glebes, vineyards and what may be hop fields. Surrounding all this is the new, fourteenth-century town wall, furnished with several towers and gatehouses. In the background there are also several villages and some of the area's important buildings. Throughout the image are dozens of human figures. Below right, for instance, a student scurries after a professor.

The monumental woodcut is not a faithful record of what the city looked like at a given moment but a recognizable representation of Leuven as a major intellectual centre. Presumably, it was a private and commercially motivated creation that was sold to students and professors as a souvenir of their stay in Leuven, or to interested merchants, travellers and proud citizens of the university town.[4] It was widely imitated in the illustrated chorographic works of the sixteenth century, one such copy being the simplified depiction of Leuven in Guicciardini's *Descrittione di tutti i Paesi Bassi* of 1567, which is obviously based on the monumental cityscape, indicating a certain distribution and appreciation of the frieze.

The portrait of Leuven was first mentioned by Passavant, who ascribed it to Anton Woensam von Worms.[5] However, the differences in design and style clearly make this attribution unsustainable. But the city view does display marked similarities to the woodcuts of Monogrammist AP, an anonymous printmaker who was active in the Netherlands around 1536.[6] One of the printers who published the woodcuts of Monogrammist AP was Willem Liefrinck in Antwerp, so it may be that the creation of this frieze can also be assigned to that city.

JL

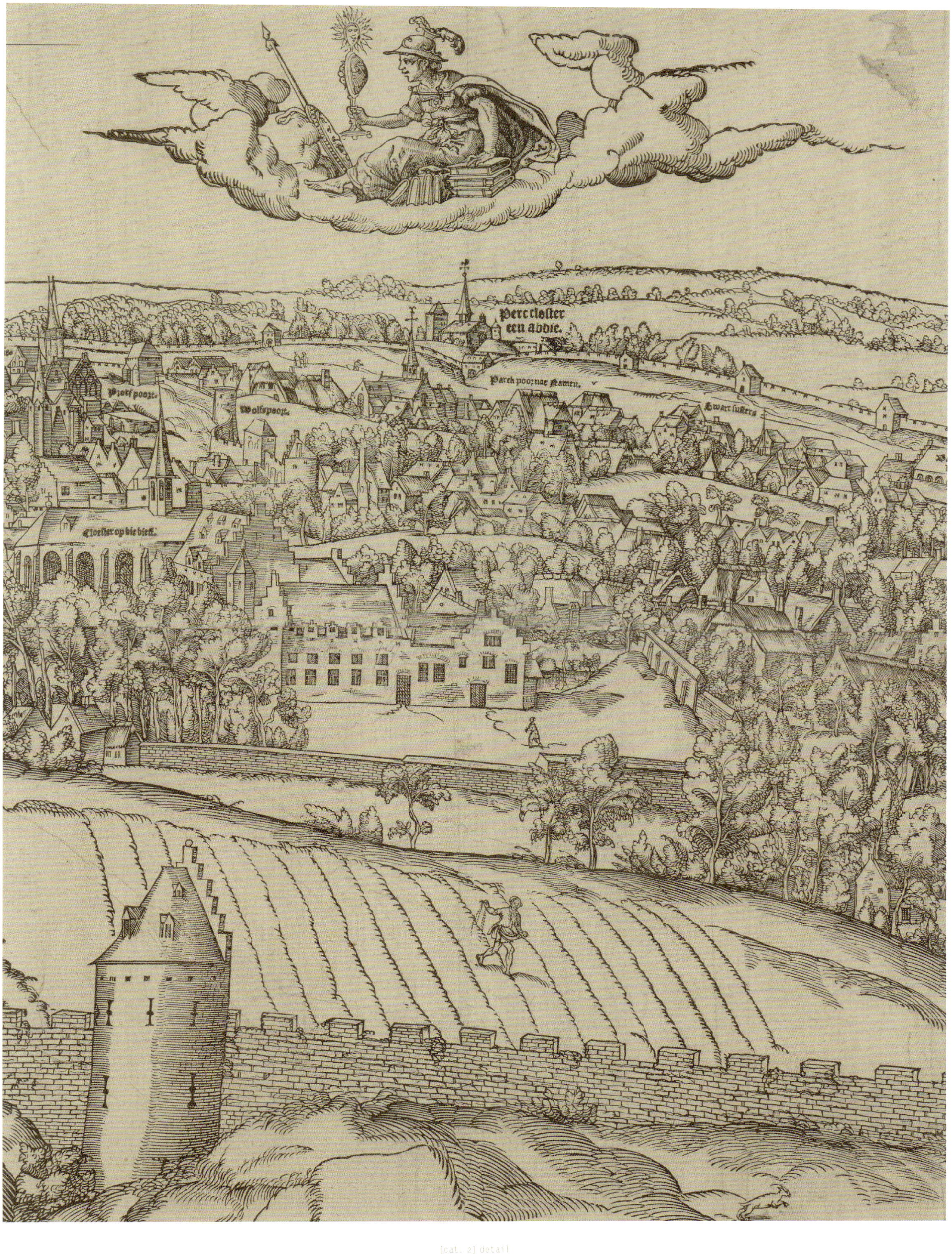

[cat. 2] detail

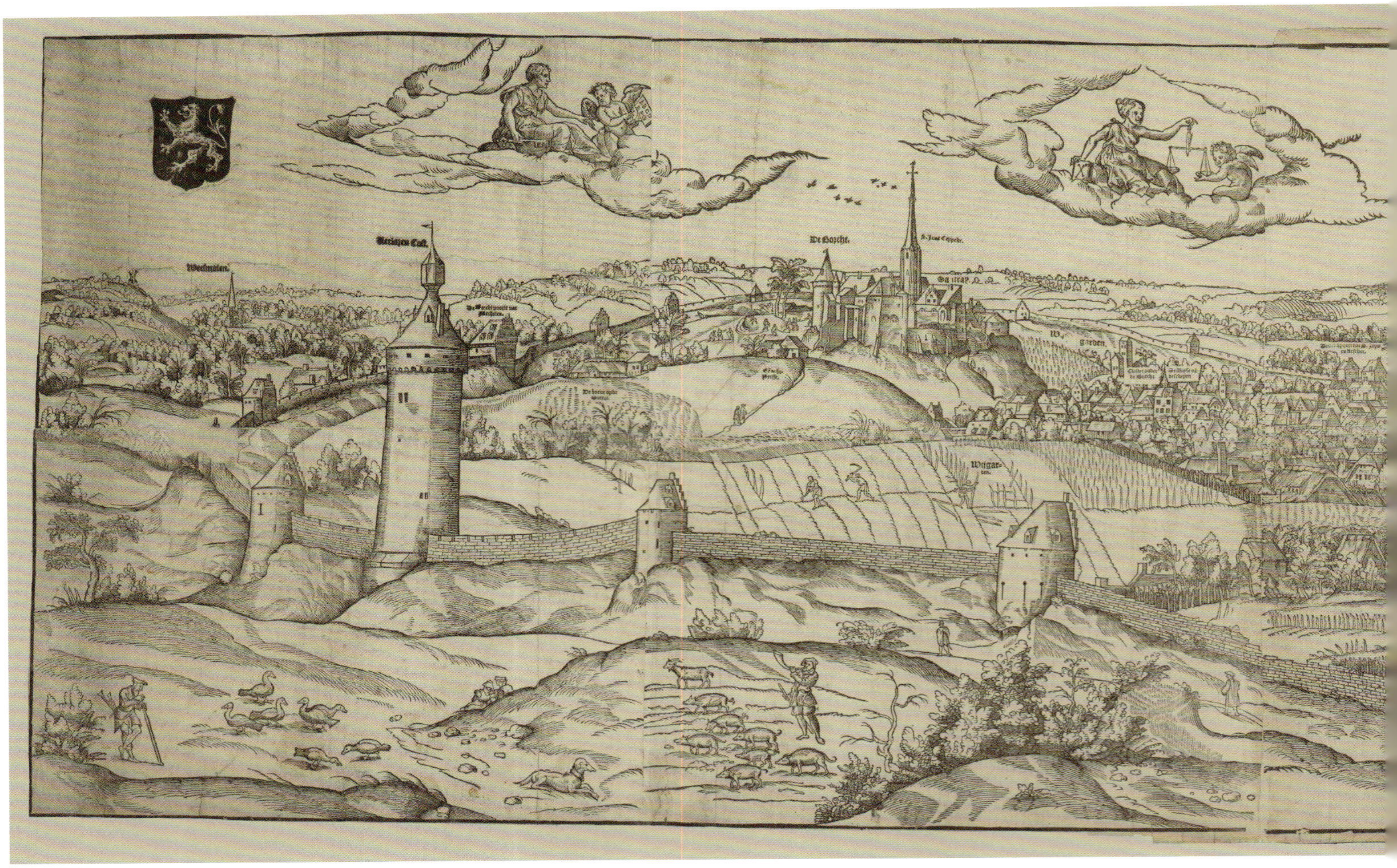

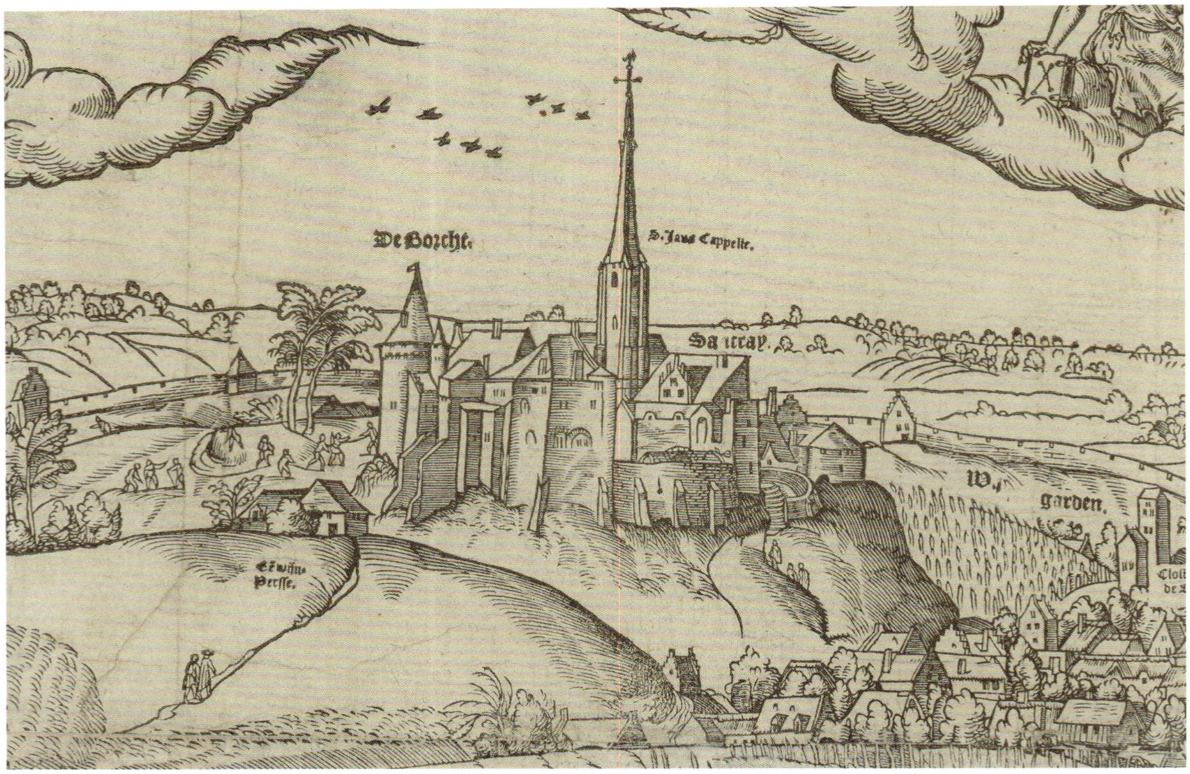

[cat. 2] detail

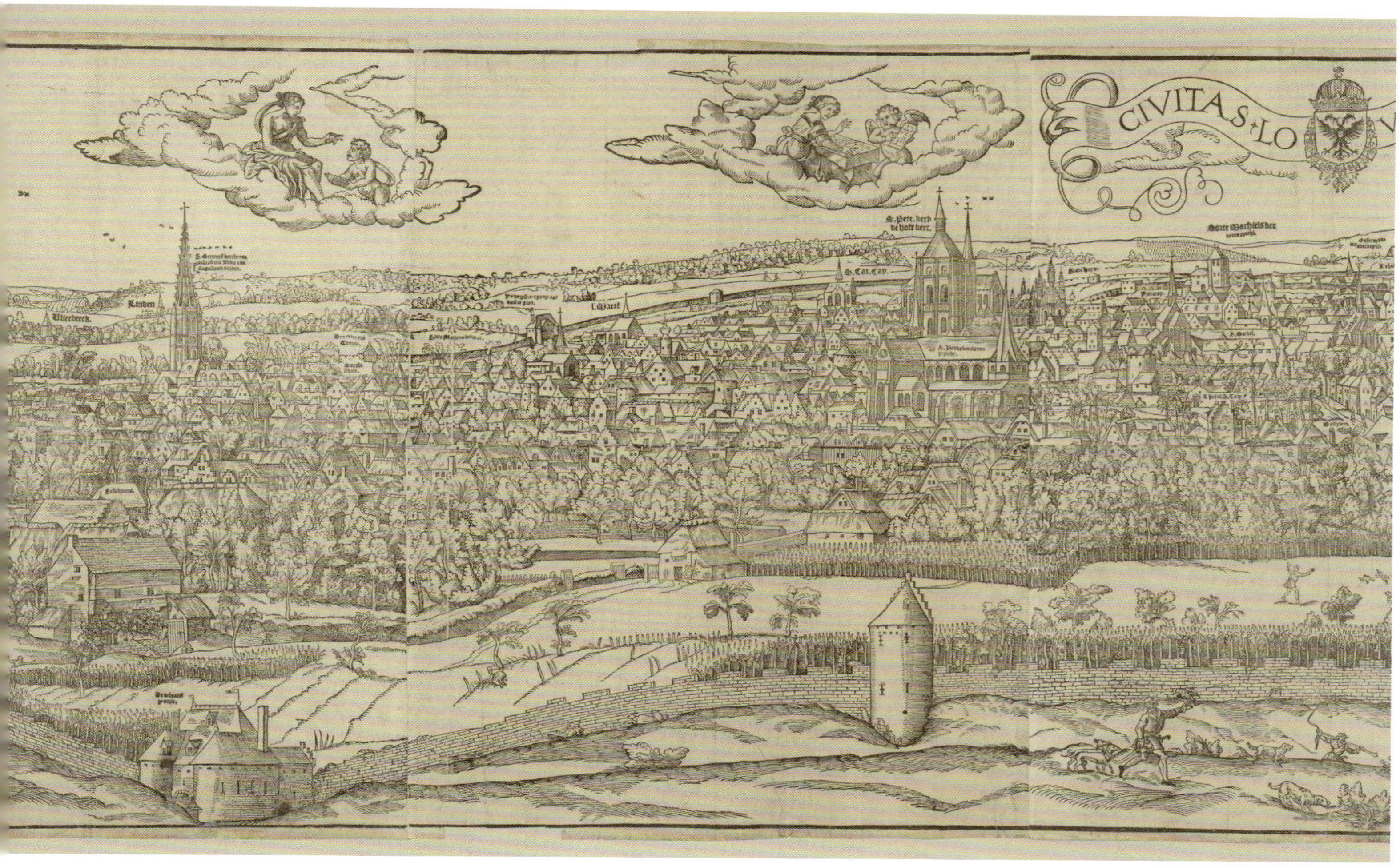

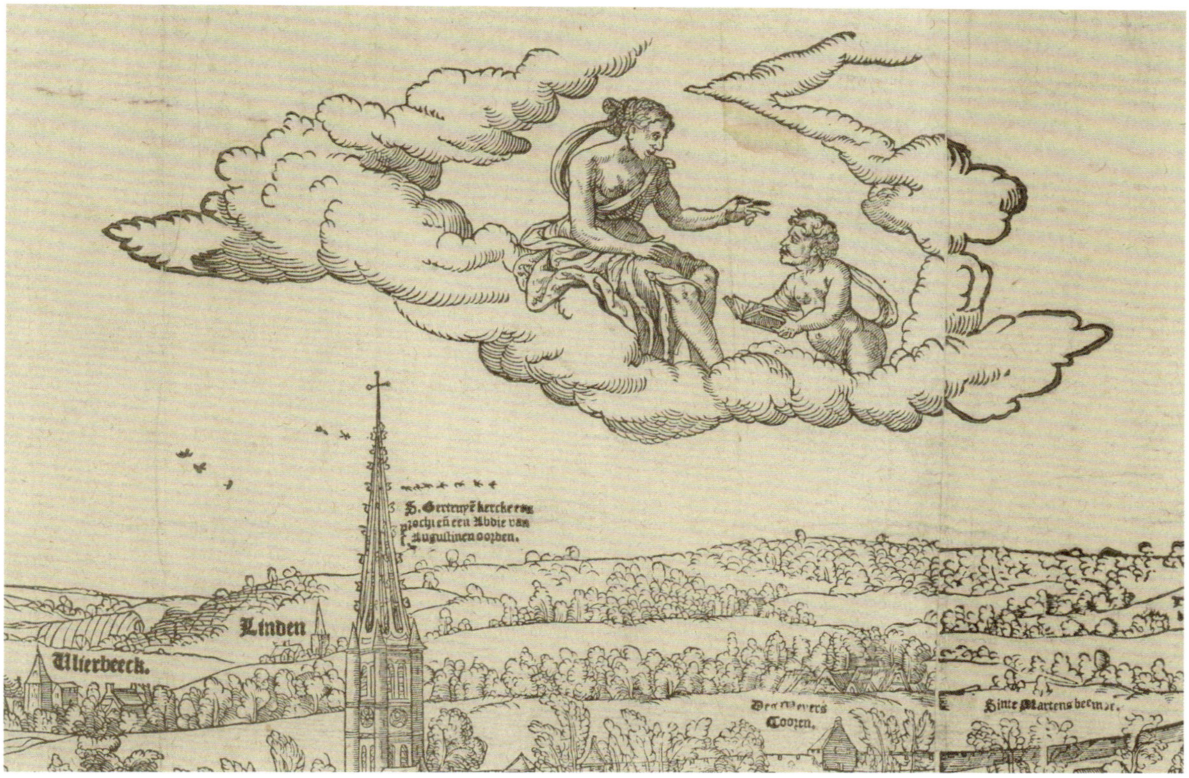

[cat. 2] detail

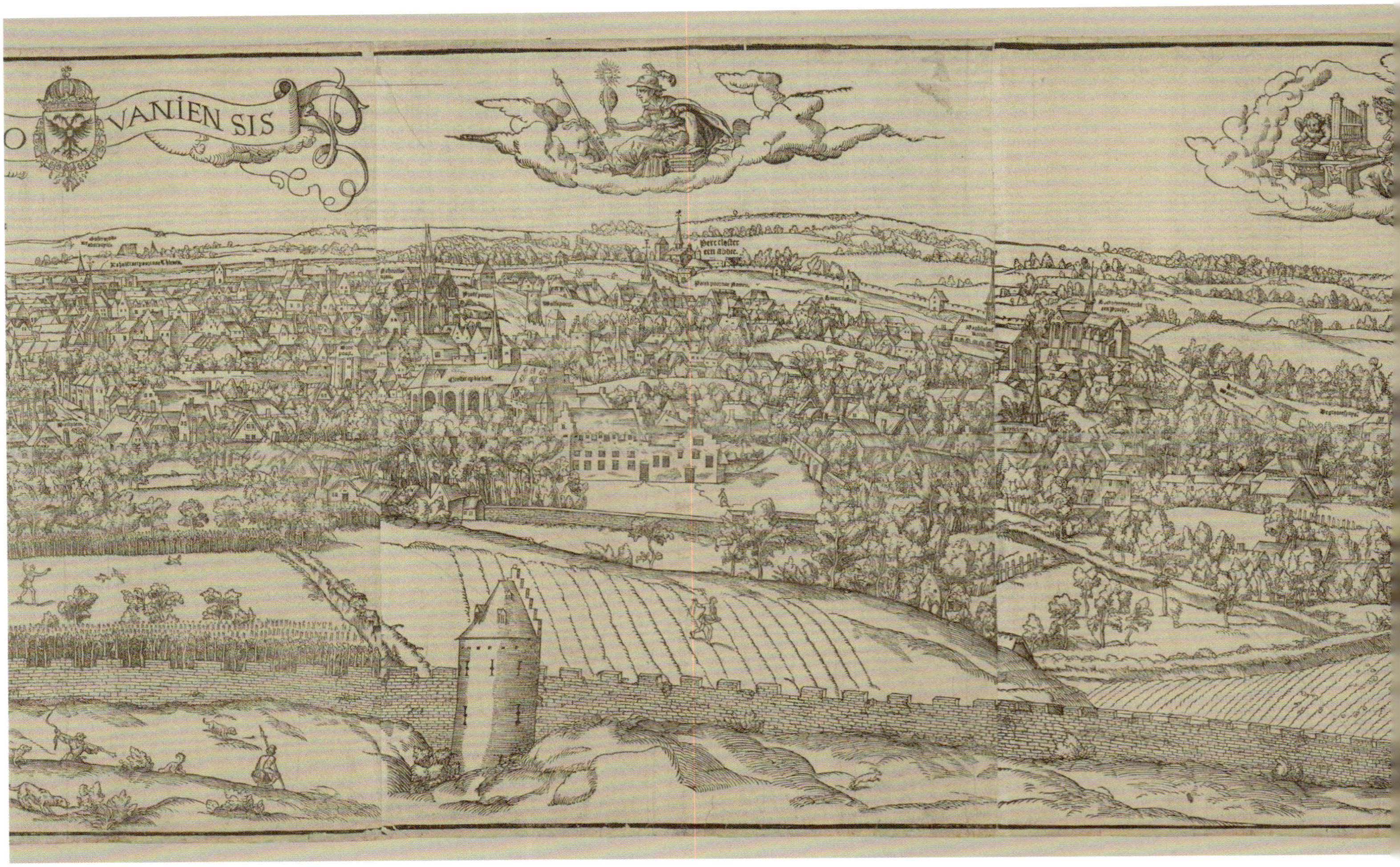
VANIEN SIS

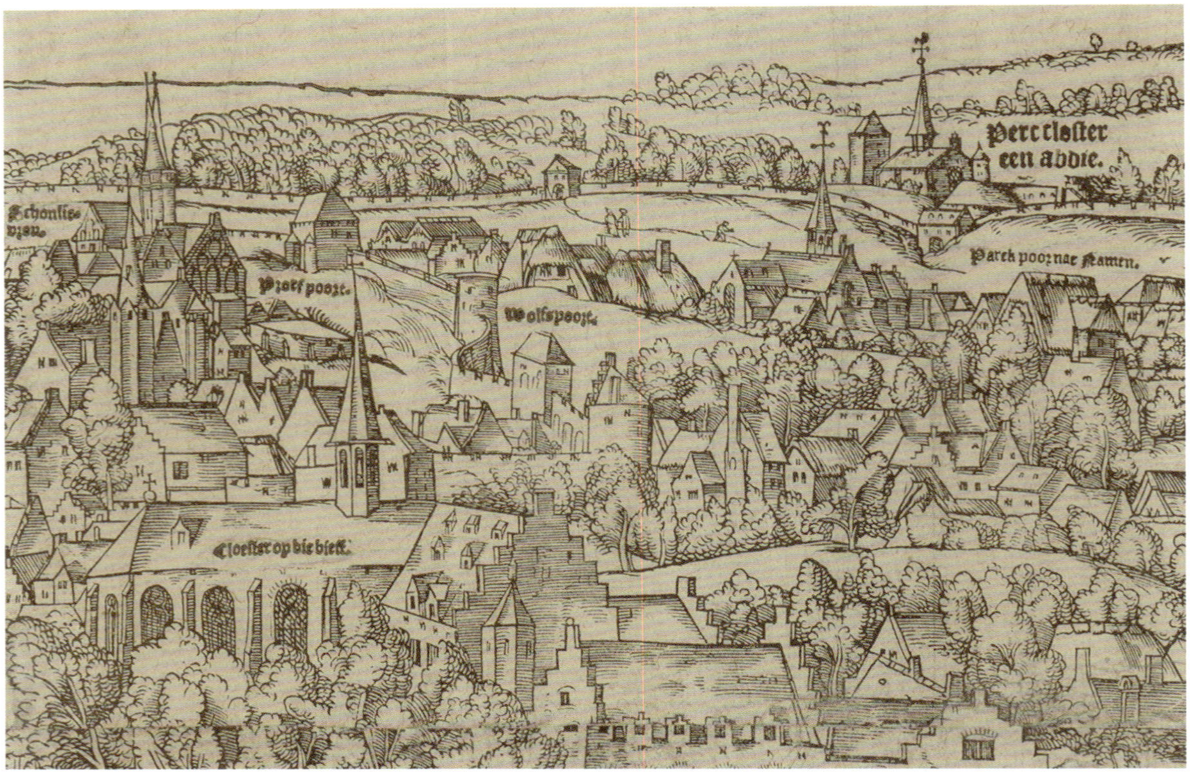
Perc closter
een abdie.
Parck poort nae Namen.
Proefpoort.
Wolfspoort.
Cloester op die biest.

[cat. 2] detail

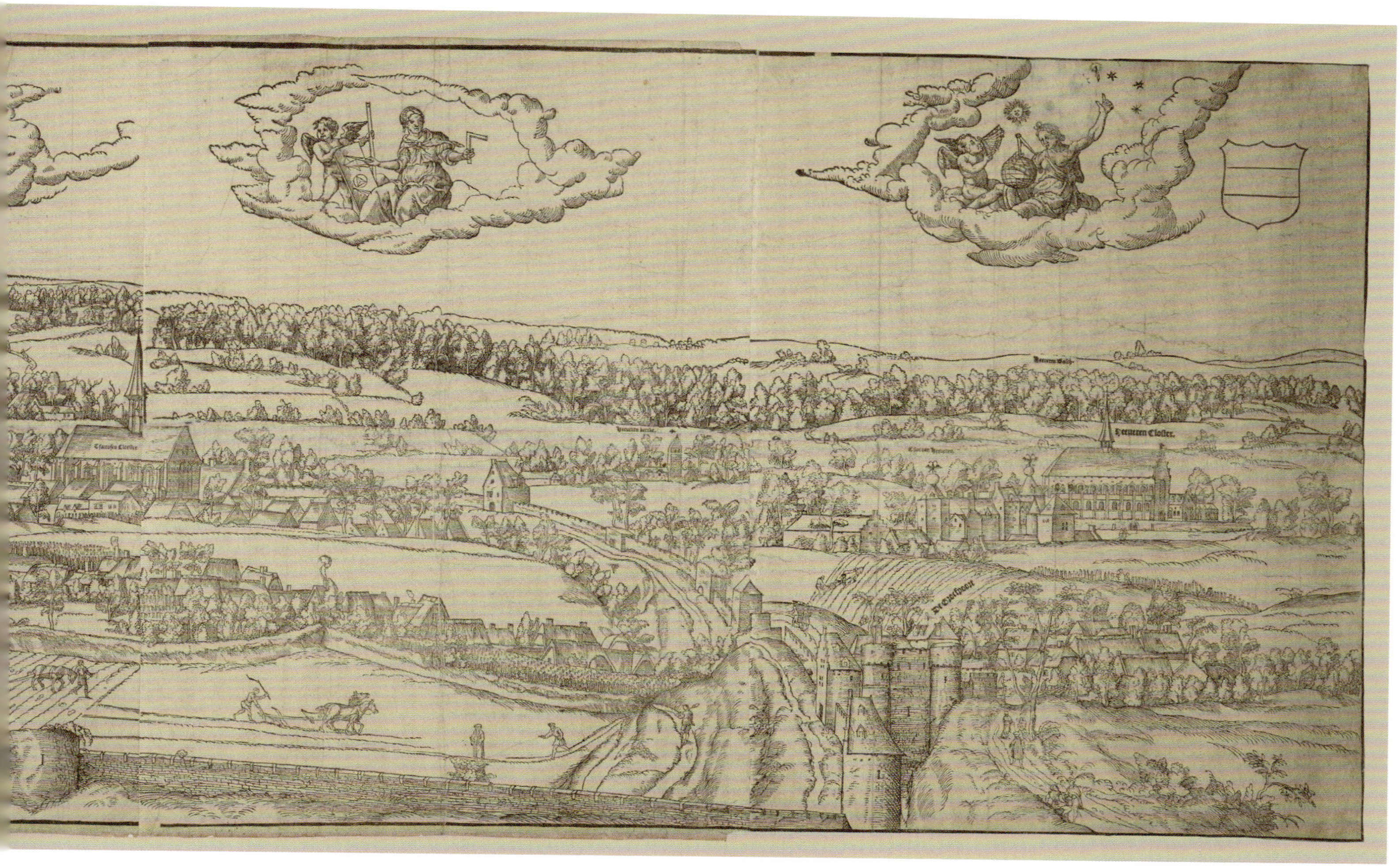

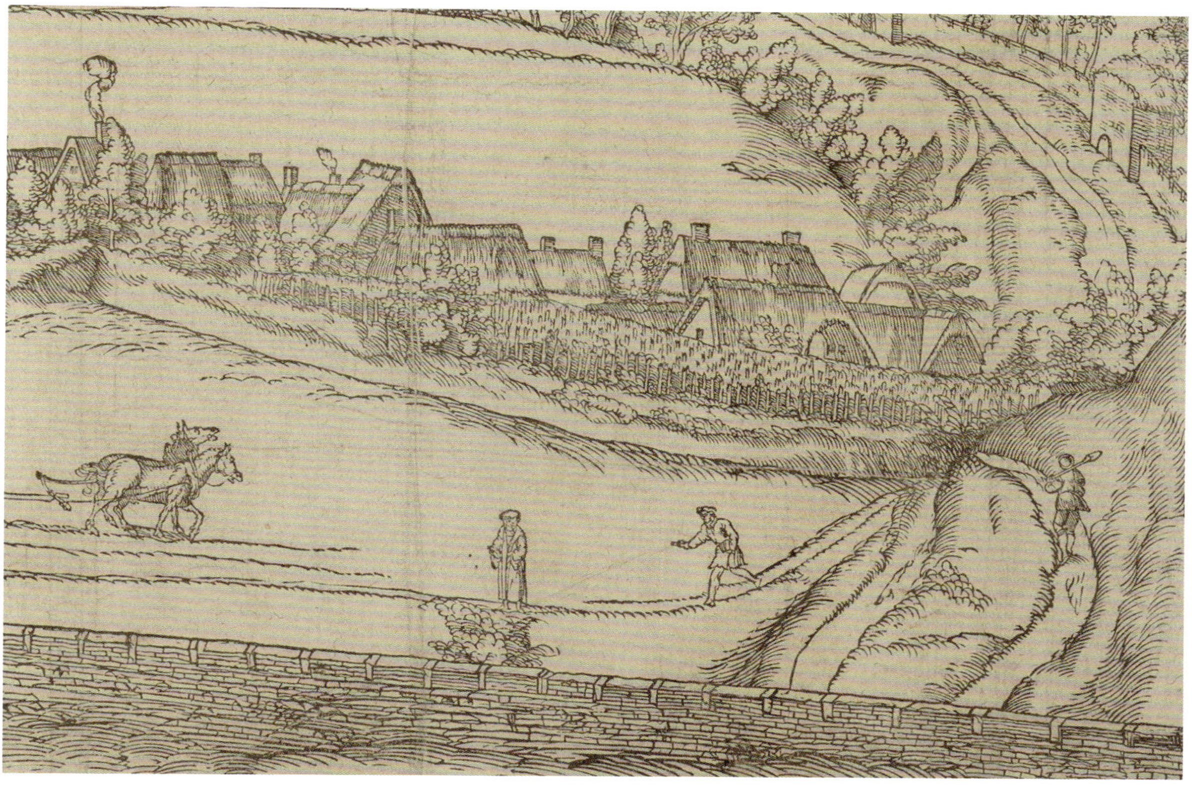

[cat. 2] detail

3

Hans Holbein the Younger, copy after *Portrait of Thomas More*

The date 1527 is copied from the original
Oil on limewood, 74.9 x 58.4 cm
London, National Portrait Gallery, inv. 4358

PROVENANCE

Francis Russell, 5th Duke of Bedford (1765-1802); sold Christie's, London, 30 June 1827 (83); afterwards in the collection of William II, King of Holland (1792-1849); bought in 1964 for the National Portrait Gallery.

BIBLIOGRAPHY

London 1890: 45, no. 127; Morison 1957; Morison 1963: 11-15; Rotterdam 1969: no. 66; Strong 1969: 228-229; Trapp & Schulte Herbrüggen 1977: 30-31, no. 26; Wells 1981: 61; Rowlands 1985: 71-72, 132-133; MacLeod 1999: 10; Foister 2004: 117.

This is an accurate, full-size copy of Holbein's original (fig. 4).[1] The limewood panel, the chalk ground, the reddish priming and the use of azurite as well as smalt indicate that it was painted in Germany in the sixteenth or the first half of the seventeenth century.[2] It is reckoned to be the best of many copies (Rubens's copy now in the Prado being a freer interpretation) and was probably made directly from the original, although the location of the original at that period is not established. *The collar of Esses*, with two portcullises and a rose, is a livery chain signifying More's loyalty to the Tudor dynasty. In the portrait, it is mordant gilded and the contour of the rose has been incised with compasses.

The face here and in the original resembles closely Holbein's damaged drawing at Windsor (fig. 3.1).[3] There are many similarities between the original and the figure of Thomas in the versions of the lost group portrait of the More family, probably painted at much the same time (cat. 5). The curtain and inscribed chair-arm are very like the curtain and parapet in Holbein's *Laïs Corinthiaca* (Basel), dated 1526, though More's curtain has a fringe and is restrained by a red cord. Holbein, according to his usual practice, has slightly distorted More's appearance. The face is enlarged in proportion to the head, the features, particularly More's left eye, are enlarged in proportion to the face and their asymmetries are exaggerated. The tear-duct of More's right eye is enlarged and turned at a very strange angle to his eye-socket; while the areas of white visible beneath both irises make him look slightly withdrawn. The very strong contrasts of light and shade are unusual in Holbein's work.

It is interesting to compare Holbein's portrait with the pen-portrait by Erasmus in a letter of 1519 (cat. 6). The stubble of his beard is not heavy; particularly when he walks, his right shoulder seems rather higher than his left … he takes no great care of his appearance. He dresses plainly and never wears his gold chain if he can avoid it …[4]

1
Oil on oak, 74.9 x 60.3 cm, New York, Frick Collection, 1912.1.77.

2
On this topic, see inter alia: Bätschmann 2013.

3
Parker 1945: 36, no. 3. 'Sir Thomas More', inv. 12268.

4
Trapp & Schulte Herbrüggen 1977: 134.

5
Reed 1926: 334.

6
Stapleton 1928: 209.

7
Trapp & Schulte Herbrüggen 1977: 119, 139–140.

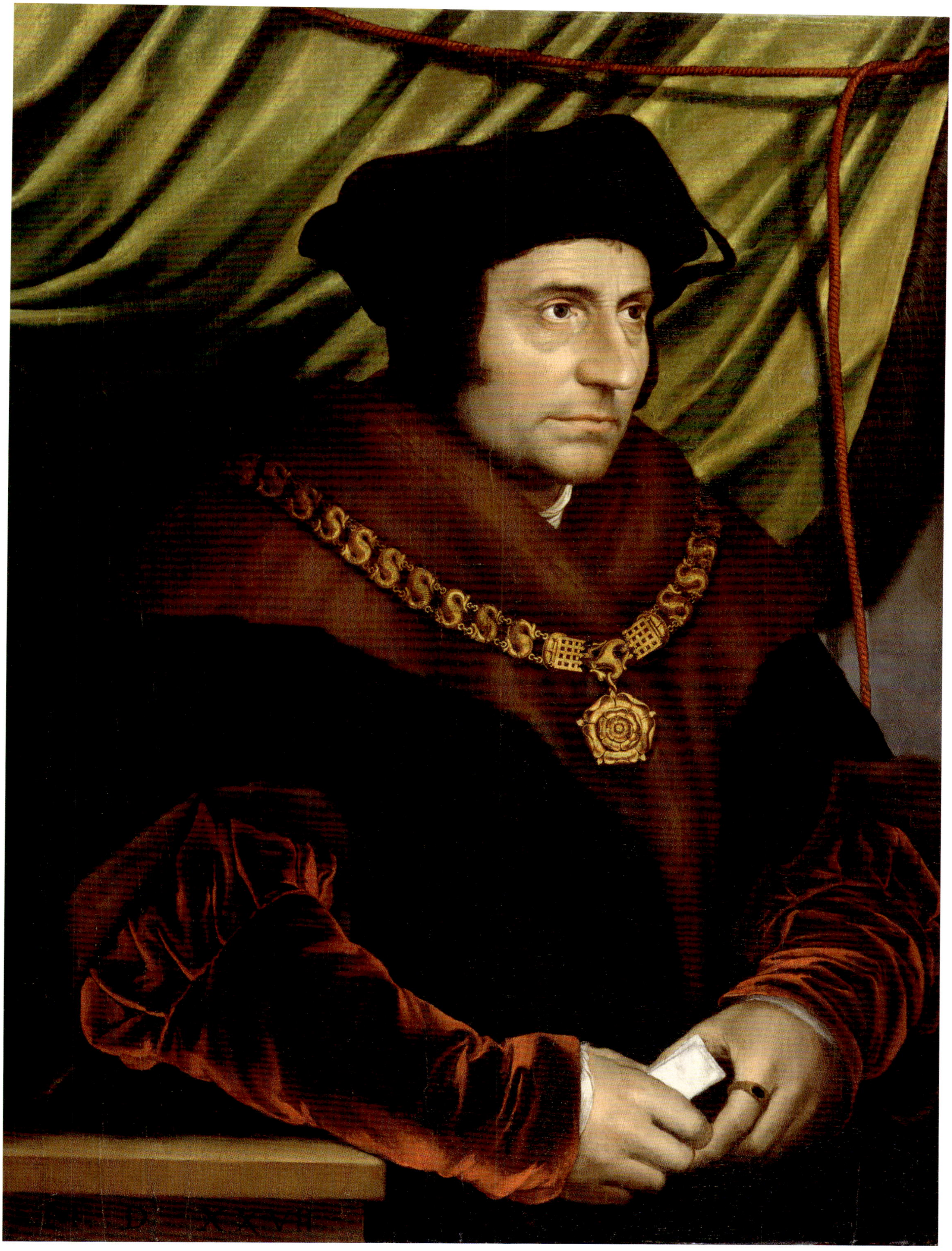

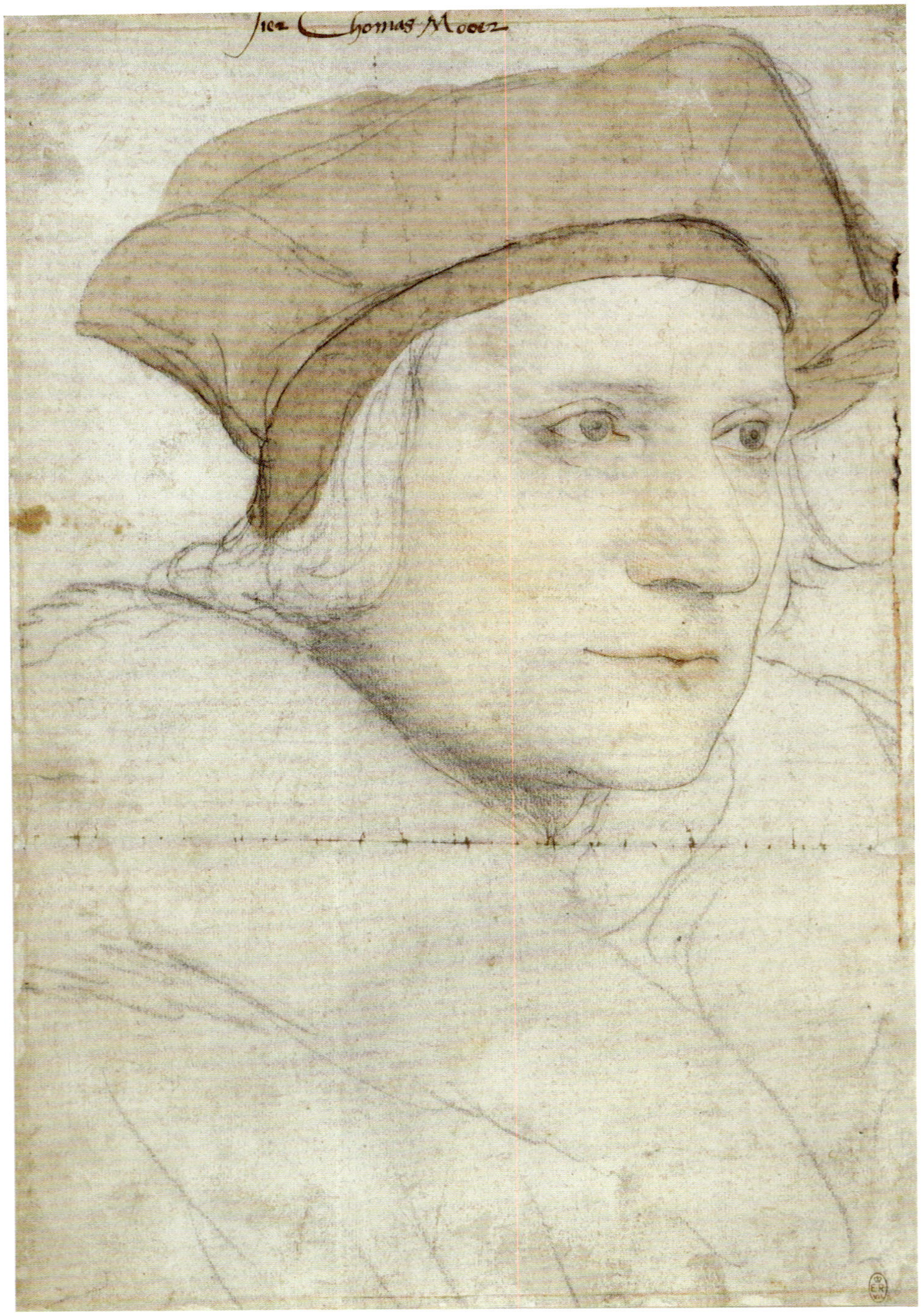

[fig. 3.1] Hans Holbein the Younger, *Portrait of Thomas More*, drawing, *c.*1526-1527, Royal Collection Trust, Windsor Castle.

More's adopted daughter Margaret Giggs owned 'a table [panel-painting] of Sir Thomas More's face', valued in 1555 at £2.[5] It was perhaps a version of Holbein's portrait. Margaret also owned 'a life-like image, made with great skill, of More going out to the place of execution …'[6] Both images were reproduced on reliquaries and as devotional images.[7]

LC

4

Hans Holbein the Younger, workshop
Portrait of Lady Alice More, Second Wife of Thomas More

London

*c.*1530

Oil on panel, 36.9 x 26.7 cm

London, Weiss Gallery

PROVENANCE

Frederick, 2nd Baron Methuen (1818-1891), Corsham Court, Wiltshire; thence by descent to Paul, 4th Baron Methuen (1886-1974); sold August 1958 to Dr Hans Schaeffer, Schaeffer Gallery, New York for £27,000; acquired (via Paul Herzogenrath) by Rudolf August Oetker (1916-2007), Bielefeld, Germany.

BIBLIOGRAPHY

London 1865: 99, no. 1146; Wornum 1867: 356; Chamberlain 1913: 307-308; Ganz 1936: 149; Borenius 1939: no. 147; Borenius 1943: 285-288; Rogers 1947; Ganz 1950: 232, no. 42, pl. 76; Salvini & Grohn 1971: no. 48; Wells 1981: 55-71; Norrington 1983; Rowlands 1985: 70, 231, no. 227; Sander 2005: 306 n. 16.

In 1511, barely a month after the funeral of his first wife, Jane Colt, Thomas More remarried.[1] His new wife, born Alice Harpur (*c.*1474–1551), was a highly desirable catch, being at that time the widow of the wealthy merchant John Middleton. Her character has been and still is difficult to describe because sources often seem to emphasize her sharp tongue and lack of manners. Nevertheless, it is a fact that she quickly showed herself to be a second mother to More's children, a loyal companion to her husband and a firm and pragmatic mistress of his household.

Today this portrait is the only known, contemporary painted likeness of Lady Alice More.[2] She does not remotely resemble the 'harpy's crooked beak' that one of More's friends once described.[3] Dressed according to her status, she seems t be mulling over an amusing thought. Her pose and clothing clearly correspond to those in the preliminary drawing (fig. 5.3) for and the copy after the famous portra of the More family (cat. 5). It is more than likely that the portrait shown here derives from the lost family portrait.[4]

An alert observer will note a faint black band under Lady More's brow. Infrared reflectology has recently revealed that beneath the paint layers is an unfinished portrait of Erasmus (fig. 4.1). This overpainted portrait shows the humanist in three-quarter profile, looking to the left. His hands rest on the pages of an open book. Portraits of Erasmus of this type were produced in great number in Holbein's workshop (fig. 4.2). Most date from the early 1530s, which may be a *terminus post quem* for the portrait of Alice More.

MB

1
In 1535 Father Bouge wrote to Dame Catherine Manne: 'I buried his [Thomas More's] first wife. And within a month after, he came to me on Sunday late at night, and there he brought me a dispensation from Cuthbert Tunstall, to be married the next day with no banns asking.' Brewer 1862–1932: 357–358, no. 1024.

2
Besides the drawing for the lost family portrait there is also a portrait miniature of Lady More by Holbein. It was once in the collection of J. Heywood Hawkins and was exhibited in London in the mid-nineteenth century. London 1865: 99, no. 1146; Wornum 1867: 356.

3
Once Alice had assumed her position as head of the household, several of More's humanist friends found it difficult to remain in his house any longer. For example, Andrea Ammonio soon relocated to London's St Thomas's College. Nor did he fail to proclaim his candid opinion of the 'guilty party', the new Lady More. Ferguson & McConica 1975: 179–182, no. 236. Ammonio to Erasmus (27 October 1511).

4
According to Ganz the portrait of Lady More shown here would be a preparatory study painted by Holbein for the family portrait. Because the work can sooner be attributed to a workshop hand than to the master himself, this is more than likely a workshop replica. Thus Ganz's proposed date of 1526–1528 is also not tenable. Borenius 1943: 286; Ganz 1950: 232, no. 42.

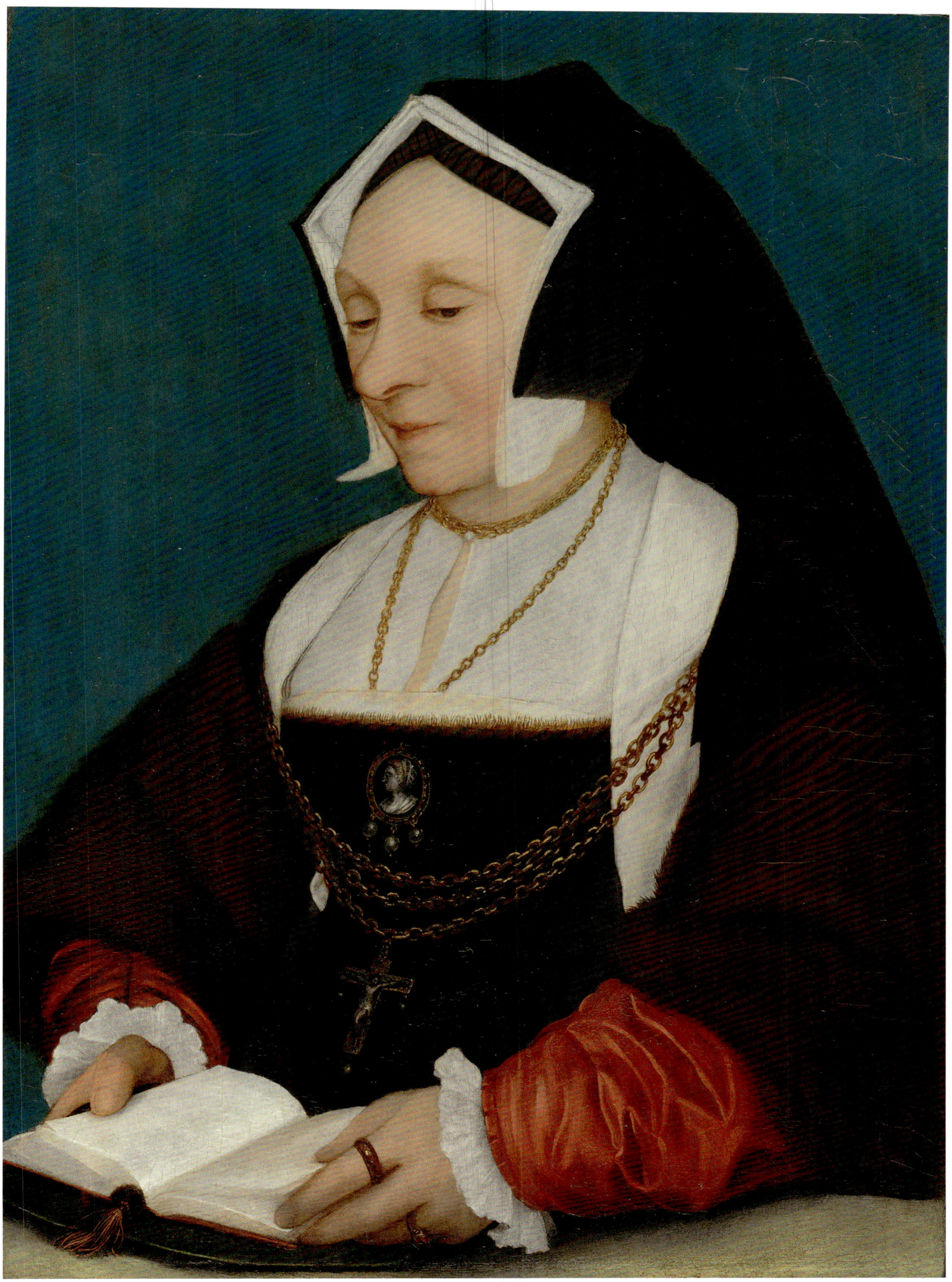

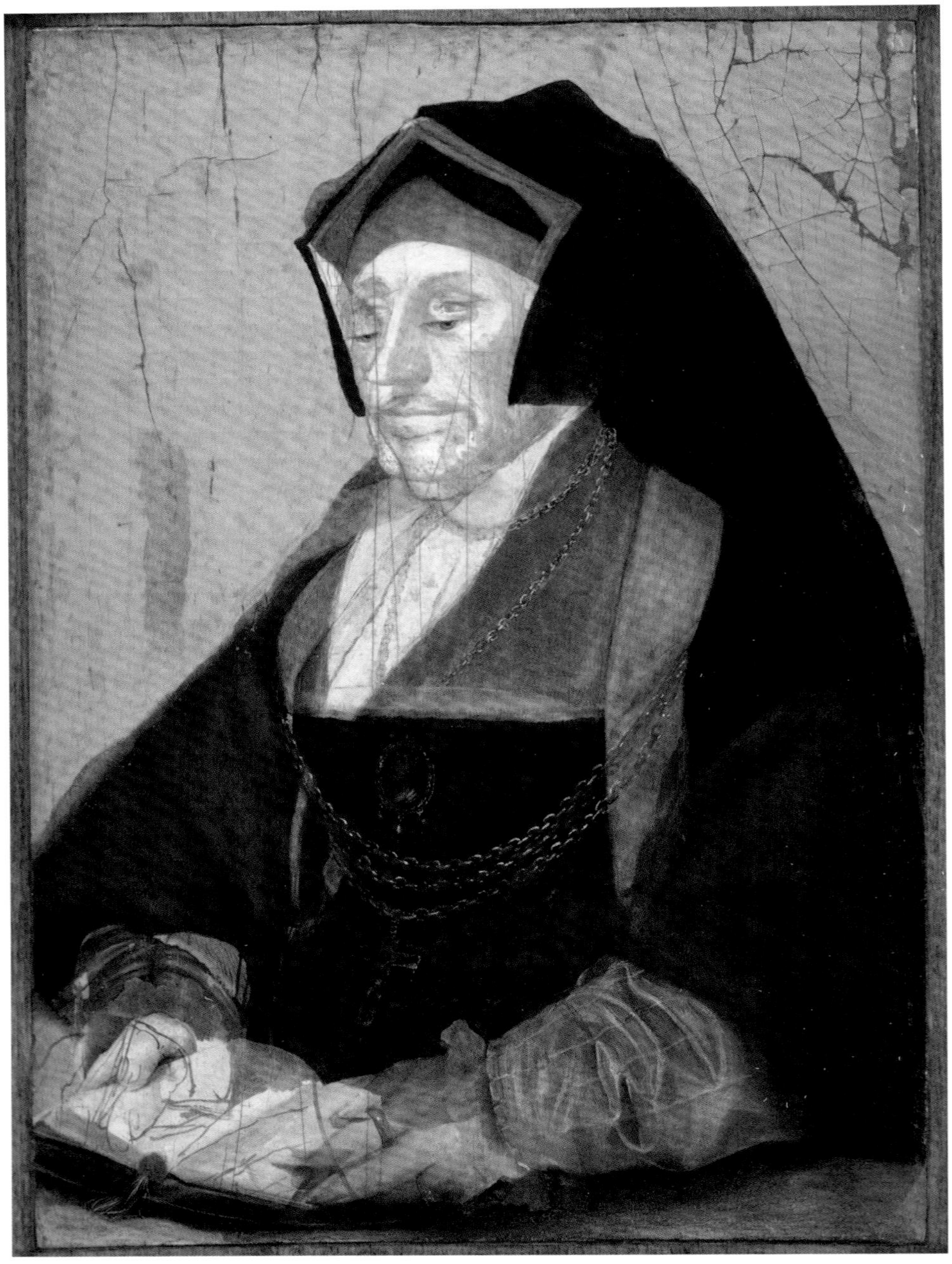

[fig. 4.1] Infrared reflectogram

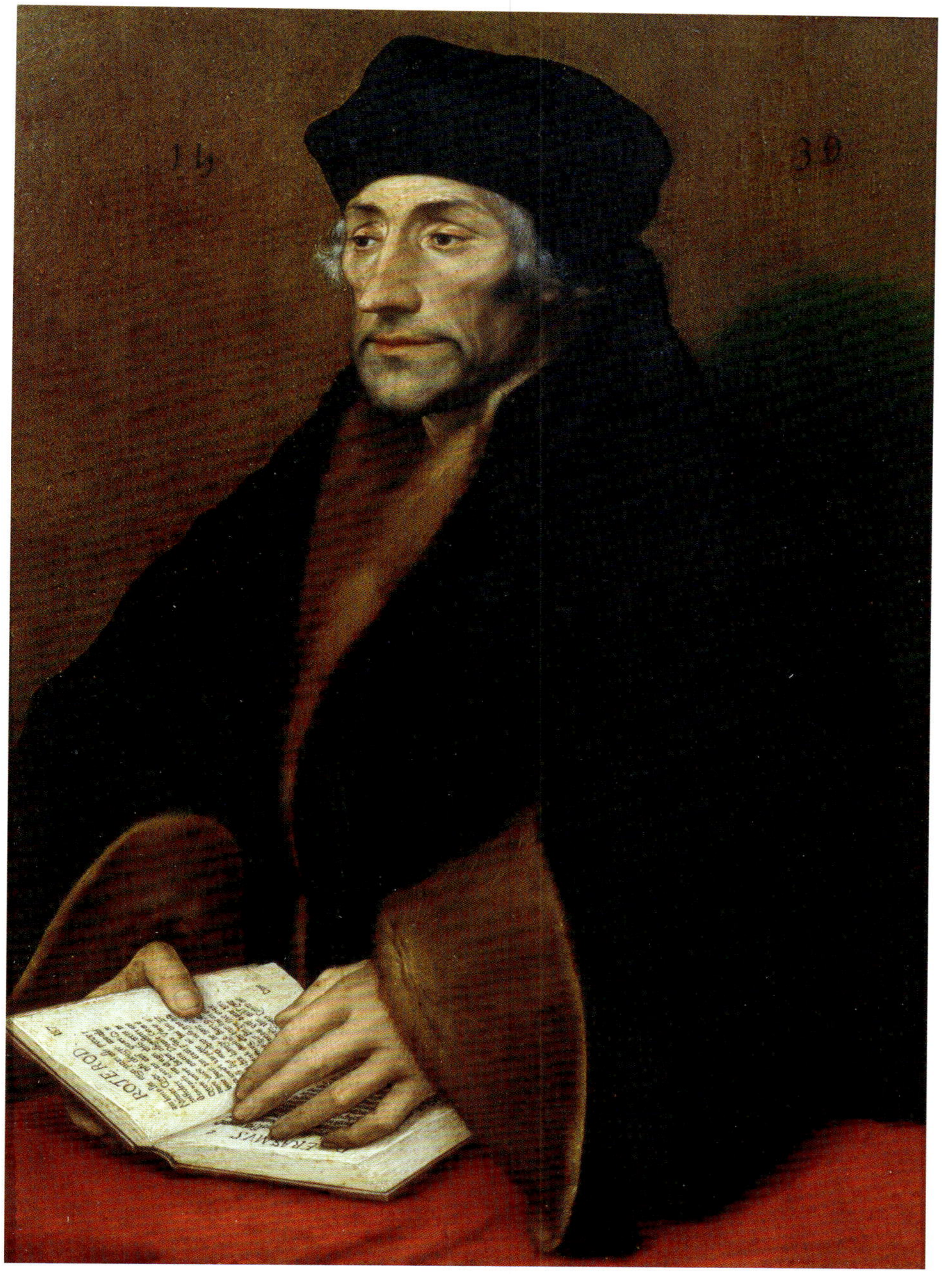

[fig. 4.2] Hans Holbein the Younger, *Portrait of Erasmus of Rotterdam*, 1530, Parma, Galleria Nazionale.

5

Rowland Lockey, partly after Hans Holbein
Portrait of Thomas More and his Family

1593

Oil on canvas, 227.4 x 330.2 cm

London, National Portrait Gallery, inv. 2765

PROVENANCE

Seen in the artist's workshop in London by the antiquarian William Burton (1575-1645); commissioned by Thomas More II; borrowed from his son Cresacre More by William Lenthall (1591-1662); retained by the Lenthall family until 1808; thereafter in various collections; acquired in 1935 by the National Portrait Gallery.

BIBLIOGRAPHY

Morison 1963: 22-25; Rotterdam 1969: no. 68; Strong 1969: 345-351; Trapp & Schulte Herbrüggen 1977: 17-18, no. 1; Wells 1981: 60-61; Rowlands 1985: 69-71, 322-323; Honigmann 1990: 80-82; Campbell 1990: 142-145; MacLeod 1996: 12; Lewis 1998: 2, 30-35; Smith 2005: 490, 494; Müller & Kemperdick 2006: 370-377; Mitjans 2007: 33-36; Cooper 2008: 38-39.

The figures are, from left to right: Sir John More (*c.*1451–1530), Thomas's father, in judge's robes; Anne Cresacre (1511/12–1577), Thomas's ward and afterwards (1529) the wife of his son John; Thomas himself; his only son John (1509–1547); Cecily (1507–?), his third daughter, wife of Giles Heron; Elizabeth (1506–1564), his second daughter, wife of William Dauntessy; Margaret (1505–1544), his eldest daughter, wife of William Roper; John More (1577–*c.*1599/1600), his great-grandson; Thomas More II (1531–1606), his grandson; Cresacre More (1572–1649), his great-grandson and biographer; Elizabeth Scrope (1534–1607), wife of Thomas More II.

Above Cresacre More hangs a portrait of his grandmother Anne Cresacre, here represented as an older woman. The coats of arms (once eight, now two) refer to the persons beneath them; the Latin inscription in the lower left corner explains their relationships.[1]

The seven figures on the left are copied from Holbein's lost portrait of Sir Thomas and his family.[2] This was painted in 'watercolour', apparently glue-size, on a very large canvas (about 275 x 360 cm); the figures were life-size. It was painted during Holbein's first stay in England (1526–1528), when, recommended by Erasmus, he lodged for a time with More. Completed in or about 1527, it can be reconstructed from a drawing by Holbein (fig. 5.3), annotated by Nikolaus Kratzer with the names and ages of the persons represented and sent by More to Erasmus, who wrote letters of thanks in September 1529.[3] It records a preliminary idea for the composition, later changed after consultations between More and Holbein. Holbein's drawings for seven of the heads are preserved. Van Mander, whose text was published in 1604, recorded that the group portrait had belonged to Andries de Loo (d. 1590) in London and that it had been acquired from his estate by Thomas More II, grandson of Sir Thomas.[4]

1
The long Latin inscription is transcribed and translated and the coats of arms are described and identified in Strong 1969: 346–348.

2
On this topic see inter alia Bätschmann 2013.

3
Hans Holbein the Younger, *Study for the More Family Portrait*, pen and brush in black over a start in black chalk, with inscriptions and separate motifs in brown ink, 389 x 524 mm, Basel, Kunstmuseum, Kupferstichkabinett, inv. 1662.31.

4
Van Mander 1994: fols. 221, 221v, 223.

5
Cited by Strong 1969: 350.

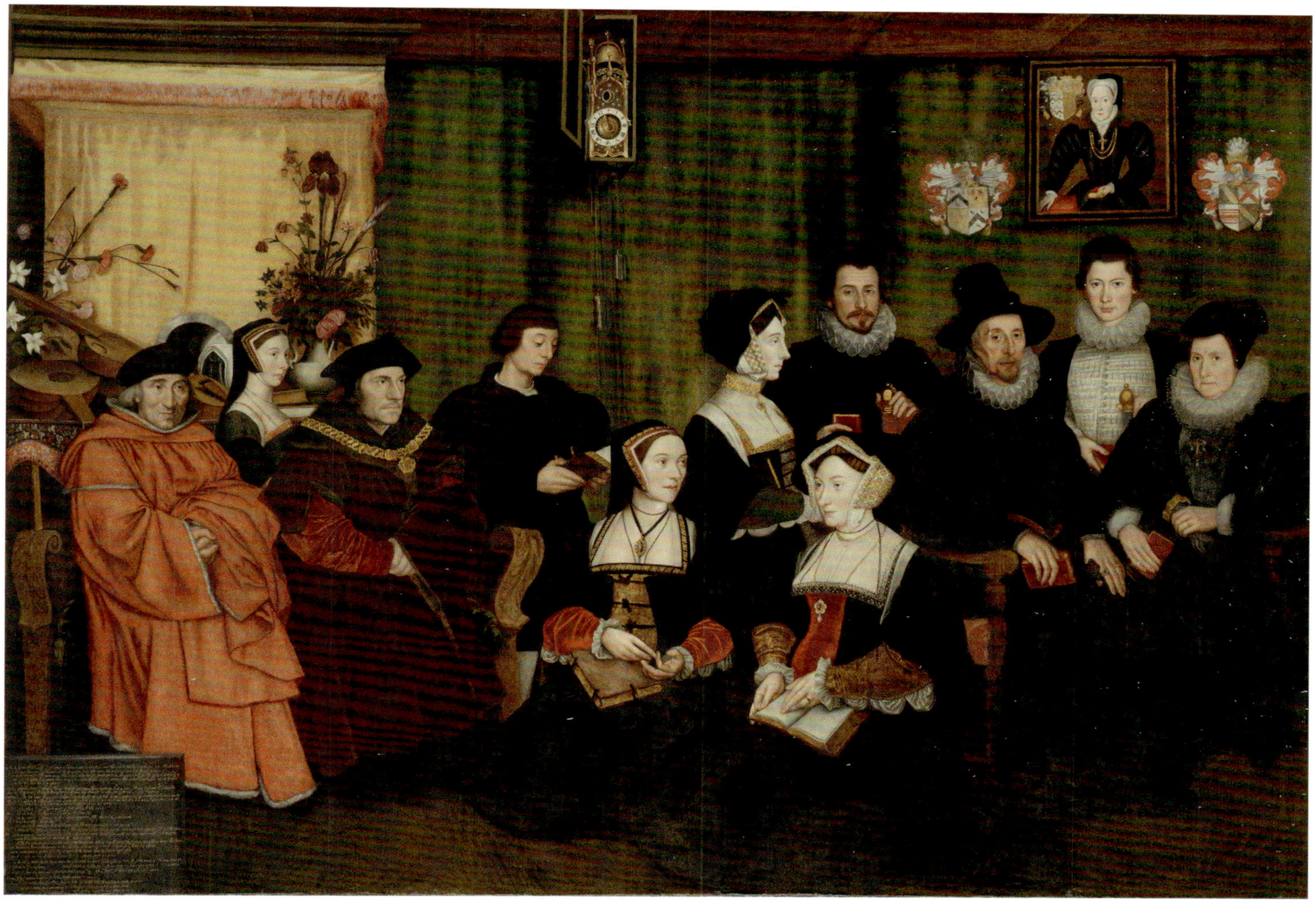

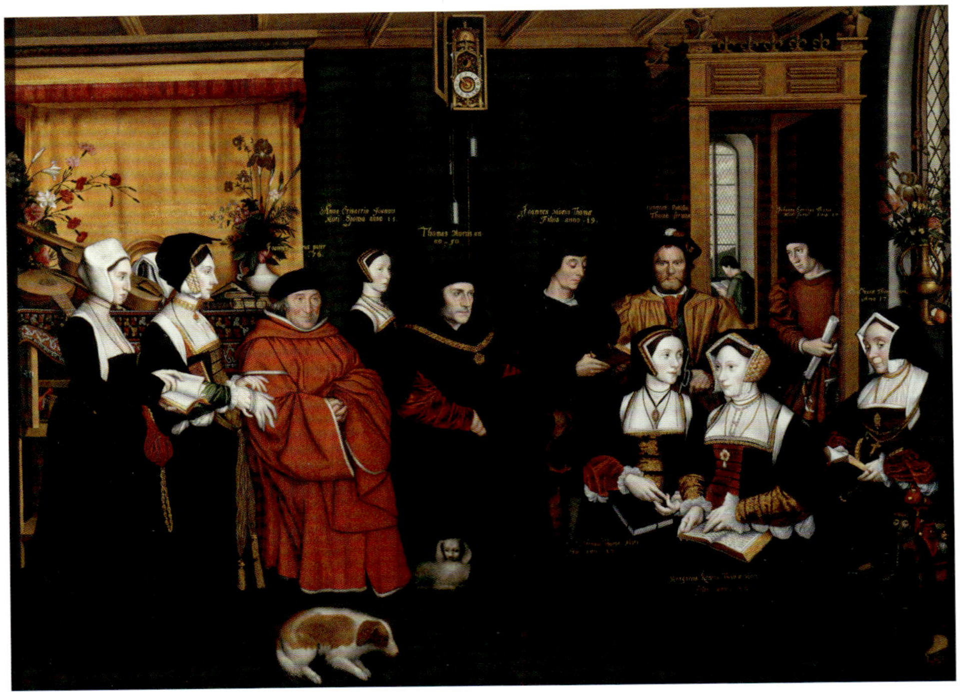

[fig. 5.1] Rowland Lockey after Hans Holbein the Younger, *Portrait of Thomas More and his Family*, 1592, West Yorkshire, Nostell Priory, National Trust.

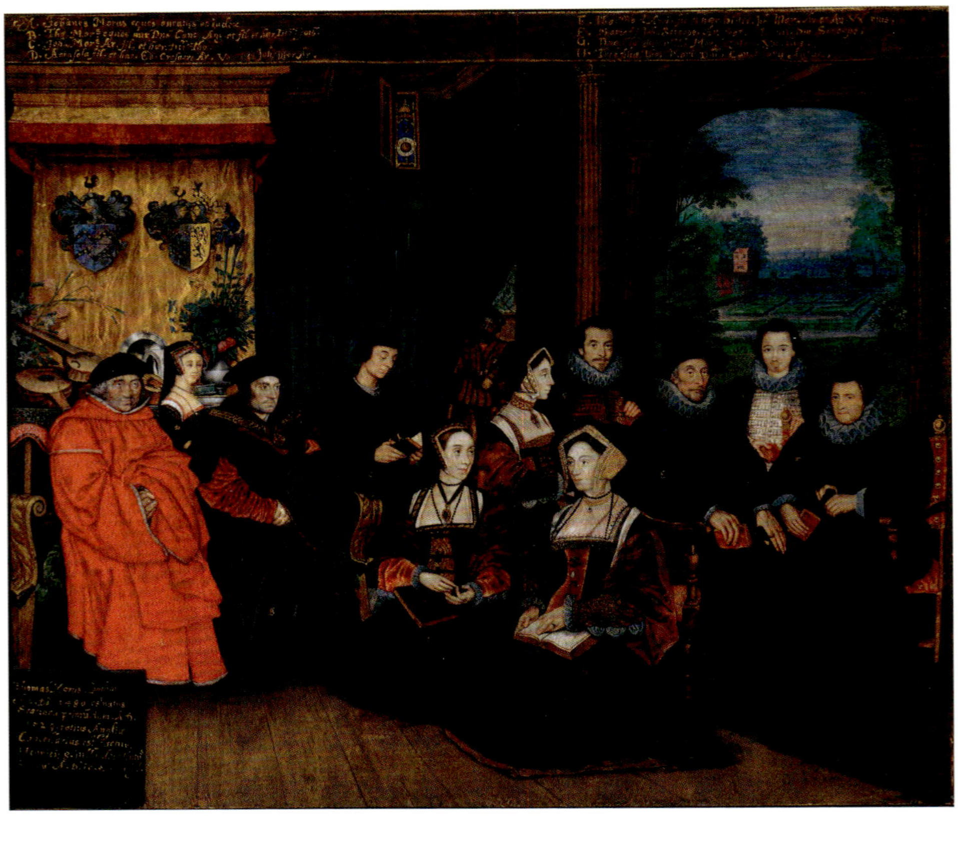

[fig. 5.2] Rowland Lockey after Hans Holbein the Younger, *Portrait of Thomas More and his Family*, 1593-1594, London, Victoria and Albert Museum.

He or his cousin Thomas Roper (1533–1598) commissioned in 1592 from Rowland Lockey a copy that descended in Roper's family and is now at Nostell Priory, Yorkshire (canvas, 249 x 343 cm; fig. 5.1). In 1593 Thomas More II commissioned from the same Lockey the altered and updated version exhibited here.

Behind Sir Thomas and his father are vases of very large flowers. They, the musical instruments, the table and its carpet and the curtain and tester appear to have been taken from Holbein's picture. There the clock seems to have hung above Thomas but here it has been moved to be close to his son John. It may have referred to the good regulation of the More household.

Burton in 1593 described the painting as a 'neat piece [...] containing in one table the picture of Sir John More [...] of Sir Thomas More [...] and of all the lineal heirs male descended from them [...] unto the present year'.[5]

The lost Holbein showed Sir Thomas with his father, his second wife, his four children, his adopted daughter Margaret Giggs, his ward Anne Cresacre, his fool Henry Patenson, his secretary John Harris and another unidentified man seen from the back. It did not include Thomas's grandchildren, his three sons-in-law, his step-mother, his step-daughter, Alice Alington, or John Clement, the husband of Margaret Giggs. No attempt was made to direct attention to Jane Colt (d. 1511), Thomas's first wife and the mother of his children. It was not a family tree. In an earlier state, recorded in the Basel drawing, there were references to prayer; but in the state recorded in the Nostell Priory copy there are no such allusions and the painting is a 'conversation piece'.

LC

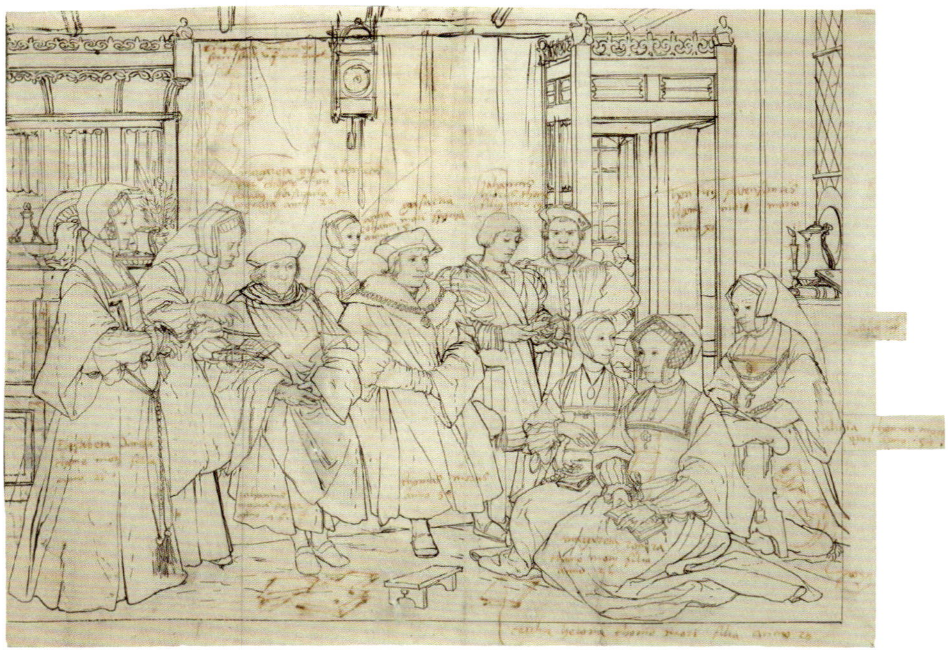

[fig. 5.3] Hans Holbein the Younger, *Portrait of Thomas More and his Family*, Basel, Kunstmuseum, Kupferstichkabinett.

6

Erasmus
Description of the Appearance and Character of Thomas More

Letter from Erasmus to Ulrich von Hutten, dated 23 July 1519

In: Desiderius Erasmus, *Farrago nova epistolarum*, Basel, Johann Froben, 1519: 329–336

Ghent, University Library, inv. BIB.ACC.020333

BIBLIOGRAPHY

Allen 1906: nos. 365, 778, 874, 923, 951, 961, 967, 745, 968, 986, 919, 999, 1233; CWE 7: 15-25; Kaegi 1924; Jardine 1993: 16-22, 160.

Ulrich von Hutten's (1488–1523) parents had intended their son for the priesthood. He soon left the monastery at Fulda, however, and went to study at the universities of Cologne, Erfurt and Frankfurt an der Oder. In October 1515 he wrote his first letter to Erasmus, lauding him as the 'German Socrates'.[1] In turn Erasmus referred to Hutten in glowing terms in letters to prominent humanists such as Guillaume Budé and Helius Eobanus Hessus. Although Hutten's polemical writings against Rome did not affect his cordial relations with Erasmus the Luther affair would ruin their friendship. Shortly before that, however, on 23 July 1519, Erasmus sent Hutten his famous epistle describing Thomas More. The letter, which allowed Erasmus to linger awhile in contemplation of a friend, as he put it, was written in Leuven during a peaceful period and subsequently sent from Antwerp. It presents a 'verbal portrait' – a counterpart to the visual *monumentum*. Erasmus sought to sketch his friend More in a narrative miniature:

> 'In shape and stature More is not a tall man, but not remarkably short, all his limbs being so symmetrical, that no deficiency is observed in this respect. His complexion is fair, his face being rather blonde than pale, but with no approach to redness, except a very delicate flush, which lights up the whole. His hair is auburn inclining to black, or if you like it better, black inclining to auburn; his beard thin, his eyes a bluish grey with some sort of tinting upon them. This kind of eye is thought to be a sign of the happiest character, and is regarded with favour in England, whereas with us black eyes are rather preferred. It is said, that no kind of eye is so free from defects of sight. His countenance answers to his character, having an expression of kind and friendly cheerfulness with a little air of raillery. To speak candidly, it is a face more expressive of pleasantry than of gravity or dignity, though very far removed from folly or buffoonery. His right shoulder seems a little higher than his left, especially when he is walking, a peculiarity that is not innate, but the result of habit, like many tricks of the kind. In the rest of his body there is nothing displeasing, only his hands are a little coarse, or appear so, as compared with the rest of his figure…[2]

1 On this topic see inter alia Klawiter 1977.

2 Citations from *Erasmus 1962*, vol. 3: 387–401.

ſor,ſed huc feſtinantem Roma abſorbuit. Non opinor quenq; apud nos tam impudentem eſſe,qui uelit in iſto theatro ſui ſpecimen ædere,& Gallus aut Germanus cum Italis,imò cum Muſuri poſteris inire certamē,quid niſi ſibilos ac riſum lucrifacturus? Pergratum eſt Petri Alcyonij per te nouiſſe nomen,utinã & lucubrationes hominis uidere liceret, præſertim quæ uertit ex Ariſtotele. Cupio totum hominem propius noſſe,præſertim cum noſtra non aſpernetur,quod haud faceret ni foret ingenio longe candidiſſimo:laceſſam eum aliàs literis meis,nunc uiro meis uerbis dices ſalutē. Eſt apud Britãnos uir undiquaq; doctiſſimus Thomas Linacrus,tibi iã olim ex Aldi noſtráq; prædicatione cognitus,uæ,pfeſſionis,ac ni fallor ætatis,is idem agit quod tu:multis annis elimata lucubrationes ſuas uiciſſim ædit in lumen. Prodijt Galenus περὶ τῶν ὑγιεινῶν,tanta fide tanta luce,tanto Romani ſermonis nitore redditus,ut nihil uſq; deſyderet lector Latinus, Imò nihil non melius reperiat,quàm apud Græcos habeatur:ſucceſſerunt libri Therapeutices,quos ſcis quales antehac habuerimus. Expectamus proxima fœtura libros Ariſtotelis Meteorologicos,nõ æſtimandis ſudorib⁹ primum emẽdatos,deinde feliciſſime uerſos. Sunt illi permulta in ſcrinijs,magno uſui futura ſtudioſis. His monumentis uir ille conſulit immortalitati ſui nominis,his ornat ſuam Angliam,his aulã regiam illuſtrat, atq; ipſum in primis principē,cui medicus eſt primarius. Apud Gallos item magnus ille Guilhelmus Copus nouis ſubinde libris rem medicã nobis aſſerit. Budæus per iurecõſultorum libros obambulans, Augeæ ſtabulum repurgat. Gaudēt omnes admoneri,ſoli Theologi nolumus meliora diſcere. Facualetudinem tuam Ambroſi doctiſſime tueare diligenter, quo & diu proſis optimis ſtudijs, & multis annis ſenex ſene frui poſſim. Nam ipſe præcurrētē te ſum aſſecutus,totus fermè canus. Aldus in familiaribus colloquijs nõ ſine uoluptate ſolitus eſt imitari ſenis decrepiti balbutiem, qua olim putaret futurũ,ut alter alterũ cõſalutarem⁹. Quõ uales,inqt,dñe Eraſme:deinde uoce æque balba,ſed exiliore,me uidelicet agens, reſpõdebat: Si uales,ego ualeo. Hæc iucũde qdē ſomniabat,præſertim ſi ſermo accederet de extruẽda noua Academia,ſed fidem non præſtitit. Reliquit nos ante balbam illã ætatem: etiamſi me multo grandior erat Aldus,annis ni fallor, plus minus uiginti. Nam ipſe nunc annum quinquageſimũ ſecundũ aut ad ſummũ tertium ago. Iuuat plura tecum confabulari,optime Leo,ſed aliò me uocãt negocia. Tuum nomen ſemper extabit in meis prouerbijs,ſi tamē his ipſis continget extare. Neq; uiciſſim in tuis lucubrationibus grauaberis Eraſmi meminiſſe. Optimum Baptiſtam Egnatium meo nomine ſalutato diligēter:item Aſulanum cũ ſua familia,nominatim Manutiolũ qui nobis iſthic puellus ludebat, cum adeſſem. Bene uale Louanij Id. Octob. ANNO M. D. XCIX.

Petrus Alcyonius.

T. Linacrus.

Guilhelmus Copus Baſilienſis.

Budæus.

Aldus Manutius.

Aetas Eraſmi

Baptiſta Egnatius.

Eraſmus

ERASMVS ROTERODAMVS CLARISSIMO EQVITI VLRICHO HVTTENO. S. D.

QVOD Thomæ Mori ingenium ſic deamas,ac penè dixerim deperis,nimirum ſcriptis illius inflammatus,quibus, ut uere ſcribis,nihil eſſe poteſt neq; doctius, neq; feſtiuius,iſtuc crede mihi clariſſime Huttene, tibi cum multis cõmune eſt, cum Moro mutuum etiam. Nam is uiciſſim adeo ſcriptorum tuorũ genio delectatur,ut ipſe tibi propemodum inuideã. Hæc uidelicet eſt illa Platonis omnium maxime amabilis ſapiẽtia,quæ longe flagrantiores amores excitat inter mortales,quàm ullæ quamlibet admirabiles corporum formæ. Non cernitur illa quidē oculis corporeis,ſed & animo ſui ſunt oculi,ut hic quoque uerum comperiatur illud Græcorum, ἐκ τοῦ ὁρᾶν γίνεται ἀνθρώποις τὸ ἐρᾶν, per hos fit aliquoties,ut ardentiſſima charitate conglutinentur,inter quos nec colloquium,nec mutuus conſpectus interceſſit. Et quemadmodum uulgo fit,ut incertis de cauſis alia forma alios rapiat: ita uidetur & ingeniorum eſſe tacita quædam cognatio,quæ facit ut certis ingenijs impenſe delectemur. Cæterum quod à me flagitas,ut tibi totum Morum uelut in tabula depingam, utinam tam abſolute præſtare queam,quàm tu uehementer cupis. Nam mihi quoque non iniucundum fuerit, interim in amici multo omnium ſuauiſſimi contemplatione uerſari. Sed primum, οὐ παντὸς ἀνδρὸς ἰδεῖν omnes Mori dotes perſpexiſſe. Deinde haud ſcio an ille laturus ſit, à quolibet artifice depingi ſeſe. Nec enim arbitror leuioris eſſe operæ Morum effingere,quàm Alexãdrum magnũ,aut Achillem,nec illi quàm hic noſter immortalitate digniores erant. Tale argumẽtum prorſus Apellis cuiuſpiam manum deſyderat: at uereor,ne ipſe Fuluij Rutubæq; ſimilior ſim quàm Apellis. Experiar tamen tibi totius hominis ſimulachrum deliniare uerius quàm exprimere, quantum ex diutina domeſticáque conſuetudine uel animaduertere licuit,uel meminiſſe. Quod ſiquãdo fiet ut uos aliqua legatio committat,tum demum intelliges quàm non probum artificem ad hoc negocij delegeris, uereórque planè, ne me aut inuidentiæ incuſes, aut cæcutiẽtiæ, qui ex tam multis bonis tam pauca, uel uiderim lippus, uel commemorare uoluerim inuidus. Atque ut ab ea parte exordiar, qua tibi Morus eſt ignotiſſimus,

ſtatura

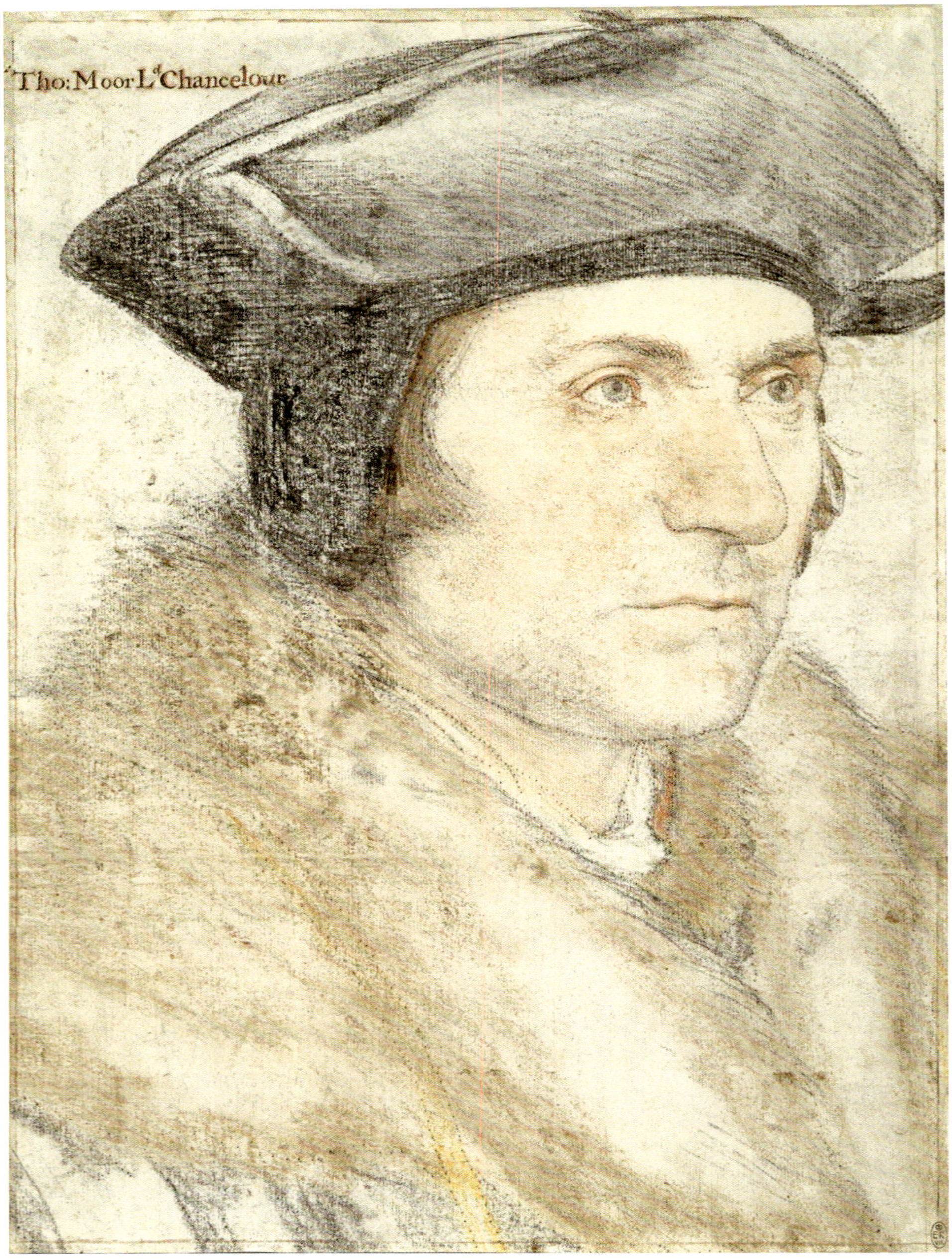

[fig. 6.1] Hans Holbein the Younger, *Portrait of Thomas More*, c.1526-1527, drawing, Royal Collection Trust, Windsor Castle.

Yet above all Erasmus wanted his portrait to show More's innate qualities – his austerity, his passion for social justice, his delight in recreation and his sense of duty, his intellectual and religious depth, his decision to marry because 'he was resolved to be a chaste husband rather than a licentious priest.' If More was an exemplary husband and father, as a friend he was a faithful refuge and rock. Moreover:

> He seems to be born and made for friendship, of which he is the sincerest and most persistent devotee. ... If anyone requires a perfect example of true friendship, it is in More that he will best find it.

In the story of a life (*vita*) set against the background of what was an almost realized European Republic of Letters, the meaning of More's life as a humanist man of letters was sought. The reader of More's *vita* is invited to find in it a catalyst for imitation and emulation. It was not merely for the pleasure of it that Erasmus twice added to his portrait of More in the sketches of More's family he sent to Budé (1521) and Johannes Faber (1533). After the first publication of the *Moriae Encomium* in 1511, which was dedicated to More himself, Erasmus included his 'letter portrait' as a preface to every edition of the *Praise of Folly*.

JP

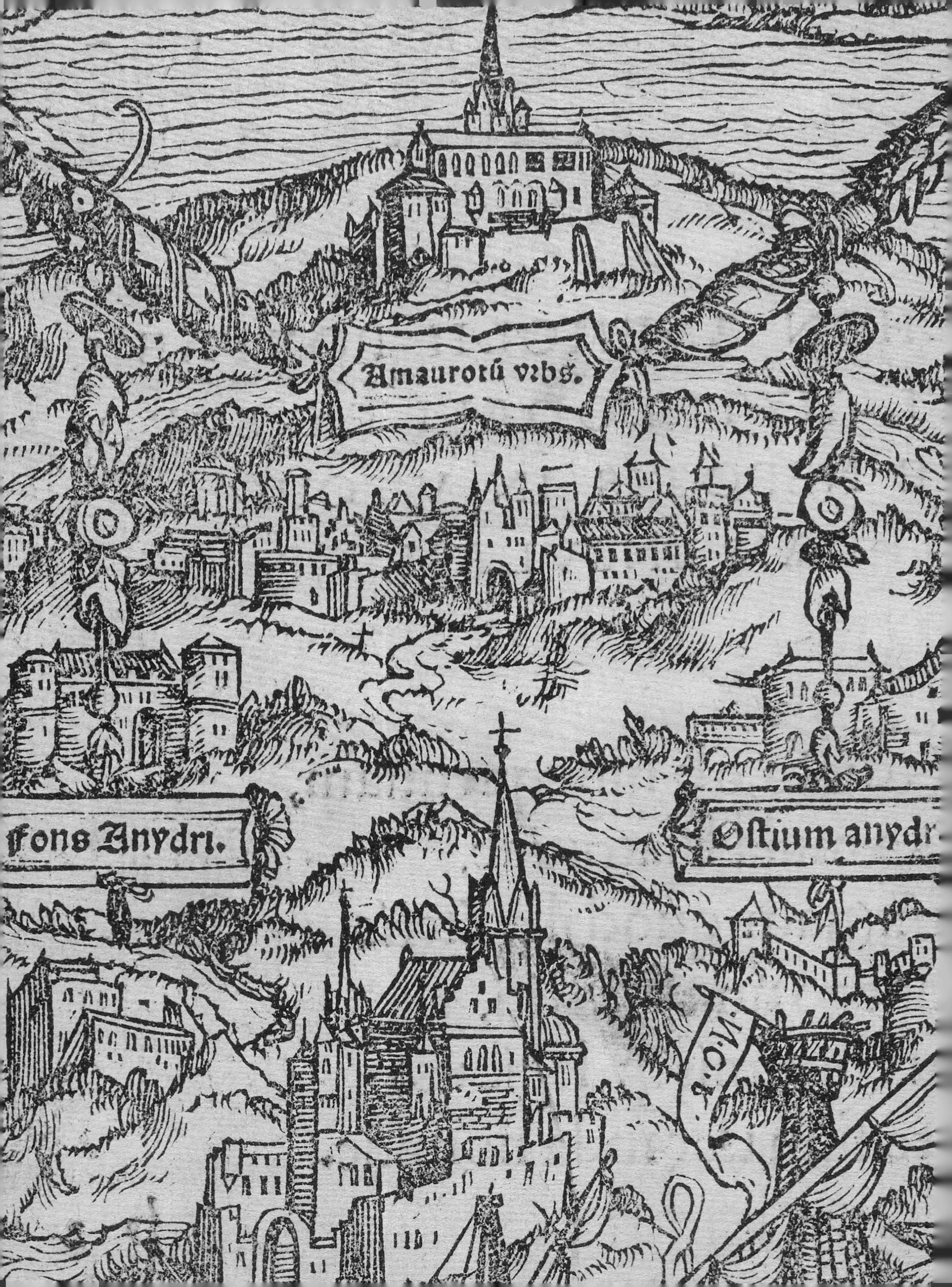
Amaurotū vrbs.
fons Anydri.
Ostium anydr

Editions and Translations of *Utopia* 1516–1750

MARCUS DE SCHEPPER

FIRST EDITION: Leuven, Dirk Martens (1516). In More's lifetime: Paris (1517), Basel (twice in 1518), Florence (1519).

LATER LATIN EDITIONS: Leuven (1548), Cologne (1555), Basel, Leuven (1565-1566), Wittenberg (1591), Frankfurt (1601, 1670, 1689), Hanau (1613, 1619), Milan (1620), Amsterdam (1629, 1631), Oxford (1663), Helmstedt (1672), Glasgow (1750).

FIRST TRANSLATIONS: German (1524), Italian (1548), French (1550), English (1551), Dutch (1553: cat. 9), Spanish (1637).

LATER TRANSLATIONS: German in Leipzig (1612), Frankfurt (1704), English in London (1624, 1639, 1684, 1685), Dublin (1737), Glasgow (1743), French in Amsterdam (1643, 1717, 1730), Leiden (1715: cat. 12), Dutch in Hoorn (1629, 1630, 1634), Rotterdam (1677), Amsterdam (1700).

DETAIL CAT. 7.1

Before 1750 over fifty editions and translations of Utopia appeared. The rhythm of their publication is uneven and happens in waves. The Latin text got off to a tremendous start, with five editions in three years. The *editio princeps* ('first edition') of this 'new' and 'golden' book was printed by Dirk Martens in Leuven in the second half of December 1516, in an era in which voyages of discovery were following one another in quick succession. The many paratexts reinforced the credibility of the tale and belief in the existence of the Isle of Utopia. Next, in early October 1517, an unauthorized edition appeared in Paris, brought out by Gilles de Gourmont. In March 1518 a third edition was printed in Basel by Johann Froben, swiftly followed by a new edition in November of that same year, the last edition to be corrected by More. In July 1519 *Utopia* was reprinted in Florence by Giunta and bound with More's Latin translations of Lucian.

After More's death in 1535 another Latin edition appeared in Antwerp and Leuven (1548: printed in Leuven by Servaas van Sassen for Birckmann, a Cologne publisher with a branch in Antwerp), Cologne (1555: Birckmann), Basel (1563: Episcopius, humanist compilation with More's *Lucubrationes*), Leuven (1565-1566: Joannes Bogardus for himself and for Petrus Zangrius, More's *Opera*, as a collection of the Latin texts of a great Catholic author) and Wittenberg (1591: Johann Kraft).

During More's life only one edition appeared in a vernacular, that being a German translation of Book II by Claudius Cantiuncula, published in 1524 by Johann Bebel in Basel. The Italian translation by Ortensio Landi was printed in Venice in 1548 for Anton Francesco Doni. The French translation by Je(h)an le Blond was printed in Paris in 1550 by Charles L'Angelier. An adaptation of it by Barthélemy Aneau came out in Lyon in 1559, published by Jean Saugrain. In 1585 Gabriel Chappuys published his translation of Book II at Pierre Cavellat in Paris, where it was reissued by Regnault Chaudière in 1598. A year after the French translation Abraham Vele published the English version by Ralph Robinson in London, followed by two reprints, in 1556 (Vele) and 1597 (Thomas Creede). The 1553 Dutch translation brought out by Hans de Laet in Antwerp was reissued in 1562 (cat. 9). Not until 1637 did a Spanish translation become available, in Córdoba.

Between 1600 and 1750 the book appeared chiefly in large publishing centres in Germany (Frankfurt book fair), England (London) and the Netherlands (the *Magasin de l'univers*). The Latin text was republished in Frankfurt (1601, 1670 and 1689), Hanau (1613 and 1619), Milan (1620), Amsterdam (two editions in 1629: Amsterdam and 'Cologne' and one in 1631), Oxford (1663), Helmstedt (1672) and Glasgow (1750).

German translations appeared in Leipzig (1612) and Frankfurt (1704), English translations in London (1624, 1639, 1684, 1685), Dublin (1737) and Glasgow (1743), French translations in Amsterdam (1643, 1717, 1730) and Leiden (1715), and Dutch translations in Horn (1629, 1630, 1634), Rotterdam (1677) and Amsterdam (1700).

Over the years, the title 'golden handbook' was displaced by 'Utopia' and the work was adapted to a Catholic or Protestant public.

7.1

Ambrosius Holbein
The (Fictitious) Voyager Raphael Hythlodaeus Tells the Story of the Isle of Utopia to Thomas More

In: Thomas More, *De optimo reip. statu, deque noua insula Utopia, libellus uere aureus*, Basel, Johann Froben, November 1518: 12–13
Woodcut on paper, 178 x 119 mm
Leuven, KU Leuven, University Library, Tabularium, inv. R 3 A 12099

BINDING

Modern parchment binding with turned edges, title in ink on spine.

PROVENANCE

'O.R.E.' ('Office de Récupération Économique': German reparations to Leuven after 1918, front pastedown).

BIBLIOGRAPHY

Gibson 1961: 10-12, no. 4; Prévost 1978; Hoyer 1981: 237-254; Bezzel 1983: no. M 6300; Schulte Herbrüggen 1997b: 215-230; Bishop 2005: 107-112; Roggen 2008: 14-31, 273-275; Salberg 2008: 32-46, 149-169; Vielle 2009; Heard, Whitaker et al. 2011: 186-187, no. 86; Scafi 2013: 160-171; Vielle 2013: 203-222.

In the German-speaking regions More's work was in great demand. On 19 February 1518, for instance, Martin Luther wrote to his friend Johann Lang, 'Capito said that *Utopia* is in the press in Basel – buy one for me at the Frankfurt fair – *Utopiam Morinam sitio*' ('I hunger for More's *Utopia*'). In March 1518 the Basel printer Johann Froben had already published a new edition of *Utopia*, edited by Beatus Rhenanus (1485–1547). The main title is framed in a woodcut by Hans Holbein (1497–1543), with a 'Vera Icon' in the upper border and in the lower border 'Lucretia kneeling before Tarquinius', which had been used in some of Erasmus's earlier publications. The start of Utopia has a border with putti. The text of this edition was reprinted in Florence in 1519 and was the basis for the modern standard edition by E. Surtz and J.H. Hexter in 1965.

It was soon obvious that one edition was not enough and the adroit Froben at once set to work on a second. A number of corrections were also made in it. It was the last edition in which More himself was in any way involved. By late November the book was ready, and in December 1518 it appeared, together with More's and Erasmus's *Epigrammata*. This time the title frame and border are the same, with putti and Froben's emblem. This edition was the model for the text in most later editions, particularly those in Leuven, Cologne and Basel.

Both Froben editions are much handsomer than the Leuven one and certainly more so than the unillustrated Paris edition of 1517. The Basel printer added woodcuts and elegant historiated initials. His editions also contain the Utopian alphabet. Most striking of all is the map of Utopia, this time designed by Ambrosius Holbein (*c*.1494–1519), elder brother of Hans. It is a modified copy, in reverse, of the map in the 1516 Leuven edition. The most significant innovation is the addition of a

1 A distorted projection or drawing which appears normal when viewed from a particular point.

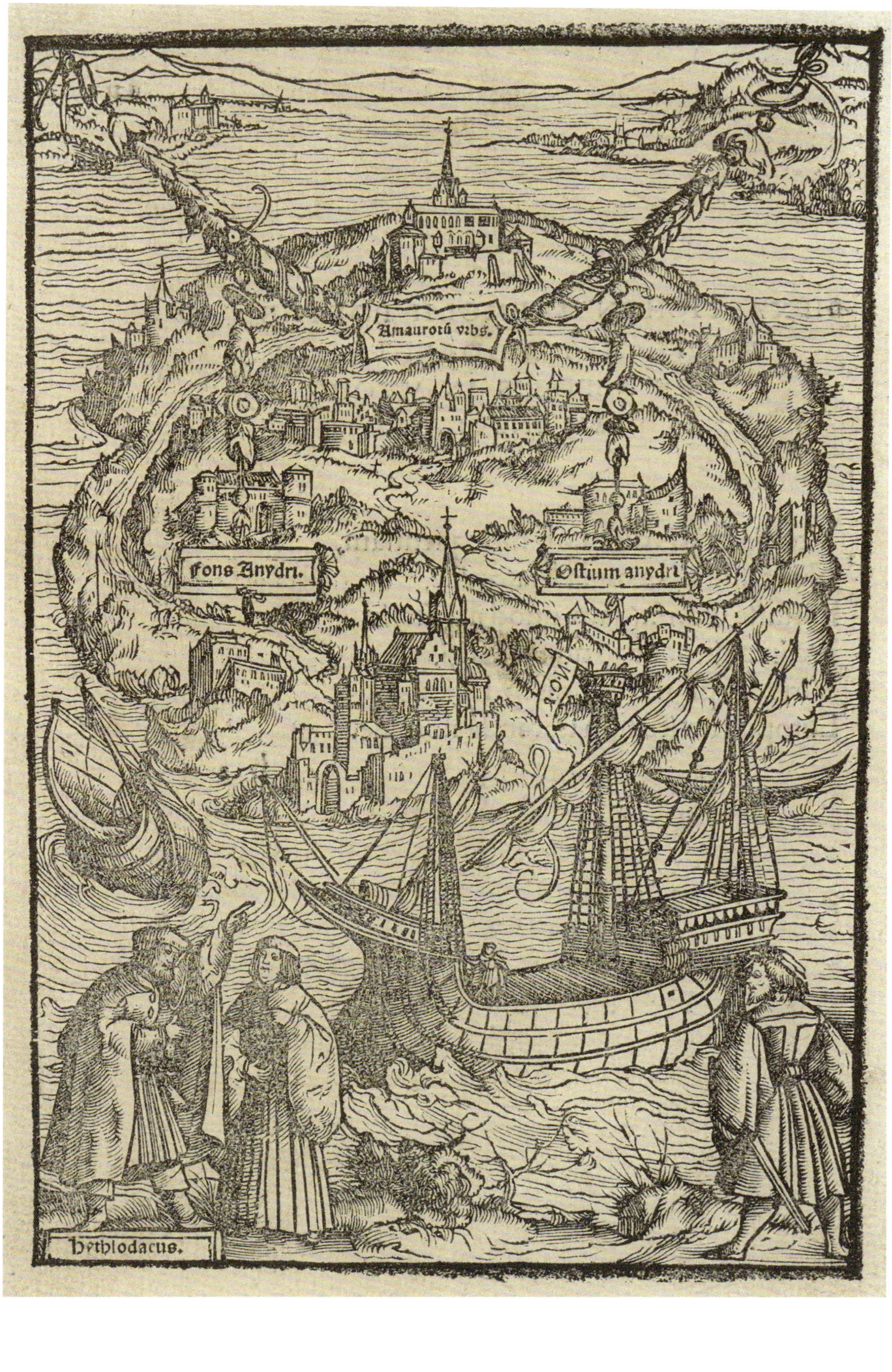
Amaurotū vrbs.
Fons Anydri.
Ostium anydri
Hythlodaeus.

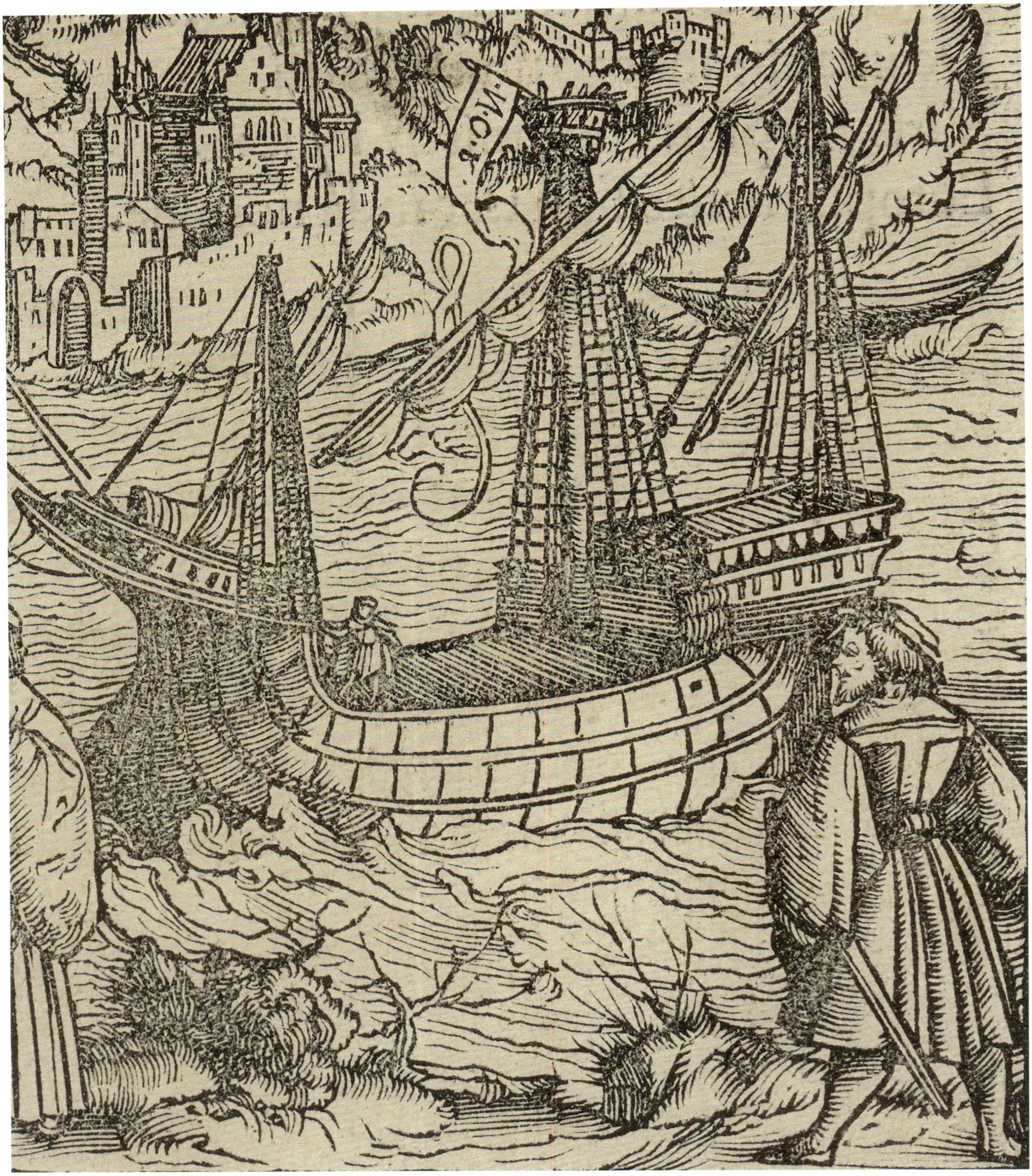

[cat. 7.1] detail

piece of land in the foreground from which three characters, including the narrator Hythlodaeus, look towards the island. On the ship's pennant an 'R' can be seen as well as the 'N' and the 'O', but it makes the meaning no clearer. The festoons of foliage, attached at the top of the image are typical of Holbein. There is one change, however, that is based on an erroneous interpretation of the text: the watchtowers have been replaced by churches. This 'decorated' edition was clearly intended for the more select and wealthy buyer. More's status was increased, but that still didn't make *Utopia* a 'popular' book.

There may be more to this map than meets the eye. Bishop has recently contended that Ambrosius Holbein was already playing with anamorphosis[1] fifteen years before his brother Hans famously painted *The Ambassadors* (1533). The shape of the island of Utopia seems to hint at a skull with the harbour mouth (*ostium*) as the mouth and the ramparts with their (church) towers as the jawbone and teeth. The 'skull' would be an allusion to the author (*mors* and *Morus*) and thus would be a *memento mori/Mori*, a visual pendant to the *beatae vitae exemplar* ('model of happy life') that characterizes the lifestyle of the Utopians.

MDS

7.2

Ambrosius Holbein
The (Fictitious) Voyager Raphael Hythlodaeus Tells the Story of the Isle of Utopia to Thomas More in Pieter Gillis's Garden

In: Thomas More, *De optimo reip. statu, deque noua insula Utopia, libellus uere aureus*, Basel, Johann Froben, November 1518: 24–25
Woodcut on paper, 63 x 104 mm
Leuven, KU Leuven University Library Tabularium, inv. R 3 A 12097

BINDING

Modern parchment binding (Italy, nineteenth century?), title in ink on spine, blue speckled fore-edge (Italianate style, eighteenth century?).

PROVENANCE

'Francisci Xaverii Guicciardi' (eighteenth century, canon in Chur/Coira?); 'O.R.E.' ('Office de Récupération Économique': German reparations to Leuven after 1918, front pastedown).

Froben's edition is 'modern', with More's letter to Gillis in the shape of an hourglass and the main text beginning with a woodcut and a playful initial. The Italianate setting shows Hythlodaeus (as a sage), More and Gillis (both wearing the typical scholar's cap), and More's assistant John Clement (*c.*1500–1572) who married More's adopted daughter Margaret Giggs in around 1526.

MDS

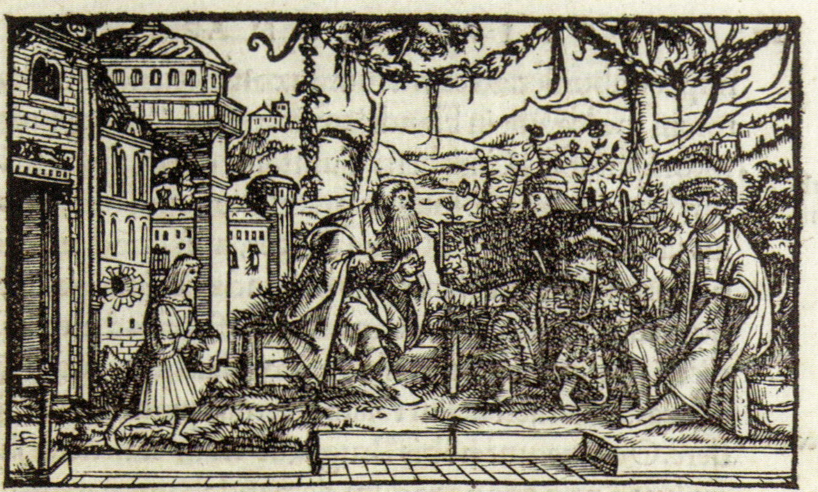

Io.Clemens. Hythlodæus. Tho.Morus. Pet. Aegid.

SERMONIS QVEM

RAPHAEL HYTHLODAEVS VIR EXIMIVS, de optimo reipublicæ statu habuit, liber primus, per illustrem uirũ Thomam Morum indytæ Britanniarũ urbis Londini & ciuem, & uicecomitem.

QVVM NON EXIGVI MOmenti negocia quædam inuictissimꝰ Angliæ rex HENRICVS eius nominis octauus, omnibus egregijs principis artibus ornatissimus, cũ serenissimo Castellæ principe CAROLO controuersa

8

Thomas More at his Writing Desk

In: Thomas More, *La description de l'isle d'Utopie ou est comprins le miroer des republicques du monde, & l'exemplaire de vie heureuse*, Paris, Charles L'Angelier, 1550, fol. 1v–2

Woodcut on paper, 36 x 52 mm

Leuven, KU Leuven, University Library, Tabularium, inv. CAAA 2212

BINDING

Recent calfskin binding 'à l'antique' with blind fillet over boards, gilt cornerpieces, back on three raised bands.

PROVENANCE

'Di Girolamo Cat[?]re da pec[?]a' (title page); acquired at The Romantic Agony auction 48, Brussels, on 16 June 2012 (lot 1083).

BIBLIOGRAPHY

Peggram 1940: 330-340; Hallowell 1960: 86-92; Gibson 1961: 24-25, no. 19; More 1970; Scrivano 1972: 99-108; Prévost 1978: ccxxix; Hosington 1984: 116-134; Minerva 1992: 51-61; Biot 1995: 11-28; Seidel Menchi 1996: 95-119; Hester 1998: 133-148; Pettegree, Walsby & Wilkinson 2007: no. 38574; Cave 2008b: 87-103, 205-217; Gjerpe 2008: 47-66, 171-179; Roggen 2008: 14-31, 273-275; Sellevold 2008a: 67-86, 181-203; Sellevold 2008b: 53-65.

In France, a country with a strong political elite and a robust intellectual consciousness, *Utopia* has always played an important role. Quite a few literary and philosophical texts refer to it (Rabelais, Montaigne). More himself was keen on having a Parisian edition immediately after the Leuven edition of 1516 came out. The introductory letter from Guillaume Budé in that edition forwarded the debate. The French favoured texts in their own language so there was often more than one translation of much-discussed books or interesting texts. Thus, too, with *Utopia*. Je(h)an Le Blond d'Evreux (1502–1553), *poeta minor* and translator of Valerius Maximus, Johann Carion and Francesco Patrizi, was a staunch advocate of the use of French. In 1550 his translation of More's book appeared, based on the Paris edition of 1517, including Budé's letter with its emphasis on Christian laws and values. The edition contains twelve woodcuts, nine of them in Book II, accompanying the start of each chapter. The opening illustration shows a learned author (More) in his study with his dog, a clear allusion to St Jerome with his lion engaged in his translation work for the Vulgate. Connoisseurs characterize Le Blond's translation as *fidèle et élégant*. The text's message was conveyed, rather than the literal wording, the aim being *l'utilite & proffit de la republique*. Therefore difficult concepts were paraphrased and terms were used that were often outdated but clear.

In 1559 Le Blond's translation was reissued (without attribution or paratexts) in Lyon by the Protestant printer Jean Saugrain in a version revised by a certain 'M.B.A.', identified by many as Barthélemy Aneau (1505–1561), rector of the Collège de Lyon and a prolific author and translator. Here the emphasis is on the exceptional, even unique nature of Utopia and the translation shows an affinity with Aneau's

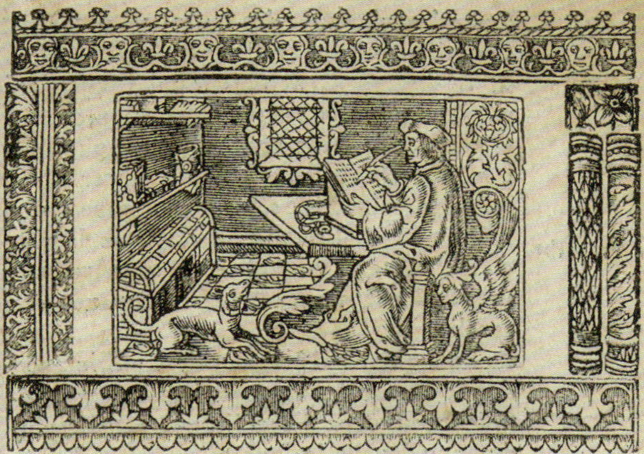

Thomas Morus.

AINSI QVE LE ttesinuincible Roy d'Angleterre Henry huictieme de ce nom, autant decoré & orné de tout ce qui apartient à vn excellent prince, qu'il eſt poſſible; de nouueau auoit different auecques Charles treſpaiſible prince de Caſtille, pour vn affaire qui ne eſtoit de petite importance, pour faire appoinctement d'iceluy, il m'enuoia en ambaſſade en Flandres en la compaignie de Cuthbert Tunſtal, homme imcomparable, lequel il auoit faict n'y auoit lõg temps ſon premier ſecretaire en la grande reſiouiſ-

Cuthbert Tunſtal ſecretaire du Roy d'Angleterre.

own utopian fiction *Alector* (1560). A very different French version of Book II appears in *L'Estat, description et gouvernement des royaumes et republiques du monde* (Paris 1585) by Gabriel Chappuys (1546–1611), a collection of texts on the government of states. Chappuys's translation appears to be a discrete adaptation of the Italian version of Book II, as included in *Del governo de i regni et delle republiche cosi antiche come moderne libri XVIII* (Venice 1561) by Francesco Sansovino (1521–1586). Chappuys's collection of political writings was intended as material for discussion at the time of the French wars of religion in the reign of Henry III. Still in 1611 there appeared a new translation of Sansovino's collection by the otherwise unknown 'F.N.D.', adapted to the new authoritarian political climate in the France of Henry IV. In 1643 an entirely new translation of *Utopia* came out in Amsterdam, written by the Protestant freethinker Samuel Sorbière (1615–1670) and aimed at a cosmopolitan readership with taste for political philosophy (Descartes, Hobbes). In France the book was later read chiefly in the *trop fameuse traduction de Nicolas Gueudeville* (Prévost) of 1715 (cat. 12).

MDS

9

Thomas More
De Utopie van Thomas Morus

8°: 87 ff.

Published in Antwerp by Hans de Laet, second edition, 1562 (first edition: 1553)

Antwerp, Museum Plantin-Moretus, inv. R 10.40

BINDING

Mottled leather, eighteenth century, gold-tooled back on five raised bands with red leather lettering label and panels with floral motif, red fore-edge.

PROVENANCE

Old ownership mark on title page ('soy de fran [...]', struck through); seventeenth-century inscription on front endpaper ('joseph de [...], geschreven den [...] january [...]', struck through); F. Dumont (on head-edge).

BIBLIOGRAPHY

Gibson 1961: 50, no. 39; Cockx-Indestege, Glorieux & Op de Beeck 1968: no. 6872; Rouzet 1975: 115-117; Smith 1981: 28; Spaans & Cave 2008: 104-109, 219-231, 284; Buyens 2009: 43-68; Pettegree & Walsby 2011: 941, no. 21639.

The name of the translator of the first Dutch edition of Utopia is unknown. Slightly more is known about its printer-publisher Hans de Laet, active in Antwerp from 1545 to 1566. He printed and published some hundred books, mostly in Dutch. His edition is marked by a no-nonsense approach: there are no paratexts, no illustrations, and the book is a handy octavo. He had received a ten-year privilege for this translation on 29 October 1550. In 1553 he brought it out on the market (with the address 'in den Salm') and just before the expiration of the ten-year period (1562), he tried again with the remaining copies, which were now provided with a new title page (with the address 'in die Rape').

Immediately conspicuous on that title page is the departure of the Dutch title from the Latin. The standard reference to *'the Best State of a Commonwealth'* has disappeared; attention is immediately focused on 'Utopia'. Most striking of all, however, is an addition in the title: *een boeck / seer profijtelijc ende vermakelijck om lesen, / bysondere den ghenen die heensdaechs / een Stadt ende ghemeynte hebben te / regeren, daer hy meestendeel toe / dienende is* ('Very useful and entertaining to read, especially those who have a town or borough to govern, for which it [the book] is mostly used'). *Utopia* is now a practical handbook with handy tips for a town's office-bearers.

The Netherlands were the cradle of Utopia. The text was conceived there (in Bruges and Antwerp), the frame narrative takes place there (Antwerp), and the finished book was published there (Leuven). In that prosperous and literate region it would certainly have found an attentive readership. In 1548 Utopia appeared in Leuven again – the first edition after More's death in 1535. The printer, Servaas van Sassen, was even related to Dirk Martens, having married Martens's daughter

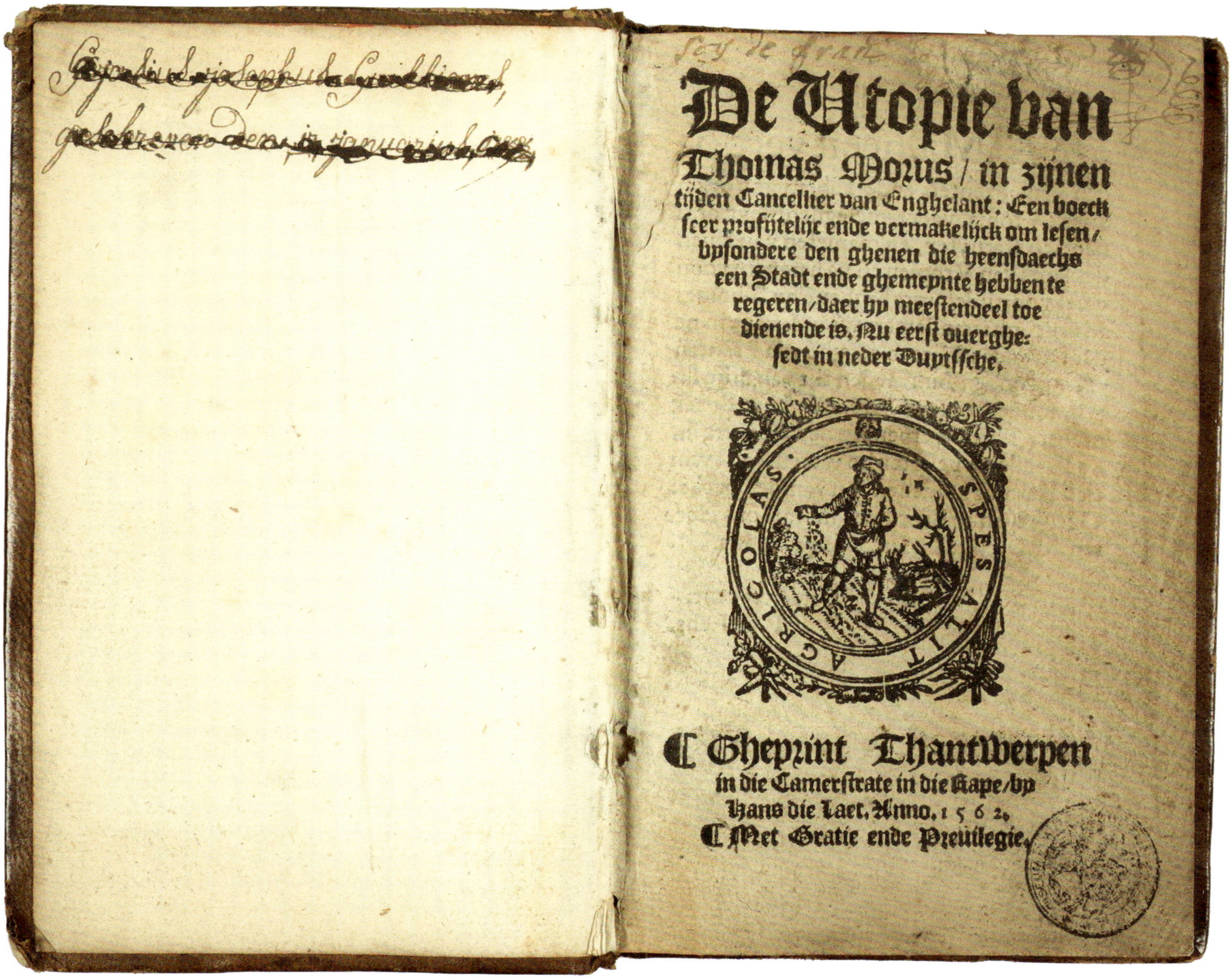

De Utopie van Thomas Morus / in zijnen tijden Cancellier van Enghelant: Een boeck seer profijtelijc ende vermakelijck om lesen / bysondere den ghenen die heensdaechs een Stadt ende ghemeynte hebben te regeren / daer hy meestendeel toe dienende is. Nu eerst ouerghe-sedt in neder Duytssche.

¶ Gheprint Thantwerpen in die Camerstrate in die Rape / by Hans die Laet. Anno. 1562.
¶ Met Gratie ende Preuilegie.

Barbara. After her death (before 1527), he married Anna Amersvoet, sister-in-law of Cologne publisher Franz Birckmann. Following the split between the northern and southern provinces of the Netherlands (1585), Utopia was no longer reprinted in the South. The North, strongly focused on the international book market, quickly out brought pocket-sized Latin editions (Blaeu and Janssonius), as well as a new and once again anonymous translation. In Hoorn, several editions of this version appeared between 1629 and 1634, all a small size. The text was tailored to Protestant receptivity: More's reservations about certain religious rituals were gingered up and 'Catholic' elements were toned down or removed altogether. Finally, in 1677, an entirely new, fluent translation by printer and professional translator Frans van Hoogstraten (1632–1696) appeared in Rotterdam, aimed at an even wider readership.

The first Dutch translation of 1553/1562 was never reissued and barely studied. Besides which, only a small number of either edition survives.

MDS

10

Hans Lützelburger, after a design by Monogrammist N.H.
Battle of Naked 'Utopians' and Clothed Peasants

Second state, *c.*1525

Woodcut, 256 x 295 mm

Watermark: a high crown with a diadem, cross and star (cf. Briquet 4968)[1]

Dresden, Staatliche Kunstsammlungen Dresden, Kupferstichkabinett, inv. A 125,1

BIBLIOGRAPHY

Passavant 1862: 445, no. 2; His 1870: 169; Woltmann 1874: 193; Lippmann 1899-1900: 40; Dörnhöffer 1904: 61; Dodgson 1907: 319-322; Dodgson 1911: 195-198 n. 1, 200-201, no. 1; Musper 1943: 171-179; Hollstein 1953: 64, no. 73; Winkler 1964a: 54-56, fig. 7; Geisberg & Strauss 1974: 888, no. G.954; Andersson & Talbot 1983: 318-319, no. 180; Falk 1984: 281-282, no. 3 (II); Parshall 1989: 30-43; Landau & Parshall 1994: 212-216, 394-395.

Following the death of Maximilian of Austria (1459–1519) the Augsburg printing industry plunged into crisis. The number of projects visibly dropped off and several illustrators were forced to tout for new customers. This was the context in which the first state of this woodcut, with its remarkable iconography, was created (fig. 10.1). With quasi-classical naked men in a setting with German peasants and drooping silver firs, designer and blockcutter tried to price themselves back into the market.[2] The sheet was effectively an advert for their artistic skills.[3] In two small supplementary woodblock panels below the composition were the name of the *Furmschnider*, Hans Lützelburger (d. *c.*1526) and a sample alphabet, declaring his expertise as a cutter of letters.[4]

Not long after 1522 the block was reused for the second state of the woodcut, which is the one shown here. The two promotional panels with Lützelburger's name and the sample of his letter cutting have gone, making way for a thirty-four-line poem in three columns. Composed in the Augsburg dialect, the rhyme gave an additional story to the iconography.[5] Because most sheets are trimmed to the margins of the picture, the whereabouts of only two prints of this second state are known today.[6]

In the middle of a forest is a group of men locked in combat. Some are naked, some are clothed; all are equipped with swords, flails, pitchforks and/or scythes. According to the first lines of the poem this is an event that must have taken place in the far distant past on the Island of Utopia. The poet himself seems not wholly convinced of the story.[7] In any case he turns almost immediately to an ode to the arts with a list of antique painters.[8] At certain points the text also refers derisively to the print. The two figures on the far left of the composition, for instance, could be identified with the designer and the blockcutter.[9] The designer, the unidentified monogrammist N.H., points with a stick to his own initials.[10] He holds aloft a flagon whose eccentric shape alludes – according to the poet – to his drinking problem. The other character with the tankard points to something outside the print.[11] Perhaps this is the *Furmschnider* Hans Lützelburger, calling the viewer's attention to the little cartouche displaying his name that can be seen on the first state.

1
Andersson & Talbot 1983: 318–319, no. 180.

2
Andersson 1968: 155–167; Landau & Parshall 1994: 394–395, no. 96–97.

3
Parshall 1989: 32; Landau & Parshall 1994: 212–213.

4
The complete signature in the cartouche at bottom left reads: 'HANNS/ LEVCZELLBVRGER / FVRMSCHNIDER/ 1522'.

5
Woltmann 1874: 193.

6
The known prints are held by the Kupferstichkabinett in Dresden and the Kunsthalle in Karlsruhe. A third print, once sold in Berlin (Boerner, 14–15 November 1933, lot 488), has now disappeared without trace. This is probably the same sheet that is described in Hollstein as being in the former Gotha collection. See Falk 1984: 282, no. 3.

7
Ain Insel haiszt Utopion / Die leyt nit ferr von Morian / Da g[e]schach ain sollichs schlagen / Hundert tausent hort ich sagen / Doch ist es eben vil der Jar / Das ich gelaub es sey nit war / Wann wie wolt ain nackend man / Ain angelegten pawren b[e]stan.

8
Der maister der das hat erdicht / Der hat sein kunst dahin gericht / Das man erkennen mug da bey / Wasz hoher kunst im maalen sey / Ausz maalen kumpt gar vil zu druck / Durch formen schneyden manig stuck. The poet continues with a list of artists who were already described in Pliny's *Naturalis Historia.*

9
Dodgson 1907: 319–322; Parshall 1989: 35; Landau & Parshall 1994: 213–214.

10
The German artist Nicolaas Hogenberg (*c.*1500–*c.*1539) has been mentioned several times in connection with the identification of 'N.H.'. Born in Munich, he was active in Augsburg before moving to Mechelen around 1527. Hogenberg's style has a rather more Netherlandish character, however, which undermines the attribution of this composition to him. See Landau & Parshall 1994: 394 n. 95.

11
Darumb man billich leben soll / Den, der sein kunst beweiset wol / Als diser auch ain maister was / Doch ist jm lieber das wein glasz / Das braucht er für ain langen spiesz / Er thue jms nach, den das verdriesz.

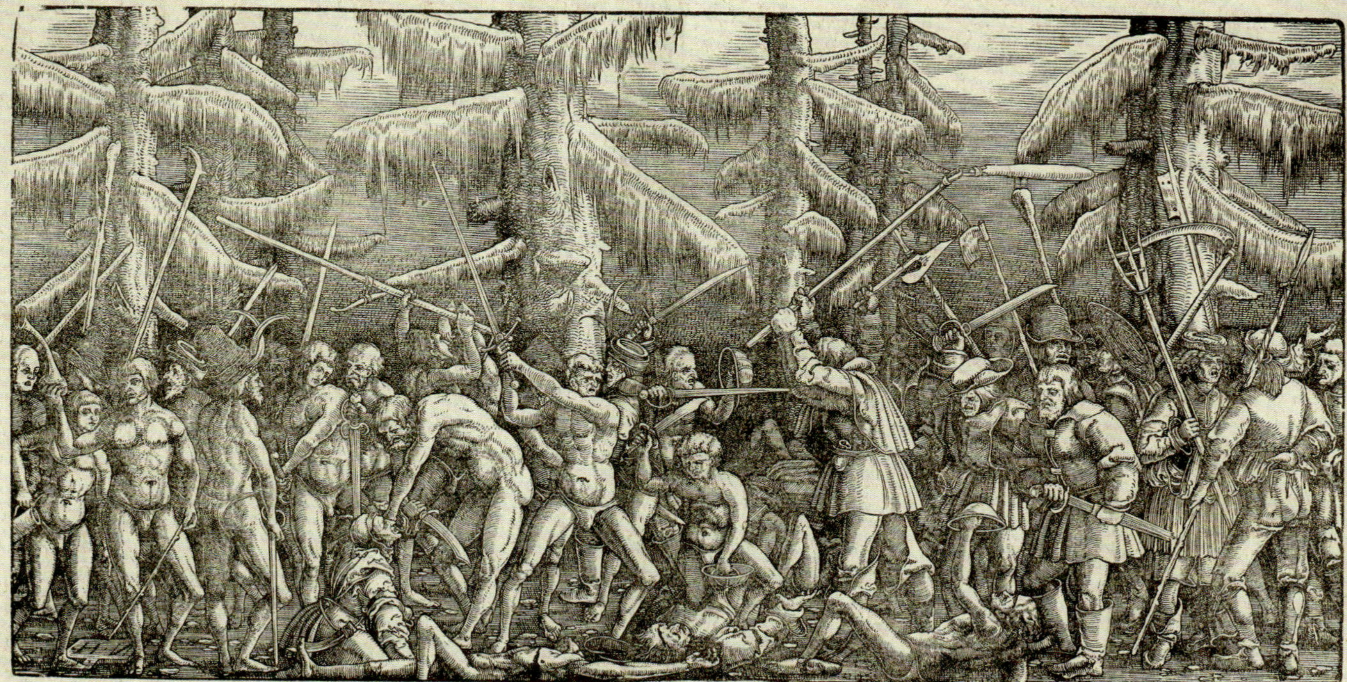

Ain Insel haißt Vtopion
Die leyt nit ferr von Morian
Da gschach ain sollichs schlagen
Hundert tausent hort ich sagen
Doch ist es eben vil der Jar
Das ich gelaub es sey nit war
Wann wie wolt ain nackend man
Ain angelegten pawren bstan
Der maister der das hat erdicht
Der hat sein kunst dahin gericht
Das man erkennen müg da bey
Waß hoher kunst im maalen sey
Auß maalen kumpt gar vil zů druck
Durch formen schneyden manig stuck
Vnd welcher sollich kunst nit waiß
Der maalt ain Esel für ain gaiß
Vnd ain Teüfel für ain Engel
Ain wurst für ain pretzenstengel
Maalen kumpt von alter her
Maalen hat groß lob vnd eer
Bey den alten offt erworben
Das zeügen die vor seind gestorben
Pythis Micon Timagor
Poligrot Appollodor
Der auch den ersten pensel fand
Cleophant gab farb dem gwand
Gyges nach dem schatten maalt
Prothagein die kunst nit faalt
Den doch Appelles über wand
Mit klůger lyni die er fand
Zensis die weinber maalt so fein
Das darzů flugen die vögelein
Vnd wolten daruon essen
Parrhasis sich nit vergessen
Der maalt ain tůch von klůger art
Daruon Zensis betrogen ward
Wer maalen will soll sein gelert
In künsten vil/als drinn begert
Die rechten kunst der maalerey
In perspectiua/Geometrey
Er soll auch gar mit nichte
Versaumen der Poeten dichte
Mit sinnen sey er gschwind vnd bhend
Den sinnen volgen nach die hend
Das er herfür künd bringen
Maß vnd gstalt in allen dingen
Mit jr natur vnd aigenschafft
Vil farben macht er auß dem safft
Grün/plaw/praun/wie mans haben will
Wer maalen will der darff noch vil
Erfarung land vnd leüte
Berg vnd tal/hoch gebürg vnd weyte
Fisch/vögel/aller thier gestalt
Er kennen soll/wie manigfalt
Gott hat sy all gezieret
Den hymel vmbher fieret
Mit liechter vil entprennet
Dabey man wol erkennet
Den schöpffer hoch ins hymels throne
Der alle gaben tailt gar schone
Mit rechter maß wer es begerte
Deß maalens seind nit vil gelerte
Darumb man billich loben soll
Den/der sein kunst beweiset wol
Als diser auch ain maister was
Doch ist jm lieber das wein glaß
Das braucht er für ain langen spieß
Er thůe jms nach/den das verdrieß.

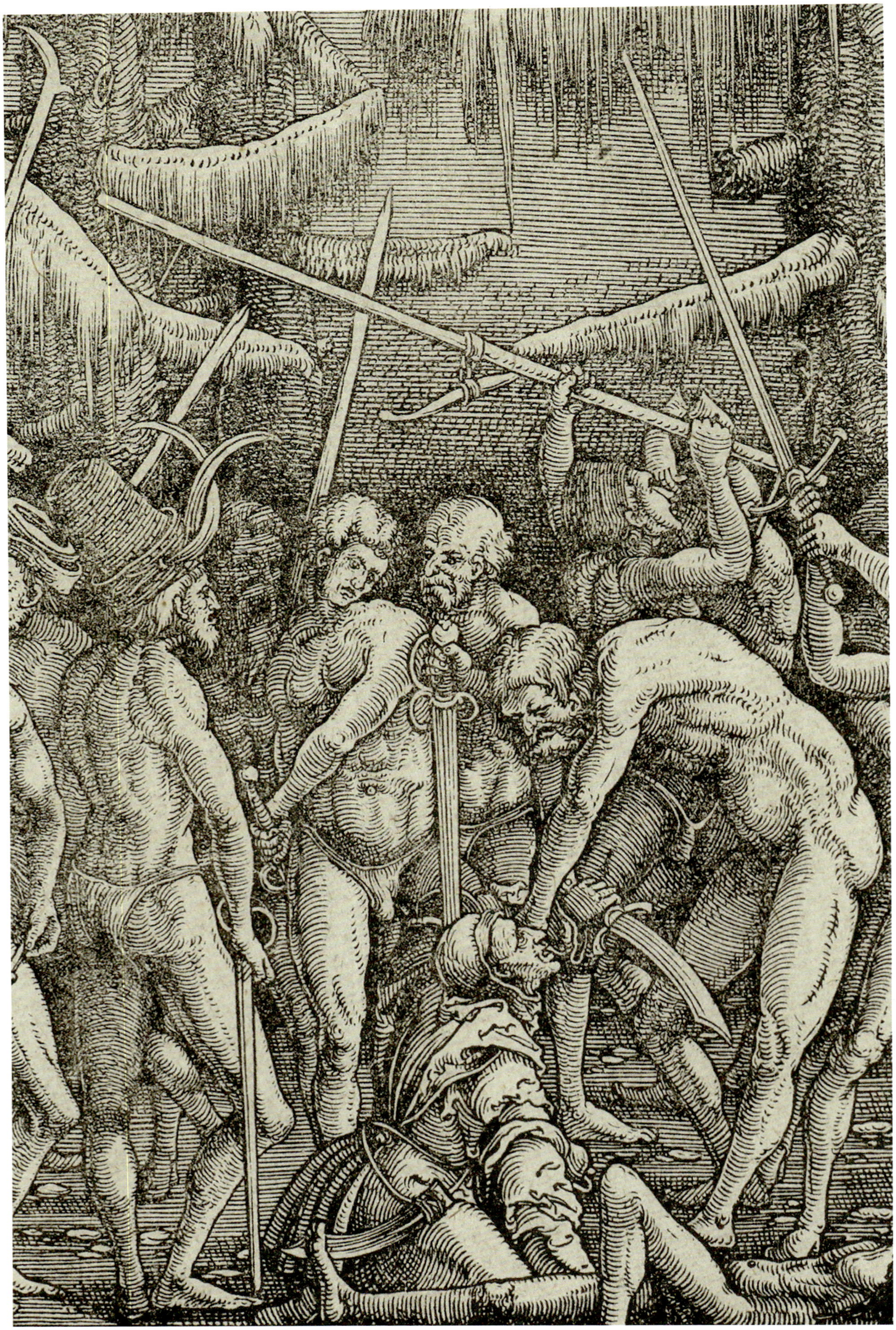

[cat. 10] detail

That the reference to Utopia should be so short – just a few lines – is in no way extraordinary. Today this woodcut is the only known contemporary reference to allude in an artistic way to Thomas More's work outside the discourse of his book. Even so, the exact context in which this second state originated is shrouded in mystery. But the fact that in the last years of his life Hans Lützelburger went to work for Hans Holbein the Younger in Basel does not seem entirely coincidental. It happens to have been Ambrosius Holbein, the brother of Hans the Younger, who provided the illustrations for the 1518 Froben reprint of More's *Utopia* (cat. 7.1–7.2).

M B

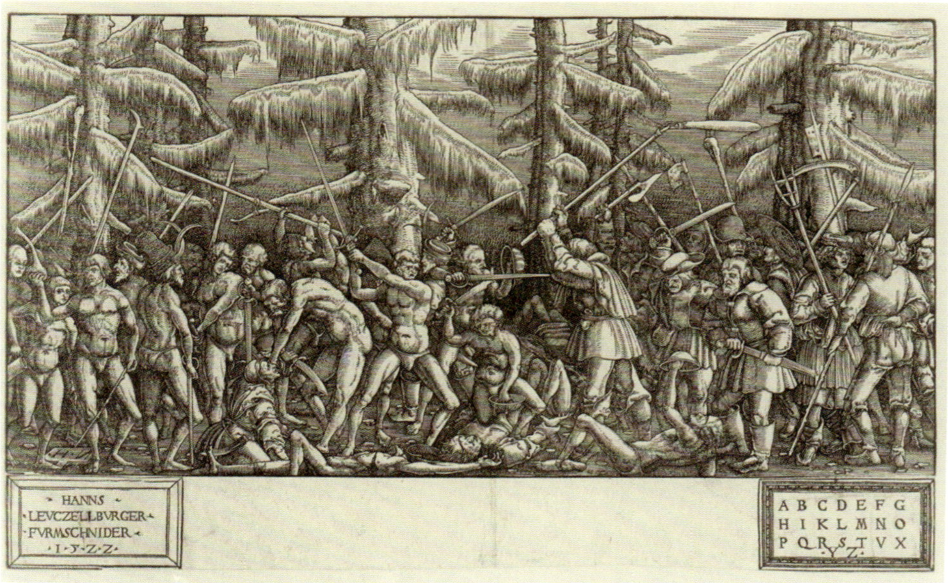

[fig. 10.1] Hans Lützelburger after the Monogrammist N.H., *Battle of Naked 'Utopians' and Clothed Peasants*, first state, 1522, woodcut, Basel, Kunstmuseum, Kupferstichkabinett.

11

Abraham Ortelius
Map of the Island of Utopia

1596

Burin engraving, 38 x 47.5 cm (intaglio)

Private Collection

BIBLIOGRAPHY

Kruyfhooft 1981: 10-14; Marc'hadour 1982: 103-105; Leslie 1998: 37-38; Meurer 1998: 151; Schaer, Claeys & Sargent 2000: 100; Stoffers 2001: 1-50; Van den Broecke 2004: 89-94; Van den Broecke & Ormeling 2005: 132-133; Van den Broecke 2011: 684-687; De Palmenaer & Parmentier 2015: no. 50.

The cartographer Abraham Ortelius (1527–1598) gained world fame as the compiler of the first modern atlas. The publication in 1570 of his *Theatrum Orbis Terrarum*, a collection of logically arranged maps with descriptive comments, immediately opened up the knowledge of the then world to a wide audience. Ortelius was also a humanist, art collector and historian and was in close contact with the scholars of his time.

In creating his map of Utopia Ortelius has closely followed the image of the island that appeared in the first editions of Thomas More's book. He has put the south at the top, which is surprising for a late-sixteenth-century map. Be that as it may, the map shows a tranquil isle with forests, vineyards, ripe cornfields and numerous rivers and towns. Many ships sail the surrounding sea, suggesting a thriving maritime trade. The map is embellished with three elegant cartouches, each framing a text. The title cartouche gives the island's name and refers to Raphael Hythlodaeus as the narrator, Thomas More as the writer and Abraham Ortelius as the map's designer. Below left, in a cartouche weighty with extravagant scroll work, Ortelius dedicates the map to his dear friend Johannes Matthaeus Wacker von Wackenfels (1550–1619), a scholar and diplomat in the service of Emperor Rudolf II, to name but one of his patrons. In the cartouche below right, the glories of Utopia are described.

Ortelius was prompted to make this map by Wacker von Wackenfels, who in a letter dated 1 August 1595 wrote to Jacobus Monau (1546–1603), likewise a friend of Ortelius, that he would also provide the map-maker with the names of rivers and towns.[1] This Utopia does not have fifty-four towns as described by More but fifty-five, for Ortelius had to add the place called Favolius to gratify a certain Johannes Baptista Favolia, an Antwerp scholar.[2]

Wackenfels supplied place names in ten different languages. The capital is Amaurotum which can mean 'foggy city' and is possibly an allusion to London. In addition to 'Utopian' the following languages were used: Latin, Greek, Italian, Spanish, French, German, Dutch, Polish, and Arabic (?). The towns and rivers have absurd and sometimes comical names like Horsdumonde, Keinstadt, Sottenbroeck, Sanspoisson, Fischlos...

Apparently Ortelius was not altogether happy with this map. In 1596 he wrote to his nephew Jacob Cole, who was living in London, that he had in fact made

1 Hessels 1887: 274.

2 Hessels 1887: 286.

3 Hessels 1887: 286.

the map of Utopia unwillingly, at the insistence of James Monau and some other friends.[3]

It appears from a letter from Cole to Ortelius written in October 1596 that twelve copies of this map were sent to London. The map of Utopia is the same size as an atlas map but it has never been found in Ortelius's *Theatrum* nor in his *Parergon*, a collection of maps of the classical world. At present the map of Utopia shown here is the only known copy.

CK

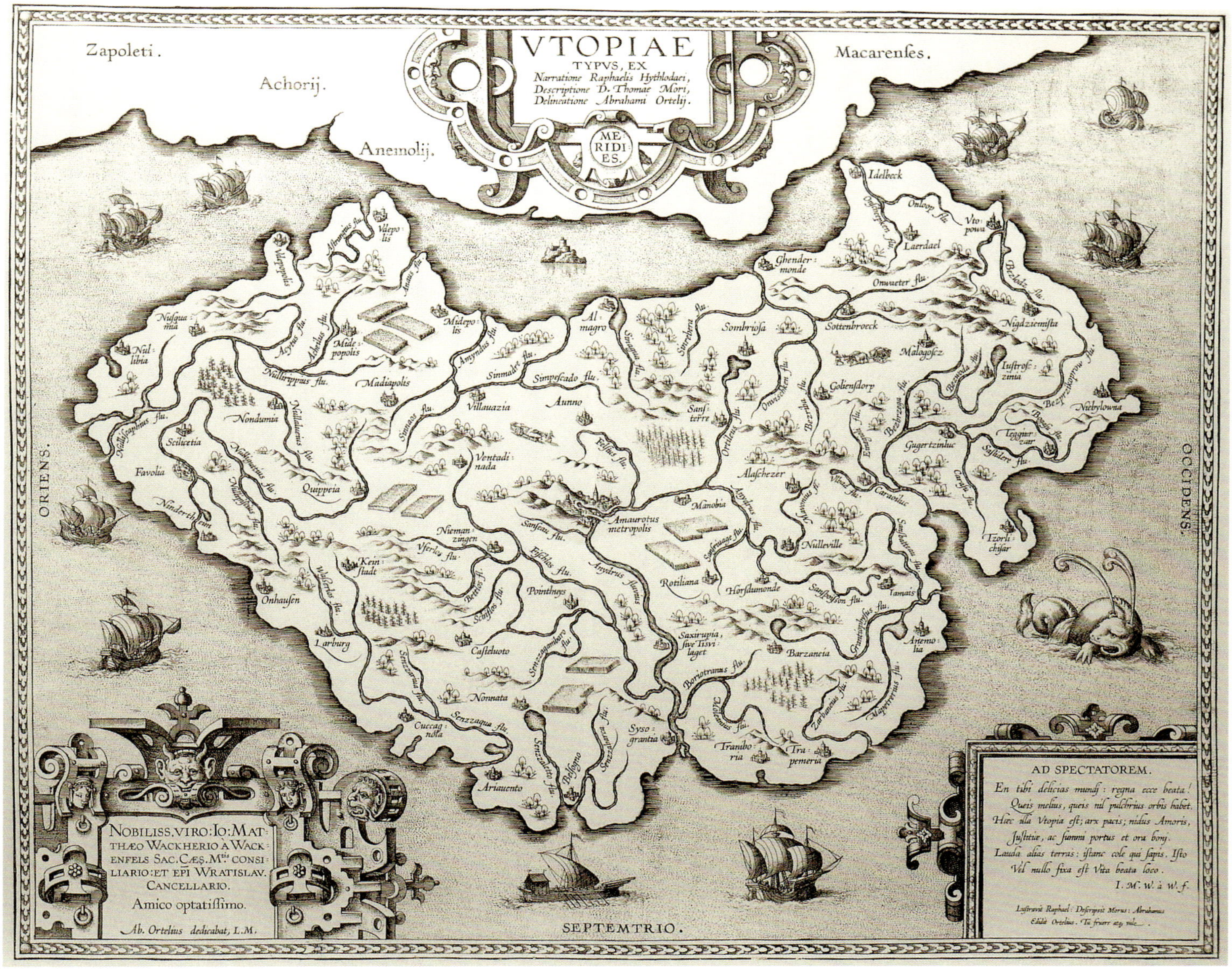

12

Thomas More, translated by Nicolas Gueudeville
L'Utopie de Thomas Morus, Chancelier d'Angleterre: Idée ingenieuse pour remedier au malheur des hommes & pour leur procurer une felicité complete

Published in Leiden by Pierre van der Aa in 1715
Engravings by Frans van Bleyswyck
Leuven, KU Leuven, University Library, Tabularium, inv. CAAA 2343
This book is not in the exhibition. It is on display in the exhibition *Utopia & More* (KU Leuven, Central Library)

BIBLIOGRAPHY

Rosenberg 1982: 208-213; Meyer 1984: 289-290; Streng 1990: 111-136.

Utopia became known to the French-speaking world primarily by way of the translation by Samuel Sorbière (1615–1670).[1] In 1642 that French philosopher settled for a short time in Leiden as a physician. His translation of Thomas More's work appeared in Amsterdam in 1643. In 1650 Sorbière returned to France, converted to Catholicism and became a historiographer in royal service.

In many respects, Nicolas Gueudeville, who produced a new translation of *Utopia* in 1715, took the opposite route. A one-time Benedictine monk, Gueudeville had lost his faith in the early 1680s. He escaped from Jumièges Abbey in Normandy and in 1688 sought shelter in the Dutch Republic as a teacher of Latin among the scholars of the diaspora of Huguenots who had fled France in great numbers on the revocation of the Edict of Nantes (1685). He became a radical freethinker and passionate opponent of absolutism, and it is in that spirit that his translation of *Utopia*, which was published in Leiden by Pierre van der Aa in 1715, must be understood. Gueudeville acknowledges in his preface that his translation of More's work (cat. 1) is a free one, glossed his own views – in fact the translation is almost twice as long as the original. According to Gueudeville, More's Utopia must be interpreted first and foremost as a plea for an egalitarian society and a policy in which the common weal takes precedence over the individual. His commentary to this core idea is intent on radicalization and clearly aimed at the absolutist regime in France, where corruption in Church and State, luxury and unbridled ambition come in for particular criticism.

At the same time the statement that 'Le monde ne jamais s'utopiera' demonstrates Gueudeville's scepticism, which is also characteristic of the early Enlightenment. The ideal state remains an unattainable ideal, incompatible with human nature. The Enlightenment, which was partly rooted in the Huguenot diaspora, was sometimes seen by those in its own ranks as a 'utopian' movement per se. In eighteenth-century France the idea of 'Utopia' flourished as never before as a narrative vehicle for enlightened ideals. But what could be evoked in literature via the imagination as a 'possible world' was not yet achievable in practice. In political philosophy there was also a much more pragmatic and almost anti-utopian countermovement. In France, unlike the English-speaking world, it would be a long

1 For the history of Utopian literature, see Racault 1991; Trousson 1999; Fortunati & Trousson 2000. On the topic of 'utopias and utopianism', see inter alia Israel 2001: 177–179, 183, 591–593, 596–598.

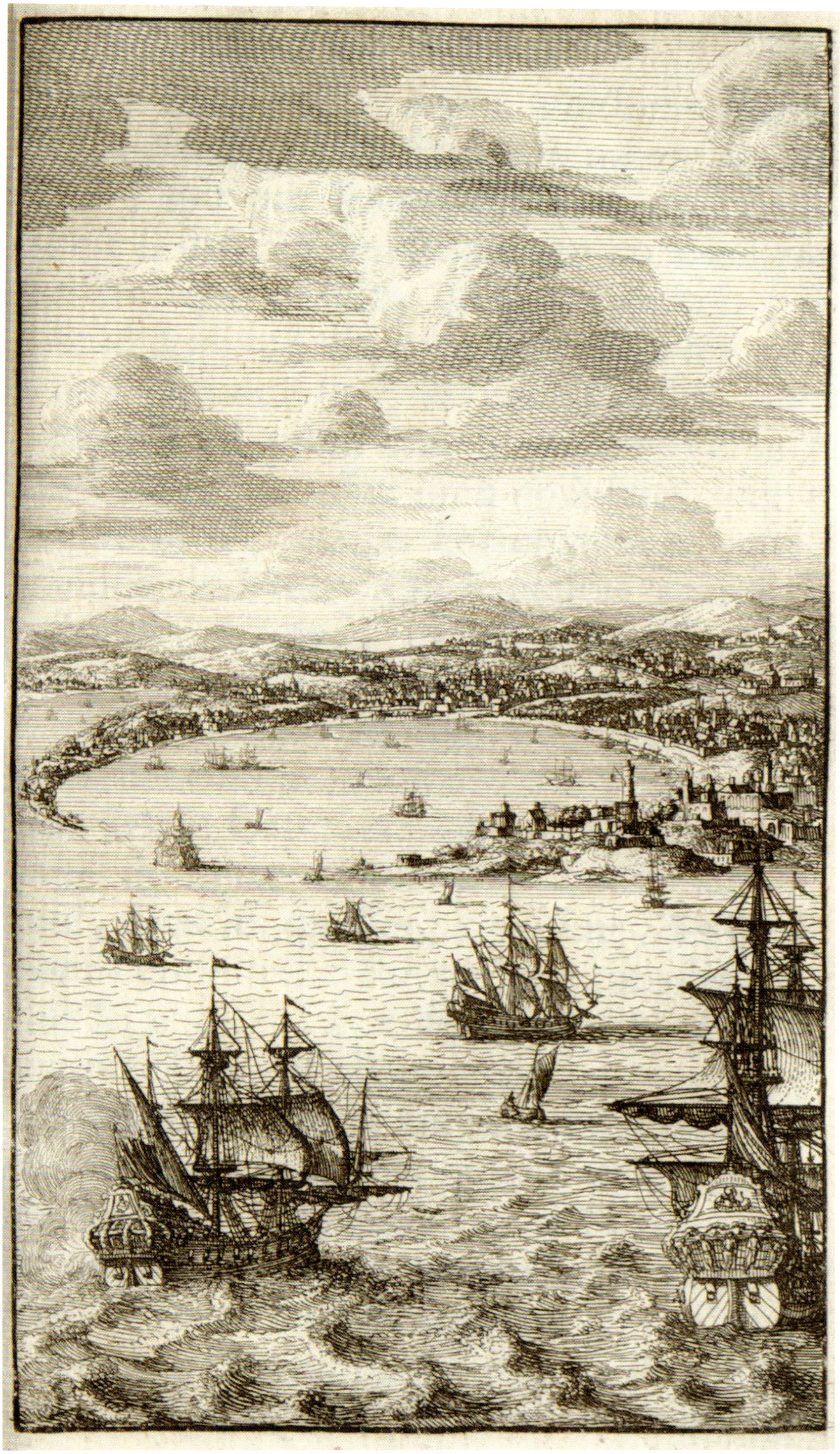

[cat. 12] The Island Utopia

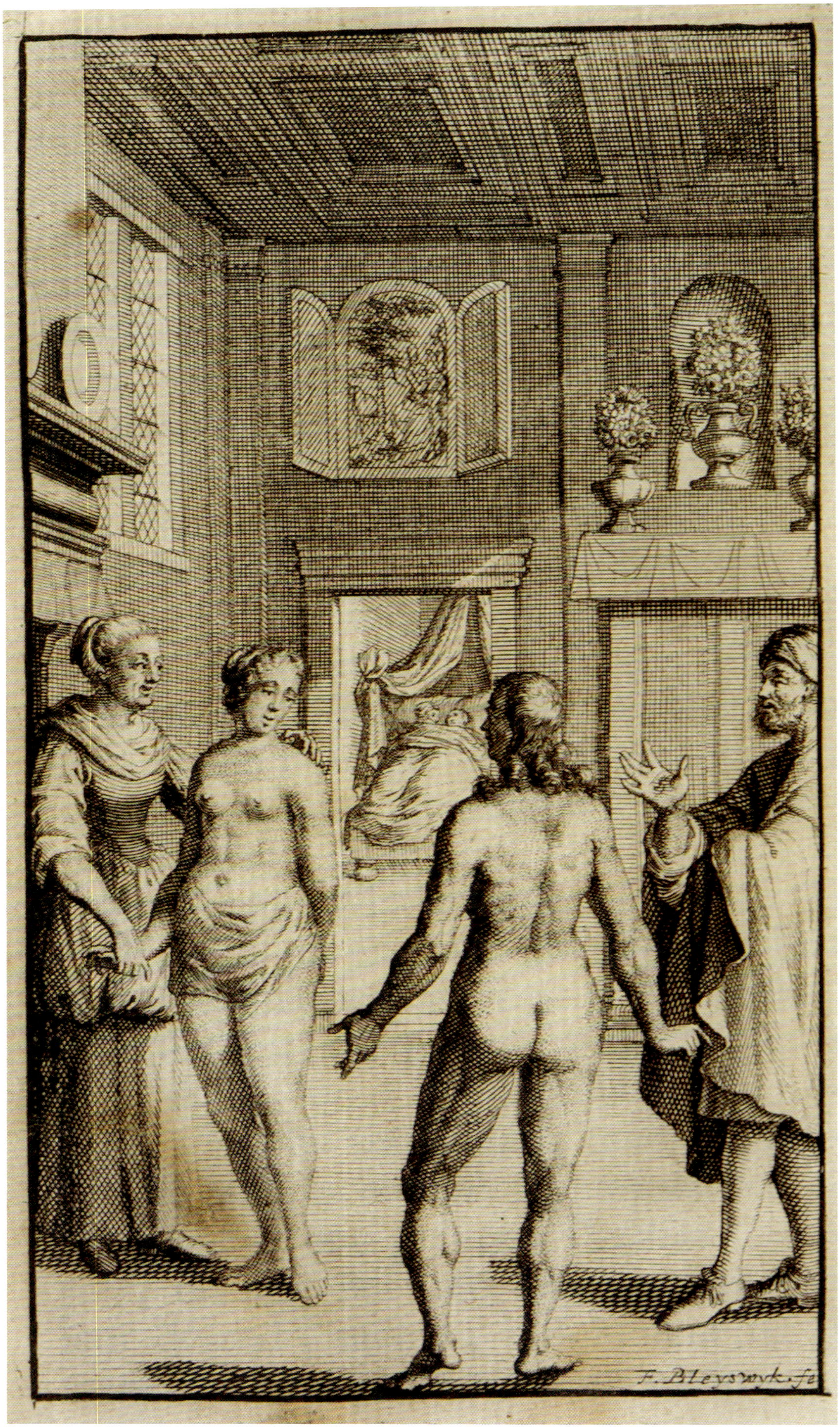

[cat. 12] The suitors are shown naked to each other prior to marriage.

time before the term 'Utopia' referred to anything but Thomas More's book. Not until 1752, in the *Dictionnaire de Trévoux*, does it enter the lexicon as a proper noun relating to a literary genre, particularly a 'plan d'un gouvernement imaginaire, à l'exemple de la *République* de Platon'.

Gueudeville's translation is the first edition to be fully illustrated. This publication, which was reprinted in 1717 and 1730, contains sixteen plates by the Leiden engraver Frans van Bleyswyck (1671–1746) who spent virtually his whole career as an illustrator. Van Bleyswyck's style is strongly influenced by French classicism, but shows a certain influence from the Holland style, particularly in the interiors.

JH

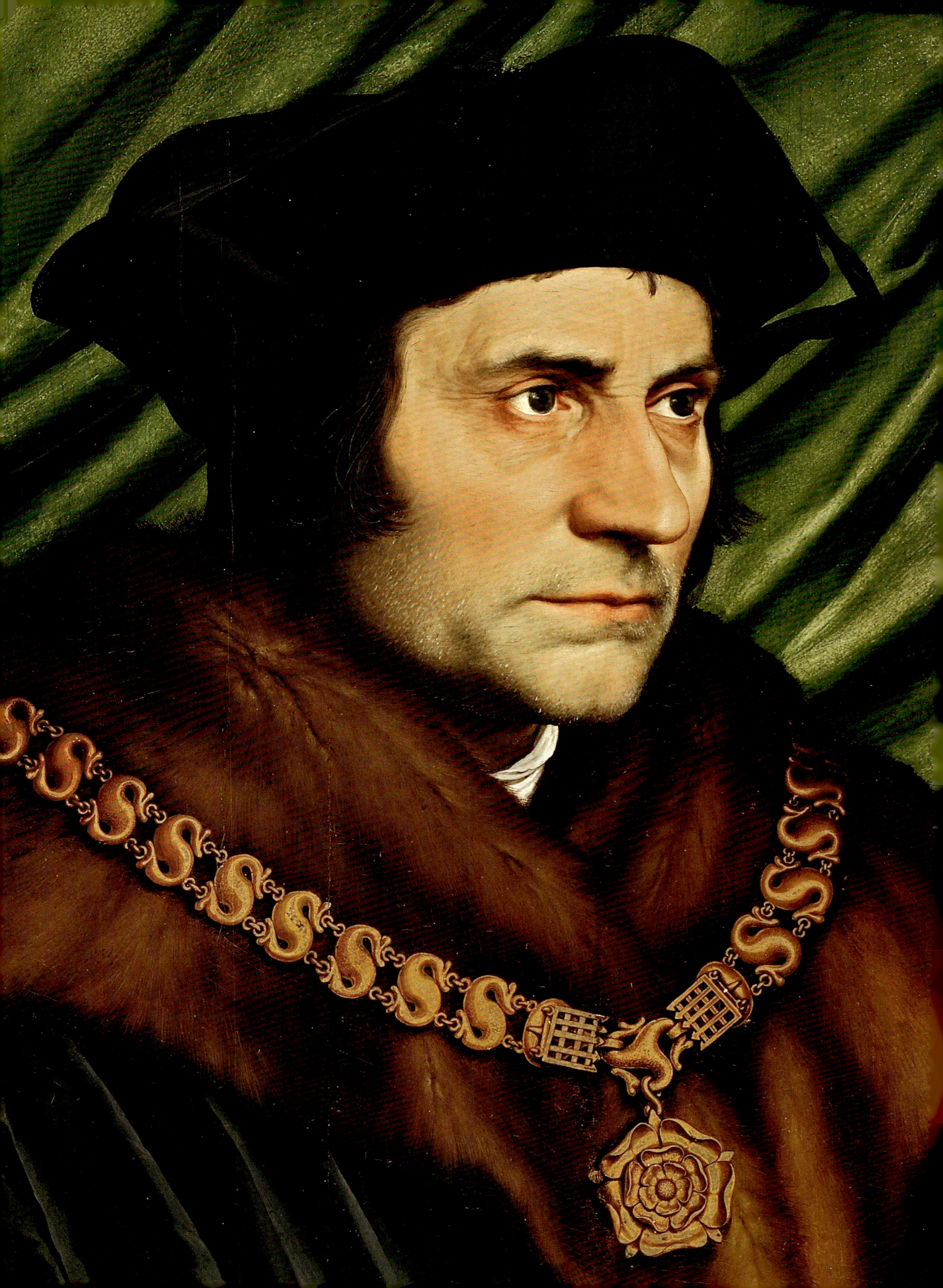

Utopia and European Humanism

JAN PAPY

Humanists were pedagogues *par excellence*. Their new Renaissance vision of what people should and could be – a well-balanced fusion of *humanitas*, based on antique culture, and Christian *pietas* – had already called for a new and adapted sort of education in fifteenth-century Italy. Yet the sedulous study of classical languages and elegant discourse along the lines of the antique authors was not at all unworldly. Young people should be able to grow into developed and cultivated adults, open-minded intellectuals who were not only familiar with the wealth of antique knowledge but could also put that knowledge to use in ways that were meaningful for their own society. That humanist upbringing and new education ought to result in *virtus*: not simply 'virtue' or 'virtuousness' but a life with the greatest potential that nature has to offer. Thus the foundations for man's complete cultural, moral and social development are already laid in childhood. Manuals on the upbringing of the young and mirrors for young princes followed each other in quick succession. Often, their core idea reprised Plato's view that the state should be ruled by a 'philosopher-king', a monarch who has been properly brought up and educated, who embraces the philosophical insights of Antiquity and humanism, and who has rid himself of short-sighted imperialism and unchristian greed. Because power without goodness is tyranny, without wisdom it is a disaster. The welfare of society therefore depends on the education of the monarch and his subjects. Only thus can a socially just society, a world of peace and cultural flourishing be guaranteed.

The pacifist Erasmus and 'utopian' More were at one in these ideals of Christian humanism. Both felt the same sense of indignation with their world; both sought their role as humanist scholars and political advisors in order to improve that world. They found a basis for that *societas perfecta* in the political thought of Plato, Aristotle, Xenophon and Cicero. But More's London humanist friends, men such as John Colet, had all studied in Italy and it was they who brought both More and Erasmus into contact with Pico della Mirandola's lofty *Oration on the Dignity of Man* and Marsilio Ficino's Neoplatonist philosophy. They also directed them to the authoritative educational treatises of Maffeo Vegio and Battista Guarino and showed them the innovative political works of Poggio Bracciolini (*Momus sive de principe*, 1450) and Filippo Beroaldo (*De optimo statu*, c.1497). Surprisingly enough, More was unaware of Machiavelli's *Il Principe* (written c.1513 to 1515, though published only in 1532). Perhaps even more surprising is that same decade should see the birth of a masterpiece on Realpolitik and a classic on the Ideal State.

Yet it is chiefly Erasmus's own *Panegyricus* for Philip the Fair (1504), his *Moriae Encomium* (*Praise of Folly*, 1511) – dedicated to More as if to demonstrate their complete congeniality – and his *Institutio Principis Christiani*, written in 1515 and dedicated to Prince Charles in May 1516, that have left the most traces in More's ideal state. If the *Institutio* was Erasmus's positive political programme for a healthy and wisely governed state, the antique sources of the *Institutio* and *Utopia* are also largely parallel. Both works are variations on the same theme: the ideal state as a tight family bond founded on the pillars of morality in which the citizens participate in the power. Compared to Erasmus's *Moriae Encomium*, More's *Utopia* as a 'Praise of Wisdom' is the literary-satirical and politico-philosophical complement to the *Praise of Folly* dedicated to him five years earlier. Both contrast, as in a mirror, the ideal with the shortcomings of their own world. Both offer a satirical stepping stone to an earnest contemplation of truth.

13A

Quinten Metsys
Portrait of Desiderius Erasmus

Antwerp, 1517
Oil on oak panel, 50.5 x 45.2 cm
Royal Collection Trust / Her Majesty Queen Elizabeth II, inv. 405759

In the open book, Erasmus has written on the recto the opening words of his Paraphrase of Saint Paul's Epistle to the Romans. On the facing page, he has written 'GRATIA', the key word of the Epistle. On the shelves behind him, three of the books are inscribed across the page-ends and are the New Testament, the works of Lucian (labelled in Greek capitals) and the works of Saint Jerome. Another book bears a damaged inscription 'M …', perhaps meant for *Encomium Moriae*, Erasmus's *Praise of Folly*, which was dedicated to Thomas More.

13B

Quinten Metsys
Portrait of Town Clerk Pieter Gillis

Antwerp, 1517
Oil on oak panel, 61.4 x 47 cm
Collection of the Earl of Radnor, Longford Castle (not in the exhibition)

The books on the shelves are labelled as by Plutarch, Suetonius and Quintus Curtius Rufus and as *Archontopaideia* (a Greek rendering of *The Education of a Christian Prince*, one of Erasmus's most famous works). Under Gillis's right hand is Erasmus's *Antibarbari*, again labelled in Greek.

PROVENANCE

The portraits formed the 'double picture' painted by Quinten Metsys in 1517 and given to Thomas More. It is not clear what happened to the portraits after More's execution in 1535 but by 1627 they belonged to Vincenzo II Gonzaga, Duke of Mantua, and were acquired in 1628, with much of the Mantuan collection, by Charles I, King of Great Britain. Both bear on their reverses the crowned CR brand that was used to mark Charles's pictures. Presumably separated from the *Gillis* when Charles's pictures were sold, the *Erasmus* was recovered for Charles II and has remained in the Royal Collection. Meanwhile the *Gillis* was in the collections of Thomas Rawlinson (1681-1725) and Richard Mead (1673-1754), at whose sale in 1754 it was bought by Jacob Bouverie (1694-1761), Viscount Folkestone, ancestor of the present owner.

BIBLIOGRAPHY

Nichols 1873; Campbell, Phillips, Schulte Herbrüggen & Trapp 1978: 725; Campbell 1985: 86-89; Hand, Metzger & Spronk 2006: 116-121, 284-285, 315; Shawe-Taylor & Scott 2007: 58-61; Heard, Whitaker et al. 2011:40-41, no. 6.

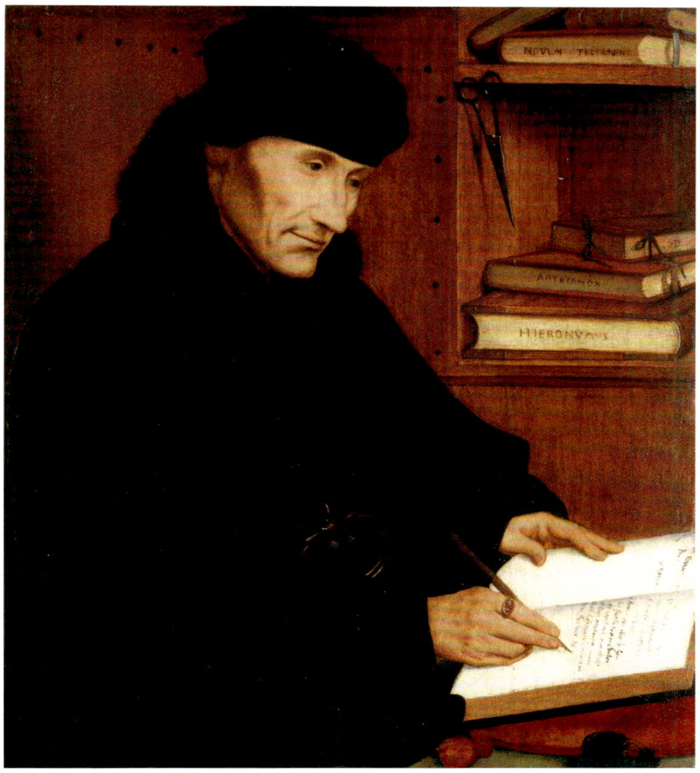
HIERONVMVS

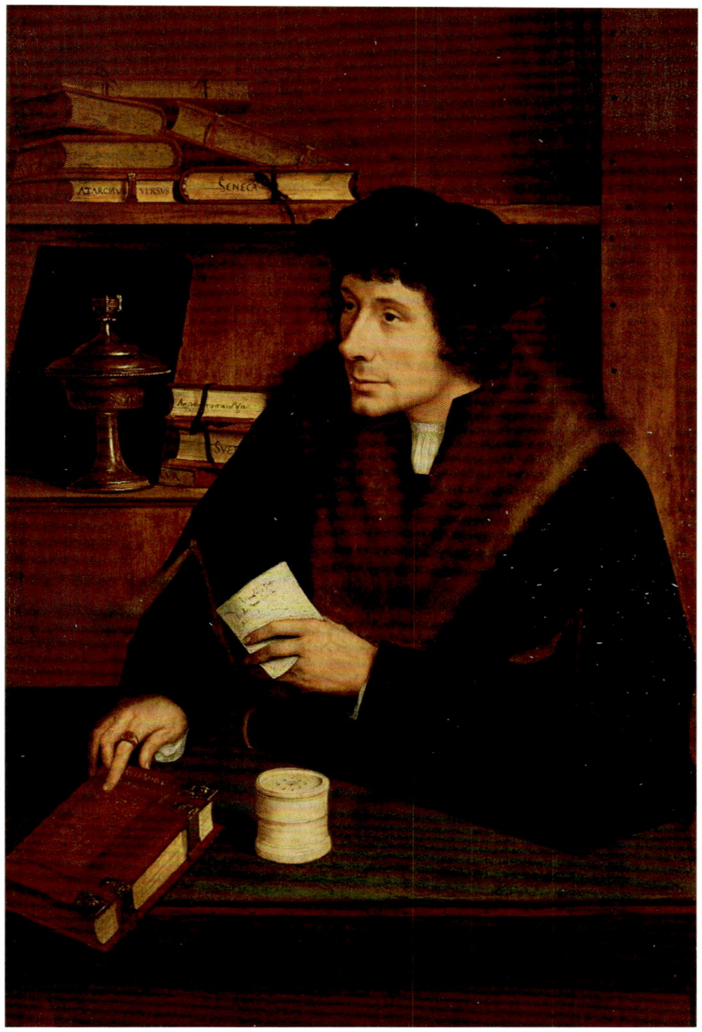
SENECA

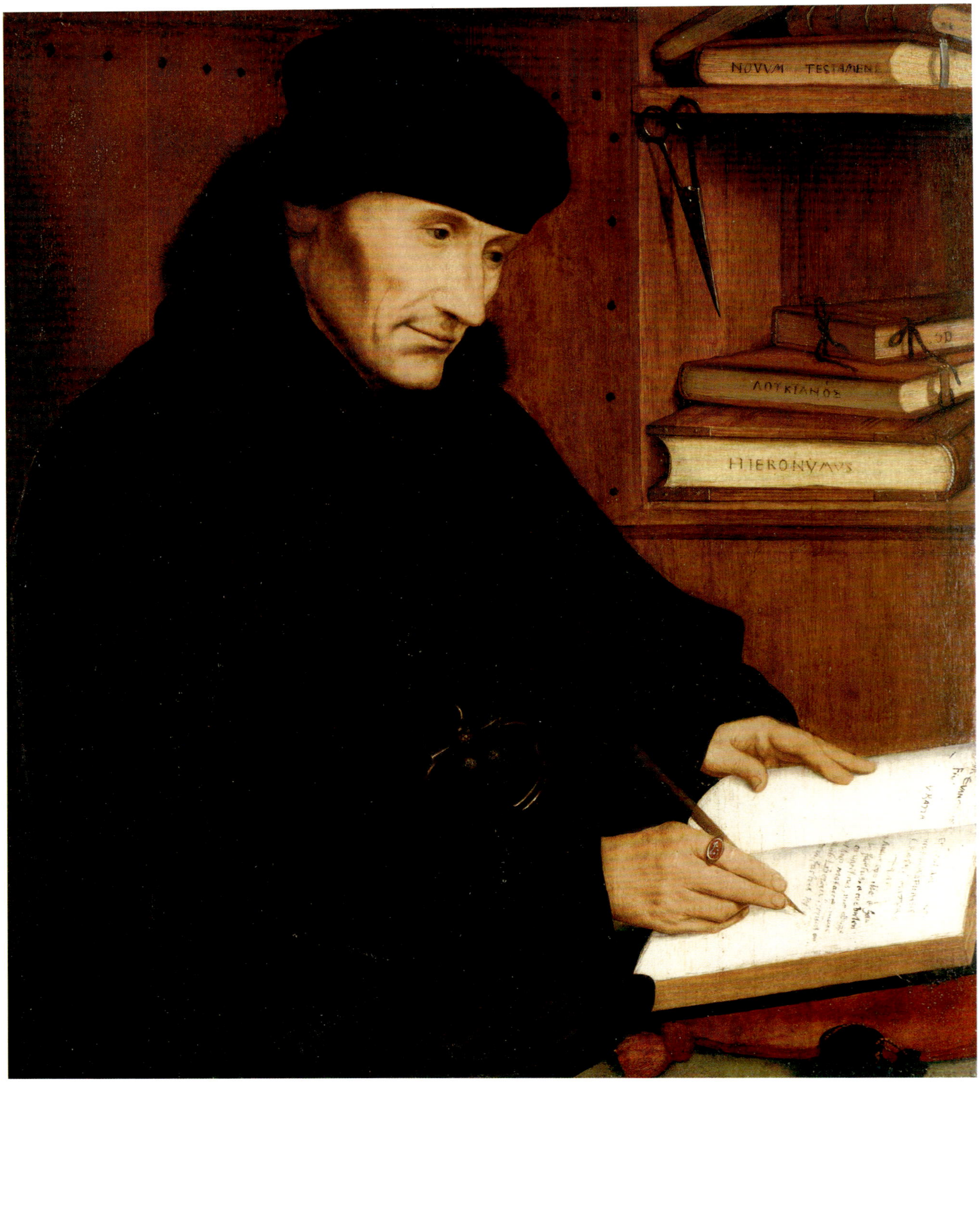
NOVVM TESTAMENT
ΛΟΥΚΙΑΝΟΣ
HIERONYMUS

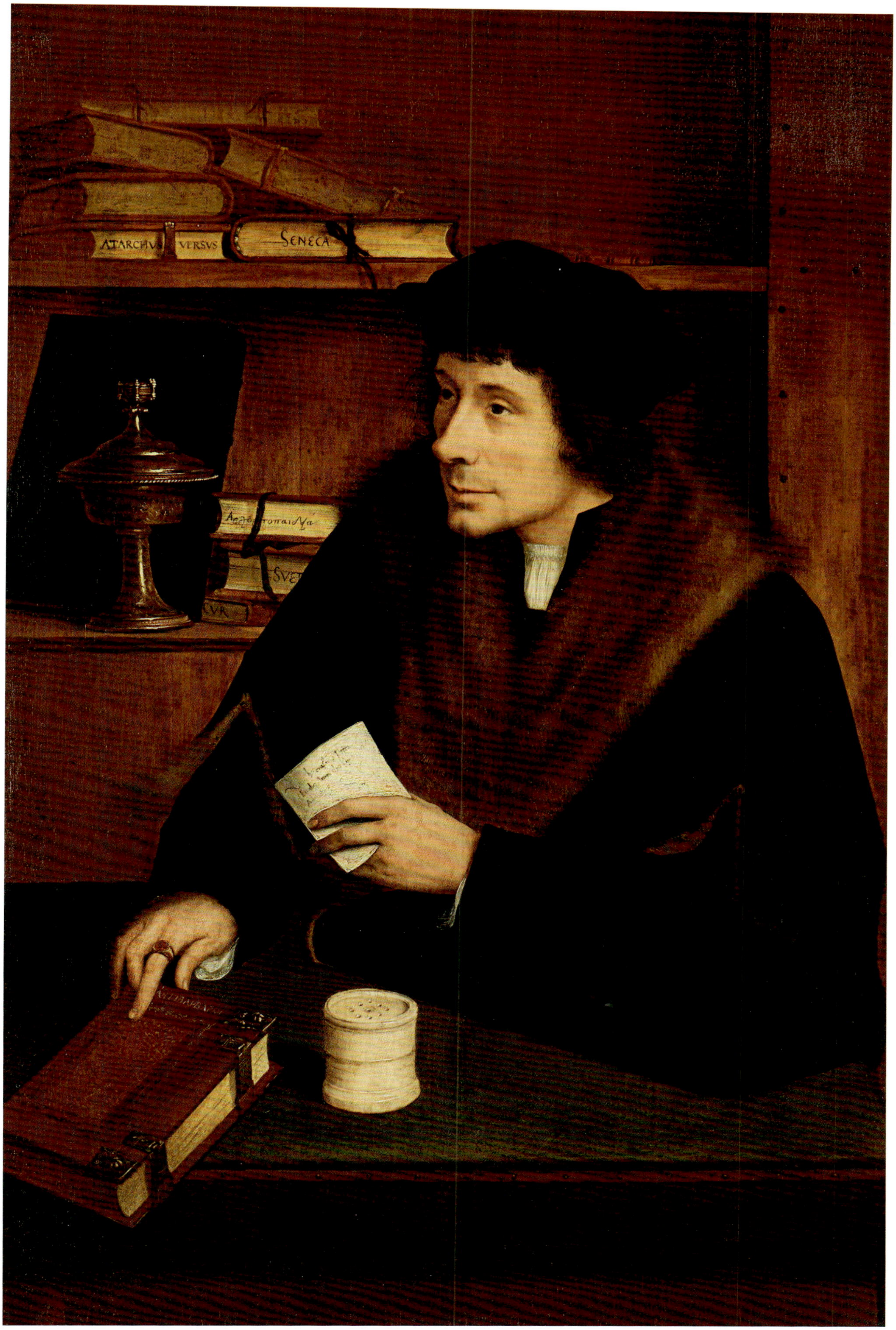
VERSVS
SENECA

14A

Quinten Metsys, studio replica (?)
Portrait of Desiderius Erasmus

Antwerp
After 1517
Oil on canvas, transferred from panel, 58.4 x 46 cm
Rome, Galleria Nazionale d'Arte Antica, inv. 1529

PROVENANCE

By 1807 in the collection of Count Aleksandr Sergeyevich Stroganov (1733-1811), St Petersburg; by inheritance, apparently through his grand-daughter Natalia Pavlovna Stroganova, who married her third cousin Sergei Stroganov, to their son Count Grigory Stroganov (1829-1910); given in 1912 by his heirs (his daughter Maria [1857-1920] married Alexei Grigorievich Shcherbatov [1848-1912]) to the Galleria Nazionale (Rome).

BIBLIOGRAPHY

Campbell, Phillips, Schulte Herbrüggen & Trapp 1978: 725; Hand, Metzger & Spronk 2006: 116-121, 284-285, 315.

14B

Quinten Metsys, studio replica (?)
Portrait of Town Clerk Pieter Gillis

Antwerp
After 1517
Oil on oak panel, 61.4 x 47 cm
Inscribed on the book in the lower left corner: '[QUERELA PA]CIS ERAS[MI] R[OTERODAME]'
Antwerp, Koninklijk Museum voor Schone Kunsten, inv. 198

PROVENANCE

Florent Joseph van Ertborn (1784-1840), Antwerp; his bequest to the museum was received in 1841.

BIBLIOGRAPHY

Hand, Metzger & Spronk 2006: 116-121, 284-285, 315.

At Antwerp in 1517, Erasmus and Pieter Gillis commissioned from Quinten Metsys (*c.*1465/66–1650) portraits of themselves as gifts for Thomas More. He received the 'double picture' at Calais that October. The progress of the commission is reported in letters from Erasmus to More. Gillis is represented holding a letter addressed to him by More; More, in his letter of thanks to Gillis, mentioned Metsys's skill in imitating More's handwriting and asked if he would return the letter. 'Placed next

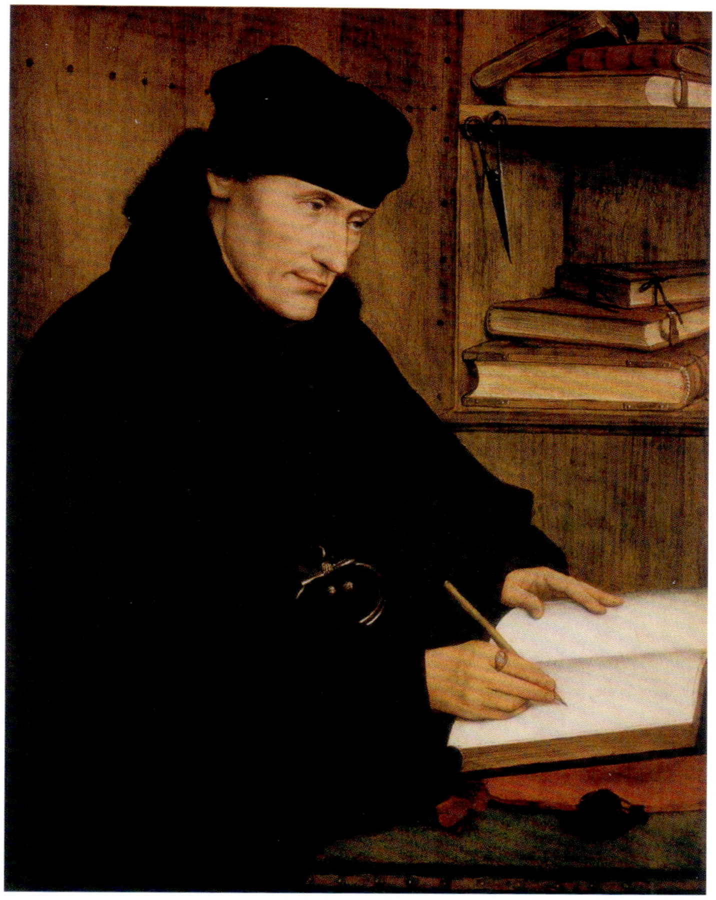

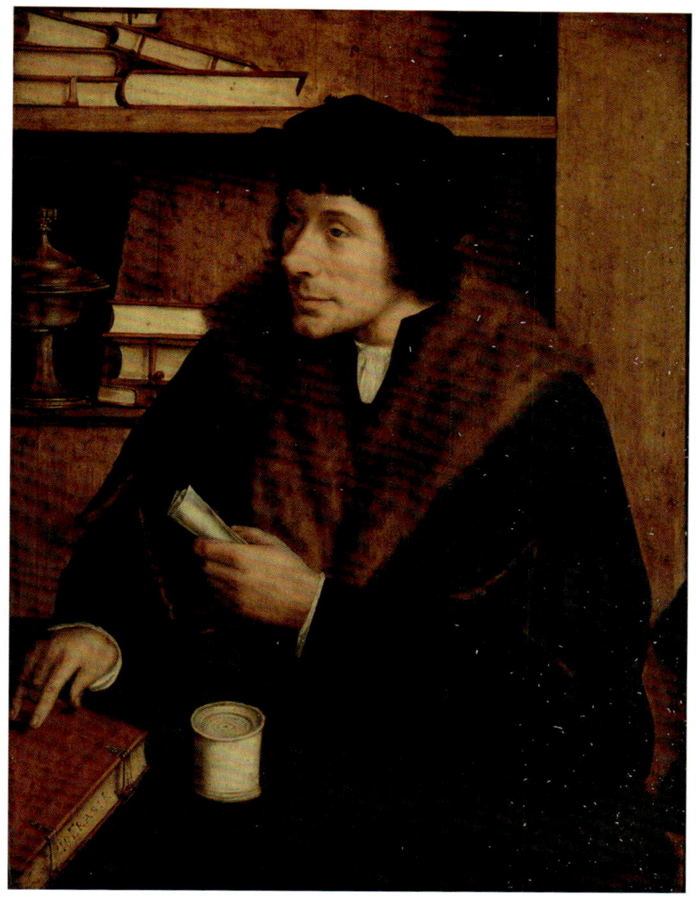

to the picture, it will make it seem more wonderful than ever.' If the original letter could not be spared, or found, More would try himself to imitate his own writing. He must have meant that he would make a counterfeit of the first letter, which he would exhibit next to the portrait.[1]

The accuracy of the likenesses had already been touched on in a letter from Erasmus to More, in which he excused the delays in sending the promised gift. He had foolishly followed a doctor's advice and taken some purgative pills, with the result that, at his next sitting, the painter said that he no longer had the same countenance. Work was delayed until Erasmus was more fully recovered.[2] More wrote from London on 16 July that he was impatient to see the portraits and his letters of thanks are full of emotion.[3] His letter to Gillis includes an epigram in elegiac couplets in which the picture speaks and expresses the feelings of all three men:

> I bear witness that Erasmus and Gillis are as dear friends as were, years ago, Castor and Pollux. More is sad to be parted from them, though affection joins them to him as closely as anyone may be to himself. In turn they are sad that their absent friend should miss them. An affectionate letter brings to More their inward thoughts but I bring their outward appearance.[4]

More explained to Erasmus that he found it difficult to say how immeasurably the pictures had delighted him. He stated that the 'likenesses are delineated and expressed with such art that they fairly excel all the old painters', and that:

> I triumph in this boast, that you esteem me so far as to make known by a memorial of such beauty that there is no one else whose affection you prefer. Perhaps with some arrogance, but with no less confidence, I assuredly interpret this present as one sent to me in order that the remembrance of you should be renewed to me not merely every day but every hour … If the truth be told, I cannot shake off this urge to boast, with which I feel both wonderfully and delightfully titillated, whenever it occurs to me that I am commended to posterity as the friend of Erasmus: that is demonstrated by letters, by books, by pictures and in every way. I wish that I had the powers to show by some evident proof that I was not unworthy of the love of a man so greatly surpassing those of his own century and those of future ages.[5]

The two panels have been cut. In 1648 they were described as 'half ovals' and may have had arched tops. Subsequently they were cut to their present, rectangular shapes. Cat. 13A measures 50.4 x 45 cm and had been reduced to that size by the time of Charles II. The *Gillis* (cat. 13B) has been cut to 50.6 x 44 cm and then enlarged on all four sides, apparently between 1734 and 1754, to 74.5 x 53.4 cm, to make it into a false pair for Holbein's portrait of Erasmus dated 1523, which measures 74.8 x 52 cm.

The Rome portrait of Erasmus (cat. 14A) and the Antwerp portrait of Gillis (cat. 14B) are about the same size and may have been painted as a pair. The last ring in the two oak boards used to make the panel on which the *Gillis* is painted was formed in 1497. The dendrochronological evidence would favour a date after 1514 for the execution of the painting.[6] The Antwerp *Gillis* is not an absolutely accurate copy after the original, since Gillis holds a letter in the original but a rolled paper in the Antwerp version; the hands differ accordingly. It is possible that the Rome and

1 'Quare nisi aut ille in suum aliquem vsum aut tu in tuum eam seruas epistolam, remitte rogo ad me: duplicabit miraculum apposita cum tabella. Sin aut perierit aut vobis vsui erit, ego experiar mee manus imitatorem ipse rursus imitari.' Allen, Allen & Garrod 1906–1958, III: 107).

2 'Nos belle valebamus, sed nescio quomodo medico venit in mentem vt purgandae bili iuberet me pilulas aliquot sumere, et quod ille stulte suasit, ego stultius feci. Iam pingi coeperam; verum a pharmaco sumpto cum ad pictorem redirem, negauit eundem esse vultum. Dilata est igitur picture in dies aliquot, donec fiam Paulo alacrior.' (Allen, Allen & Garrod 1906–1958, II: 576).

3 'Tabellam quae mihi tuam Petrique nostri imaginem referat non credas quam auide expectem …' (Allen, Allen & Garrod 1906–1958, III: 11–12).

4 'TABELLA LOQVITVR. Quanti olim fuerant Pollux et Castor amici, / Erasmum tantos Egidiumque fero. / Morus ab his dolet esse loco, coniunctus amore / Tam prope quam quisquam vix queat esse sibi. / Sic desyderio est consultum absentis, vt horum / Reddat amans animum littera, corpus ego.' (Allen, Allen & Garrod 1906–1958, III: 106).

5 ' … quae quam impense me delectent, procliuius cuius est a suo sensu consequi quam dicendo mihi … tanto artificio deliniatis atque expressis imaginibus vt quae veteres facile pictores omnes prouocauerint? … ego vehementer hac triumpho gloria, quod tanti abs te estimor vt tam egregio monumento declares non alium esse quenquam a quo tu amari malis. Nam sic ipse certe – satis pol superbe, sed tamen sic – interpretor, quod istud abs te missum est mihi quo non in dies modo sed in horas etiam tui apud me memoria renouetur … Sed tamen, vt verum fatear, vnum hunc pruritum glorie excutere profecto non possum, quo mirum quam suauiter titillor, quoties animum subit sere demum posteritati me Erasmi amicitia litteris, libris, tabulis, omnibus denique modis contestata commendandum. Vtinam ea mihi facultas esset quae insigni aliquo documento possit efficere ne indignus amore tam impenso viri, cum suo seculo, tum secuturis etatibus incomparandi, videar!' (Allen, Allen and Garrod 1906–1958, III: 104).

6 Peter Klein, cited by Hand, Metzger & Spronk 2007: 284.

Antwerp portraits were executed in the workshop of Metsys for Erasmus. Metsys's compositions were evidently known to both Dürer and Holbein and influenced Dürer's engraving and Holbein's paintings of Erasmus. It was perhaps Erasmus who showed them versions of Metsys's two portraits.

LC

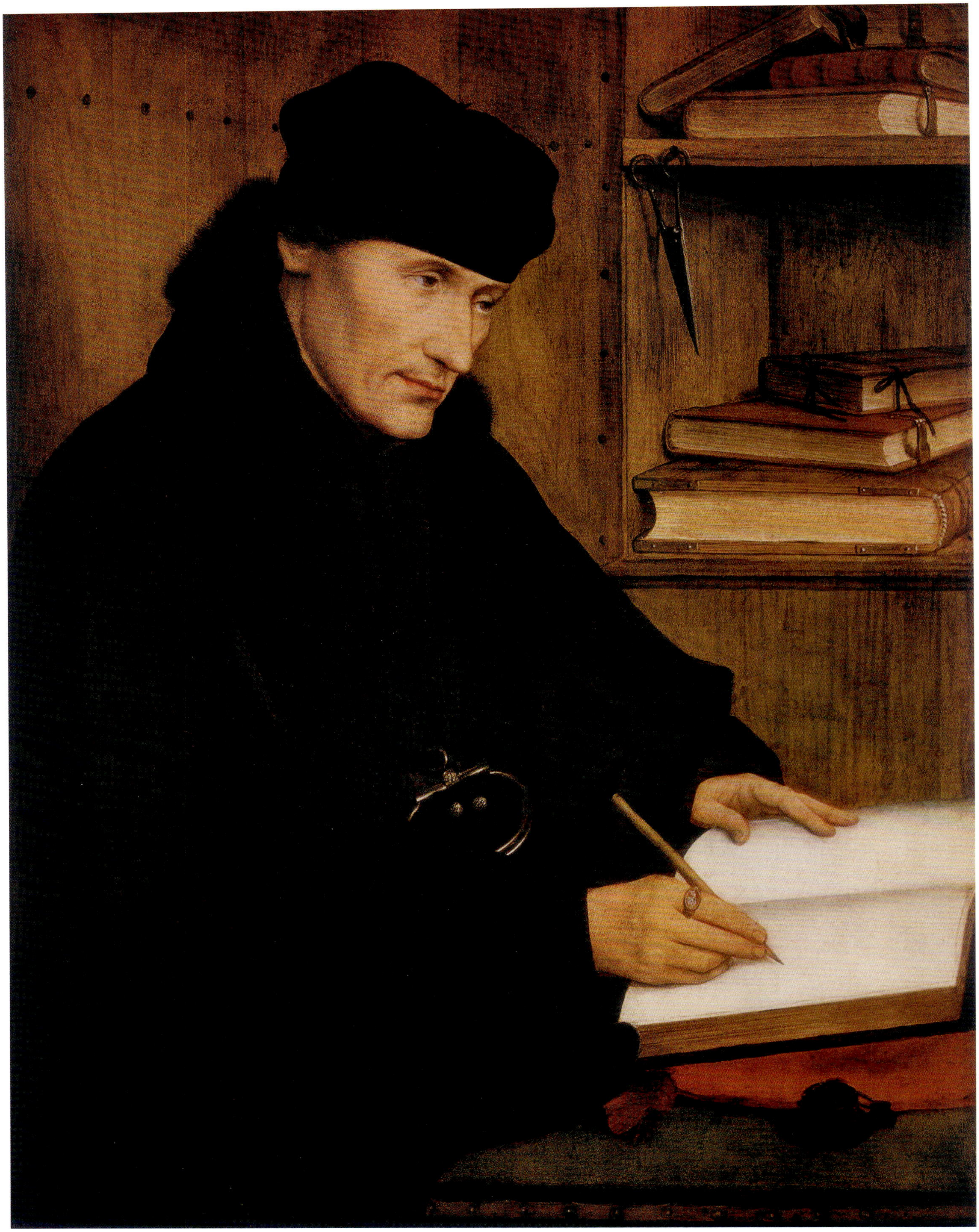

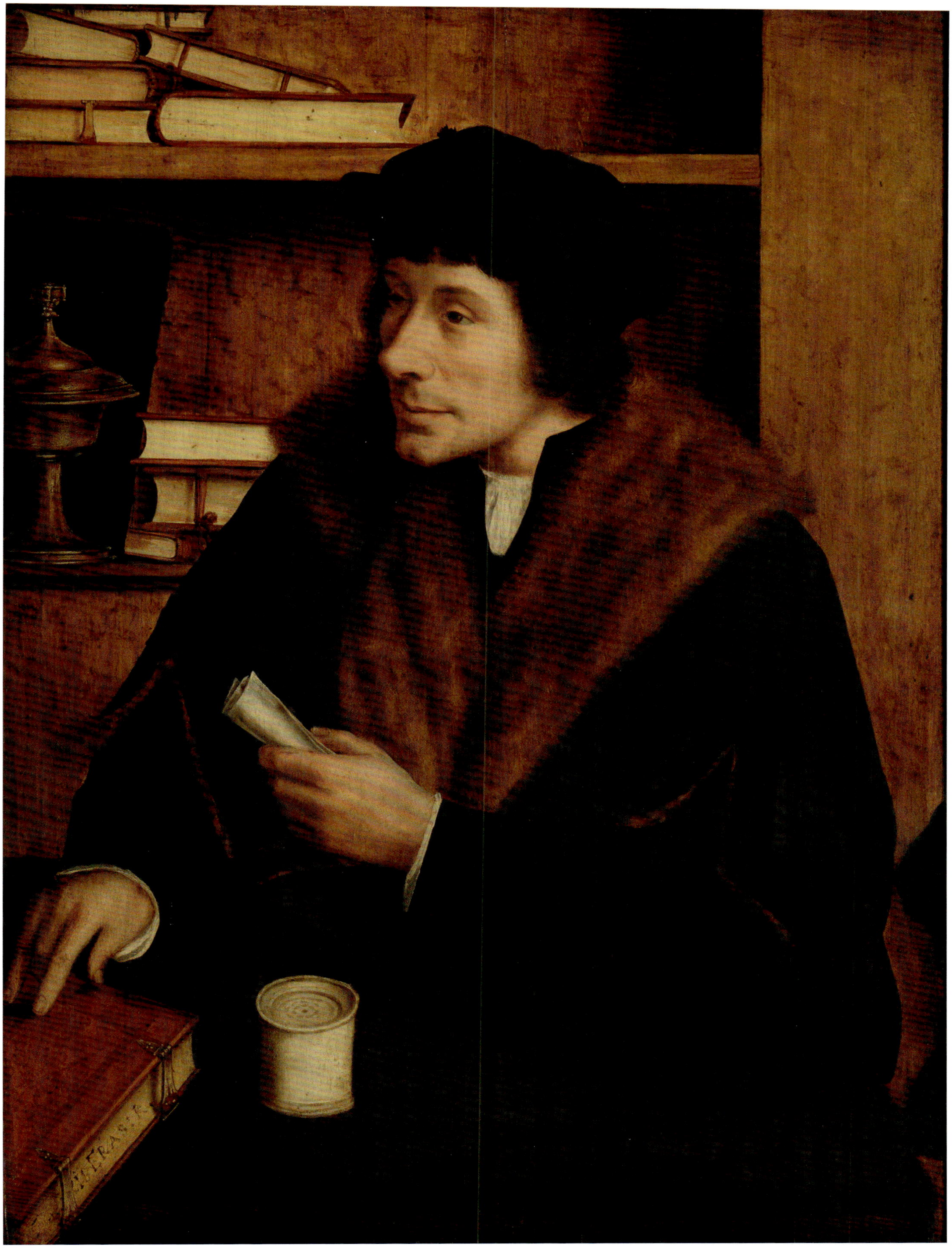

15A–B

Quinten Metsys
Portraits of the Antwerp Town Clerk Pieter Gillis and his Wife Cornelia Sandrin (diptych)

Antwerp (?)

c.1514

Oil on oak panel, 32.7 x 27.5 cm and 32.8 x 27.8 cm respectively

Oldenburg, Landesmuseum für Kunst und Kulturgschichte, inv. 15.565 and 15.566

PROVENANCE

Until 1804 in the collection of the painter J.H.W. Tischbein; acquired by Duke Peter Friedrich Ludwig of Oldenburg, 1804-1919 Großherzogliche Gemäldegalerie Oldenburg; 1922 transferred to the Landesmuseum für Kunst und Kulturgeschichte Oldenburg.

BIBLIOGRAPHY

Böttiger 1800: 70, 72 ff.; Parthey 1863: 603, no. 86; Bode 1888: 81; Bredius & Schmidt-Degener 1906: 28, no. 14A; Hartlaub 1912: 74; Schäfer 1912: 6; Oncken 1913: 542; Friedländer 1916: 95, 183; Winkler 1924: 206; Holtze 1927: 292 ff.; Friedländer 1924-1937: VII, 65, 122, no. 49-50; Köhler 1947: no. 103; Van Wachem 1954: 13, 71; Müller-Wulckow 1961; Keiser 1967: 42; Friedländer 1967-1976: VII, 34, no. 49-50; Nuremberg 1971: 117 ff., no. 205-206; De Bosque 1975: 239 ff.; Anzelewsky, Mende & Eeckhout 1977: 105 ff., no. 187-188; Silver 1984: 113, 192, 234 ff., no. 56; Van der Stock 1993: 207, no. 59; Deuter 2001: 30-31; Oldenburg 2012: 28 ff.; Woodall 2014: 46.

In 1804 the Duke of Oldenburg acquired the entire painting collection of the classicist artist Johann Heinrich Wilhelm Tischbein. It included this pair of portraits. In the nineteenth century they were thought to be by German artists such as Hans Holbein or Barthel Bruyn. Friedländer was the first to connect them with Quinten Metsys.[1] It was the resemblance between this sitter and the portrait of Pieter Gillis in the 'double picture' Metsys painted in 1517 (cat. 13B) that allowed the identification to be made.[2] Just as he does in the 'double picture', Gillis wears a darkish coat with a light brown fur collar, which Metsys has painted with loose brushstrokes. Also typical of Pieter Gillis is the dimple just at the tip of his nose. The light falls from the left onto the sitter's face as he turns his head to the right to look at the woman at his side. Her face is fully lit. Her white headdress reflects the light and surrounds her entire face in starched pleats. A delicate pattern has been worked into the fabric of the veil. Unlike her husband, the woman lowers her eye and has a subdued expression.

The portraits were probably painted on the occasion of a marriage. In July 1514 Gillis wed Cornelia Sandrin. After her death he married again, in 1526, his second wife being Maria Denis Adriaensdr.[3] Because Gillis looks older in this wedding portrait than in the 'double picture' of 1517, it has repeatedly been asserted that he is shown here with his second wife.[4] Silver, however, noted that the strong contrast between light and dark in the portraits is similar to handling of light in *The Money-Changer and his Wife*[5] painted by Metsys in 1514, the same year as Gillis's marriage to Cornelia Sandrin. He ascribed the difference in Gillis's appearance in

1 Friedländer 1916: 95.

2 Van Wachem 1954: 71.

3 Gillis married a third time in 1530. Quinten Metsys died in that same year.

4 Müller-Wulckow 1961: 7; De Bosque 1975: 240.

5 Musée du Louvre, Paris, inv. 1444.

6 Silver 1984: 235; Köhler 1947: 30 May 1517, letter from Erasmus to Thomas More. See also cat. 13 in this catalogue

7 Dendrochronological analysis and report, especially for this exhibition, by Peter Klein, 1 August 2015.

8 Erasmus published the poem in 1524 in the revised edition of the *Colloquia*.

1514 and 1517 to the fact that Gillis was ill when this diptych was painted. Erasmus mentions Gillis's malady in a letter to Thomas More.[6]

Recent dendrochronological analysis of the two panels has led to greater certainty. Both Baltic oak panels came from the same tree and would have been available for use from around 1513.[7] The dating of the panels adds support to the hypothesis that the portraits were produced around 1514, to mark the marriage of Gillis and Sandrin, a union that was celebrated by Gillis's friend Erasmus in the form of an *epithalamium* and by their mutual friend Quinten Metsys with this pair of portraits.[8]

M F

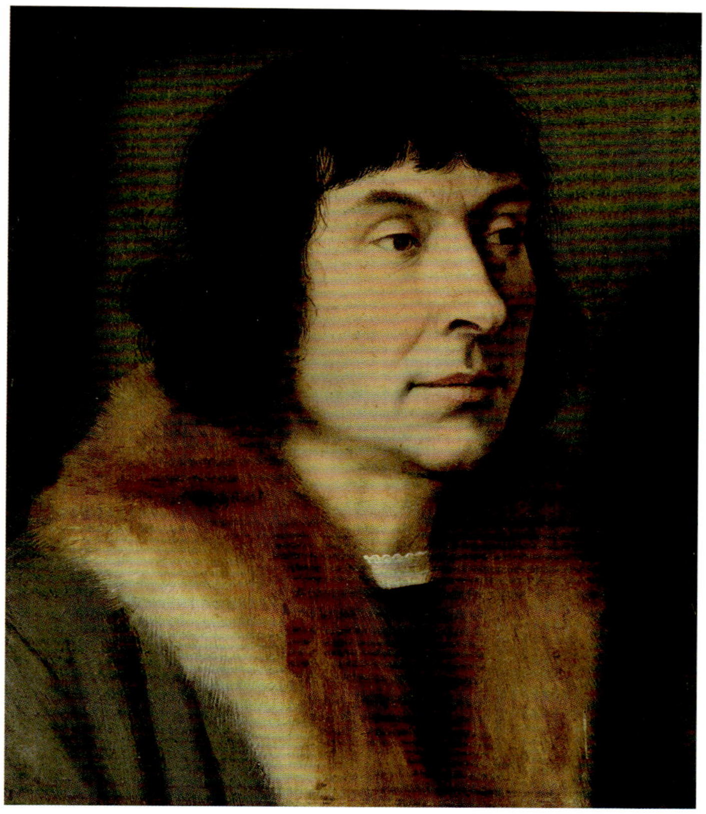

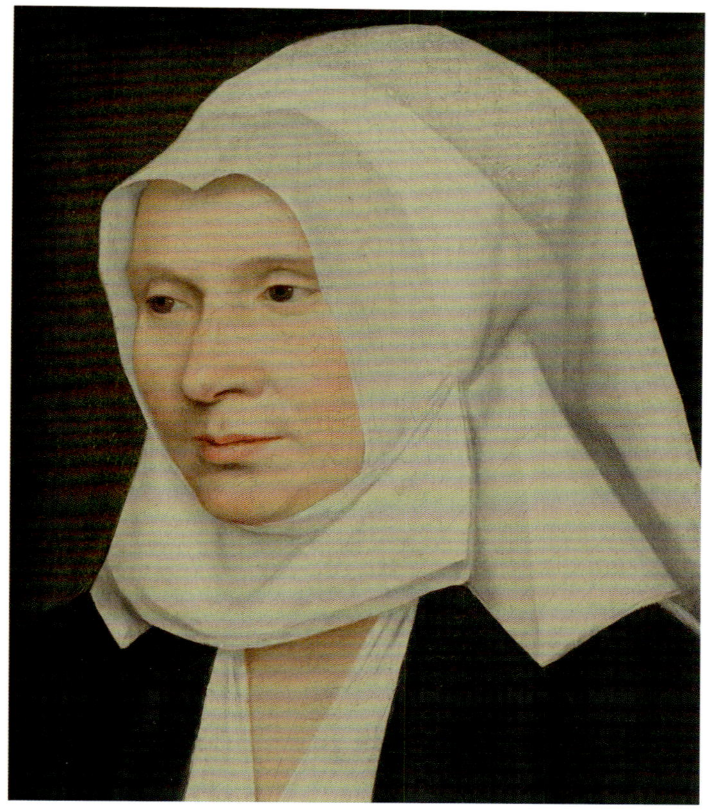

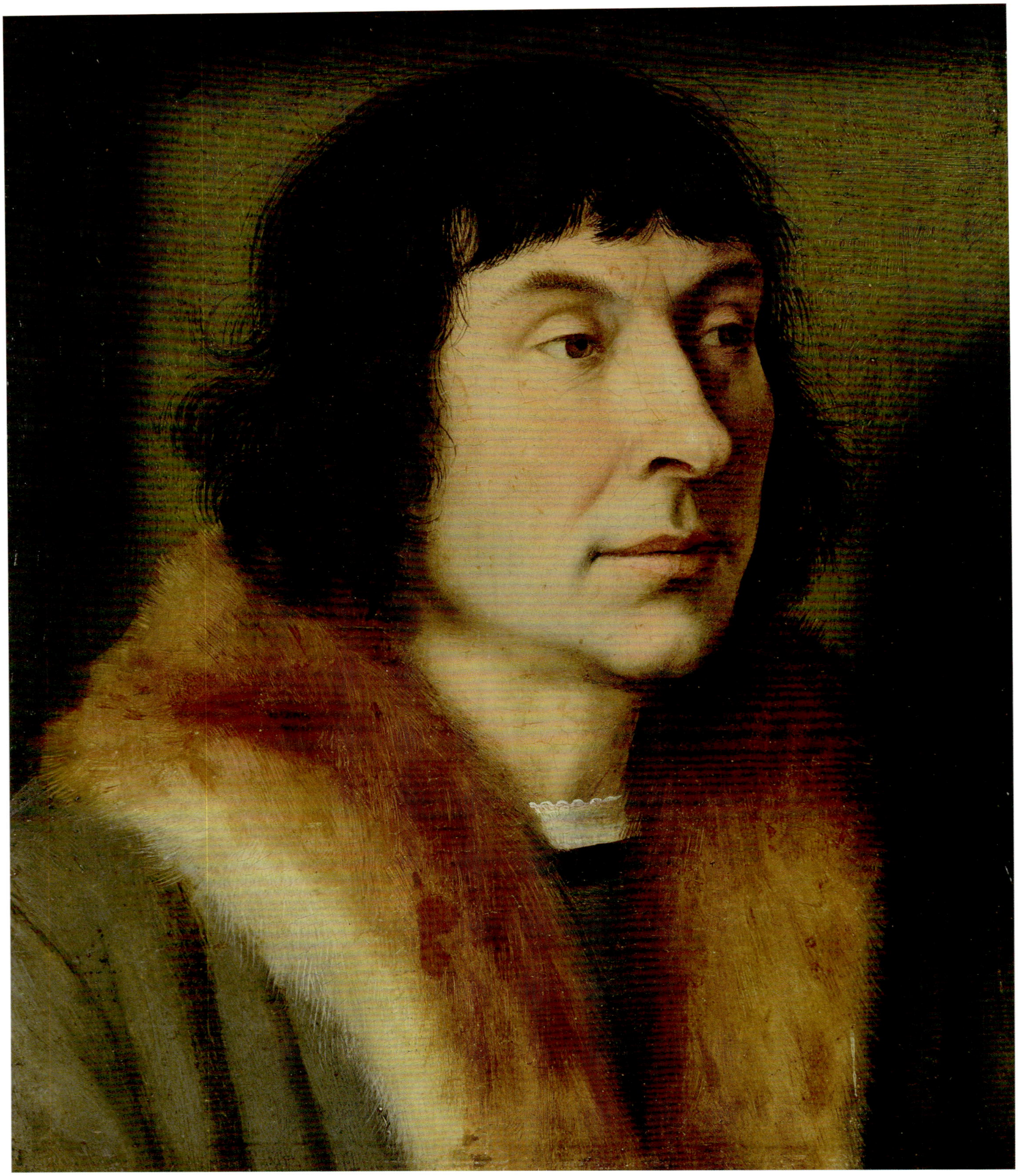

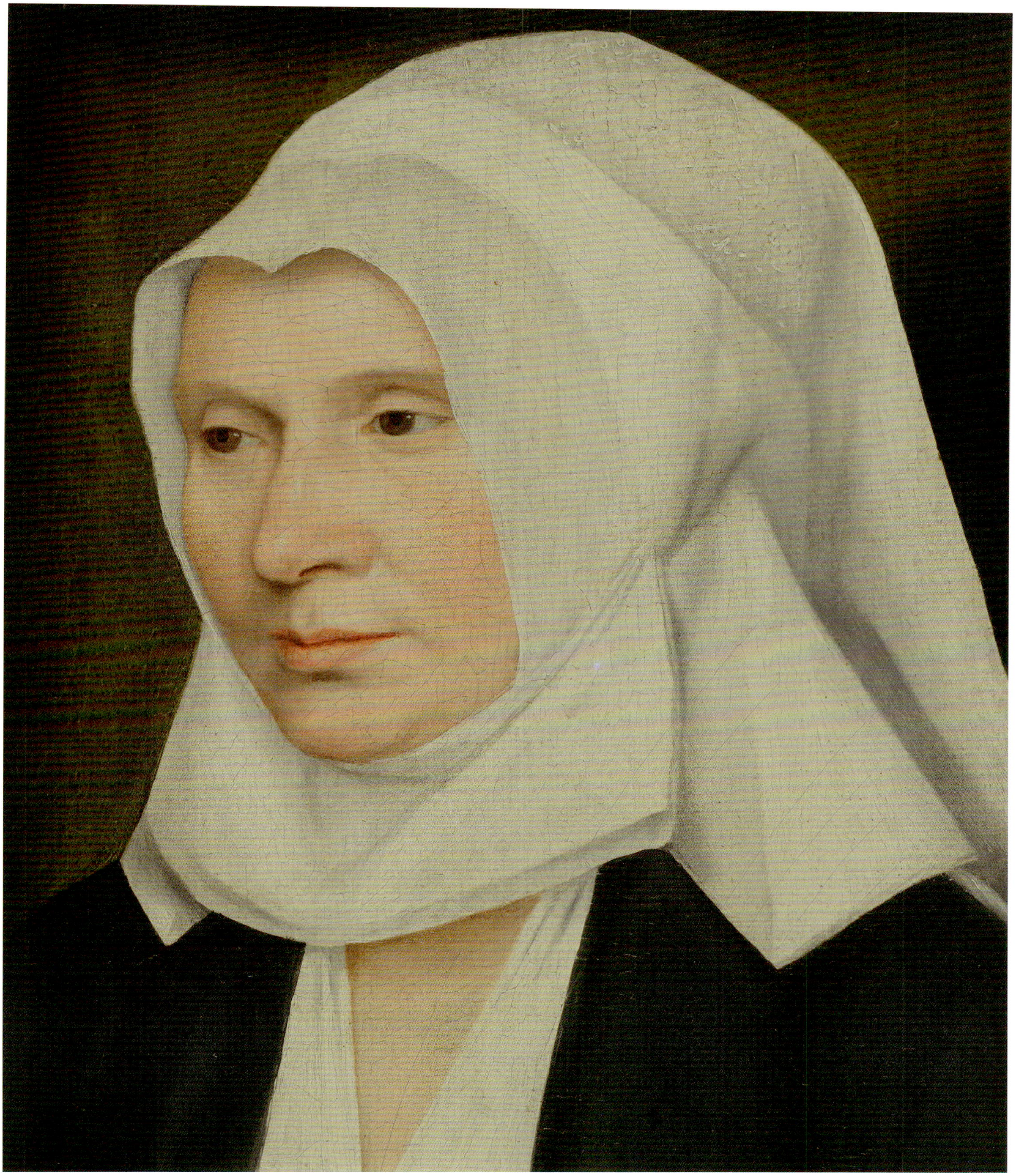

16

Quentin Metsys
Portrait of a Scholar

Antwerp
*c.*1522–1527
Oil on oak panel, 68.8 x 53.3 cm
Frankfurt am Main, Städel Museum, inv. 766

BIBLIOGRAPHY

De Bosque 1975: 262-265, fig. 77; Silver 1984: 61, 114, 165, 166-167, 237-239, no. 61, fig. 146; Sander, Knobloch & Klein 1993: 282-294, fig. 20; Sander 2002: 15, 282-294, 500 ff., 511 ff. (with bibliography); Büttner 2011: 66; Seelig 2015: 501.

Quinten Metsys's portrait of an unknown scholar in the Städel Museum in Frankfurt is not only one of his greatest achievements as a portrait painter but also one of the finest examples of humanist portraiture. Around ten years earlier, in his 'double picture' of Erasmus of Rotterdam and Pieter Gillis (cat. 13A–B), Metsys had suggested the relationship between the two learned friends with the utmost subtlety. This time, however, he brings the unknown scholar into convincingly direct contact with the viewer. The man has paused in his perusal of the book before him and removed the spectacles that are meant to correct his short-sightedness while he turns over in his mind the words he has just read. We seem to be witnessing the very moment that light breaks and causes his involuntary gesture. Seldom has that dazzling moment of intellectual insight been so persuasively portrayed. Sadly, Metsys leaves us in the dark about the subject of the scholar's sudden enlightenment for the text in the book is nothing more than illegible quasi-gothic script.

The monumentality of the portrayal, the deliberate absence of attributes intended to glorify the sitter, and the resolute focus on the scholar's physiognomy and gesture, indicate that the painting must have been produced in the last five years of Metsys's life. Compared to his earlier portraits, influences from Italian masters of the genre, such as Pietro Perugino and Andrea Solario, are much more marked in creations that are, nonetheless, entirely his own invention.

As yet the portrait's scholarly sitter has not been identified. According to Silver, who argues his case at some length, he is Cornelius Grapheus, humanist and town clerk of Antwerp. But Silver's suggested comparison with a copper engraving in Philips Galle's 1572 *Virorum doctorum effigies* raises serious doubts about the sitter's identity. Conversely, the weak portrait in the Louvre (inv. 1346; fig. 16.1), now attributed to Jan Provost, undoubtedly shows the man we see in the Städel portrait. Whether the work in the Louvre is a portrait painted from life or a modified repeat of the one shown here is uncertain. Even so, it offers the best chance of identifying the scholar, as there is a coat of arms and a motto on the reverse – though whom they belong to has still to be discovered.

When the work came to the Städel Museum from a collection in Augsburg in 1829 an inscription on the now lost frame identified it as a portrait of Berend

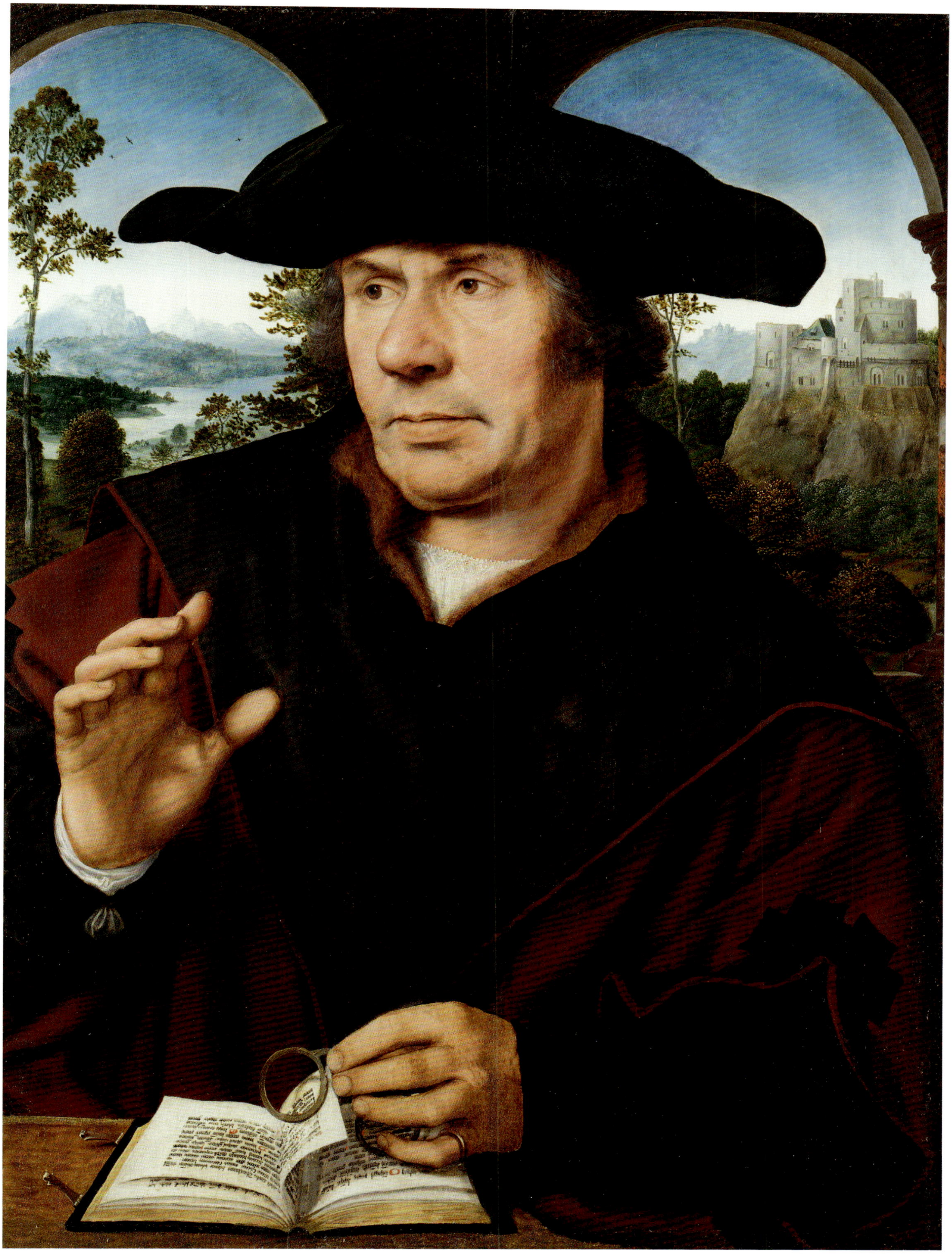

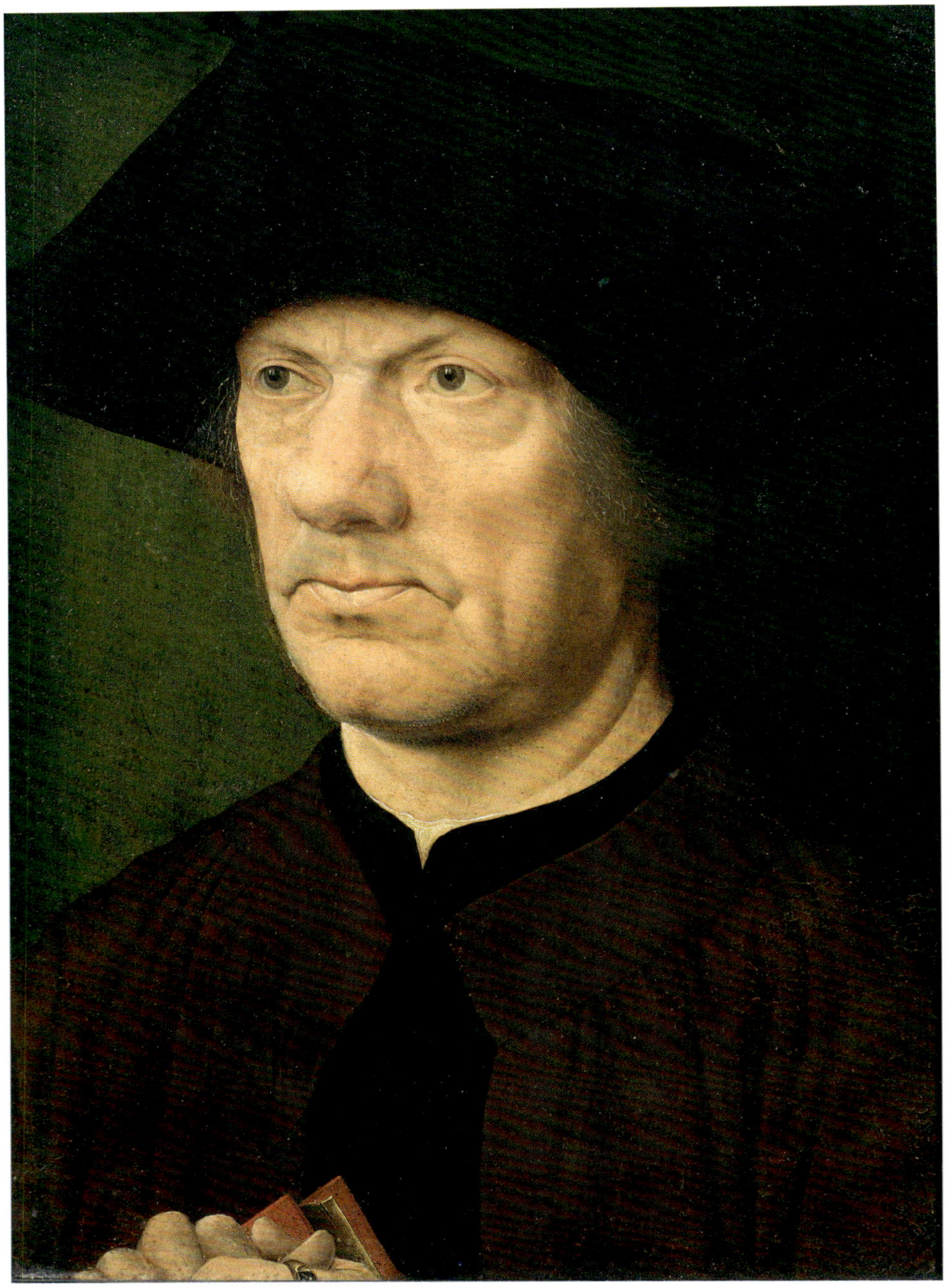

[fig. 16.1] Attributed to Jan Provost, *Portrait of a Man*, c.1520, Paris, Musée du Louvre.

Knipperdolling, leader of the Münster Anabaptists, ostensibly painted by Metsys in 1534. This is hardly possible as by that date Metsys had been in his grave for four years. In 1734 the correct attribution to Metsys and incorrect identification of the sitter as Knipperdolling were already cited in the catalogue for the sale of Amsterdam burgomaster Willem Six's collection. And there are even earlier references to the painting. It can be seen in Willem van Haecht's picture, dated 1628 and now in the Rubenshuis, of the visit of the Archdukes Albert and Isabella to the Antwerp art collection of Cornelis van der Geest on 23 August 1615. From 1626 until his death in 1637 Van Haecht was responsible for Van der Geest's art collections, and he repeatedly included works from them in his own depictions of art cabinets. For instance, the present portrait can also be seen in an undated painting of the workshop of Apelles (now in the Mauritshuis in The Hague; inv. 266). Van der Geest sought to preserve Metsys's fame for posterity, even to the extent of having the painter's tombstone transferred to the west front of Antwerp Cathedral as an early monument to an artist.

JS

17

Albrecht Dürer
Portrait of a Humanist (Sebastian Brant?)

1520–1521

Silverpoint on prepared paper, cut out along the contours (191 x 147 mm) and mounted on a new sheet of paper (194 x 148 mm)

Berlin, Staatliche Museen zu Berlin, Kupferstichkabinett, KdZ 2229

PROVENANCE

Purchased in 1881.

BIBLIOGRAPHY

Kalkoff 1905: 474-485; Janitsch 1906a: 75-78; Janitsch 1906b: 5-18; Friedländer 1908: 79-80; Veth & Muller 1918: vol. I, 31; Bock 1921: vol. I, 34 no. 2229; Flechsig 1928-1931: vol. II, 238; Tietze & Tietze-Conrat 1938: no. 754; Winkler 1936-1939: no. 817; Panofsky 1943: no. 1073; Tietze 1951: 34; Musper 1952: 276; Winkler 1957: 305; Strauss 1974: vol. IV, no. 1520/42; Anzelewsky & Mielke 1984: 110, no. 107; Siewert & Bartetzky 1994: 45-46, no. 5; Unverfehrt 2007: 45.

This silverpoint drawing embodies Dürer's mastery of portraiture. A few fine lines indicate the wrinkles on the man's face. Hair-fine yet confident contours define the eyelids, lips, bags under the eyes, the shape of the nose and the left cheek. With only the play of wavy parallel lines, Dürer evokes the rich fur lining of the coat. For the depiction of the chest and arms, he uses coarser but no less confidently executed lines and hatching. The hands are only schematically indicated. The structure of the composition reveals that the portrait probably came about during an informal sitting. The man's relaxed posture distinguishes it from Dürer's other portrait drawings. The loving care with which the portrait was composed, as well as the sitter's posture and intelligent gaze, seem to suggest a considerable amount of trust between artist and model. Still, the man's identity cannot be determined with any certainty. The drawing would have been created during Dürer's sojourn in the Netherlands (1520–1521).[1] Perhaps it is a loose-leaf sheet that Dürer gave the sitter as a token of friendship.[2] In his travel diary, Dürer made note of events and encounters, and especial of his income and expenses. He mentions around 140 drawings, most of them portraits. Not all of these drawings have survived, while those that have rarely match the ones mentioned. There are clear indications that Dürer did not note down all of his encounters, and also that countless portraits were never mentioned.

Already in 1887, Janitsch pointed out the striking similarity between the man depicted on this sheet and several known portraits of the German humanist and satirist Sebastian Brant.[3] Dürer had met Brant in Basel sometime before, and designed some of the illustrations to his renowned work *Das Narrenschif* (1493).[4] Brant was moreover in the Netherlands in 1520 as a representative of the city of Strasbourg. On 6 August the new emperor, Charles V, received him in Ghent. On 8 August he continued on to Antwerp, where Dürer had arrived six days earlier.

1
The drawing may have formed part of a larger sketchbook, although the figure is a few centimetres too large to fit in what remains of the silverpoint sketchbook that the artist took with him on his trip. Fifteen, often double-sided sheets in silverpoint (*c.* 133 x 194 mm) have survived. They are today dispersed in collections in Berlin, London, Chantilly, Vienna, Frankfurt, Nuremberg and Bremen. Dürer himself first mentions this sketchbook ('mein Büchlein') during his stay in Aachen in 1520. Joachim von Sandrart describes what is probably the same, still intact sketchbook in the imperial collections in Vienna in 1675. See Unverfehrt 2007: 99, 208. On the art of drawing in silverpoint and metal stylus, see Sell & Chapman 2015.

2
The silverpoint portrait by Lucas van Leyden in Lille has approximately the same dimensions and was evidently also conceived as a loose-leaf sheet commemorating the friendship between two artists. See Strauss 1974: no. 1521/26.

3
First in the *Schlesische Zeitung*, 1887, no. 861; convincingly argued in Janitsch 1906 and Janitsch 1906b.

4
Brant describes how 109 fools depart in a rudderless ship for the imaginary island of 'Narragonia'. The fool, for Brant, represented the foolish and morally deficient person. Humanity in his view was a rudderless ship populated by fools. Sebastian Brant, *Das Narren schyff*, Johannes Bergman von Olpe, Basel 1494. Today, 73 of the 114 illustrations are attributed to Dürer. See Schoch, Mende & Scherbaum 2004: no. 266.

5
According to our observations, the perforated part corresponds perfectly with the contours of the figure in the engraving, although the figure in the latter is differently framed and slightly tilted. The engraving in Berlin (162 x 122 mm; Kupferstichkabinett, inv. 154–9) was first associated with the drawing by Bock 1921. Nagler also mentions an exemplar in Paris and identifies the monogrammist erroneously as Christoph Amberger; see Nagler 1858: vol. I, no. 1123.

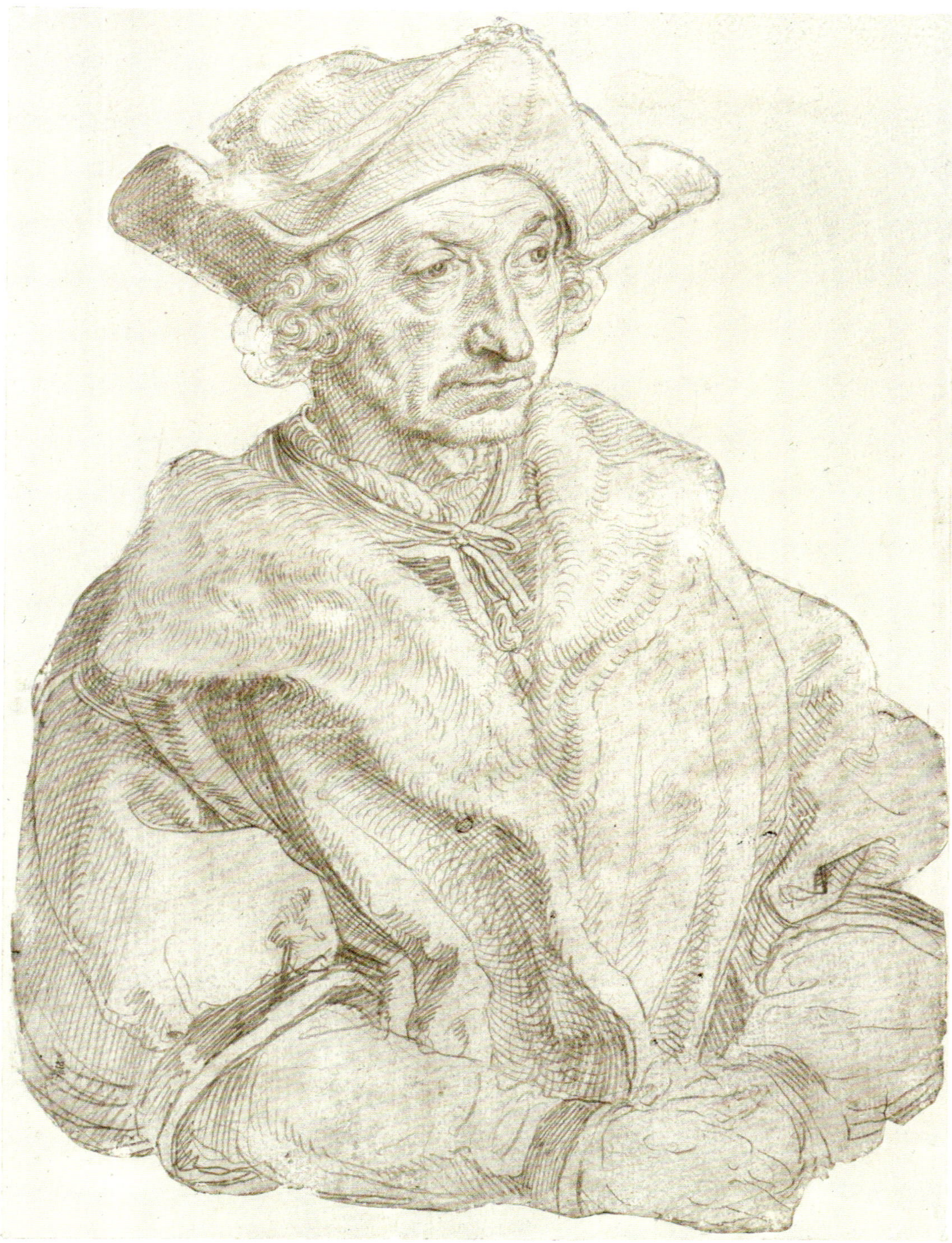

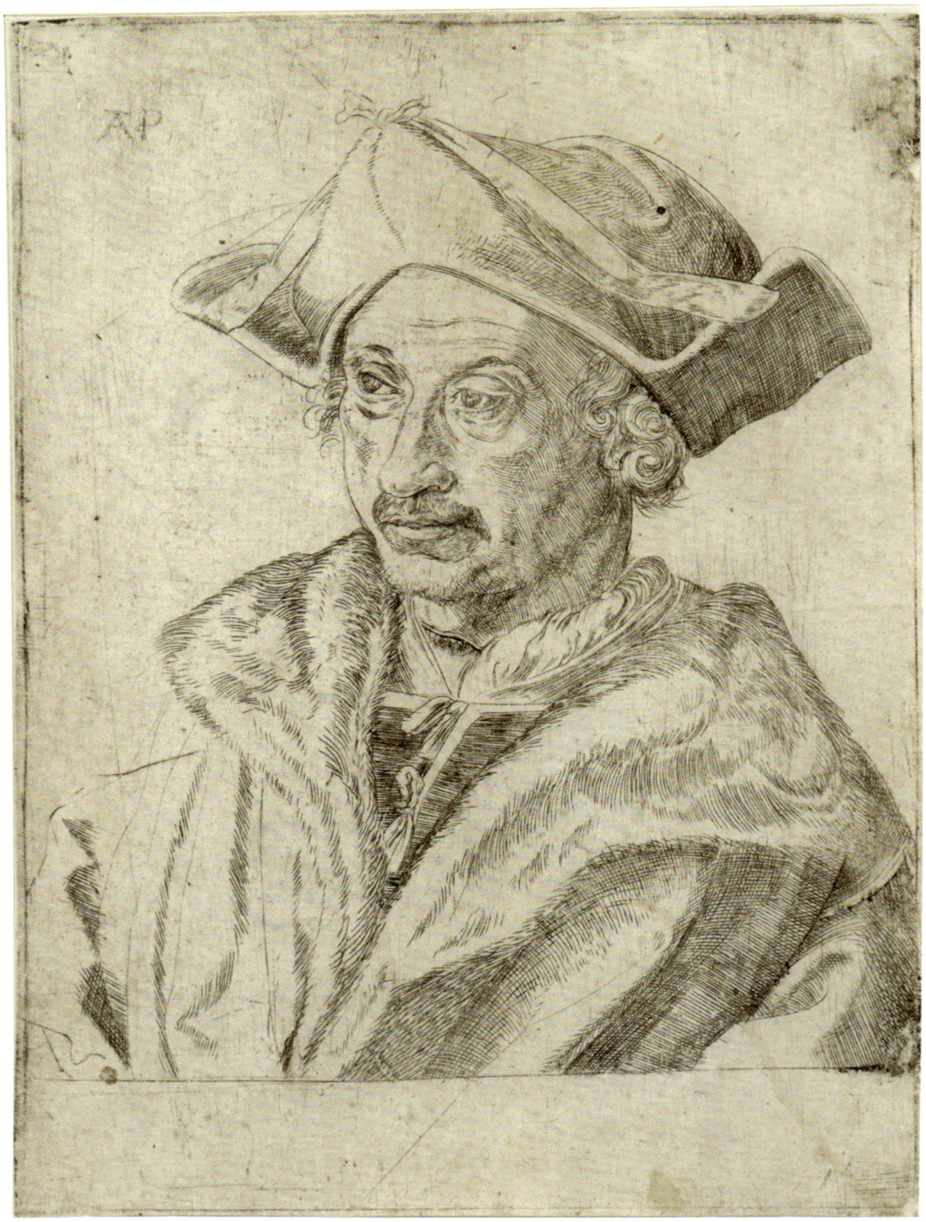

[fig. 17.1] Monogrammist AP, *Portrait of a Man*, engraving, Berlin, Staatliche Museen, Kupferstichkabinett.

It seems likely that the German artist portrayed him then. In his diary, however, Dürer does not mention any such meeting with Brant, or a portrait. Given that Brant was already back in Strasbourg by 22 August, this means his portrait would have been made in the second week of the month at the latest. Yet the sitter in this spontaneous and informal looking portrait wears formal and above all very warm clothing. This could mean that he was drawn in the winter of 1520–1521, in which case he cannot be Sebastian Brant. Another possibility is that Brant deliberately had Dürer portray him in a fur coat and cap in August, but rather as preparation for a formal portrait to be executed as an engraving or oil painting. In that case, the clothing is an appropriate status symbol for the renowned humanist. That it is a portrait of a well-known personality is also evident from the fact that the man's head and shoulders have been indented by a stylus for transfer to a copperplate. The only known impression of the resulting engraving in reverse, by the Monogrammist AP, unfortunately bears no inscription that can help us identify the sitter (fig. 17.1).[5]

JVG

18

Marinus van Reymerswaele
St Jerome in his Study

*c.*1540

Oil on oak panel, 74 x 109 cm

Douai, Musée de la Chartreuse, inv. 234

PROVENANCE

Collection of Marquis d'Aoust; collection of Marlier, painter, Douai; in 1836 acquired by Dr Escallier from Marlier's widow; in 1857 acquired by the Musée de la Chartreuse, Douai, from the Escallier estate.

The painting has been specially conserved for this exhibition by Anne Perrin.

Conservation work was funded by KU[N]ST Leuven vzw.

BIBLIOGRAPHY

Demmler 1918: 63 no. 304, 116 no. 304; Held 1931: 85, no. 3; Leroy 1937: 43, no. 209; Hoogewerff 1941-1942: 459; Douai 1953: 9, no. 16; Marlier & Reis-Santos 1954: 132, no. 27; Van Werveke 1955: 129-130, no. 113; Guillouet 1956: no. 28; Marlier 1963: 163-164, no. 212; Foucart 1965: 202-203, no. 247; Gillouet 1967: 108, no. 55; Chatelain 1967: no. 18; Recklinghausen 1967: no. 16; Friedländer 1975: 106-108, no. 162b, pl. 92; Anzelewsky, Mende & Eeckhout 1977: 115, 160-161, no. 343; Milkers 1985: 174-176; Baligand 1999: 55-56, fig. 49; Baligand 2001; Heck 2005: 272-278, no. 36; Borchert 2010: 205, no. 67.

With piercing eyes St Jerome (*c.*342–420) stares at the viewer from his study. A skull lies on the table in front of him: he points to it with an almost claw-like admonitory finger, a stark reminder of the transience of earthly life. From the early sixteenth century this was a popular *topos* in literature and the visual arts. In Jerome the humanist milieu found the ideal ally for the propagation of their ideas.[1] His origins in the classical world helped to justify a theological culture in which secular knowledge and piety coalesced. For the Greek doctor of the church represented a bridge between the two currents that have shaped humanism: the literary legacy of the classical world and the patrimony of Christianity. His phenomenal activity, not least as a translator of the Old Testament from Hebrew into Latin (*Editio Vulgata*, the first 'popular' translation) and a as writer of commentaries on the Bible, makes Jerome the ultimate 'intellectual', the perfect symbol of the *studia humanitatis*.[2] In response to the medieval scholastic study of the divine, the focus now shifts to the development of man. It is this aspect of him that Marinus van Reymerswaele (1490/5–1546/67) portrays time and time again.[3]

It is generally assumed that Van Reymerswaele's portraiture (*c.*1540)[4] is based on the melancholy half-length picture of Jerome in his study by Albrecht Dürer (1521) (fig. 18.1).[5] There are also some noticeable similarities with the paintings of Quinten Metsys and his son Jan, including the horizontal format and the grotesque physiognomy.[6] Van Reymerswaele puts his own slant on this stylistic characterization, transforming Jerome from a venerable saint into a gaunt and emaciated old man who exudes the finiteness of human life. Iconographically

1
Milkers 1985: 38–40; Hand 1990: 3–5; Cavalli-Björkman 1993: 49; Pabel 2002: 470–471, 473. On the iconographic aspect of humanism's appreciation of Jerome, see Rice 1985: 104–113.

2
Pabel 2002: 470–471. In his *Omnia Opera Divi Eusebii Hiëronymi Stridonensis* (Basel 1516) – nine volumes including a revised *vita* of Jerome –Erasmus even presented the saint as his own alter ego (Pabel 2002: 472, 474–475; Pabel 2013: 134).

3
That limited iconographic repertory is typical of Van Reymerswaele's oeuvre. It encompasses only six variously repeated subjects. Sixteen versions of 'St Jerome in his study' are known (Hendrikmann & Filedt Kok 2009b). For a concise summary of his core oeuvre, see Ackroyd, Billinge, Campbell & Kirby 2003: 50; for a comprehensive study with a revised biography, see Mackor 1995.

4
The date is problematic, given that it is based on stylistic connoisseurship of a changing core oeuvre and the erroneous reading of the year recorded on the few signed works (Ackroyd, Billinge, Campbell & Kirby 2003: 50, 59 n. 3). For the attribution and dating of the present *Jerome*, see Heck 2005: 274–276.

5
On the considerable influence of Dürer's *Jerome* on Southern Netherlandish painting, see Held 1931: 87, 139–140.

6
According to Silver and Milkers the lost original by Quinten Metsys is known from copies in the Kunsthistorisches Museum, Vienna (inv. 691) and the Alte Pinakothek, Munich (inv. 5320/2888). Silver even dates the original before 1520, which means that this painting, not Dürer's, should be regarded as the prototype (Silver 1984: 218, no. 28; Milkers 1985: 40–42, 40 n. 4, 41 n. 1, 93 n. 2).

7
Van der Haagen 1837: 819–874; Milkers 1985: 44–45.

8
Milkers 1985: 47–48, 170; Johns 1989: 47; Cavalli-Björkman 1993: 49–50; Jeon 2005: 210–211, 212 n. 905; Borchert 2010: 205.

9
Eccl. 1: 2.

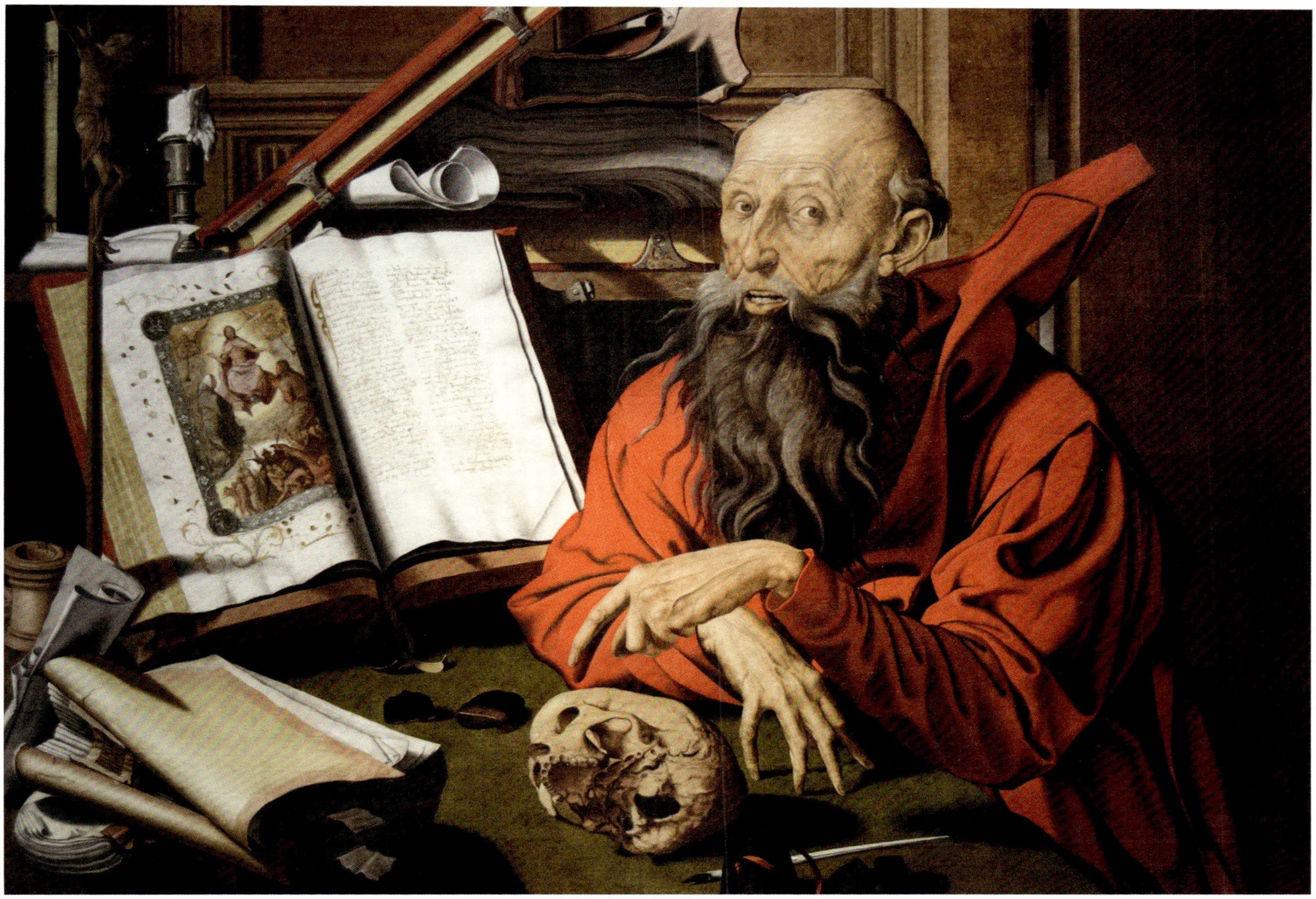

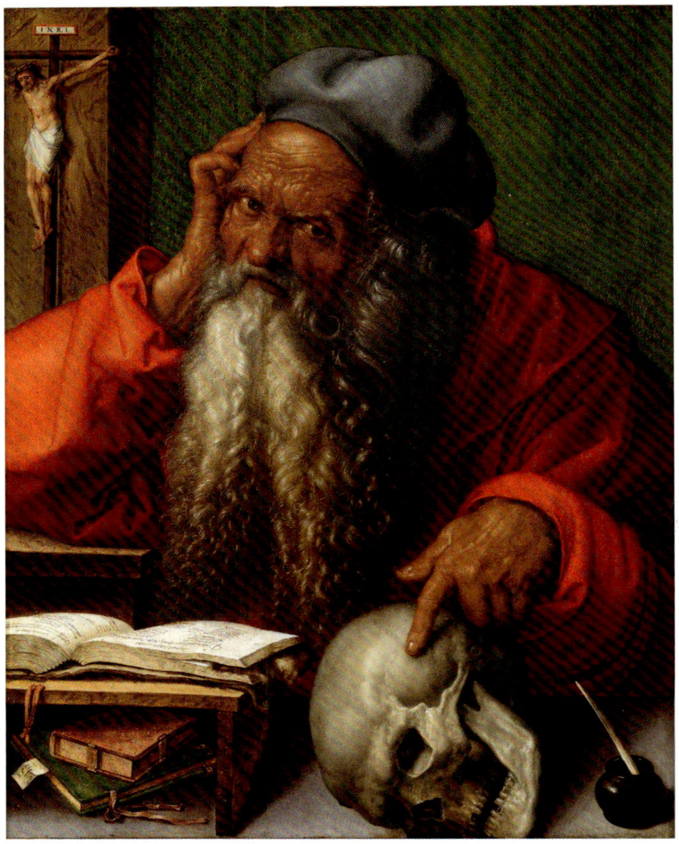

[fig. 18.1] Albrecht Dürer, *St Jerome*, 1521, Lisbon, Museu Nacional de Arte Antiga.

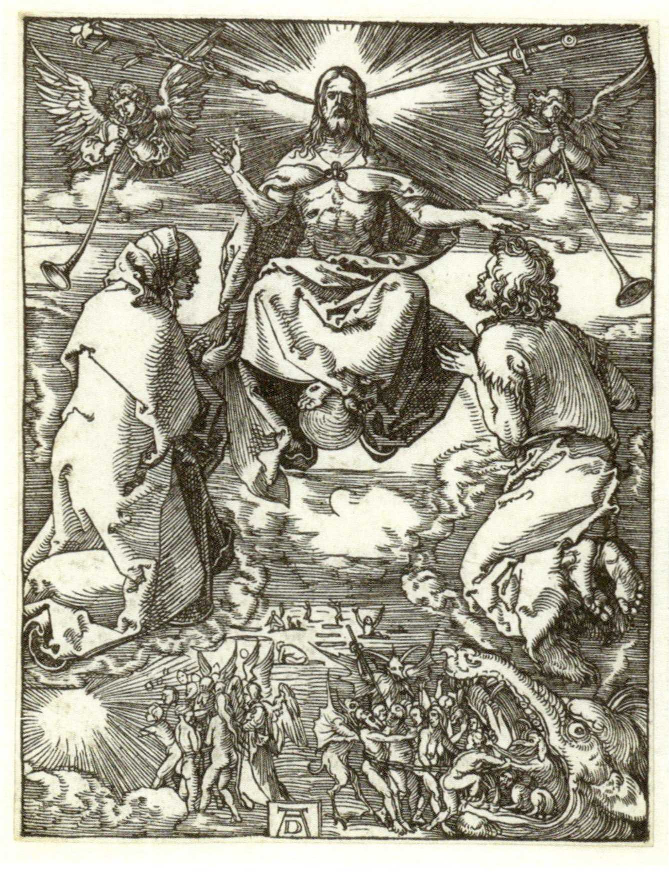

[fig. 18.2] Albrecht Dürer, *Last Judgement*, woodcut from the *Small Passion* series, Amsterdam, Rijksmuseum, Rijksprentenkabinet.

speaking, images of St Jerome fall into two categories, in which he either does or does not look at the viewer. Meeting that penetrating gaze creates a sense of involvement. Both the torments he endured in his restless period of wandering, which contemporary viewers would have been sufficiently familiar with, and his contemplative frame of mind, are manifest in this work. The painting invites self-reflection and so gains the status of a cautionary image. Jerome confronts us personally with the prospect of ultimate accountability to God.[7]

The notion of transience in Van Reymerswaele's *Jerome* is entirely in keeping with the spirit of the time. From the sixteenth century his study becomes filled with vanitas attributes that underline the empty and fleeting nature of earthly existence.[8] The skull moves into the foreground and is portrayed as chillingly as possible, the occiput turned to the viewer. This open reference to the idea of *memento mori* confronts the viewer with his own mortality. The need for inner reflection in the light of human finitude is also evinced in the Bible open at the miniature of Christ at the Last Judgment (Mt 25, 31–46), which is based on Dürer's woodcut from around 1510 (fig. 18.2).The crucifix, the extinguished candle, the worn books and curling parchment scrolls that represent the transience of spiritual riches, reinforce the painting's message: *Vanitas vanitatum, et omnia vanitas* (Vanity of vanities! All is vanity).[9]

W W

19

Bust of Charles of Habsburg (from 1519 Emperor Charles V)

Southern Netherlands (?)
Sixteenth (?) or nineteenth century replica of a lost original of c.1515–1517
Polychromed plaster, height: 53.5 (with cap), 51 (without cap) x 62 x 31 cm; cap removable
Ghent, Museum voor Schone Kunsten, inv. 1983-E

PROVENANCE

Collection of L. Gonthier, Brussels; collection of J. Willems, Brussels; collection of J. Ladot, Brussels; 1983 acquired by the Friends of the Museum voor Schone Kunsten Ghent.

BIBLIOGRAPHY

Devliegher 1960: 240-241; Delmarcel 1976: no. 56; Lowenthal 1976: 84; Anzelewsky, Mende & Eeckhout 1977: 126-127, no. 264; Eeckhout 1985: II, 400, no. B2; Mechelen 1987: no. 204; Van den Boogert 1993: 237, no. 173; Martens 1998: II, no. 232; Checa Cremades 1999: 32; Soly & Van de Wiele 1999: 192, no. 39; Katz, Auersperg & Zock 2000: 18; Noever 2000: 131-132, no. 37; Checa Cremades 2000: no. 33-34; Kruse 2000: 132-133, no. 38; Eichberger 2005: 89, no. 25; Eikelman & Burk 2006: 252-253, no. 69; Van der Coelen, Rümelin, Trapman et al. 2008: 133, fig. 45; De Roy 2011; Michel & Sternath 2012: 158-159, no. 16; De Roy 2013: 561-562, 564-569; De Roy & Van Audenaeren 2015: 10-13; De Roy 2016: 97-108.

This bust of Charles V (1500–1558) bears a great likeness to the portraits of him that were painted around 1515 by Bernard van Orley at the court of Margaret of Austria. In 1515 Charles was declared of age and became sovereign ruler of the Netherlands. In 1516 he became king of Castile and Aragon and in the autumn of 1517 he left the Netherlands for Spain, returning only in 1520. From 1515, therefore, his portraits in the Netherlands had a special significance, irrespective of medium. On the death of his grandfather, Emperor Maximilian I, in 1519, Charles was elected King of the Romans and subsequently Holy Roman Emperor.

The bust is cut horizontally at the bottom. Charles turns his head slightly to the left and looks ahead, his mouth open, his jaw jutting. Descriptions by contemporaries – by Antonio de 'Beatis in 1517, for example – repeatedly emphasize the same physiognomic characteristics and the dignity and import of his actions. The young prince wears a flat broad-brimmed cap with a drooping ostrich feather at the back. His jaw-length hair is cut straight. Beneath his open cloak he wears a gold-trimmed doublet over a finely-pleated shirt. Around his neck hangs the collar of the Order of the Golden Fleece, into which he was inducted not long after his birth and of which he became grand master in 1506. In 1515 – when the prototype of this bust was created – Erasmus of Rotterdam was appointed as counsellor to Charles. The following year he wrote his *Institutio principis Christiani* (*The Education of a Christian Prince*), a 'mirror for princes' in which he advises Charles on what he should strive for in his policies. His rule should be based on Christian ethics – an ideal also pursued by Thomas More in his *Utopia* as he emphasized justice, moderation and the nurturing of the common good.

1
Bruges, Gruuthusemuseum, inv. C.4.

2
Valladolid, Museo Nacional de Escultura, inv. 2700. See Checa Cremades 2000: 212–213, no. 34, and Katz, Auersperg & Zock 2000: 18.

3
De Roy 2013; De Roy & Van Audenaeren 2015; De Roy 2016.

4
Sterk 1980: 45, 234 (Duurstede, 1529 inventary).

5
Michelant 1872: 361.

6
On the authorship of the bust, see Duverger 1934: 47, according to whom, Leone Leoni could not have been the maker of the sculpture at that time.

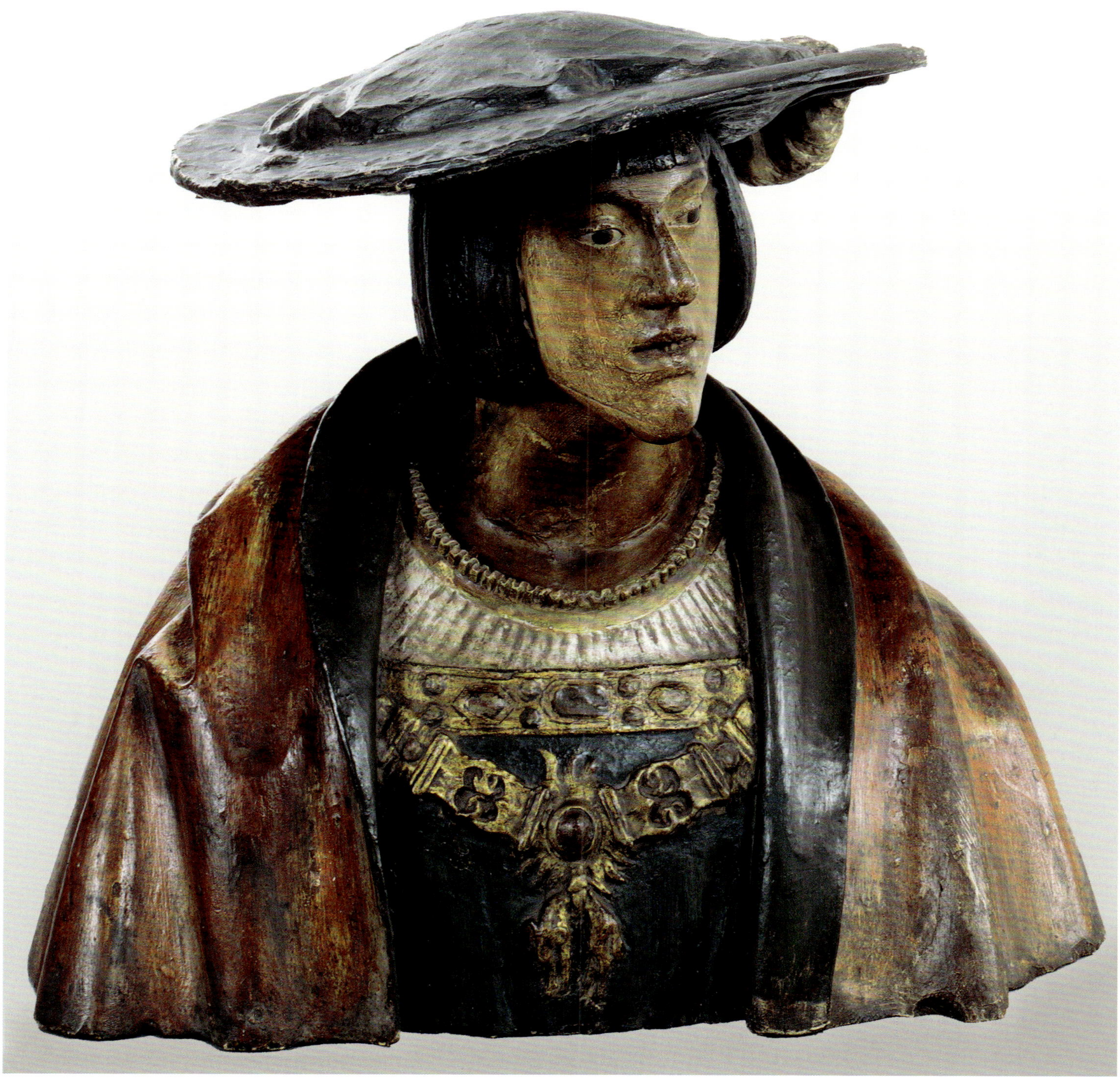

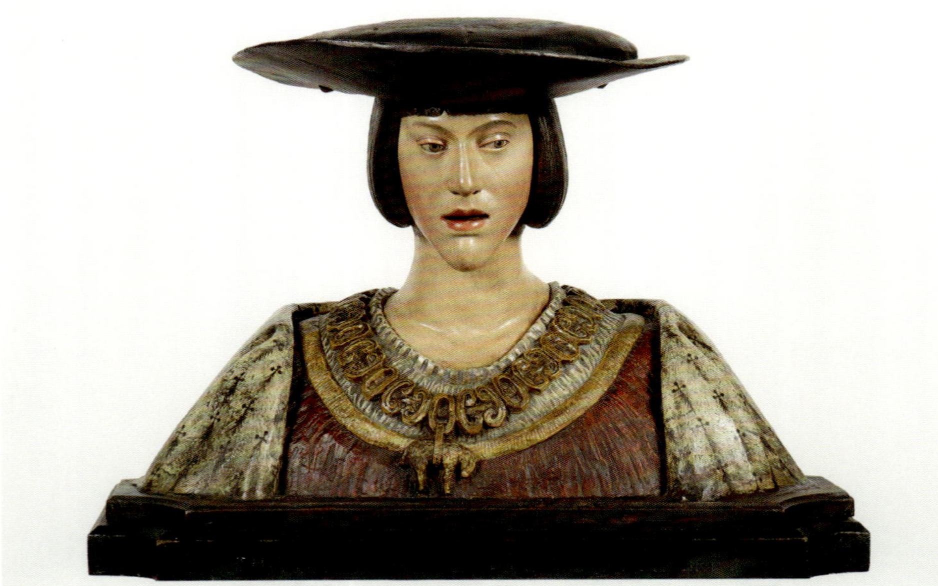

[fig. 19.1] Southern Netherlands (?), *Portrait bust of Charles V*, c.1520, Bruges, Museum Gruuthuse.

In addition to the bust in Ghent, two others survive: one in the Gruuthuse-museum in Bruges (fig. 19.1)[1] and one in the Museo Nacional de la Escultura in Valladolid.[2] Yet even after recent research it is not clear how the three busts relate to each other.[3] Until some decades ago there was also a fourth bust that adorned Middelburg Town Hall, was destroyed during the Second World War, and is now only known from photographs. That it was made in 1522 is attested by the inscription on it. There is nothing to support the earlier attribution of the bust to Conrat Meit. Other artists in the Netherlands are known to have made busts in terracotta at that time. In 1518, for instance, the Frenchman Claude de Chartres was commissioned by Philip of Burgundy, Bishop of Utrecht, to make a set of fifteen such sculptures for his castle in Duurstede.[4] Charles V himself also had a bust of this kind. His Brussels inventory of 1536 mentions *La resemblance de l´Empereur en terre cuycte, faicte par ung nommé Mre Lyon*'.[5] That bust was therefore certainly not made by Conrat Meit, and although 'Lyon' may sound a little like 'Leoni', given the early date it cannot be one of the bronze portrait busts of Charles that Leone Leoni once made and which represented the pinnacle of imperial iconography.[6] We can therefore assume that there were once still more terracotta busts of Charles V in existence.

JLB

BEYOND UTOPIA

Images of Paradise and Hell

20

The Fountain of Eternal Youth

Strasbourg
c.1430–1440
Wool and silk, 90 x 115 cm
Colmar, Musée Unterlinden, inv. 161

PROVENANCE

Foltz collection until 1860.

BIBLIOGRAPHY

Jubinal 1840: 29; Burckhardt 1923: 12, fig. 15; Kurth 1926: 235, no. 124; Van Marle 1932: 433, fig. 461; Göbel 1933: 89; Rapp 1976: 88-90, 121-125, 127, 148, no. 16.; Stewart 1989: 64-88; Rapp Buri & Stucky-Schürer 1990: 23, 308-309, no. 92; Camille 1998: 81-87; Antoine 2002: 57-59, 130-131; Lecoq-Ramond 2003: 23-25.

This tapestry, produced in Strasbourg in the first half of the fifteenth century, is one of the earliest German representations of the Fountain of Eternal Youth.[1] In the middle of a paradisiacal garden stands the miraculous pool. The gothic, hexagonal construction is enclosed by a wall with two gates.[2] The arrangement recalls the religious Enclosed Garden, which is sometimes surrounded by the *Porta Ezechiel* and the tower of David (cat. 32).[3] At the left, a group of crippled, exhausted elders enters the building. Some of them walk with crutches; others are carried to the water in carts or baskets. Then they undress and enter the pool via the steps. Above their heads is a banderol with the text: 'I praise you God, I, an old man, that I have found the fountain'.[4] Above, a youth points to the old cripple and says: 'Because we will all grow old, we save our money'.[5] In the water, the elderly are transformed into their young selves. In the middle, a half-naked couple caress one another. At the right, a young man helps a rejuvenated woman from the water. Just outside the gate at the right, the young celebrate their recovered youth with a banquet. The faces of the figures were originally executed in colourful silk embroidery, but today only traces of it are still visible.

The Fountain of Eternal Youth is a legendary spring with a rich visual tradition. Representations of the unattainable source first appear in fourteenth-century French ivories and miniatures. The originally religious and courtly motif evolved over the course of the fifteenth and sixteenth centuries into a voyeuristic and erotically charged theme (cat. 21).[6]

The fascination with the imaginary and miraculous construction was particularly great and led to various stories and phantasms. Longing for the fountain was so cherished that in the course of the sixteenth century, an actual search for the source was undertaken. The expedition was allotted to Juan Ponce de León, a Spanish explorer who was charged with finding the land of *Benimy* in 1513. According to legend, he found the Fountain of Eternal Youth in the New World – more specifically, in present-day Florida. The legend illustrates the sixteenth-

1
Rapp 1976: 121–125.

2
The typology of the fountain was widespread in the fifteenth century. Comparable constructions can be found in painting, prints and the applied arts of the fifteenth and sixteenth century. The angular fountain as live-giving source is also depicted in the *Lamb of God* polyptich by Van Eyck (Ghent, St Bavo's Cathedral).

3
Antoine 2002: 57–59.

4
Freely translated from German: 'Ich lobe dich Got ich altter man das ich de burnen funden han'. Rapp Buri & Stucky-Schürer 1990: 23; Camille 1998: 83–85.

5
Freely translated from German: 'Sind wir gewesen also die altten so ist unser gelt gar wol behalten'. Rapp Buri & Stucky-Schürer 1990: 308–309; Camille 1998: 83–85.

6
Stewart 1989: 64–88; Antoine 2002: 130–131.

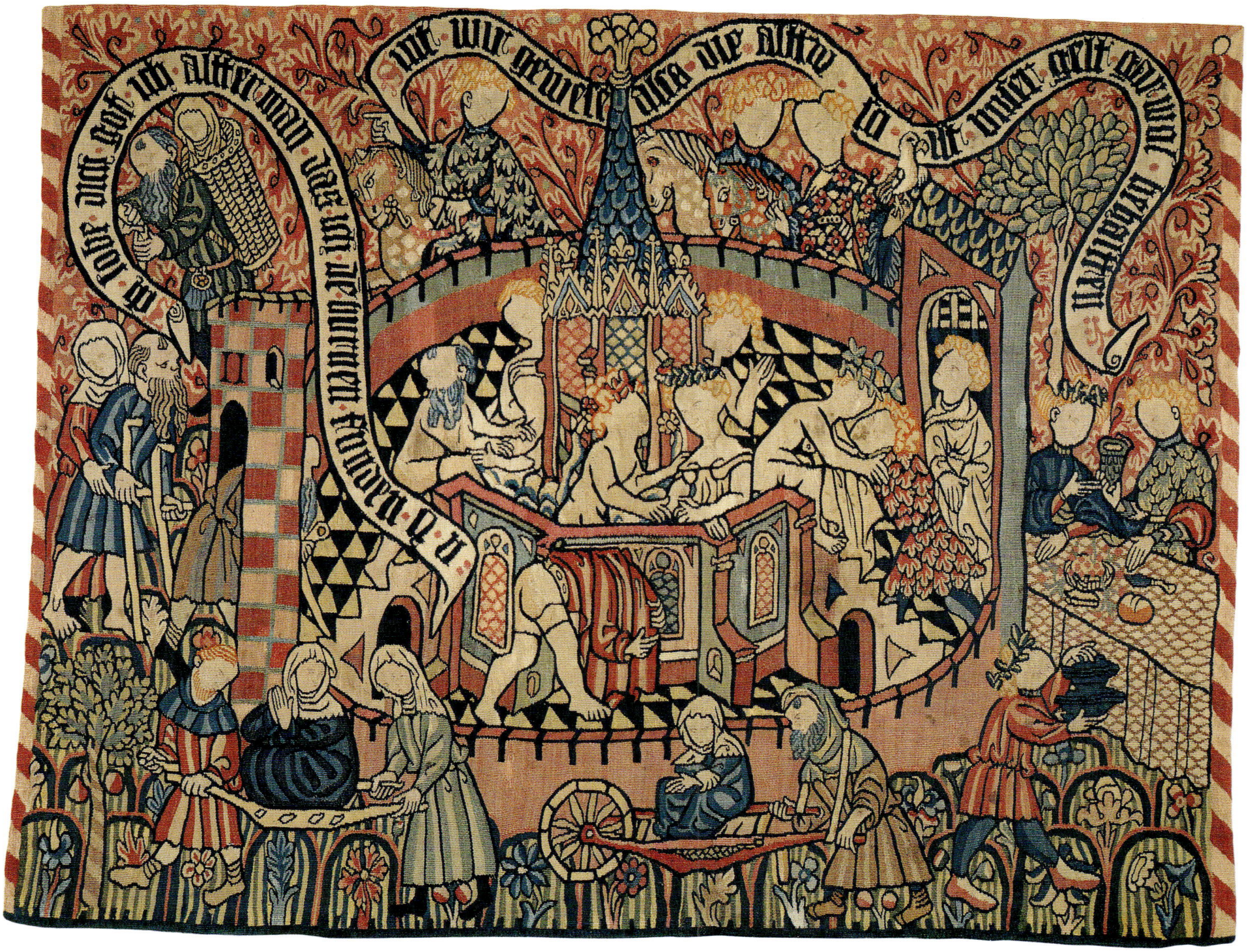

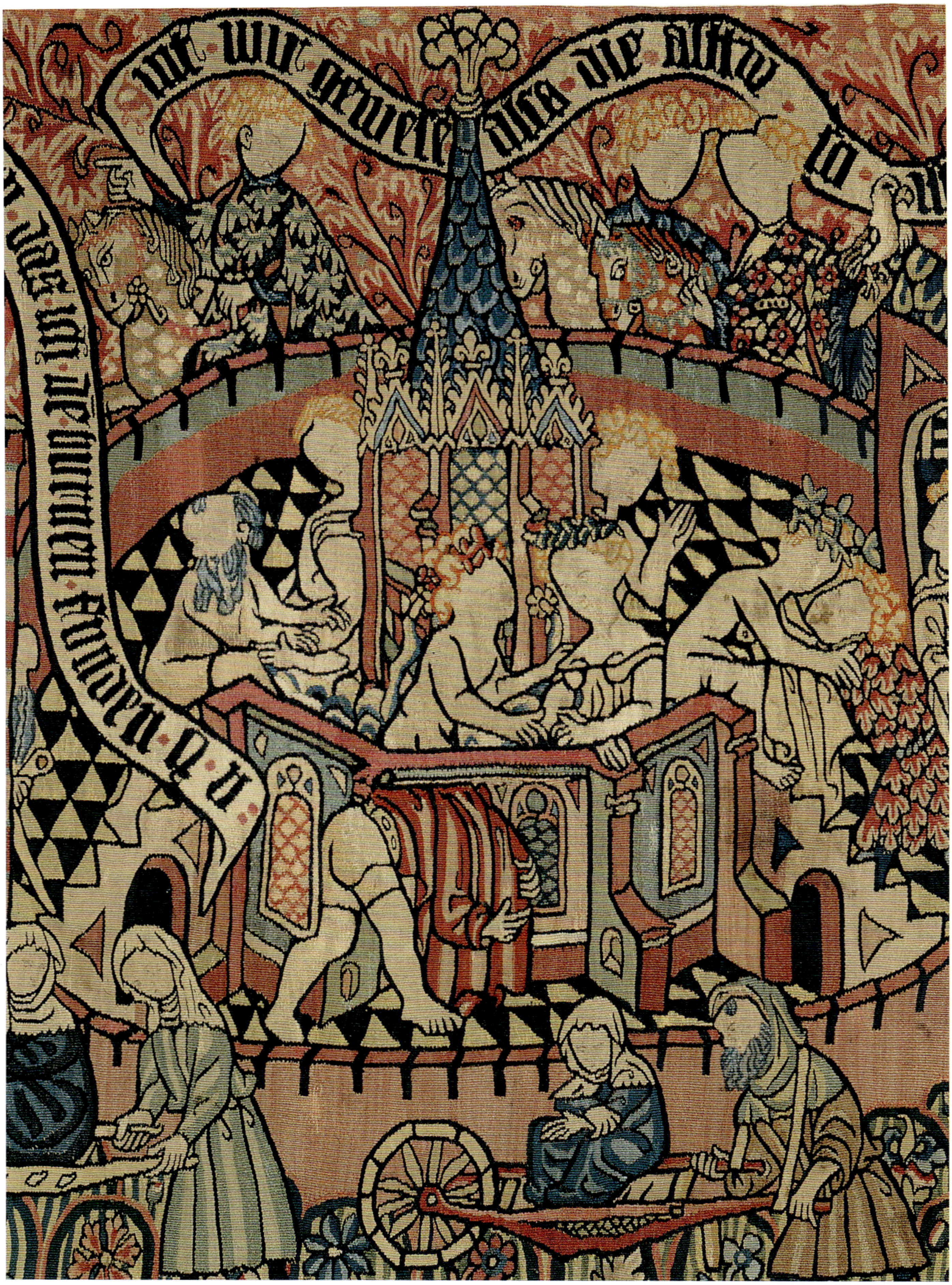

[cat. 20] detail

century tendency to look for mirages, drawn from the well of human desire, beyond the horizon and to project them upon the New World.

HI

21

Hans Sebald Beham
The Fountain of Eternal Youth

Nuremberg

*c.*1531

Woodcut coloured with brown, blue and grey, 506 x 1095 mm

Oxford, Ashmolean Museum, inv. WA1863.3079

PROVENANCE

Collection of Francis Douce (1757-1834); bequeathed to the Bodleian Library, Oxford; in 1863 transferred to the Ashmolean Museum, Oxford.

BIBLIOGRAPHY

Pauli 1901: 394-395, no. 1120; Dodgson 1933: 121; Stewart 1989: 64-88; Landau & Parshall 1994: 232-234; Stewart 2008: 73-84; Müller 2007: 309-318; Yasui 2016: 19-36.

In the Land of Cockaigne, only three miles from Christmas, behind the great Rice Pudding Mountain, lies the Fountain of Eternal Youth. Only one hour of bathing in the marble pool with warm and cold streams of water gives every greybeard a rejuvenating cure for the heart, body and soul.

The preceding excerpt is taken from the verses of the Nuremberg shoemaker Hans Sachs (1494–1576).[1] This *Meistersinger* created the world known as *Schlaraffenland*, the land of milk and honey, in which the *Jungbrunnen* was situated, in numerous of his texts. In doing so the poet joined a longstanding tradition and thereby gave it new impetus. His poems, coupled with the graphic work of the woodcutters, brought dreams of abundance and eternal youth into many a sixteenth-century middle-class home and bathhouse.

When, in the second half of the sixteenth century, the medium of large woodcuts became popular as wall decoration, Hans Sebald Beham (*c.*1500–1550) developed his own interpretation of the Fountain of Eternal Youth. His composition of four large sheets, joined together and provided with a poem of at least 230 lines, was published in 1531 by the printer Albrecht Glockendon the Younger (*c.*1500–1545).[2] The fact that the majority of these sheets were usually pasted on walls has meant that only a few have survived the ravages of time. The Oxford exemplar is unique. It is a coloured, late impression of the first state.[3]

The complexity of the woodcut can be grasped in a multi-layered reading of the whole. A first glance reveals how, at the left, the elderly are carried on stretchers to their invigorating bath. Once in the water, they display signs of rejuvenation immediately. Crutches and walking sticks are from then on superfluous and can only serve as fuel for a joyous bonfire. The right side of the print depicts a voyeuristic view into an Italianate bathhouse. Enjoying the benefits of their recovered youth, bathers fill their time with music, games and each other. Satire and eroticism form a distinct undertone here. When, in a second phase, attempts were made to place Beham's print in the visual culture of the time, it became evident just how much inspiration the artist drew from the work of Italian artists.

1
Sachs 1884: chapter 50, no. 48; Sachs 1893: 8–11, no. 4 and 321–323, no. 115.

2
In 1989 Stewart indicated Glockendon as the author of the poem on the woodcut, because his name was added after the text. This cannot be considered a conclusive argument, however, if one considers, Erhard Schöns' (1491–1542) representation of *Schlaraffenland*, for example. There the name of the publisher, Wolff Strauch, appears after the poem, while the verses can be attributed without a shadow of a doubt to Hans Sachs. The origin of the texts under Beham's woodcut is therefore more likely to have originated with Sachs or his circle. See Stewart 1989: 81.

3
Dodgson was the first to be able to place the various states of Beham's composition in the right order. Previously Pauli, and later Hollstein, erroneously described a first state with the monogram HSB. According to them, a second state bore Glockendon's address. In reality, the printing history of these two states should be reversed. See Dodgson 1933.

4
Marcantonio Raimondi after Raphael, *The Judgment of Paris*, c.1517–1520, engraving, 298 x 442 mm (B.XIV.197.245).

Several figures in the bathhouse print, for example, are literally taken over from Marcantonio's *Judgement of Paris*.[4]

MB

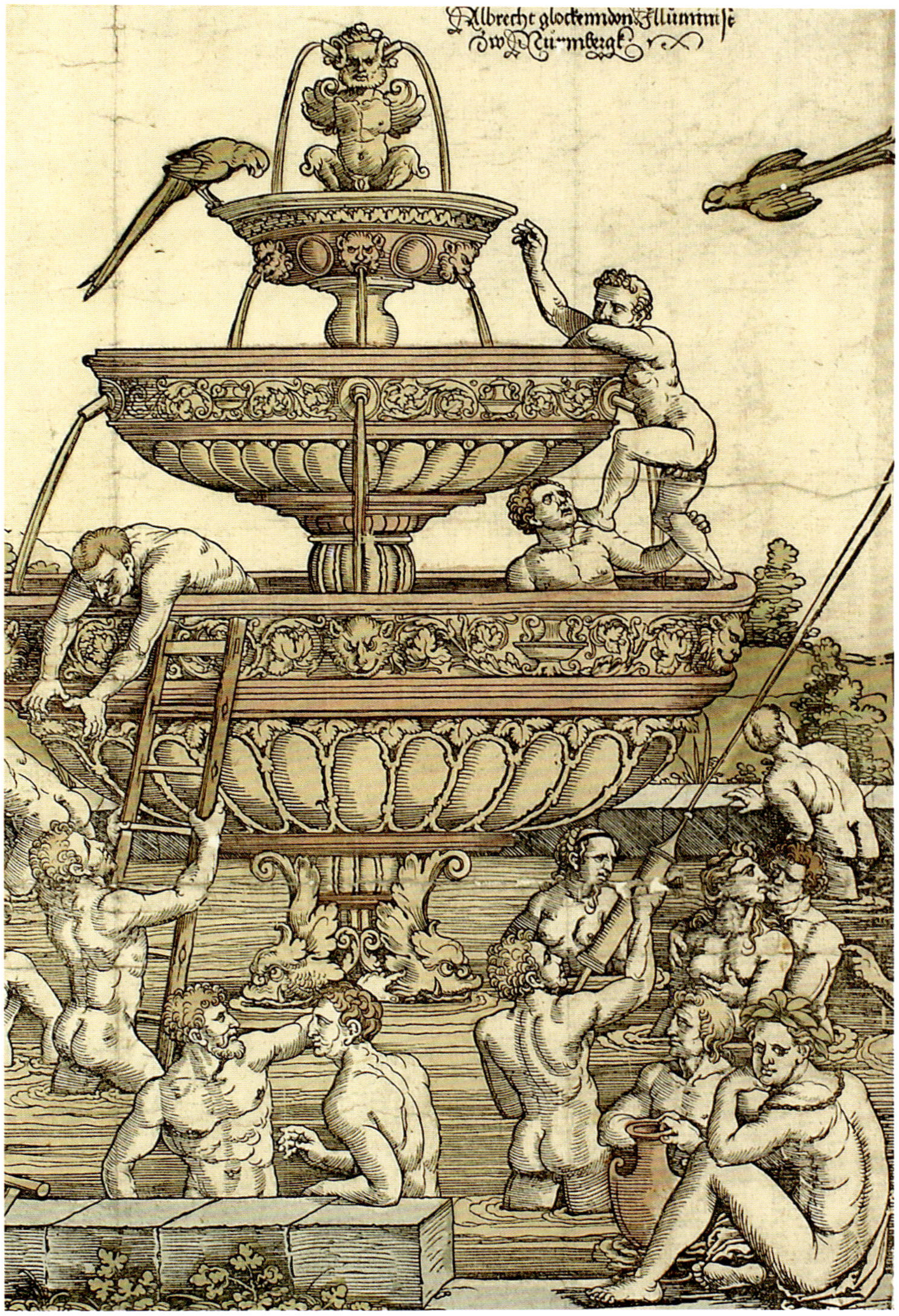

[cat. 21] detail

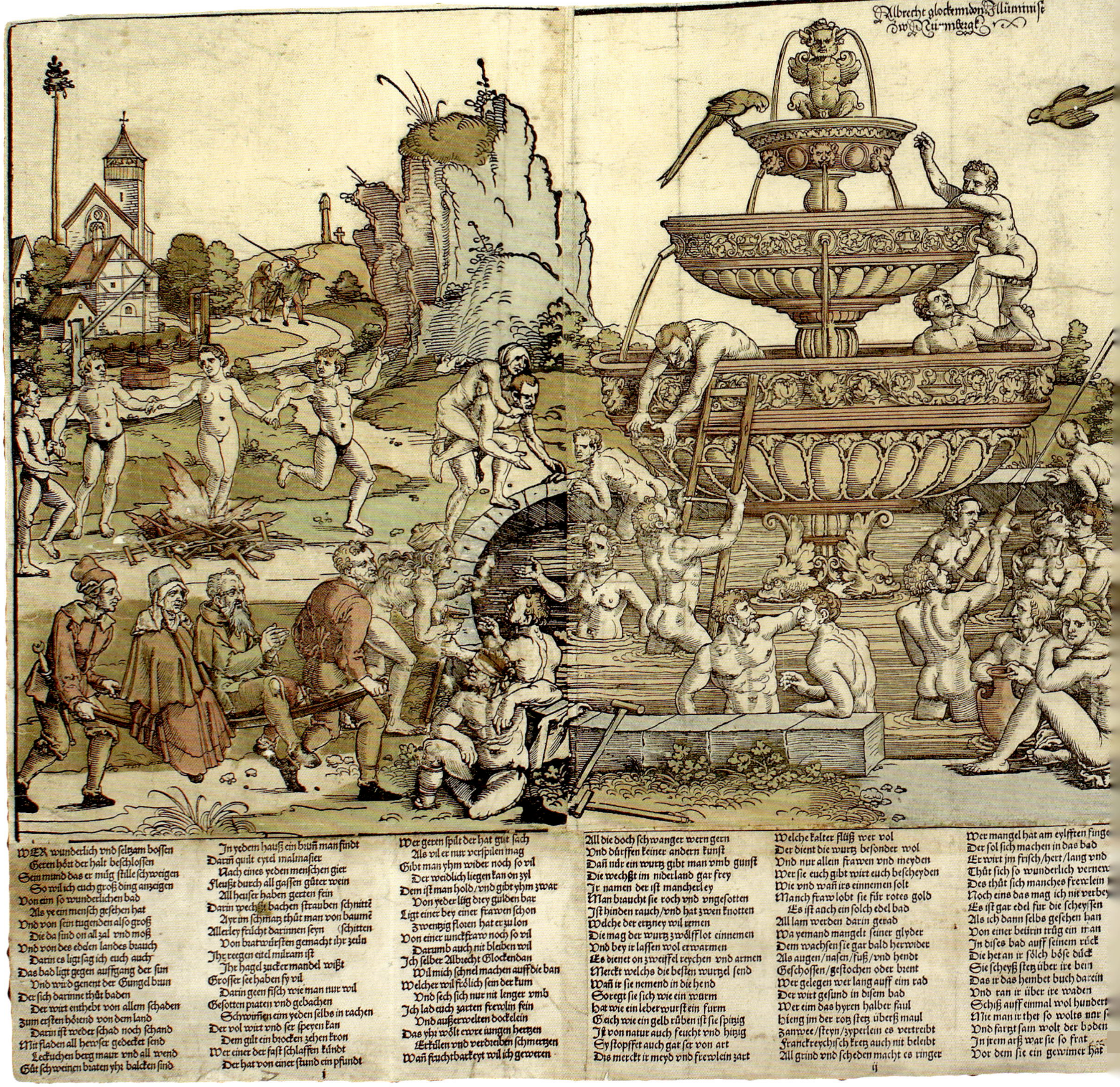

WER wunderlich vnd seltzam bossen
Geren hört der halt beschlossen
Sein mund das er müg stille schweigen
So wil ich euch groß ding anzeigen
Von ein so wunderlichen bad
Als ye ein mensch gesehen hat
Vnd von sein tugenden also groß
Die da sind on all zal vnd moß
Vnd von des edlen landes brauch
Darin es ligt sag ich euch auch
Das bad ligt gegen auffgang der sun
Vnd wird genent der Gungel brun
Der sich darinne thůt baden
Der wirt enthebt von allem schaden
Zum ersten hörend von dem land
Darin ist weder schad noch schand
Mit fladen all hewser gedecket send
Leckuchen berg maur vnd all wend
Gůt schweinen braten yhr balcken sind
In yedem hauß ein brun̄ man findt
Darin̄ quilt eytel malmasier
Nach eines yeden menschen gier
Fleußt durch all gassen gůter wein
All heuser haben gerten fein
Darin wechßt bachen strauben schnitte̅
Ayr im schmatz thůt man von baume̅
Allerley frücht darinnen seyn (schitten
Von bratwürsten gemacht ihr zeun
Ihr reegen eitel milram ist
Ihr hagel zucker mandel wißt
Grosser see haben sy vil
Darin geen fisch wie man nur wil
Gesotten praten vnd gebachen
Schwimmen eim yeden selbs in rachen
Der vol wirt vnd ser speyen kan
Dem gilt ein brocken zehen kron
Wer einer der fast schlaffen kündt
Der hat von einer stund ein pfundt
i
Wer geren spilt der hat gut sach
Als vil er nur verspilen mag
Gibt man yhm wider noch so vil
Der weidlich liegen kan on zyl
Dem ist man hold / vnd gibt yhm zwar
Von yeder lüg drey gulden bar
Ligt einer bey einer frawen schon
Zwentzig floren hat er zu lon
Von einer iunckfraw noch so vil
Darumb auch nit bleiben wil
Ich selber Albrecht Glockendan
Wil mich schnel machen auff die ban
Welcher wil frölich sein der kum
Vnd sech sich nur nit lenger vmb
Ich lad euch zarten frewlin fein
Vnd außerwelten dockelein
Das yhr wölt ewre iungen hertzen
Erkülen vnd verdreiben schmertzen
Wañ fruchtbarkeyt wil ich geweren
All die doch schwanger wern gern
Vnd dürffen keiner andern kunst
Dañ nůr ein wurtz gibt man vmb gunst
Die wechßt im niderland gar frey
Ir namen der ist mancherley
Man braucht sie roch vnd vngesotten
Ist hinden rauch / vnd hat zwen knotten
Welche der ertzney wil remen
Die mag der wurtz zwölff lot einnemen
Vnd bey ir lassen wol erwarmen
Es dienet on zweiffel reychen vnd armen
Merckt welchs die besten wurtzel send
Wañ ir sie nemend in die hend
Soregt sie sich wie ein wurm
Hat wie ein leber wurst ein furm
Gleich wie ein gelb rüben ist sie spitzig
Ist von natur auch feucht vnd hitzig
Sy stopffet auch gar ser von art
Dis merckt ir meyd vnd frewlein zart
Welche kalter flüß wer vol
Der dient die wurtz besonder wol
Vnd nur allein frawen vnd meyden
Wer sie euch gibt wirt euch bescheyden
Wie vnd wañ irs einnemen solt
Manch fraw lobt sie für rotes gold
Es ist auch ein solch edel bad
All lam werden darin gerad
Wa yemand mangelt seiner glyder
Dem wachsen sie gar bald herwider
Als augen / nasen / fuß / vnd hendt
Geschossen / gestochen oder brent
Wer gelegen wer lang auff eim rad
Der wirt gesund in disem bad
Wer eim das hyren halber faul
Hieng im der rotz stetz überß maul
Zanwee / steyn / zyperlein es vertreibt
Franckreychisch kretz auch nit beleibt
All grind vnd scheden macht es ringer
ij
Wer mangel hat am eylfften finge
Der sol sich machen in das bad
Er wirt im frisch / hert / lang vnd
Thůt sich so wunderlich verner
Des thůt sich manches frewlein
Noch eins das mag ich nit verbey
Es ist gar edel fur die scheyssen
Als ich dann selbs gesehen han
Von einer beürin trůg ein man
In dises bad auff seinem rück
Die het an ir sölch böse dück
Sie scheyß stetz über ire bein
Das ir das hembet buch darein
Vnd ran ir über ire waden
Schiß auff einmal wol hundert
Nie man ir thet so wolts nur s
Vnd fartzt sam wolt der boden
In irem arß war sie so frat
Vor dem sie ein gewimer hät

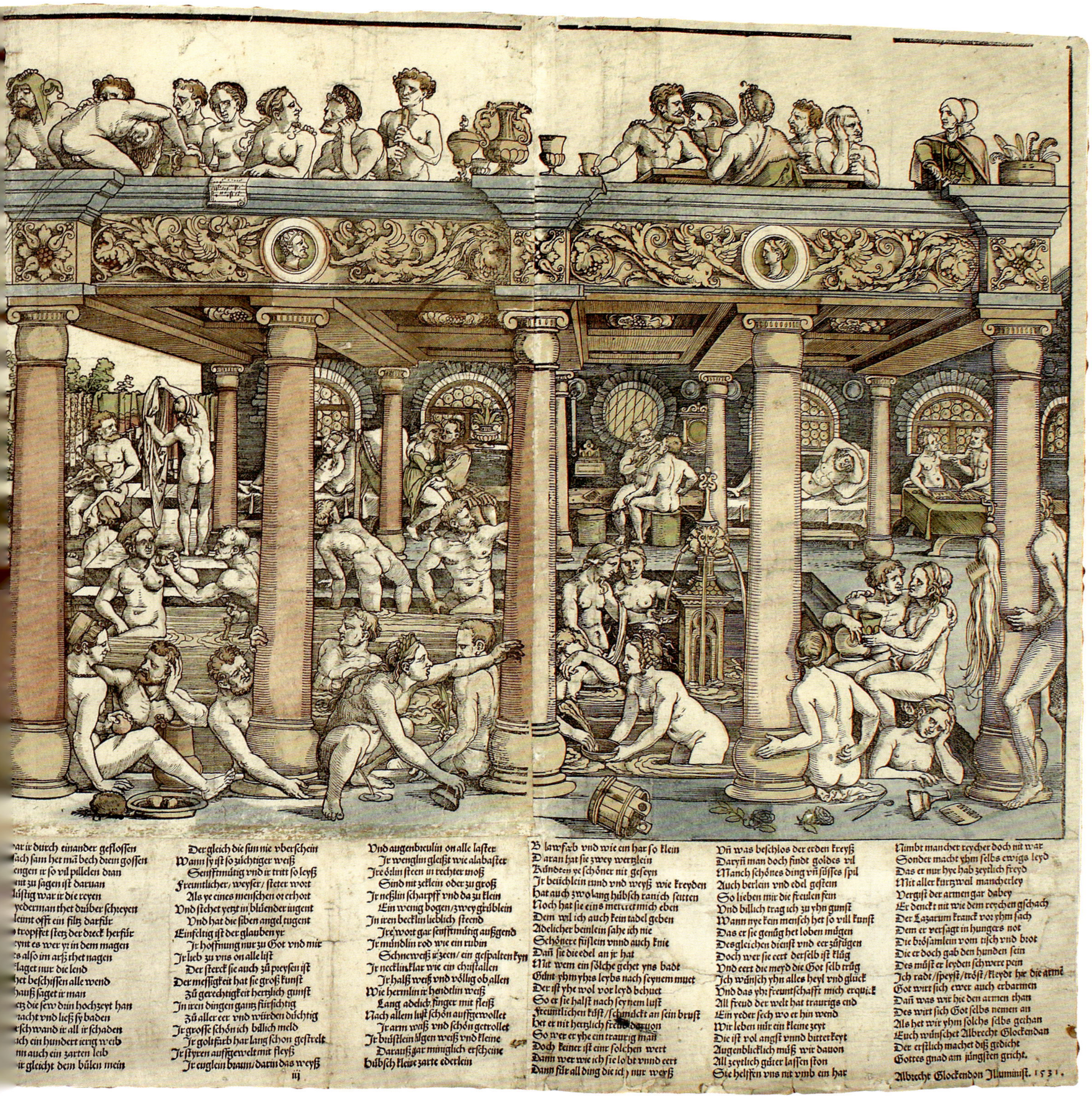

ar ir durch einander geflossen
ach sam het mã bech drein gossen
engen ir so vil pillelen dran
nit zu sagen ist daruan
lüstig war ir die reyen
yederman thet druber schreyen
leimt offt ein filtz darfür
tropffet stetz der dreck herfür
eynt es wer yr in dem magen
s also im arß thet nagen
laget nur die lend
het beschissen alle wend
hauß saget ir man
etz die sew drin hochzeyt han
acht vnd ließ sy baden
schwand ir all ir schaden
ich ein hundert ierig weib
nn auch ein zarten leib
ir gleicht dem bülen mein

Der gleich die sun nie vberschein
Wann sy ist so züchtiger weiß
Senfftmütig vnd ir tritt so leyß
Freuntlicher/ weyser/ steter wort
Als ye eines menschen or erhort
Vnd stehet yetzt in blüender iugent
Vnd hat die siben angel tugent
Einfeltig ist der glauben yr
Jr hoffnung nur zu Got vnd mir
Jr lieb zu vns on alle list
Der sterck sie auch zü preysen ist
Der messigkeit hat sie groß kunst
Zü gerechtigkeit hertzlich gunst
Jn iren dingen gantz fürsichtig
Zü aller eer vnd würden düchtig
Jr grosse schön ich billich meld
Jr goltfarb har lang schon gestrelt
Jr styren auffgewelt mit fleyß
Jr euglein braun/ darin das weyß
iij

Vnd augenbreulin on alle laster
Jr wenglin gleißt wie alabaster
Jre örlin steen in rechter moß
Sind nit zeklein oder zu groß
Jr neßlin scharpff vnd da zu klein
Ein wenig bogen/ zwey grüblein
Jn iren becklin lieblich steend
Jre wort gar senfftmütig außgend
Jr mündlin rod wie ein rubin
Schneweiß ir zeen/ ein gespalten kyn
Jr necklin klar wie ein christallen
Jr halß weiß vnd völlig ob allen
Wie hermlin ir hendtlin weiß
Lang adelich finger mit fleiß
Nach allem lust schön auffgewollet
Jr arm waß vnd schön getrollet
Jr brüstlein ligen weiß vnd kleine
Darauß gar minglich erscheine
hübsch kleine zarte ederlein

Blawfarb vnd wie ein har so klein
Daran hat sie zwey wertzlein
Kündten ye schöner nit geseyn
Jr beüchlein rund vnd weyß wie kreyden
Hat auch zwo lang hübsch ranich seitten
Noch hat sie eins mercket mich eben
Dem wil ich auch kein tadel geben
Adelicher beinlein sahe ich nie
Schönere füßlein vnnd auch knie
Dañ sie die edel an jr hat
Mit wem ein sölche gehet yns badt
Gunt yhm yhrs leybs nach seynem muet
Der ist yhe wol vor leyd behuet
So er sie halst nach seynem lust
Freuntlichen küst/ schmuckt an sein brust
Het er nit hertzlich freüd daruon
So wer er yhe ein traurig mañ
Doch keiner ist eins solchen wert
Dann wer wie ich sie lobt vnnd eert
Dann für all ding die ich nur weyß

Vñ was beschlos der erden kreyß
Daryñ man doch findt goldes vil
Manch schönes ding vñ süsses spil
Auch berlein vnd edel gestein
So lieben mir die freulen fein
Vnd billich trag ich zu yhn gunst
Wann nye kein mensch het so vill kunst
Das er sie genüg het loben mügen
Desgleichen dienst vnd eer züfügen
Doch wer sie eert derselb ist klüg
Vnd eert die meyd die Got selb trüg
Jch wünsch yhn alles heyl vnd glück
Vnd das yhr freuntschafft mich erquick
All freud der welt hat traurigs end
Ein yeder sech wo er hin wend
Wir leben nür ein kleine zeyt
Die ist vol angst vnnd bitterkeyt
Augenblicklich müß wir dauon
All zeytlich güter lassen ston
Sie helffen vns nit vmb ein har

Nimbt mancher reycher doch nit war
Sonder macht yhm selbs ewigs leyd
Das er nur hye hab zeytlich freyd
Mit aller kurtzweil mancherley
Vergist der armen gar dabey
Er denckt nit wie dem reychen gschach
Der Lazarum kranck vor yhm sach
Dem er versagt in hungers not
Die brösamlein vom tisch vnd brot
Die er doch gab den hunden sein
Des müst er leyden schwere pein
Jch radt/ speyst/ tröst/ kleydt hie die armẽ
Got wirt sich ewer auch erbarmen
Dañ was wir hie den armen than
Des wirt sich Got selbs nemen an
Als het wir yhm solchs selbs gethan
Euch wünschet Albrecht Glockendan
Der erstlich machet diß gedicht
Gottes gnad am jüngsten gricht.

Albrecht Glockendon Illuminist. 1531.

22

Brussels Master after Hieronymus Bosch
Garden of Earthly Delights

Before 1560 (after a cartoon of c.1528–1540)
Gold, silver, wool and silk, 292 x 492 cm
Madrid, Patrimonio Nacional, Real Monasterio de San Lorenzo de El Escorial, inv. 10004013 (Tapicería, Serie 36/IV. – inv. A 265-7837)

PROVENANCE

In 1566 the series of tapestries after Bosch hung in the residence of Cardinal Granvelle in Brussels; at his death in 1586 an inheritance - among which the tapestry of *Garden of Earthly Delights* - went to his nephew Jean-Thomas, lord of Maîche in Franche-Comté; in 1600 sold to Emperor Rudolph II; its subsequent history is unclear; between *c.*1610 and 1660 transferred to Madrid; subsequently Valencia de Don Juan in the collection of the Spanish crown since Philip IV (1621-1665); mentioned in the inventory of the Palacio Real in Madrid from 1666 and 1701; in 1694 mentioned in the Pardo among tapestries in need of restoration.

BIBLIOGRAPHY

Kurz 1967: 150, 152-153, 156; Steppe 1967: 28-37; Schneebalg-Perelman 1971: 263-264, 289-290; Vandenbroeck 1981; Junquera de Vega & Herrero Carretero 1986: 263-267; Vandenbroeck 1989: 9-210; Vandenbroeck 1990: 9-192; Volckaert & Delmarcel 1993: 94-95; Delmarcel 1999: 22, 362; Silva Maroto 2000; Vandenbroeck 2001: 87-92; Vandenbroeck 2009: 212-269.

In Flemish tapestry of the sixteenth century, there is one exception to the unwritten rule that all series of tapestries are devoted to a particular theme or narrative, of which diverse aspects or episodes are depicted. This unusual series without any thematic unity whatsoever was woven in Brussels around 1530–1540 after compositions by Hieronymus Bosch. The representations were assembled *on the basis of their common author*. As far as we know, this has never been the case with any other series of tapestries. This attests to Bosch's unique status in the sixteenth century: as an artist, he created a universe that was so idiosyncratic that it could stand alone 'without a story'. That this was also understood as such in the sixteenth century is demonstrated by this series of tapestries. In an inventory of the art collection of Francis I from 1542, which survives in a copy from 1552, they are summed up under the title 'the concepts of Hieronymus'.[1] Thus, the title itself does not indicate a subject, but brings together several of the master's concepts.

The series consisted of five tapestries with heterogenous subjects: the so-called *Garden of Earthly Delights*, the *Elephant*, the *Haywain on the sea of the world*, *Saint Anthony in the wilderness* and *Saint Martin leaving a city*. The tapestry reproduces the open triptych, albeit with each panel in reverse due to the weaving technique.

Although there were at least three versions, only one has survived – the one now in the Spanish Patrimonio Nacional.[2] The earliest document concerning the series of tapestries after Bosch isthe abovementioned inventory of the art

1 'Cinq pièces de tappicerie de divers histoires rehaulsée d'or d'argent et soye des *devys* de Hieronyme'. Schneebalg-Perelman 1971: 253–304 & spec.: 263–264, 289–290.

2 Junquera de Vega & Herrero Carretero 1986: 263–267.

3 Schneebalg-Perelman 1971: 253–304 & spec.: 263–264, 289–290: '5. Une aultre pièce en laquelle est figuré le Monde, l'Enfer et le Paradis terrestre, enchassé en ung tableau contrefaict de menuyserie comme les précéddans, sauf qu'il y a deux coulonnes par voye… la 5e 4 [aulnes] et demy quart sur 2 [aulnes] et demy [= *c.*300 x 495 cm].' These dimensions correspond, within a few centimetres, with those of the series in the Patrimonio Nacional.

4 Delmarcel 1999: 22, 362.

5 Volckaert & Delmarcel 1993: 92–112 & spec.: 94–95.

6 Steppe 1967: 28–37; Junquera de Vega & Herrero Carretero 1986: 263. Ze gaan zo ver om te beweren: 'los últimos trabajos de I. Mateo Gómez y J.K. Steppe sugieren un encargo directo de Felipe II a Jerónimo Bosco.' If Philip II (°1527) commissioned anything from Bosch (d. 1516), it can only have occurred by transcendental means.

7 For the details: Vandenbroeck 2009: 212–269 n. 210.

8 Vandenbroeck 2001: 87–92.

collection of Frans I.[3] Thus, the tapestries were certainly created before 1542. Since the tapestries of the Spanish Patrimonio Nacional are marked with a double B ('Brussels-Brabant'), they had to have been produced *after 1528*, the year in which it became obligatory to mark in this way any tapestries wider than six feet.[4] The paintings are set in Renaissance pilasters or columns in the Italian style, which place the Bosch compositions in a 'modern' frame.[5]

In an essay written on the occasion of the Hieronymus Bosch exhibition in 1967, J. K. Steppe discussed the sixteenth-century history of the tapestries. From this it was evident that Cardinal Granvelle and the Duke of Alva already had such series in their posession.[6] The exemplar of the Patrimonio Nacional is probably that of Granvelle; the cartoon on which it was based was probably *not* painted after the original, but after a copy that belonged to the collection of the Count of Pomereu and which is now in the German art trade. It is to this version that the tapestry shows the greatest number of similarities.[7]

Bosch's renowned counter-utopian painting, which hung in the Brussels palace of Nassau on the Coudenberg from or in 1517, formed the stakes of the 'political' conflict that the Duke of Alva provoked with William of Orange.[8]

PVDB

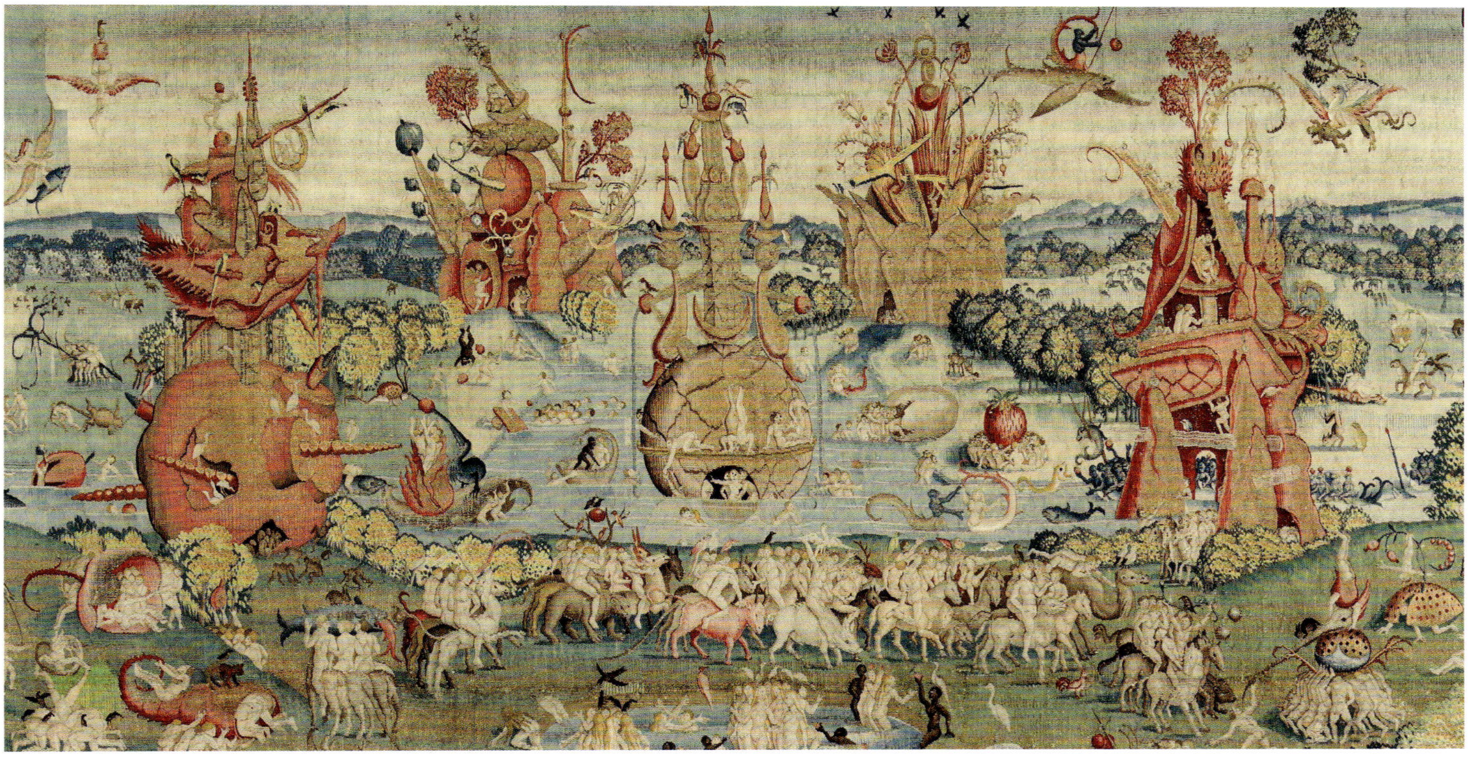

[cat. 22] detail

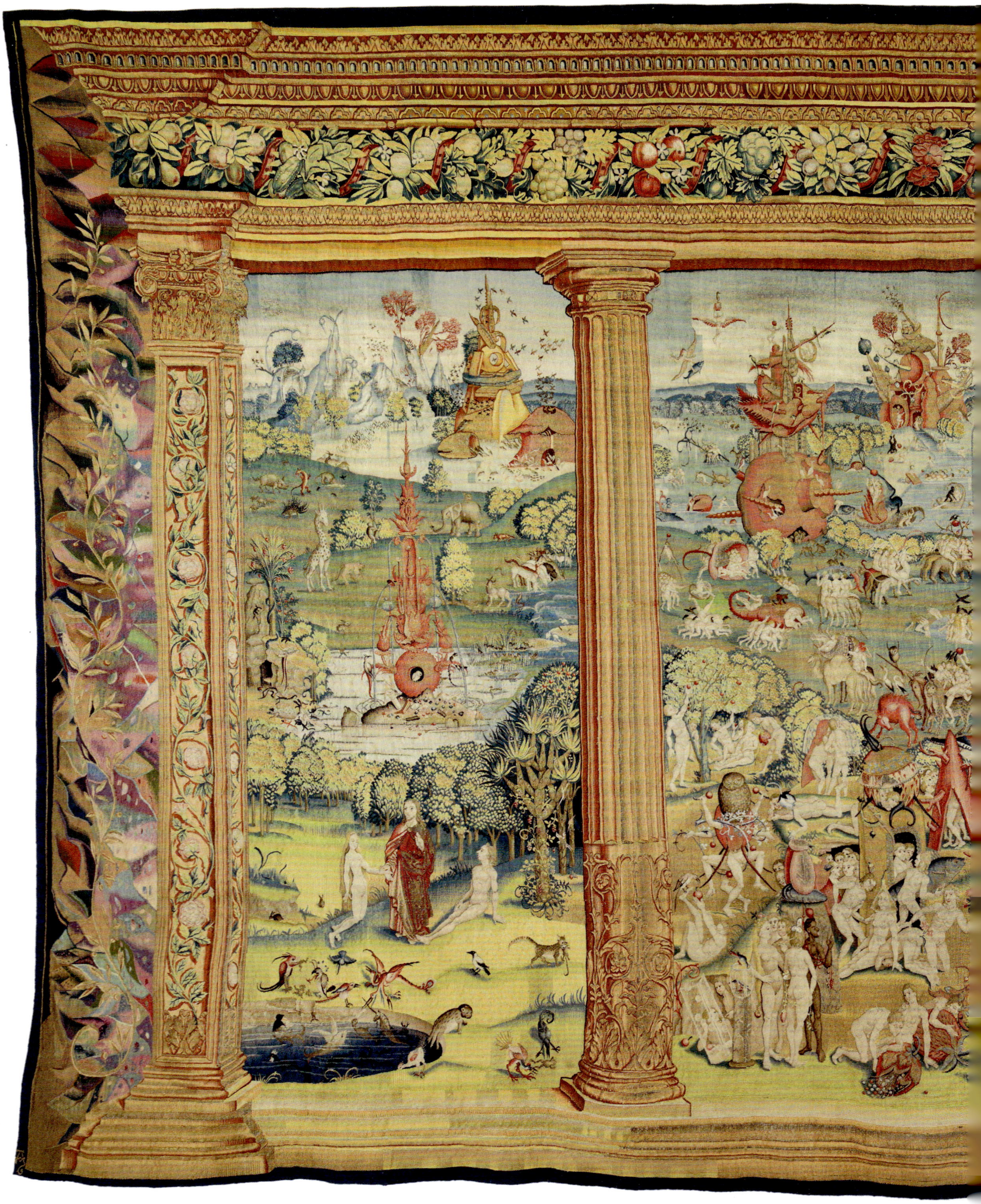

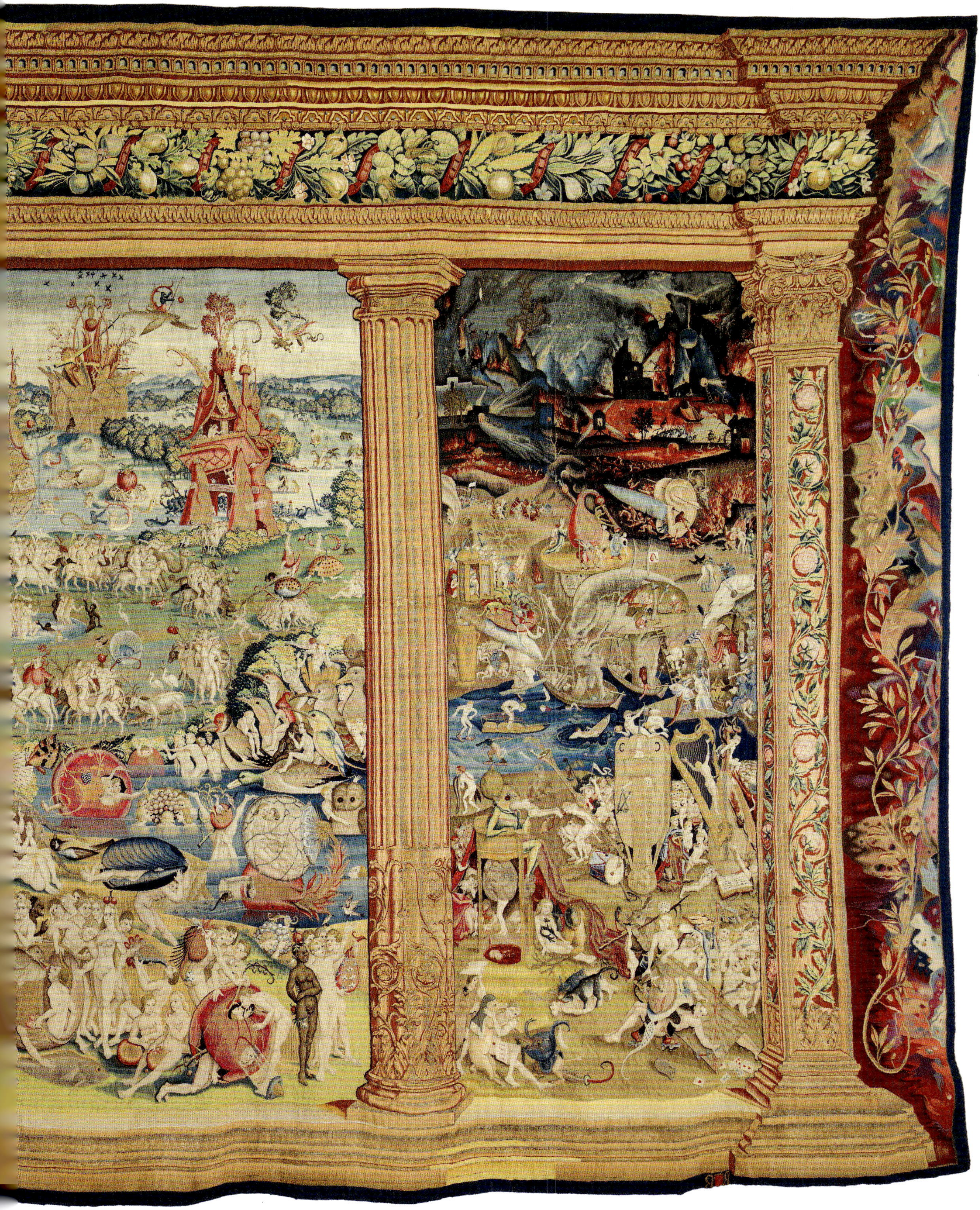

23

Master of the Prayer Books of around 1500
The Lover Entering the Garden of Delights

In: Guillaume de Lorris and Jean de Meung, *Le Roman de la Rose*, Brugge, c.1490–1500, fol. 12v.
Parchment, 395 x 290 mm (sheet), 210 x 195 mm (miniature)
London, The British Library, inv. Harley Ms. 4425

PROVENANCE

Made for Engelbert II, count of Nassau and Vianden (d. 1504); annotated in Spanish, late sixteenth-seventeenth century; in the library of Jean Antoine II de Mesmes, count of Avraux and Brie-Comte-Robert (d. 1723); in 1726 purchased by Edward Harley, second duke of Oxford (d. 1741), at the sale of the library of Valentin Esprit Fléchier, bishop of Nîmes (d. 1710); in 1753 purchased for the British Museum from Harley's widow and daughter; transferred to the British Library in 1973.

BIBLIOGRAPHY

Bourdillon 1906: 12, 28, 75, 98, 149; Winkler 1915: 338-341; Durrieu 1921: 32, 62, pl. 72; Winkler 1925: 128-129, 179; Kelly 1953-1954: no. 609; Tuve 1966: 257, 324; Fleming 1969: 28, 42, 71, 208; Hassall 1970: 343; Wright 1972: 145-146, 152, 238, 358-359, 444; Scott 1980: 196-197, fig. 135-137; Le Loup 1981: no. 87; Kren 1983: 49-58, no. 6; Backhouse 1987: 36; Simpson & Fawcett 1987: 24; Braet 1989: 183-192; Gullick 1990: pl. 23; Martin & Vezin 1990: 282; Arnould & Massing 1993: 207; Backhouse 1993: 8 n.-9 n.; Braet 1994: 110-116; Braet 1996: 491-504; Backhouse 1997: no. 190; Camille 1998: 80-81, 91-92, pl. 68, 78; Jones 1998: 55, pl. III; Korteweg 1998: no. 12; Smeyers 1998: 444, 445, fig. 37; Smeyers 1999: 444, pl. 37; Backhouse 2000: 160; Nascimento & Miranda 2000: no. 68; Antoine 2002: no. 27; Nijs 2002: 1033; Kren & McKendrick 2003: no. 120; Porter 2003: 7; Braesel 2004: 45-50; Wijsman 2006: 94-101; Scott 2007: 161-162, 164, pl. 100; Campbell et al. 2008: no. 60; De Kesel 2010: 126; Wijsman 2010a: 371, 458, 463; De Kesel 2011; Smith 2011: no. 3; De Kesel 2013.

The *Roman de la Rose* was one of the most requested texts of the middle ages.[1] More than 300 manuscripts and countless early impressions bear witness to its popularity between its completion (1269–1278) and the end of manual copying in the sixteenth century. Many European nobles had copies of the Roman made for their libraries, with fine calligraphy and an abundance of splendid illuminations. The most splendid of all is shown here.

Comprising around 20,000 octosyllabic lines of French verse, the *Roman de la Rose* narrates the dream of a young lover, in which the long quest he has undertaken ends when he obtains the much long-for rose. As is now well established, the *Roman* is a composite of two texts composed by different authors at least 40 years apart. Its second and much larger part, produced by Jean de Meung (*c.*1240–before 1305), was appended to a pre-existing text of 4,000 lines. This earlier text – originally attributed to Guillaume de Lorris (thirteenth century), although this is now disputed – is significantly different in character to its sequel. Whereas Jean de Meung's text is didactic, scholarly, pessimistic and clerically

1 See http://romandelarose.org.

2 For these events see De Lorris and De Meun 1995: 39–40, 48–53.

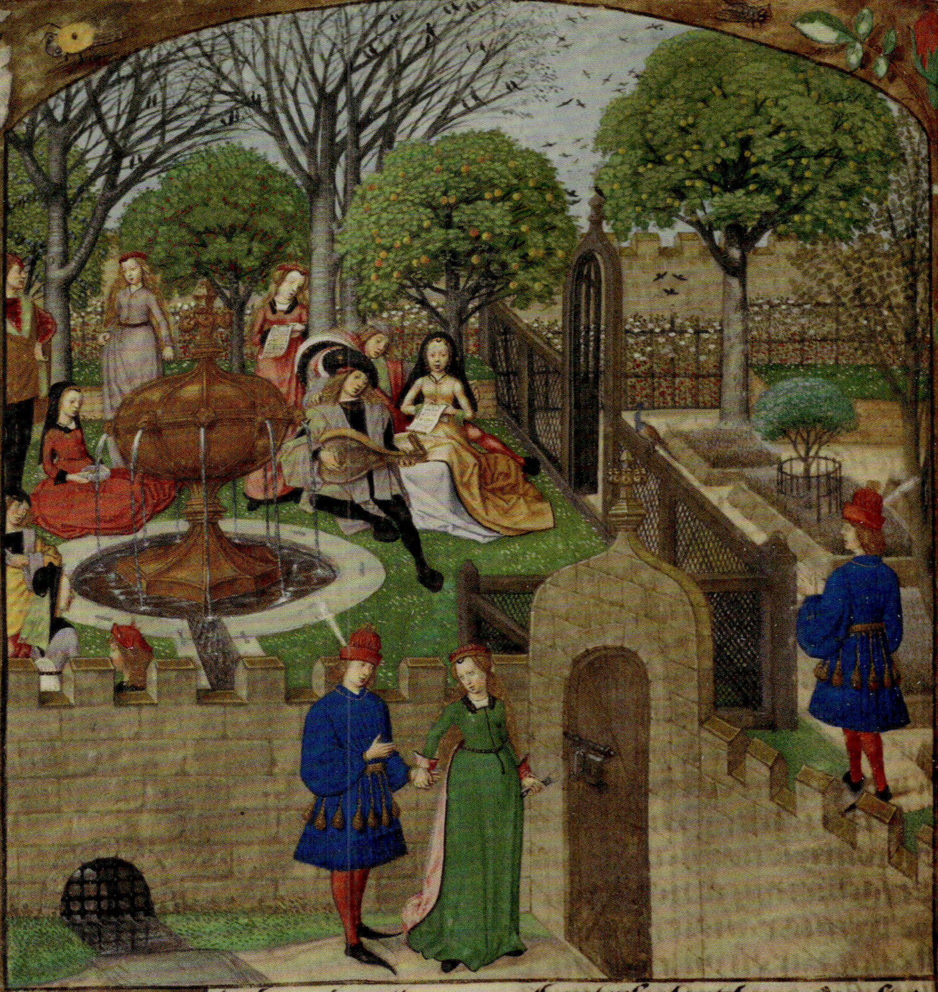

Asses y fery et
hurtay.
Et maintesfois
je escoutay
Se je orroye leans nulle ame
Le huichet qui estoit de charme
Me ouvrit une pucellette
Qui assez estoit gente et nette
Cheveulx eut blons cõe ung bassin
La char plus tendre que ung poussin
Front reluisant sourcilz voultis
Lentreoeil si nestoit pas petis
Ains fut assez grans par mesure
Le nez eut bien fait a droiture
Les yeulx eut vers cõme faulcone
Pour faire envie a tous homme
Doulce alaine eut et savouree
La face blanche et coulouree
La bouche petite et grossette
Et au menton une fossette

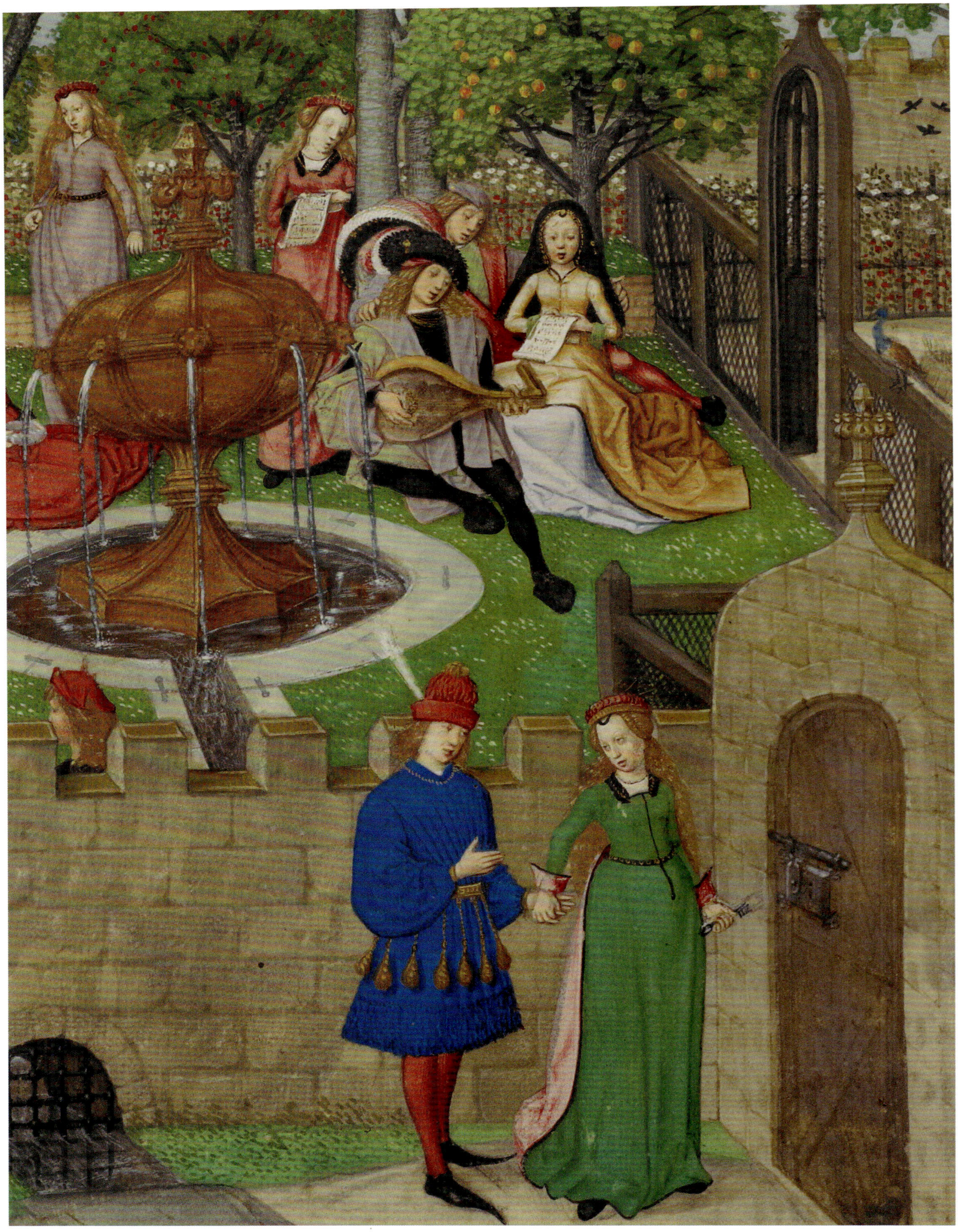

[cat. 23] detail

mysogynist, the first fifth of the *Roman* is lyrical, courtly and redolent of the world of romance.

The artists responsible for the illustration of the present manuscript were particularly responsive to the courtly and poetic aspects of the text. Although their 92 illustrations depend in number, distribution and subject on an earlier sequence devised for a printed edition of the *Roman* issued in Lyon in 1487, they diverge significantly from their model in their painterly and imaginative treatment of the narrative. Most notable are the four large illuminated pages that punctuate the original narrative assigned to Guillaume de Lorris and take us from the beginning of the lover's dream to its apparent original conclusion with the lover barred from the object of his love: the rose.

In one of these illustrations (fol. 12v) the illuminator delights in the depiction of the fabled garden of Pleasure. The lover, dressed in a blue gown, red hat and red tights, is shown being led by a young woman, Idleness (*Oiseuse*), to the door of a walled garden. Distinguished by his costume, the lover is again depicted, this time within the garden. The remainder of the illustration shows the wonders he experiences inside that *locus amoenus*. Here, 'happy, gay and full of joy' the lover imagines himself 'truly in the Earthly Paradise'; indeed 'so delightful was the place that it seemed to belong to the world of the spirit'. Within the garden he sees flocks of birds whose song seems to come from 'heavenly angels'. Passing through a formal garden the lover comes to a separate space where Pleasure and his company of 'beautiful people' like 'winged angels' are engaged in music. Within the garden are fruit trees of remarkable variety and the Fountain of Love where the lover first sees and is entranced by the rose.[2]

SMK

24

Master of Frankfurt
Utopian Gathering of the Antwerp Archery Guilds (the so-called *Festival of the Archers*)

Antwerp
1493
Oil on oak panel, 176 x 141 cm
Antwerp, Koninklijk Museum voor Schone Kunsten, inv. 529

BIBLIOGRAPHY

Friedländer 1917: 139-140, 150, no. 42; Pauwels 1963: 172-173, no. 180; Goddard 1984: 126-129, no. 1; Goddard 1985: 3, 5, 10, 15; Hand & Wolff 1986: 151; Vandamme et al. 1988: 493; Vandenbroeck 1997: chap. 1;[1] Bloom 2002: 106-217; Lowe 2005: 36; Chapmann & Woodall 2010: 70-89.

This extraordinarily complex and unclear representation is an exceptional remnant of late medieval profane painting. A crowd of people enjoy themselves in a luxuriant garden enclosed by a hedge with a gate: it is thus an Enclosed Garden. This popular late medieval motif was used in both religious and profane contexts. It was also performed as a *tableau vivant,* or stage show, at public festivities, and appeared often in the literature of the time.

The Garden was a representation of a protected, heavenly environment where human beings lack for nothing. It could merge with the motif of the 'pleasant place', or *locus amoenus*. Characteristic elements include plucking fruit from heavily laden trees, making merriment, making music, making love, luxuriant plant growth, a lady holding a flower in her hand, outsiders who try to enter. The loving couples make the garden into a garden of love. In the centre stands a young woman who holds a large apple. This fruit is a symbol of love from time immemorial. There is music: one many plays a bagpipe, and further to the left, under a tree, stands a black African who plays a tabor and drum. Two fools in their characteristic fool's cloths dance to the music. We are witness to the performance of a *moresca,* a dance that was highly beloved of aristocratic and urban society in the fifteenth century. At courtly festivities, the *moresca* was inserted into what today would be considered a scurrilous, phantasmagorical world of love castles, wild animals and dragons, goddesses, dwarves and wild men. In middle-class milieus, the Morris dance was primarily understood as an allegory of 'foolishness' of men caused by the power of love or the sex drive (cat. 25).

Measured against the stylistic norms of the time, the *moresca* was wild, outrageous and excessive. The lady who has chosen a man for herself (one of the dancers), could only mean one thing in the eyes of fifteenth- and sixteenth-century burghers: the reversal of gender roles, and hence perversion. A man chose a woman – not vice versa. A lady who made a choice from among a selection of men turned the 'natural' course of things on its head. In this way, the man is reduced to a spineless fool and power relations are disturbed. The artist makes an allusion to this (with a nudge and a wink?): he has painted himself in this utopian garden, while

1 French edition: *Vols d'âmes. Traditions de transe afro-européennes,* with extensive bibliography.

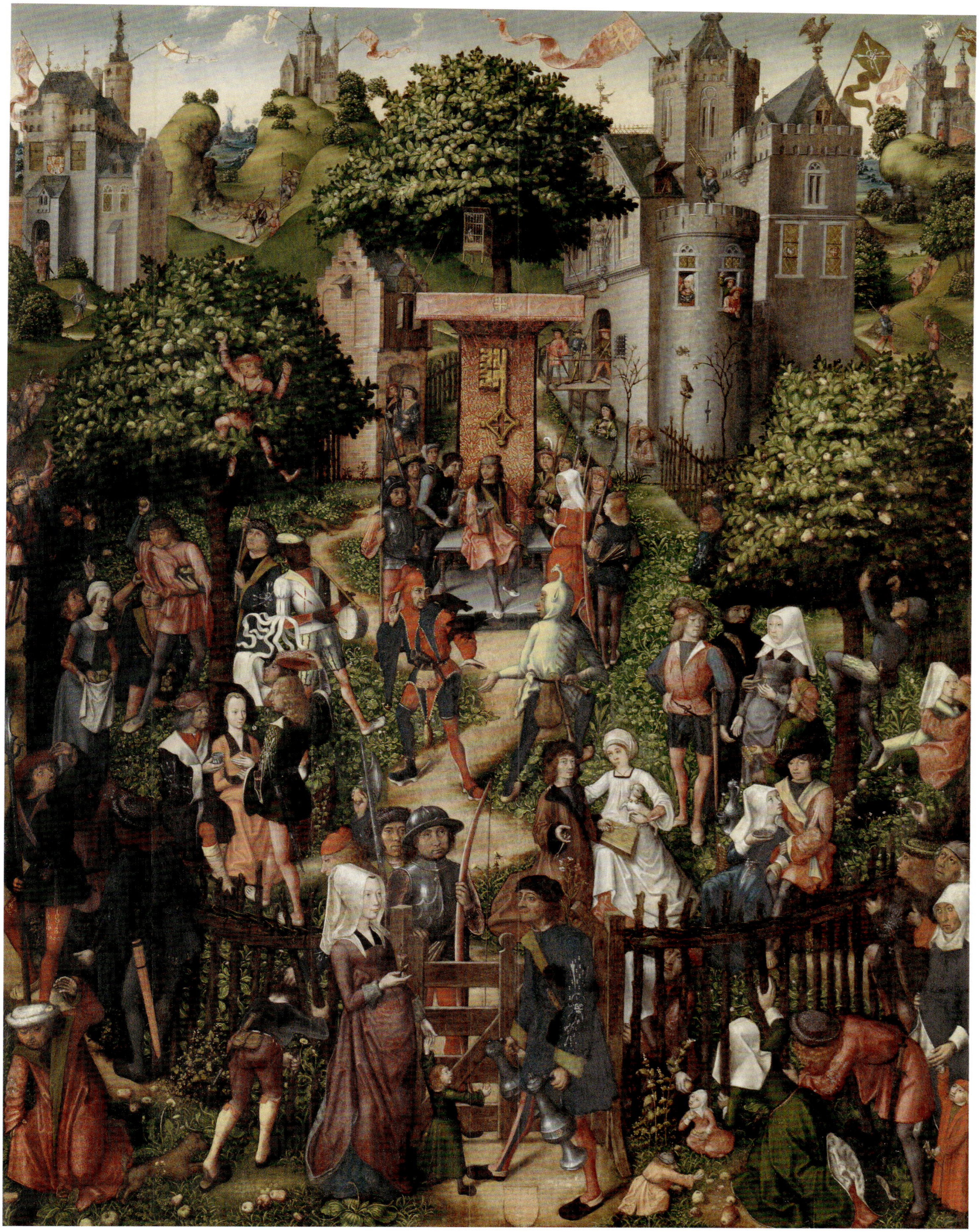

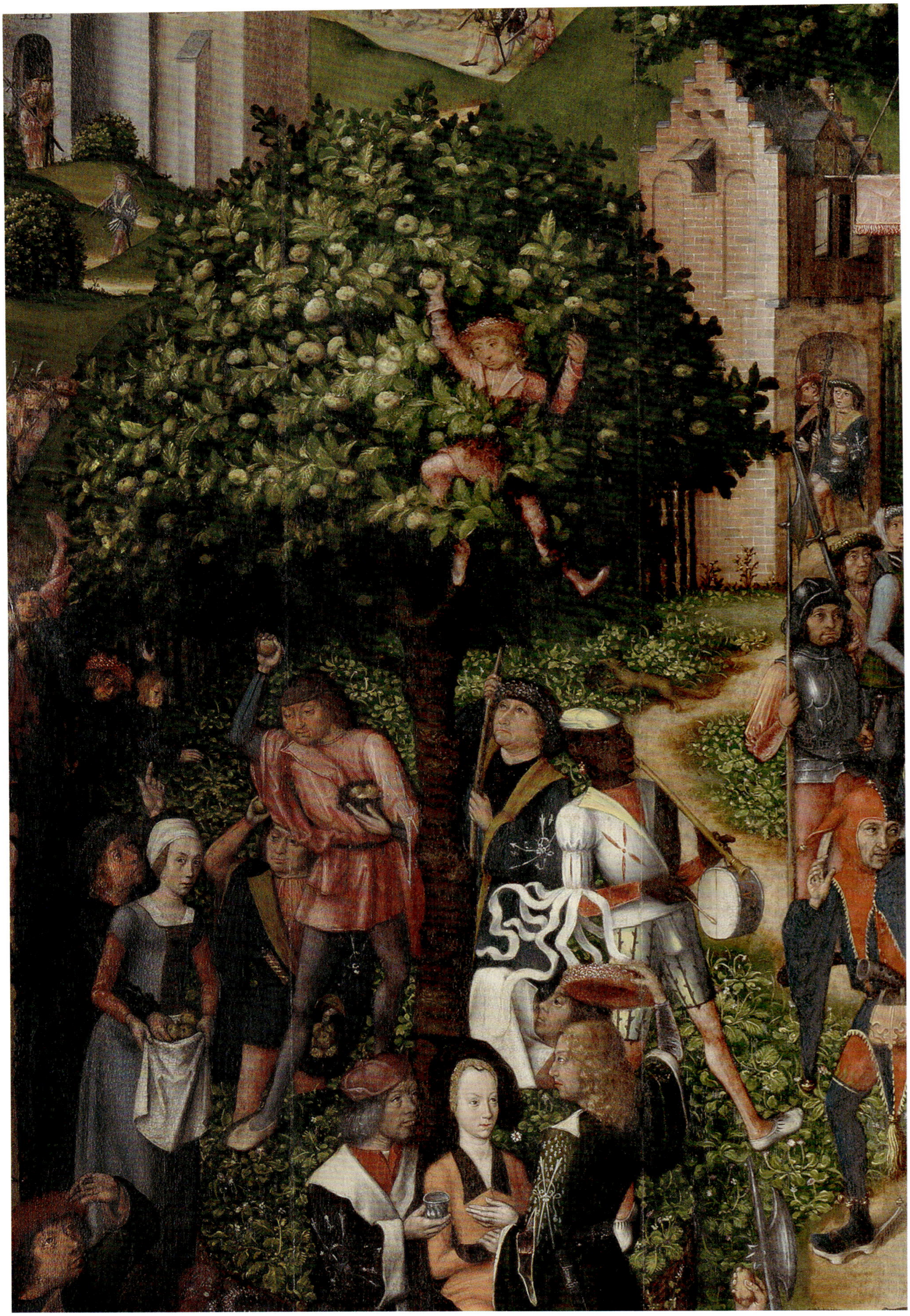

[cat. 24] detail

a beautiful gypsy – her clothing reveals her origins – suggestively drapes her arm over his shoulder as they play a dulcimer together. Making music together was an image of making love.

One could call the *Festival of the Archers* a fantastical *social utopia*: a group of city-dwellers – in this case members of Antwerp's archery guilds – imagine how they might live in harmonious joy and delight in a protected, heavenly environment.

PVDB

25

Jörg Breu the Elder, circle of
Fool's Platter (The Eternal Regeneration of Folly)

Augsburg (?)
1528
Oil on panel, 78.9 cm (diameter)
Vienna, Kunsthistorisches Museum, Kunstkammer, inv. KK 4955.

PROVENANCE

Possibly identical to the 'gar grosse hulzine gemalte schüssel in ir fürstlich durchleucht kunstchamer' mentioned in the collection of Count Hannibal von Hohenems in 1577;[1] afterwards mentioned in the estate inventory of Archduke Ferdinand II of Tirol from 1596;[2] in 1821 the platter is mentioned in the inventory of Schloss Ambras.

BIBLIOGRAPHY

Kenner 1893: 167, no. 10670; Kunsthistorisches Museum 1977: 117; Mezger 1979: 161-180; Vandenbroeck 1987: 40, 324; Mezger 1991: 64, 176-178, 325-329, 379-380, 382, 394-396, 484; Porteman 1995: 194, fig. 12; Vandenbroeck 2002: 40-42, 290-291; Seipel 2008: no. 85, 192; Mezger 2009: 178-212.

The humanist preoccupation with wisdom and reason grew in proportion to the cultural popularization of their opposite: folly. The figure of the Fool thus became a prominent theme in early modern literature, visual arts and festival culture. Among the most well known texts are the *Ship of fools* by Sebastiaan Brant (1494) and the *Narrenbeschwöring* (Exorcism of fools) by Thomas Murner (1512), in which folly is inextricably tied to sinfulness and evil from an intellectually driven ethical standpoint.[3] The *Platter of fools* from Schloss Ambras, shown here, should also be understood within the context of the early modern cultural concept of 'fool-being'. Central in this respect is the eternal regeneration of folly and the senselessness of trying to purge it from the fool.[4]

In the middle of the wooden platter the Mother of all Fools stands in an Enclosed Garden.[5] Various fools dance and make music as they circle around her. One of them is hatching eggs, from which new fools will be born. The Enclosed Garden was a popular motif in both religious and secular iconographic programmes and usually alludes to a paradisiacal environment where life takes its course peacefully and virtuously.[6] In this way, the artist may have wanted to create a pseudo-paradise, in which corrupt folly prevails over virtuous reason. This artistic strategy of social and religious inversion is also characteristic of Bosch's *Garden of Earthly Delights* (cat. 22).

The centre of the composition is taken up by a water mill in which fools are ground. To the right, a fool's head is being sharpened on a grinding stone. These 'fool mills' probably allude to the eternal return of folly. In Christianity, the symbol of the mill was associated with death and resurrection since time immemorial.[7] Interestingly, the central compositional scheme is borrowed from a drawing of *The Children of Luna* after Jörg Breu the Elder (*c.*1480–1537; fig. 25.1). The symbolism of the children of the planets is here translated ingeniously as the philosophy of fools.

1 Kenner 1893: 167, no. 10670.

2 Mezger 1979: 162, n. 7.

3 Vandenbroeck 2002: 20–44; Mezger 1991; Hess 1971.

4 Mezger 1979: 161–180; Vandenbroeck 2002: 40–42, 291; Mezger 2009: 178–212.

5 For the motif of the mother of fools, see Mezger 1991: 324–338; it is unclear whether the inscription 'Bertel Kesselschmid' at the lower right of the garden can be identified as the signature of the (as yet unknown) artist. Mezger 1979: 177.

6 See the essay by Barbara Baert in this catalogue (pp. 49–53).

7 Vandenbroeck 2002: 41; Mezger 1991: 386–402.

8 Mezger 2009: 196–200; on the drawing, see also Butts & Hendrix 2000: 206, no. 80.

9 Vandenbroeck rightly notes, however, that the *Platter of Fools* came into existence independently of Bosch's *Cutting the Stone*. Vandenbroeck 2002: 41; this in contrast to Mezger 1979: 171, who sees a direct connection between the scene on the platter and Bosch's formal language.

10 Mezger 1991: 374–386.

11 Vandenbroeck 2002: 42.

12 Cf. Porteman 1995: 194, fig. 12.

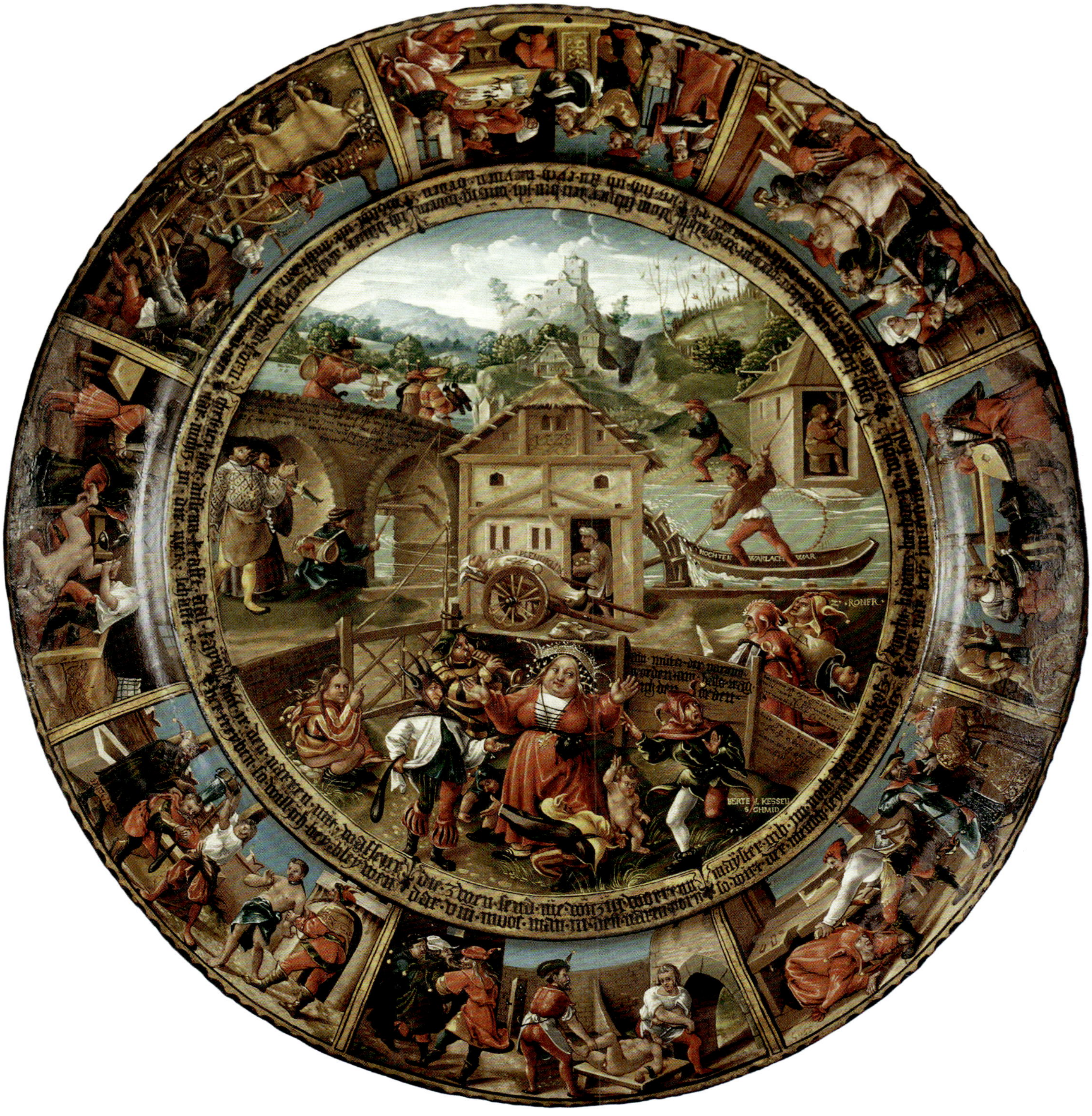

WARLACH
RONER
KESSEL
GHMID

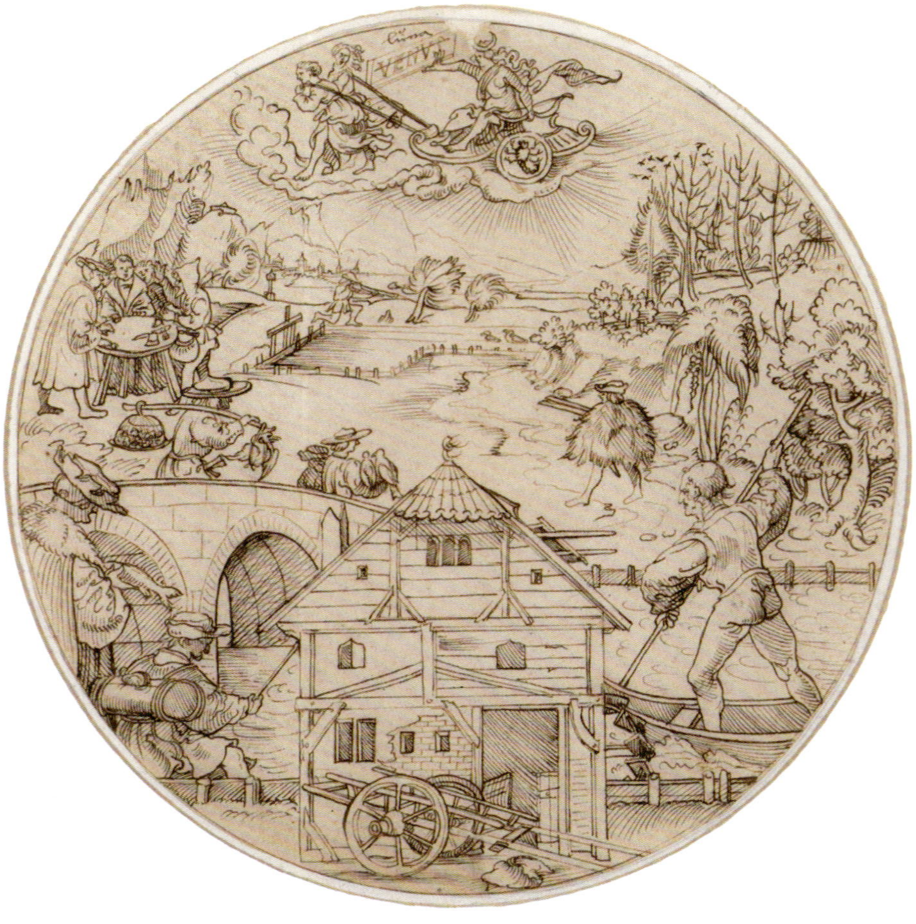

[fig. 25.1] After Jörg Breu the Elder, *The Children of Luna*, c.1512–1516, drawing, Weimar, Kunstsammlungen zu Weimar, Schlossmuseum.

Those who were influenced by the moon were held to be fickle and inconstant – subject to *lunacy* or *moonstruck* behaviour. Fishermen and sailors (both depicted on the platter), among others, were believed to be governed by this heavenly body.[8]

Around the central scene are eight vignettes that likewise deal with the endlessness of 'fool-being'. In seven different ways, people try to cure the fools of their folly: it is excised with a knife (Bosch's *Cutting the Stone*!), bound with a barrel hoop, smothered with a bellows, sanded away with a plane, removed with a drill, drowned with water and even purged with a clyster.[9] These healing processes, however, are futile: in the eighth and final scene, countless new fools are sown. Folly is like an ineradicable weed that germinates again and again.[10] Sometimes the 'doctors' themselves are dressed as fools, which underscores all the more the senselessness of their labours.[11]

The platter format of the painting is striking. It may be a reference to the *lanx satura* (the etymological origin of the word *satire*), a platter filled with a variety of sweetmeats that the fool presents to his public.[12] In this way, not only the satirical imagery but also the bearer thereof is a visual mockery of human folly.

DVH

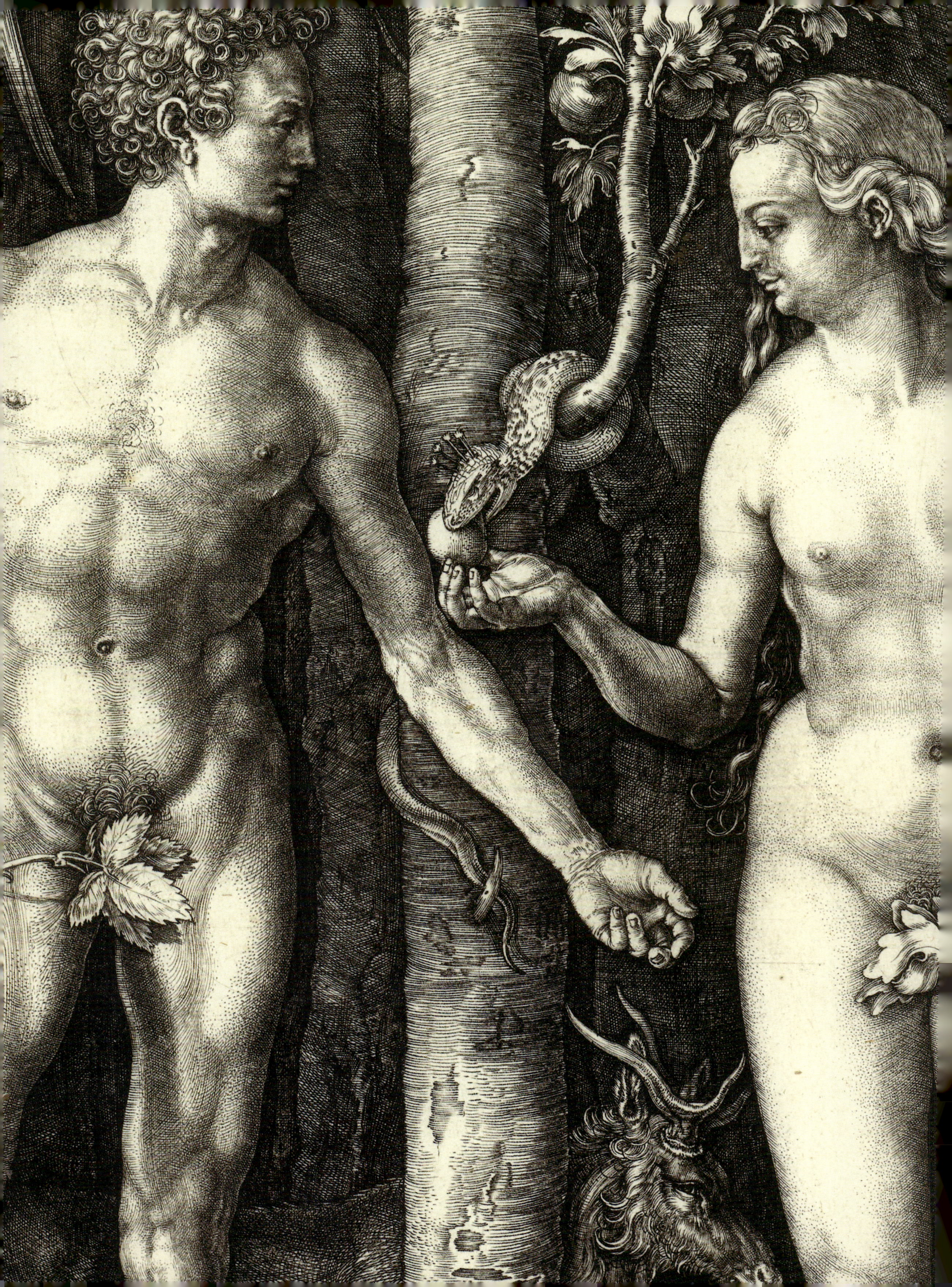

The Garden of Eden – The Earthly Paradise

WENDY WAUTERS

[CAT. 30] DETAIL

In the beginning there was chaos and darkness, a non-place (cat. 70) out of which the first utopian place would grow: the Garden of Eden (cat. 27 and 29). The character of this place is captured in the etymology of its name - Eden, derived from Semitic *'dn* - which means abundance, luxuriance. Over the centuries its paradisiacal beauty was celebrated in literature and art, in particular to emphasise the contrast with the decline of the world after original sin. In this contradiction lies the theology of Christianity.

The root of evil was already germinating in Eden. Traces of it were embedded in the nature of Nature and the essence of man. Well-known *topoi* of evil and decline are the serpent (cat. 28 and 30), the apple (cat. 31) and - centrally depicted in Bosch's *Garden of Eartly Delights* (cat. 22) - the owl. Even in this ideal society, created by God, evil was already inherently present and formed the occasion for the failure of life in Paradise. The utopian world was destined from the start to be transformed into a dystopia. In an endless struggle to recreate this paradise, humans set in motion a pendulum between utopia and dystopia. These two poles are mutually dependent, and rather than being absolute opposites, they fuel each other's existence.

According to late medieval theologians, the symbolic reading of the story of Genesis could be read in three ways: allegorically, morally or eschatologically. These variations were considered complementary to a realistic worldview. The Garden of Eden was after all more than an imaginary place - it was a historical and topographic reality. Augustine identified the four rivers of Paradise not only as an allegory of the four cardinal virtues, but also as actually existing rivers. Even though the precise location of paradise on earth was hotly debated by late medieval theologians and writers, there was clearly a consensus that this place could still be found somewhere on earth. Unfortunately it was heavily guarded and inaccessible to mortals. Its preferred location shifted through the centuries. During the great age of exploration in the fifteenth and sixteenth centuries, it shifted from the East in the direction of the New World. The desire to rediscover Paradise, along with the urge to find a utopian society in times of social and political crisis, undoubtedly formed one of the driving forces behind the expeditions. Various literary offshoots of the Genesis story originated in this period. Out of a sense of nostalgia and yearning, cartographers regularly indicated the Garden of Eden on their maps. The search for this Paradise continued until the end of the eighteenth century. It was only abandoned with the discovery of fossils and the realisation that the idea of an Earthly Paradise as the beginning of human existence was no longer tenable.

26

Enclosed Garden with Earthly Paradise

Southern Netherlands

c.1500

Mixed media, 90 x 104.5 x 16 cm

Antwerp, Koninklijk Museum voor Schone Kunsten, inv. 5094

PROVENANCE

Before 1929 in the possession of an inhabitant of the Beguinage of Kortrijk; in 1929 purchased by family of Eugeen Van Herck; in 1971 purchased by the Koninklijk Museum voor Schone Kunsten Antwerp from Dr. J. Van Herck, honorary deacon in Kontich.

BIBLIOGRAPHY

Antwerp 1948: 50, no. 355; Jansen & Herck 1948: 77, no. 228; Brussels/Mechelen 1954: no. 84; Bruges 1960: 208-209, no. 105; Mechelen 1961: 182-183, no. 839; Vandamme 1981: 143-149; Eichberger 2002: 395-399; Huvenne 2003: 58; Geelen & Steyaert 2011: 222-225; Ruby 2011: 110-118; Van de Velde & Van Hout 2013: 12; Vandenbroeck 2014: 85-87.

Both the Fall and Salvation are the basic ideas behind the iconographic programme of the Enclosed Garden. The central part of the shrine consists of several groups of carved figures that are not all related. At the lower left, Adam and Eve are driven from the Garden of Eden by the archangel Michael with upraised sword. [1] In the centre is an image of St Agnes, possibly the patron saint of an early owner. At the right are images of John the Evangelist, who is related to the Madonna in the nimbus above. In this representation, Mary is identified with the woman of the apocalypse in the Revelations of John.[2] This woman symbolizes the Salvation of mankind at the end of times. In addition, Mary is likewise associated with the verse *Fair as the moon, bright as the sun* (Song of Solomon 6: 10), in which the chastity of the virgin is praised. This Madonna in a nimbus is also surrounded by a wreath of silk flowers. In this way the theme of the *Marianum* merges with that of the Madonna of the Rosary, a symbol of eternal glory and dominion. Later, the rosary was associated with objects of daily use such as the paternoster.[3]

The figurines are surrounded by two portals that suggest a passage that begins with the Fall and ends with Salvation. This notion is confirmed by the inscription below:

Die tyt is cort, die doot is snel,
Wacht U van Sonde(n) so doet ghi wel.
Och wat vroegde(n) daer wese(n) mach,
Daer duse(n)t jaer is ene(n) dach.[4]

The group of figurines stands before a background of silk flowers and plants. In the corners above are two packets containing relics. The human remains, wrapped in silk, are accompanied by an *authentica*, an inscription that identifies the contents of the packet. The one on the right reads: *daer Sinte Joris ghedoot wort* (there where

1
From photographs it appears that this Enclosed Garden was dismantled in 1971, and when it was reassembled, several elements had changed place: Adam and Eve were originally reversed. Under the tree of good and evil is a bust of a warrior with a coat of arms, the meaning and function of which in relation to the other figurines is unclear.

2
Smeyers 1994: 283–285. 'A great portent appeared in Heaven: a woman clothed with the sun, with the moon under her feet, and on her head a crown of twelve stars.' (Revelation 12: 1–2).

3
From old photos (1960) it appears that the Madonna wore a paternoster around her neck.

4
In translation: 'Time is fleet, death is swift, you do well to turn away from sin, o what a joy to be here, where a thousand years is (as) a day'. The inscription is made up of two rhymed proverbs that have been combined. The first, 'Die tyt is cort, die doot is snel, Wacht U van Sonde so doet ghi wel', also occurs in the *Rethoricale wercken* of Anthonis de Roovere (1430–1482). This same combination also appears on a pair of iron hinges on a fifteenth-century charter cabinet from Ypres. Mulder 2004: 47–51, although there it is a hundred years and not a thousand.

5
On earth and stone as contact relics, see Rudy 2011: 107–118.

6
Geelen & Steyaert 2008: 222.

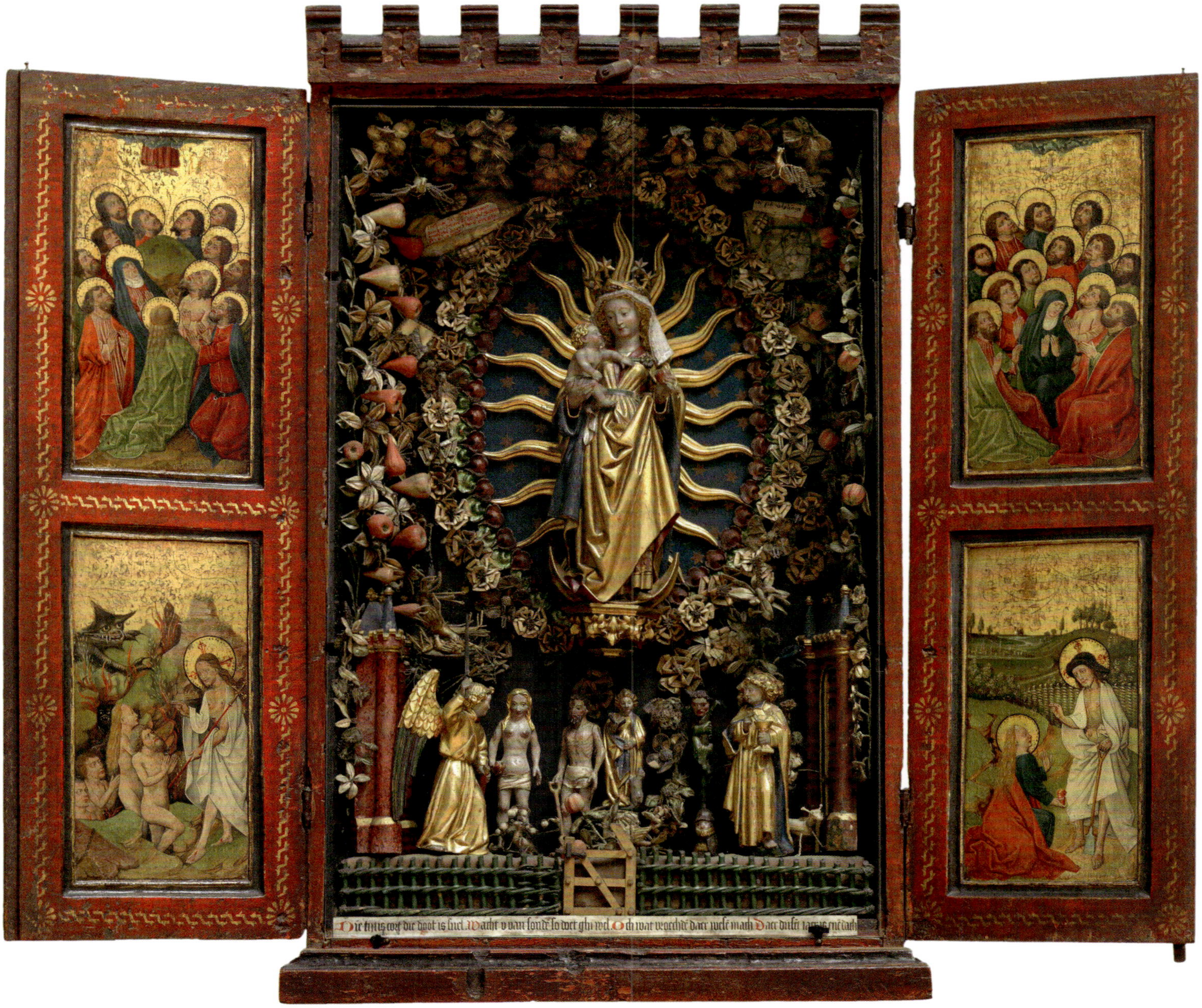
Die tijt is cort die doot is snel. Wacht u van sonde so doet ghi wel.

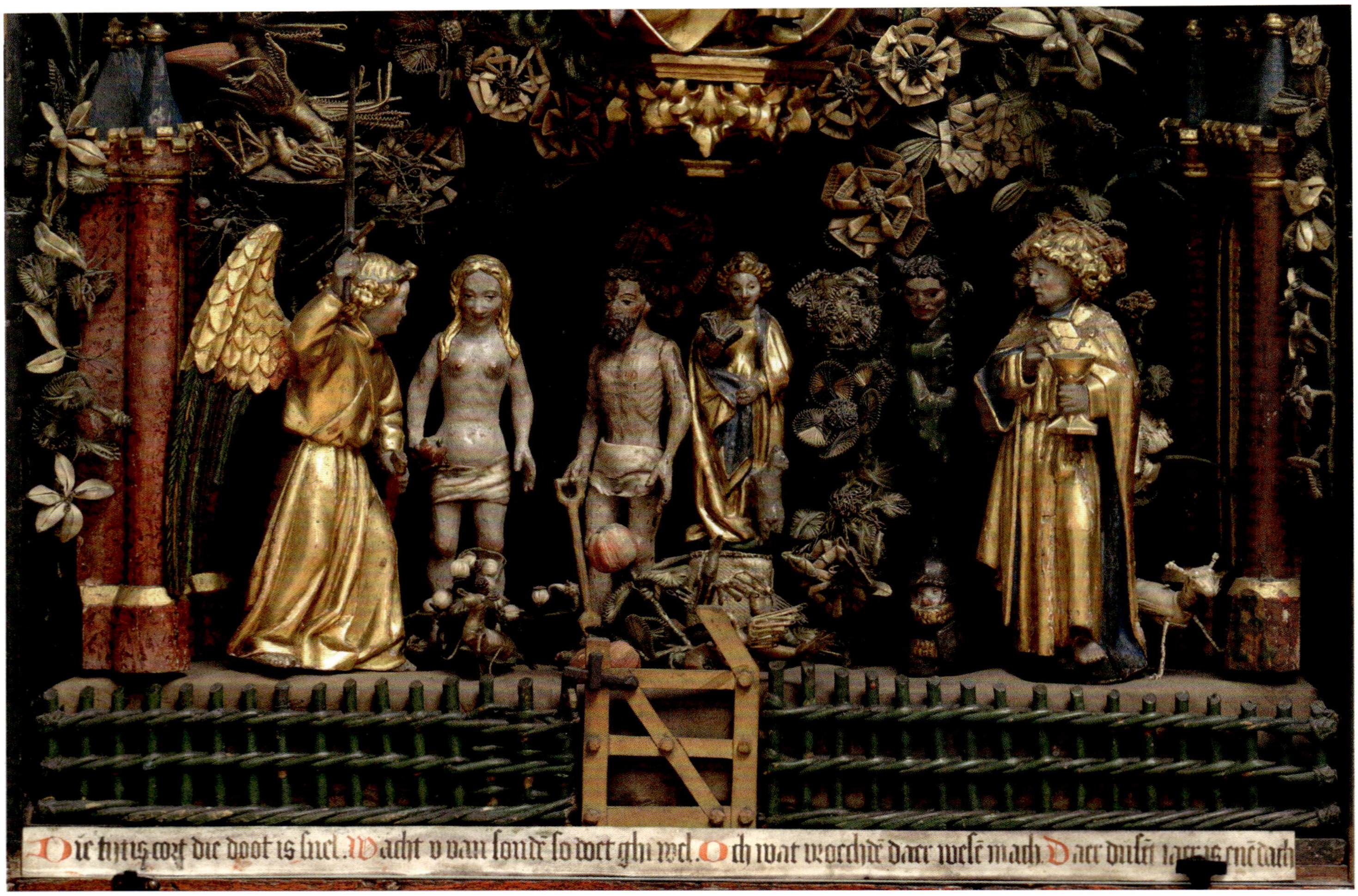

[cat. 26] detail

Saint Joris – that is, George – was killed) referring to the Promised Land, the place where St George met his death. Hence, the packets contain not only bone fragments of saints, but also stones and earth from the Holy Land.[5] The whole is surrounded by a green hedge referring to the Enclosed Garden of the Song of Solomon. Traditionally, the place where this retable was produced, given its typology, would be situated in the cloisters of Mechelen. Yet this Enclosed Garden differs markedly in both extent and composition from known Mechelen exemplars (cat. 32, 33, 34). Moreover, its figurines are very different from the typical *poupées de Malines* and none of them bear the city's seal.[6] Not only the figurines but also the wings of the shrine have a different origin. On the left wing, Christ's ascension into Heaven and his descent into Limbo are depicted; on the right, the descent of the Holy Spirit and Christ's appearance to Mary Magdalene. These wings have a pronounced German character. The case may have had another function originally and was only later recuperated as housing for an Enclosed Garden.

HI

27

Simon Marmion
The Garden of Eden

In: Jacquemart Pilavaine (scriptor), *Le livre des sept Ages du Monde,* Mons and Valenciennes (?), c.1460
Parchment, 435 x 305 mm
Brussels, Royal Library of Belgium, inv. Ms. 9047

PROVENANCE

De Croÿ family.

BIBLIOGRAPHY

Bayot 1928: no. 5; Delaissé 1959: 59-60, no. 49; Prevenier & Blockmans 1983: pl. 6, 18; Lyna & Pantens 1984-1989; Dogaer 1987: 51-55; Van den Bergen-Pantens 1989: no. 1, fig. f. 1v; vol. 3.1, 208-210, no. 299 and vol. 3.2 446, no. 299; Debae 1995: 123-125, no. 82; Lachièze-Rey & Luminet 1998: 135; Cardon, Van der Stock & Vanwijnsberghe 2002: 476, 482-483, 501, 645, 655, 657, 804; Kren & McKendrick 2003: 46-47; Wijsman 2010a: 322; Wijsman 2010b: 112; Bousmanne & Delcourt 2011: 401, no. 113.

The splendid paradisiacal miniature – attributed to Simon Marmion – appears in the only known exemplar of an illuminated world chronicle *Les Sept ages du Monde*, written by an unknown French author. The text is partly based on St Augustine's *Sex aetates mundi* (c. 400). It begins with the story of Creation and is filled in up to the year 1326. The manuscript was written in Mons (Hainaut) in the workshop of the renowned copyist Jacquemart Pilavaine for the De Croÿ family. The illumination is by the particularly talented miniaturist Simon Marmion (1425–1489), who came from a family of artists from Amiens. He was active at the Burgundian court and his work – paintings on panel as well as miniatures – exercised considerable influence on later generations of artists in the Low Countries.

The miniature shows Adam and Eve in a paradisiacal environment. Marmion reinforces the idea of paradise on earth as an endless garden and has placed the two figures in a broadly spreading landscape. The garden is an idyllic place: a hart drinks water from a river; the landscape unfolds in delicate shades of green that transition into atmospheric nuances of blue. Marmion manages the light and space of the landscape with consummate virtuosity. Above, God the Father gives his blessing, surrounded by choirs of angels in vibrant shades of red and orange. Around them are the spheres of Creation. Near god is the firmament of stars, along with the planets, the sun, and close to the horizon the moon. The different days of Creation as related in Genesis are visualised in a broad, panoramically conceived, realistic miniature.

The two nude figures of Adam and Eve are strikingly depicted: they are slender, their curves indicated by fine hatching, and they are painted in soft, transparent shades of brown and pink. The two nude bodies suggest a figure type that Marmion introduced in the third quarter of the fifteenth century. It can also be seen in the depiction of the nude Tondal in the apocalyptic story 'The Visions of the Knight Tondal' (*Les visions du chevalier Tondal*), illuminated in 1475 (cat. 37).[1]

1 Kren & McKendrick 2003: 47.

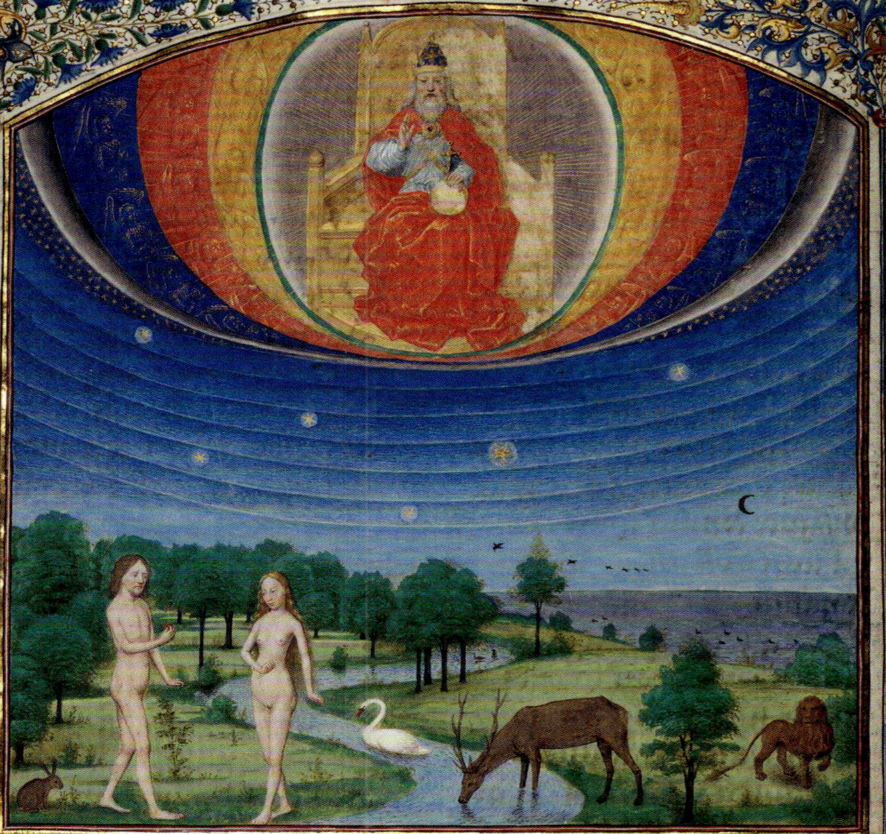

Dieu le souverain
cōmencement et
lumiere parfaite.
en quoy toutes
choses visibles et
invisibles passees presentes et ad
venir manifestement et aperte
ment apperent. Et devez savoir
que du mouvement et verbera
tion de celluy pur air et pure lu
miere naist la parolle dieu. car
parolle nest aultre chose que voix
fourmee par verberation et par
mouvement dair. Et sachiez q̄
dieu est sa parolle et sa parolle est
dieu. Et est ceste parolle complie
et parfaite par iii. souverainnes
vertuttes sans lesquelles vertuttes
la parolle dieu cest a dire la deite
ne puet estre acōplie ne parfaite
entierement. Et po͛ ce dist pla
tons que ces iii. vertuttes sont un
souverain bien principal le ql͂
est cōmencement de tous aultr̄s
biens et de qui toutes choses prē
dent cōmenchement et perfection

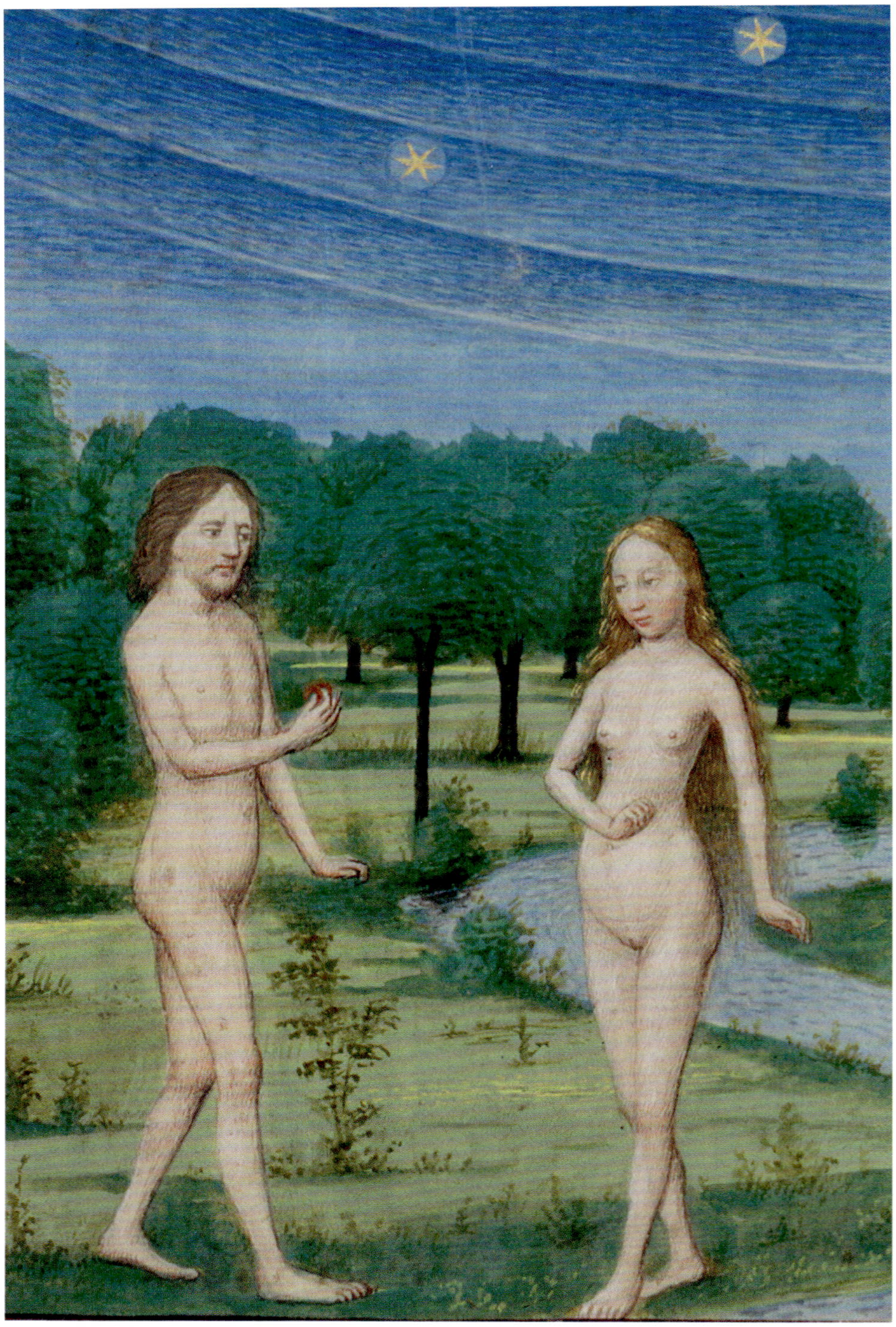

[cat. 27] detail

Marmion's influence on Hieronymus Bosch is notable: the depiction of the nude Adam and Eve on the left panel of the *Garden of Earthly Delights* – painted around 1480–1490 – shows the same physiognomy and a highly similar transparent palette of colours. The challenge of integrating the monumental and the pure in this miniature, the utopian Garden of Eden before the Fall, acquires a new dimension in Marmion's hands.

LW

28

Aert van den Bossche (?)
Adam and Eve before the Fall

Brussels
c.1490
Oil on oak panel, 47.5 x 33 cm
Ludwigshafen, Wilhelm-Hack-Museum, inv. 457/42

BIBLIOGRAPHY

Clemen 1899: 70, pl. VII; Firmenich-Richartz, Clemen & Hartmann 1904: 65, no. 146; Friedländer 1926: 134, no. 36.b; Winkler 1964b: 43 n. 4; Friedländer 1969a: 75, no. 36b, pl. 43; Wilhelm-Hack-Museum 1979: 6-10, 184-187, no. 60; Demus 1981: 192; Scherer 2001: 363-370; Gombert 2005: 32-33, fig. 24; Scherer 2007: 31-44; Steyaert 2007: 180; Steyaert 2013: 258-259, no. 55.

In a densely wooded landscape, Adam and Eve stand before the Tree of Good and Evil at the moment just before the Fall. From the distance, behind a tree, the 'serpent' looks on. In a few moments it will convince Eve to pluck the apple. In the foreground, Adam tries with a gesture, doomed to failure, to prevent his companion from carrying out the deed. The serpent is depicted here as a lizard that stands erect, with four legs and a woman's head (*virgineum vultum*). The serpent standing erect already occurs in Jewish and Christian literature from the first century. It is only later in the story that God condemns this being to crawl upon its belly. The addition of a woman's head is of later but more obscure origins. The serpent in the Garden of Eden is described in this way by Petrus Comestor (*c.*1100–1179) in his *Historia Libri Genesis* and in the *Speculum Naturale* by Vincent of Beauvais (1190–1264), among others.[1] According to Comestor, the devil chose a snake with a woman's head because Eve would be more inclined to trust a creature with 'similar' features.[2]

There are three more or less identical versions of the composition described above. Their contour lines correspond to such a degree that we may assume that the same or similar patterns were used to make the paintings.[3] In 1899 the piece was attributed to a painter from the circle of Hugo van der Goes.[4] Nearly thirty years later, Friedländer published the two other versions. In 2001 they were all three attributed to the 'Master of the Embroidered Foliage', a name derived from a certain similarity between the painted leaves of the trees and fine embroidery.[5] Recently, Steyaert attributed the work to Aert van den Bossche, who was probably a pupil or assistant of Hugo van der Goes. The painting dates from around 1490.[6] In any case the panel is related to *The Fall* by Hugo van der Goes (Vienna, Kunsthistorisches Museum; fig. 28.1).[7]

1
Koch 1965: 323–326; Kelly 1972: 301–328; Flores 1981.
2
Koch 1965: 323–326; Kelly 1972: 301–328; Crowther 2010: 28–29. This type of 'snake' was highly popular in the visual arts and literature of the period. For a thorough study of this tradition, see Flores 1981 and also Bonnell 1917.
3
The panel in Ludwigshafen was cut down above and below over the course of time.
4
Clemen 1899: 442–450. The panel was then at Slot Frens, near Cologne. This castle already existed in the fifteenth century. One might wonder whether this was the panels original location.
5
Scherer 2001; Scherer 2007.
6
Steyaert 2007: 180; Steyaert 2013: 258–259, no. 55.
7
Friedländer 1969a: 75, no. 36, pl. 43.

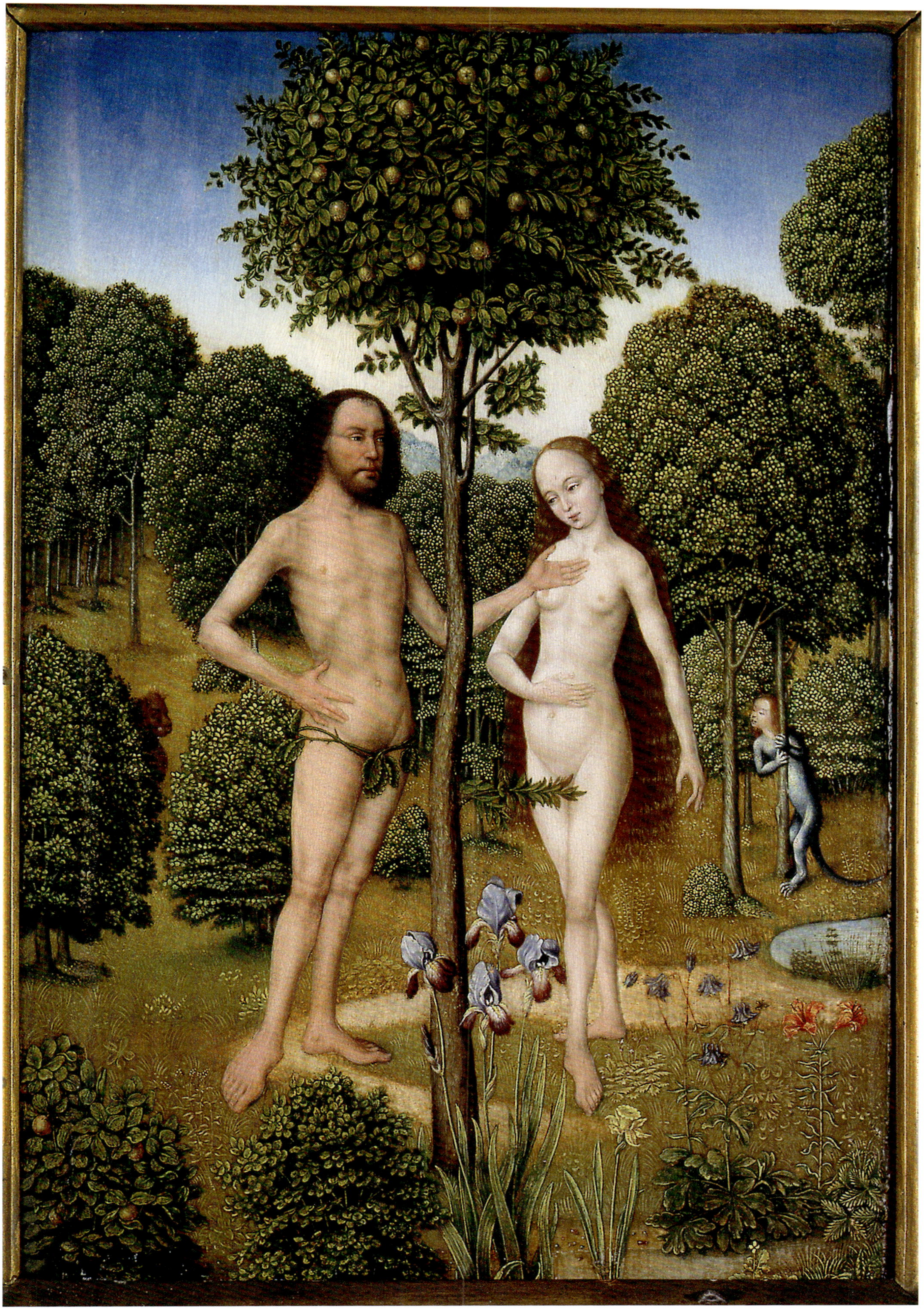

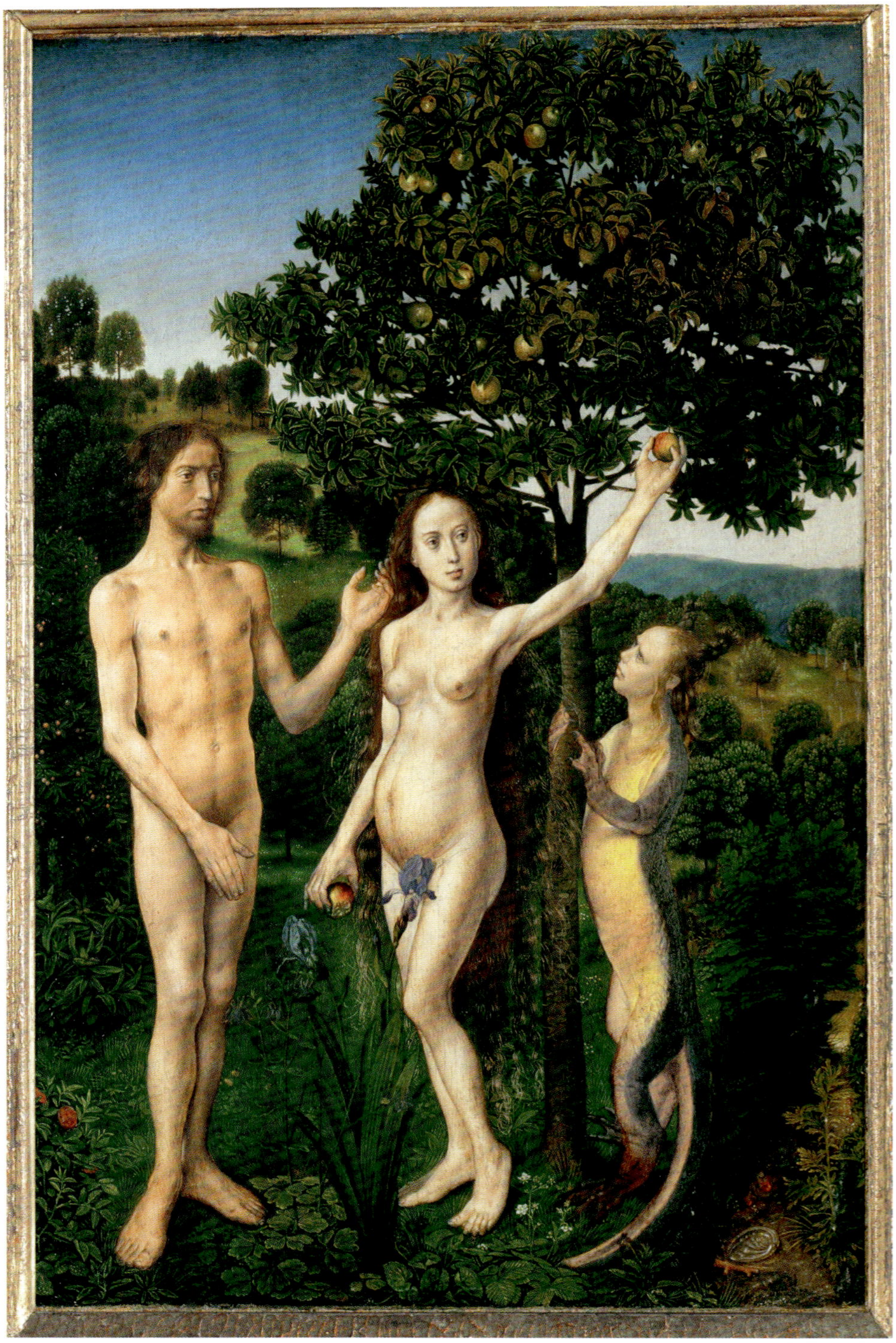

[fig. 28.1] Hugo Van der Goes, *The Fall of Man*, 1470-1475. Vienna, Kunsthistorisches Museum

29

Herri met de Bles
The Garden of Eden

*c.*1541–1550

Oil on walnut panel, 46.6 x 45.5 cm (tondo)

Amsterdam, Rijksmuseum, inv. SK-A-780

PROVENANCE

Sale, H.G. van Otterbeek Bastiaans (Deventer), Amsterdam (C.F. Roos), 31 January 1882, no. 9 (as 'Velvet Brueghel'), fl. 320, to J. Balfoort, Utrecht, for the museum; via Victor de Stuers, fl. 368, to the Rijksmuseum; on loan to the Mauritshuis, The Hague, 1963-1997; on loan to the Bonnefantenmuseum, Maastricht, 2004-2011.

BILBIOGRAPHY

Winkler 1924a: 309; Friedländer 1924-1937: XIII, 136, no. 57; Tóth-Ubbens 1968: 8-9, no. 959; Boeke 1974: 67-163; Friedländer 1967-1976: XIII, 78, no. 57; Broos 1988: 60-61; Gibson 1989: xxi, 27; Serck 1990: I, 171-189, no. 1; Serck 2000: 148-151, no. 1; Kloek 2000: 184-185, no. 76; Wied 2003: 286-287, no. 103; Bakker 2004: 123, 160; Van Wegen 2005: 122-123; Filedt Kok 2008: 104-111; Hendrikmann & Filedt Kok 2009a; Bakker 2012: 106-108; Weemans 2012: 263-312; Bakker 2015: 115-132; Kloek & Ubl 2015: 200-203.

In a sort of microcosm, this tondo depicts the biblical story of Creation and the Fall (Genesis 1–3). Concentric spheres surround the earth: outside is the dark, barren void in the form of deluge, followed by the heavens with the sun, the moon, the stars and the clouds. From this heavenly sphere, birds comoe to populate the earth. Paradise is shielded from the rest of the world by dense forests. In the centre, a fountain provides the luxuriant garden with water. In this apparently peaceful Garden of Eden, a wide variety of plants and animals grow and swarm, among them a unicorn, a lion and an ape. Only upon closer inspection does one notice the enmity between some of the animals, such as the dog and boar in the foreground, or the pair of rutting goats just above. The water to the left, populated by countless sea monsters, likewise refers to impending doom, the Fall. At the left, God creates Eve from Adam's rib; behind them, the two are admonished by Christ not to eat from the Tree of Knowledge of Good and Evil. Spurred on by the snake, Eve finally plucks the forbidden fruit and the first human couple are driven from paradise.

Through the vast landscape, artist Herri met de Bles (*c.*1500/10–before 1567), gives Paradise the form of a concrete location. In comparison to the Enclosed Gardens from Mechelen (cat. 32, 33, 34), this paradise is much more than an imaginary space. It is a hidden place that actually exists: the Garden of Eden, which is lost to humankind from the Fall until the Last Judgment. The striking fountain, of which there is no mention in the Bible, must also be seen in this context. In the first place, it does not refer to the source of the four rivers of Paradise (Genesis 2: 10). Rather, the fountain refers primarily to the 'spring of the water of life'. It is the fountain of life that refers to Christ as the new Adam, the Saviour and the new paradise in the Book of Revelations (Revelations 21: 6).[1]

The landscape in the background shows an idealised combination of wild and cultivated nature with tall, capricious mountains, hills, fields, woods and sea. The fields of

1
Bakker 2015: 119–120, 127–128.

2
Boeke 1974: 70–71; Serck 2000: 148; Kloek & Ubl 2015: 202–203 fig. 76B-F.

3
Hieronymus Bosch, *The Last Judgment*, oil on panel, 98.7 x 110.7 cm, Bruges, Groeningemuseum, inv. 0000.GR00208.I

4
Ilsink & Koldeweij 2016: 164; Borchert 2016.

5
Allart 2000: 22, 25; Limentani Viridis 2010; Weemans 2013: 8.

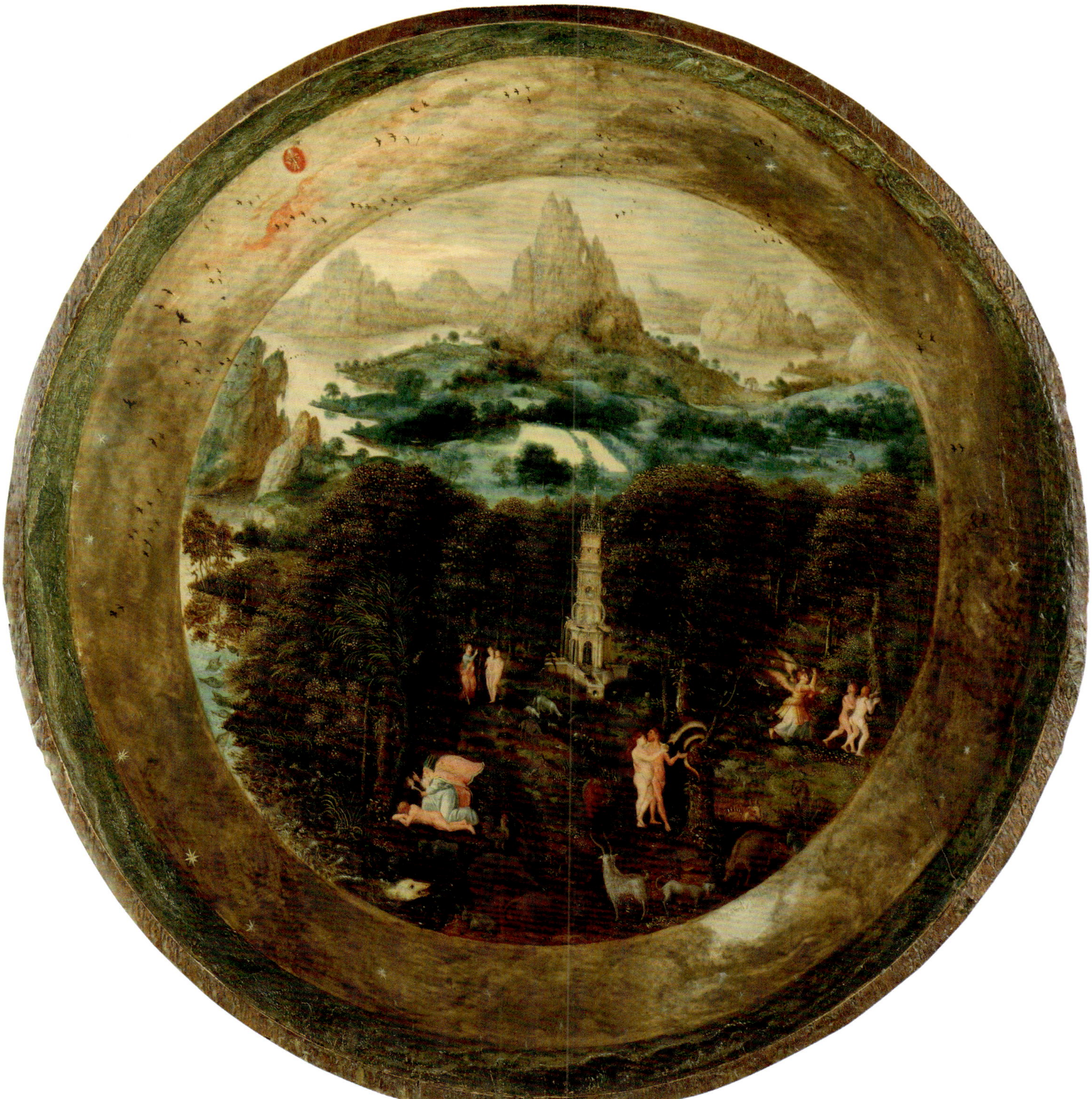

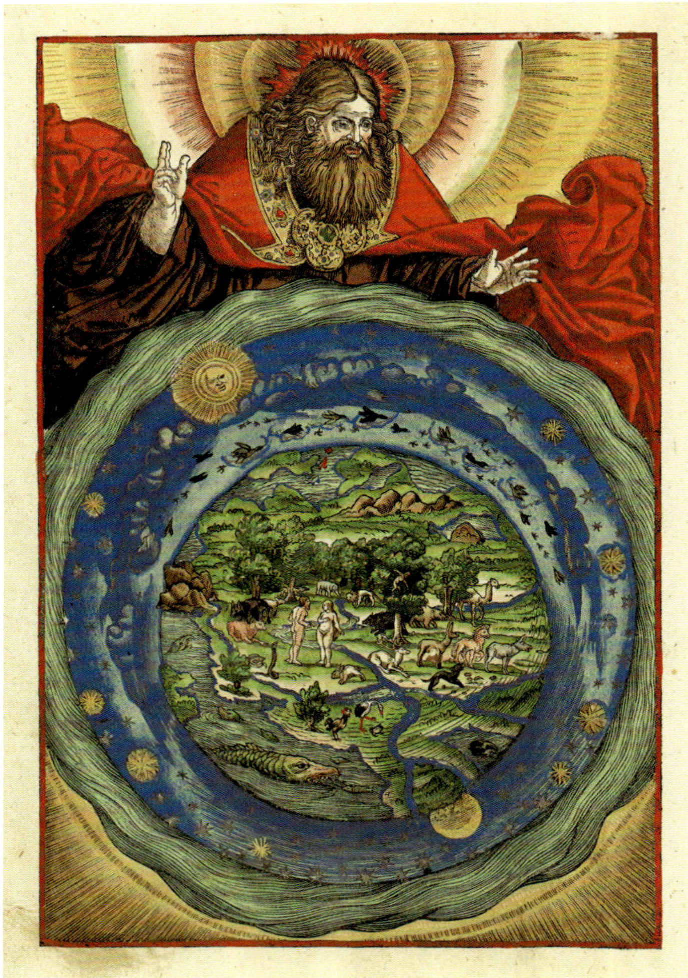

[fig. 29.1] Monogrammist MS (workshop of Lucas Cranach), *The Creation of the World*, woodcut on the frontispiece of the Luther Bible published by Hans Lufft in Heidelberg, 1534, Weimar, Herzogin Anna Amalia Bibliothek.

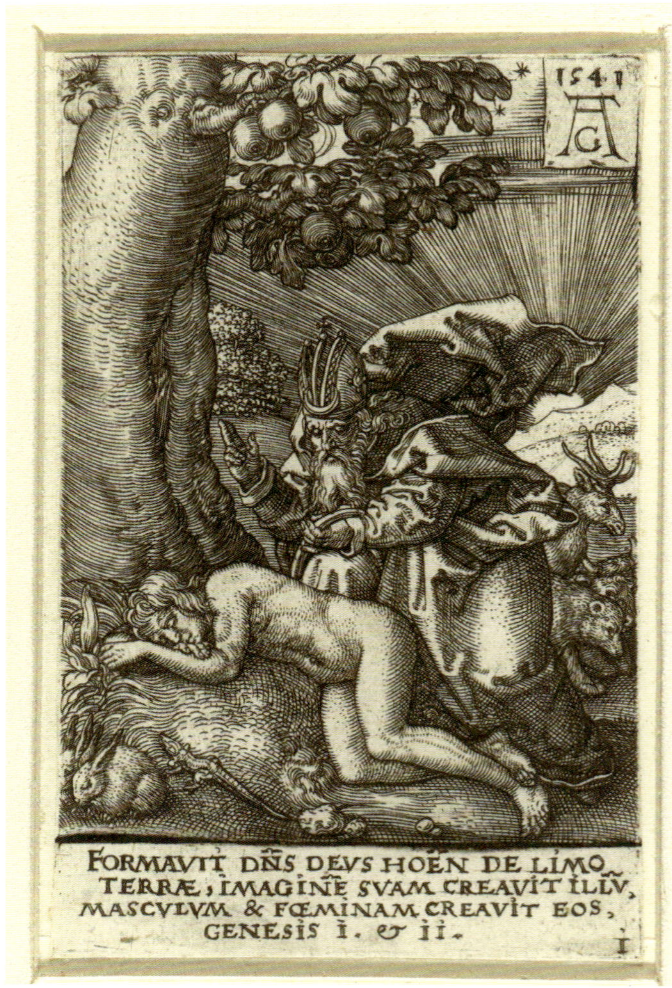

[fig. 29.2] Heinrich Aldegrever after Hans Holbein the Younger, *The Creation of Eve*, engraving, 1541, Amsterdam, Rijksmuseum, Rijksprentenkabinet.

grain, a farm and the figures labouring in the fields refer to the world after the Fall, in which man must earn his bread by the sweat of his brow.

Bles reveals himself here as a true miniaturist and narrator. In this painting, which invites close looking, the successive yet simultaneously depicted biblical narratives and especially the countless details speak to the imagination. These are not limited to the foreground: even the background is crawling with little vignettes and motifs, although many of the figures have grown transparent with time and are now scarcely visible. We see a miniscule elephant to the right of the large mountain, and just before it a bear walking erect in the meadow.

Bles based his overall composition on a woodcut by the anonymous Master MS from the first illustrated Luther Bible of 1534 (fig. 29.1). The four scenes with Adam and Eve are also borrowed from German prints: the first two from engravings by Heinrich Aldegrever from 1540 and 1541 (fig. 29.2) and the second two from prints by Albrecht Dürer from 1510. This provides the earliest possible date for the creation of the painting, 1541.[2] Bles borrowed the fountain of life from a depiction by Hieronymus Bosch, as seen in his *Last Judgment* triptych in Bruges.[3]

It is not the exceptional iconographic combinations alone that make this tondo a unique work. The use of walnut – a type of wood more commonly used in Italy, France and Spain – for the painting's support is also unusual in Northern painting,[4] and probably not without significance. It raises the question as to whether Bles, who was active in Antwerp, might also have been to Italy.[5]

MU

30

Albrecht Dürer
The Fall of Man

Nuremberg

1504

Gravure, 251 x 190 mm

Signed and dated on tablet: *ALBERT[US] / DVRER / NORICVS / FACIEBAT / [Dürermonogram]. 1504.*

Amsterdam, Rijksmuseum, Rijksprentenkabinet, inv. RP-P-OB-1156

PROVENANCE

Royal Library, The Hague, transferred to the Rijksmuseum, 1816.

BIBLIOGRAPHY

Bartsch 1803-1821: vol. 7, no. 1; Dodgson 1926: 308-311; Meder 1932: no. 1; Panofsky 1955: 68, 70, 90, 91 ff.,119 ff., 150, 157, 171, 204, 262, fig. 117, 132; Thiem 1963: 21, no. 37; Stechow 1967: 5; Levenson 1991: 290-291, no. 195; Cole & Gealt 1991: 130; Boorsch & Orenstein 1997: 30-31; Schoch, Mende & Sherbaum 2002: 110-113 no. 39 (with extensive bibliography); Smith 2001; Eikelman & Burk 2006: 182-183, no. 36; Verdi 2007: 90-91, no. 19; Bull 2009-2010: 93, no. 2, fig. 6; Borchert 2010: 425, no. 236; Vogelaar, Filedt Kok, Leeflang et al. 2011: 303, no. 98; Chipps Smith 2011: 188-189; Leemans 2012; Niessen 2012: 55, fig. 45.

This engraving is one of Dürer's most influential works. With heroic and idealised nudes after antique models, he issued a direct challenge to his Italian contemporaries. The bodies recall statues from antiquity, which in the Renaissance had one again become models for the ideal human being. Numerous drawings from the years 1500–1540 attest to his interest in theory of proportions based on the writings of Vitruvius. In a draft of the foreword to his posthumously published treatise *Vier Bücher zur Menschlicher Proportion* (1528), Dürer writes that he first encountered the principles of proportion through Jacopo de' Barbari.[1] Dürer could have met the Venetian artist during his first visit to the Republic in 1494. In 1500 De' Barbari settled for a longer period in Nuremberg, where Dürer tried in vain to pry from him the secrets of proportion. According to him, he eventually succeeded in mastering the Vitruvian rules through self-study. The extent to which the rules of proportion and construction schemes determined the figure of Adam is evident from a study sheet now in Vienna (fig. 30.1).[2] This drawing was in turn used as a model for the figure study in New York dated 1504 that served as a model for the figures in the engraving (fig. 30.2).[3]

From the large number of preliminary studies and proof impressions it is clear just how much Dürer strove in this work to surpass everything that had been made in the genre up to that time. Filled with confidence, he signed not only with his well-known monogram but also with his full name and place of birth in Latin. Here too he is following in the footsteps of his Italian models. The wealth of details and the symbolism are characteristic of the art of the north, but Dürer departs from

1 Albrecht Dürer, *Hjerin sind begriffen vier bücher van menschlicher Proportion…*, Hieronymus Andreae for Agnes Dürer, Nuremberg, 1528. See London, Sloan 5230/44; Rupprich 1956–1959: vol. I, 101–106; Schoch, Mende & Scherbaum 2004: 277.

2 Albrecht Dürer, *Construction drawing of Adam*, pen and brown ink, 265 x 167 mm, Vienna, Albertina, inv. 3080 D 80. See Strauss 1974: no. 1504/11; Schröder & Sternath 2003: 254–257.

3 Albrecht Dürer, *Adam and Eve*, 1504, pen and brown ink, brown wash, with touches of white body colour, 242 x 200 mm, New York, Morgan Library, inv. I, 257d. See http://www.themorgan.org/drawings/item/144288; Strauss 1974: 1504/17.

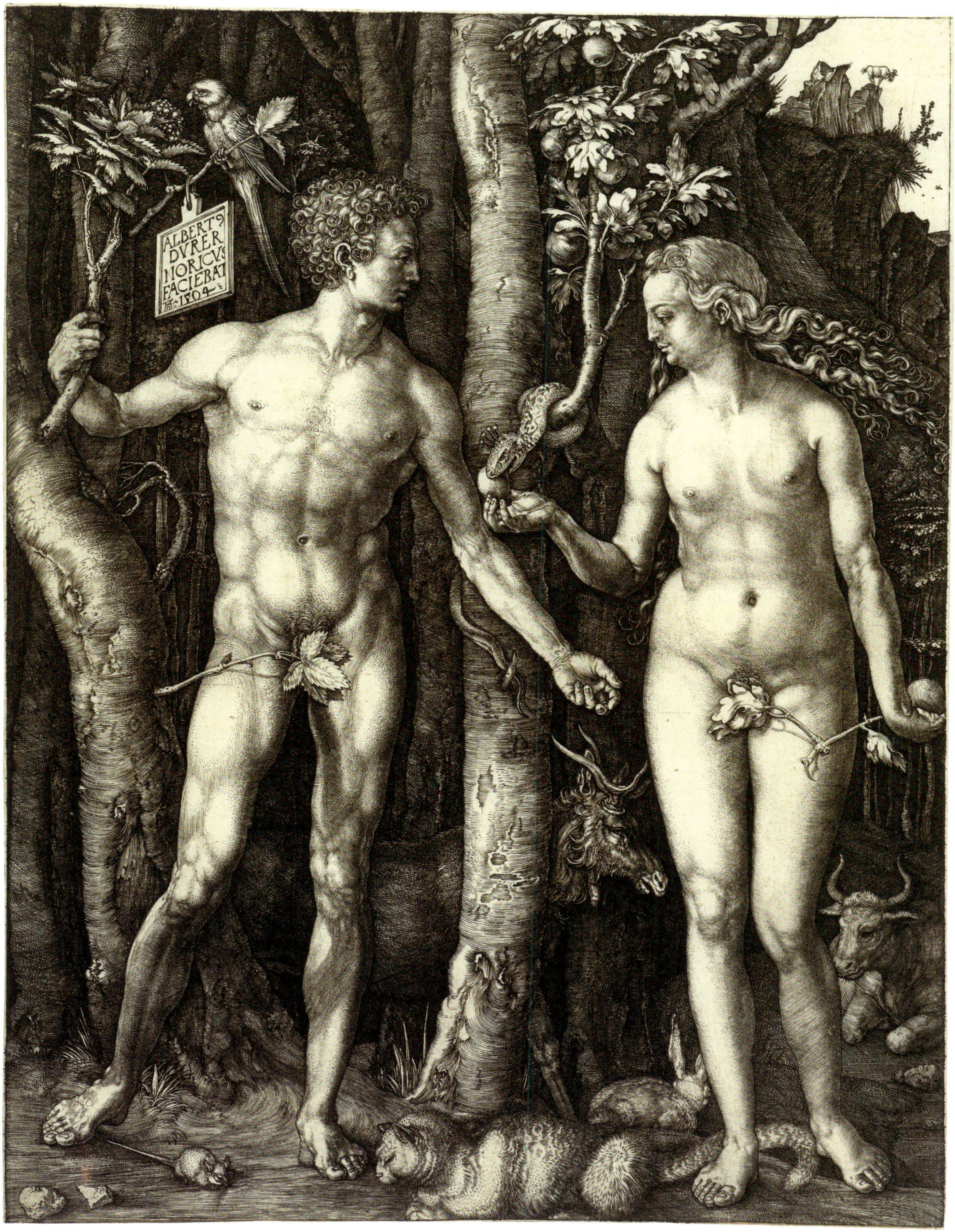
ALBERTS
DVRER
NORICVS
FACIEBAT
1504

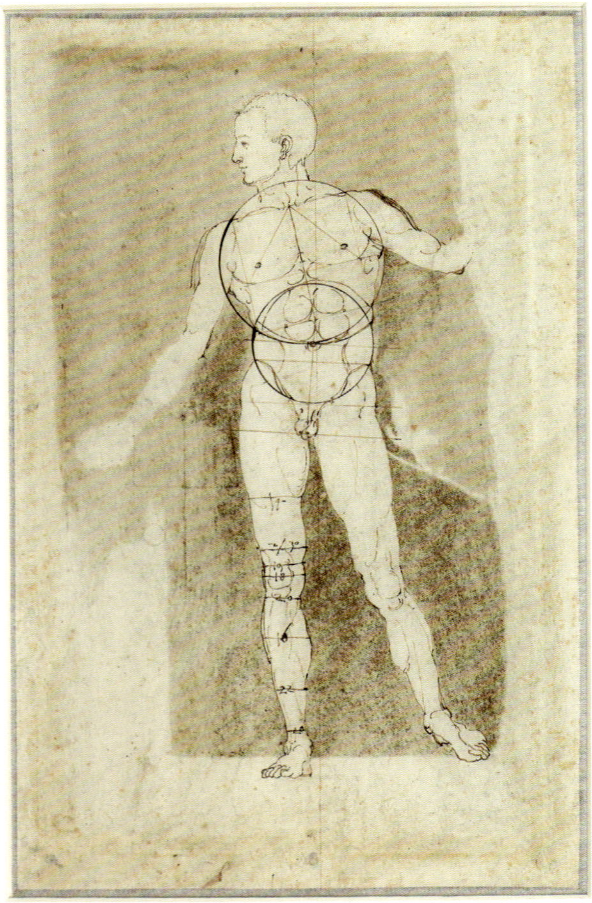

[fig. 30.1] Albrecht Dürer, *Adam*, drawing, Vienna, Albertina.

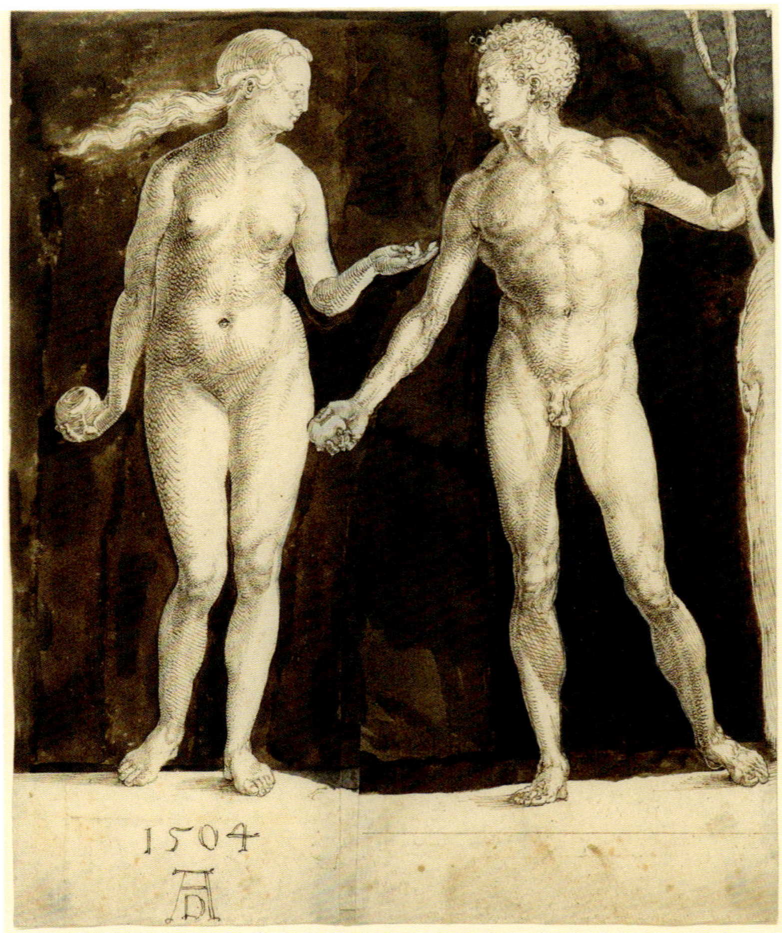

[fig. 30.2] Albrecht Dürer, *Adam and Eve*, 1504, drawing, New York, Morgan Library.

the iconographical tradition in a number of respects. The Tree of Knowledge of Good and Evil, from which Eve plucks fruit, is represented as a fig tree, by analogy with the fig leaves with which, according to Genesis, the first humans covered their shame at their nakedness. The tree Adam holds onto has been identified as the *Fraxinus ornus,* or flowering ash. This sort of tree was widely regarded in Central Europe as the Tree of Life, and was hence thought to be avoided by snakes.

The animals that populate the Garden of Eden have not been chosen by chance. The parrot was a symbol for caution (*prudentia*) and has perched on the Tree of Life. As a personification of male virtue, he stands in stark contrast to the deceitful seductiveness of the snake at Eve's side. In her state of innocence, the cat snoozes peacefully in the foreground, unperturbed by the presence of a mouse. As future predator and prey, they are a foreshadowing of the tension that will arise between the two species after the Fall. According to medieval tradition, the elk, the cow, the rabbit and the cat represent the four temperaments that were also associated with the four bodily fluids. These 'humours' would only determine human behaviour after the Fall, leading to the eternal repetition of sin.

JVG

31

Conrat Meit
Adam and Eve

Mechelen or Antwerp
c. 1530–1535
Boxwood, traces of polychromy, height: 25.5 cm (Adam) and 24 cm (Eve)
Vienna, Kunsthistorisches Museum, inv. KK 9888 (Adam) and KK 9889 (Eve)

PROVENANCE

From the property of the Augsburg art dealer Philipp Hainhofer (1578-1647); in 1609 given to Emperor Rudolf II by the city of Prague (?); Prague art inventory of Emperor Rudolf II from the years 1607-1611;[1] in 1650, 1718, 1737, 1763 and 1782 in the inventory of the Kaiserliche Schatz- und Kunstkammer in Prague Castle; Adam from *c.*1817 in the possession of Josef Daniel Böhm, director the Vienna engraving academy;[2] in 1865 at the sale of Böhm's estate purchased by the art dealer Artaria for the k.k. Österreichisches Museum für Kunst und Industrie;[3] Eve in 1869 in the estate of the painter and sculptor Hanns Gasser (1817-1868), Vienna; that same year purchased as a pendant to Adam for the k. k. Österreichisches Museum;[4] 1942/1943 transferred to the Kunsthistorisches Museum.

BIBLIOGRAPHY

Vöge 1908: 110 ff., fig. 25-27; Winkler 1924b: 46; Troescher 1927: 12, 20 ff.; Bange 1928: 64, pl. 64; Falke 1937: 111 ff.; Duverger 1934: 109; Van der Kemp 1964: 22, no. 26; Rasmussen 1973: 123 ff.; Lowenthal 1976: 91-104, 129-131, 181; Ullmann 1984: 339; Mechelen 1987: 98 ff., no. 203; Kunsthistorisches Museum Wien 1988: 177; Chipps Smith 1994: 388; Eichberger 2002: 305 ff.; Dienst 2002: 328 ff., fig. 166; Gosman, MacDonald & Vanderjagt 2003: 247; Burk 2004: 1037 ff.; Eichberger 2005: 314-315, no. 123; Eikelmann & Burk 2006: 38 ff., 80-82, no. 4; Herrmann-Fichtenau 2007: 97; Borchert 2010: 182, no. 52; Burk 2016: 537.

Carved images of Adam and Eve are typical products of the Renaissance in Northern Europe. Conrat Meit depicted the first human couple with consummate artistry, and in his sculptures incorporated references to renowned kindred spirits such as Albrecht Dürer (cat. 30) and Jan Gossaert, whom he knew personally. No wonder, then, that his Adam and Eves were considered the best in the genre.

Adam and Eve are the first and only known nudes by Meit. Although the pair of *c.*1510 that are now in Gotha (fig. 31.1), incorporate models by Albrecht Dürer and Lucas Cranach the Elder, the pair from Vienna shown here are characterised by a more muscular build and dynamic execution, and as a result by other proportions. The intimate involvement of the first pair, visible in their physiognomy and gestures, with the second pair gives way to a tension that is no longer hesitant but focused and decisive.[5] The figurines from Vienna show greater knowledge of the human body, but at the same time these nudes have something mannerist about them thanks to their pronounced musculature. All trace of the artist's late gothic early years has vanished.

That the nude concept of the Vienna Adam and Eve has become more dynamic can be connected with a similar evolution in contemporary painting in the

1 Bauer & Haupt 1976: 99; Falke 1937: 112 ff.
2 Eikelmann & Burk 2006: 82, n. 3.
3 Eikelmann & Burk 2006: 82, n. 4.
4 Eikelmann & Burk 2006: 82, n. 5.
5 Troescher 1927: 22.
6 Silver & Smith 1978: 240.
7 Silver & Smith 1978: 239.
8 Schwarz 1953: 165.
9 On the Adam and Eve paintings by Jan Gossaert, see Silver 1986: 3.
10 See Schoen 2001: 100.

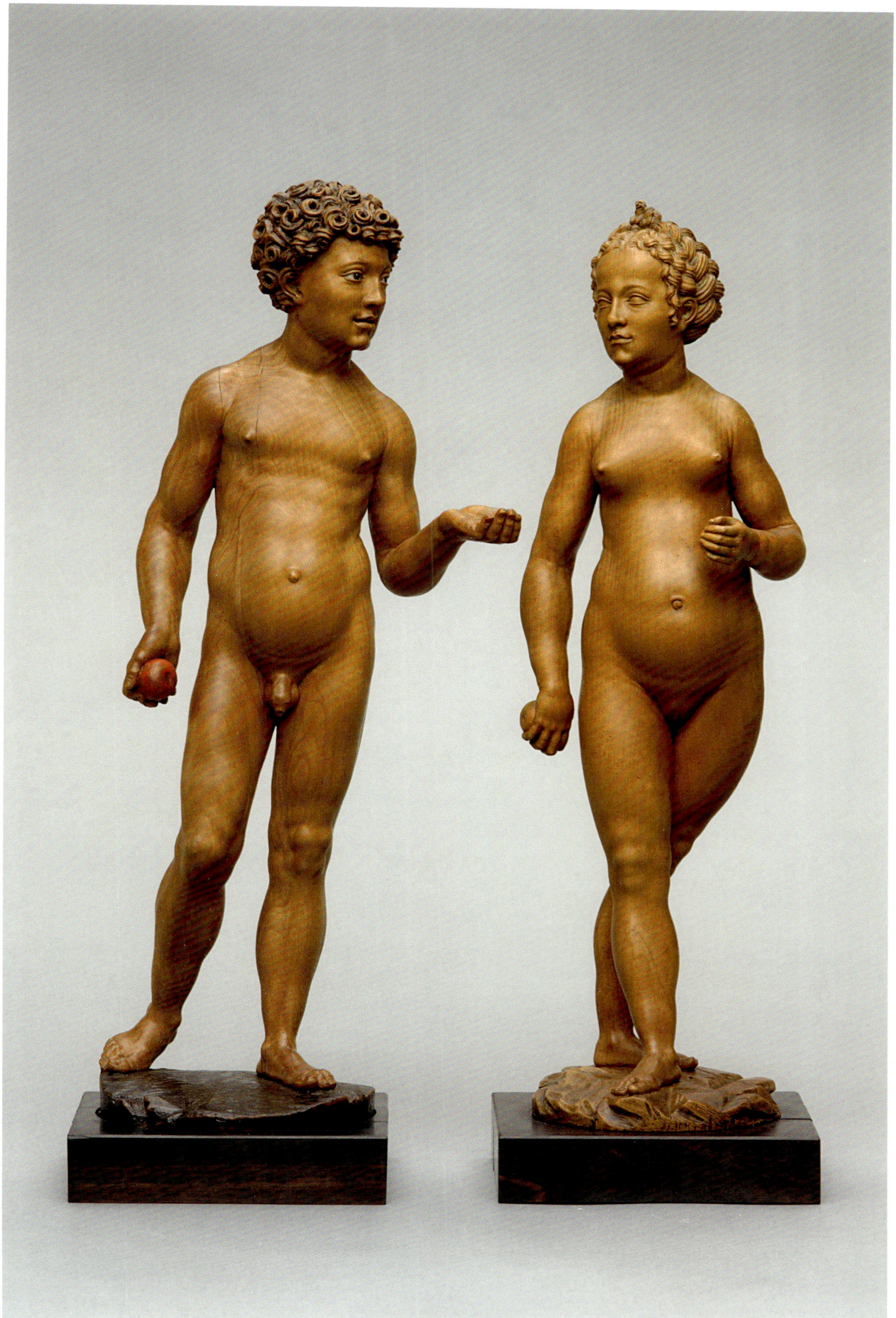

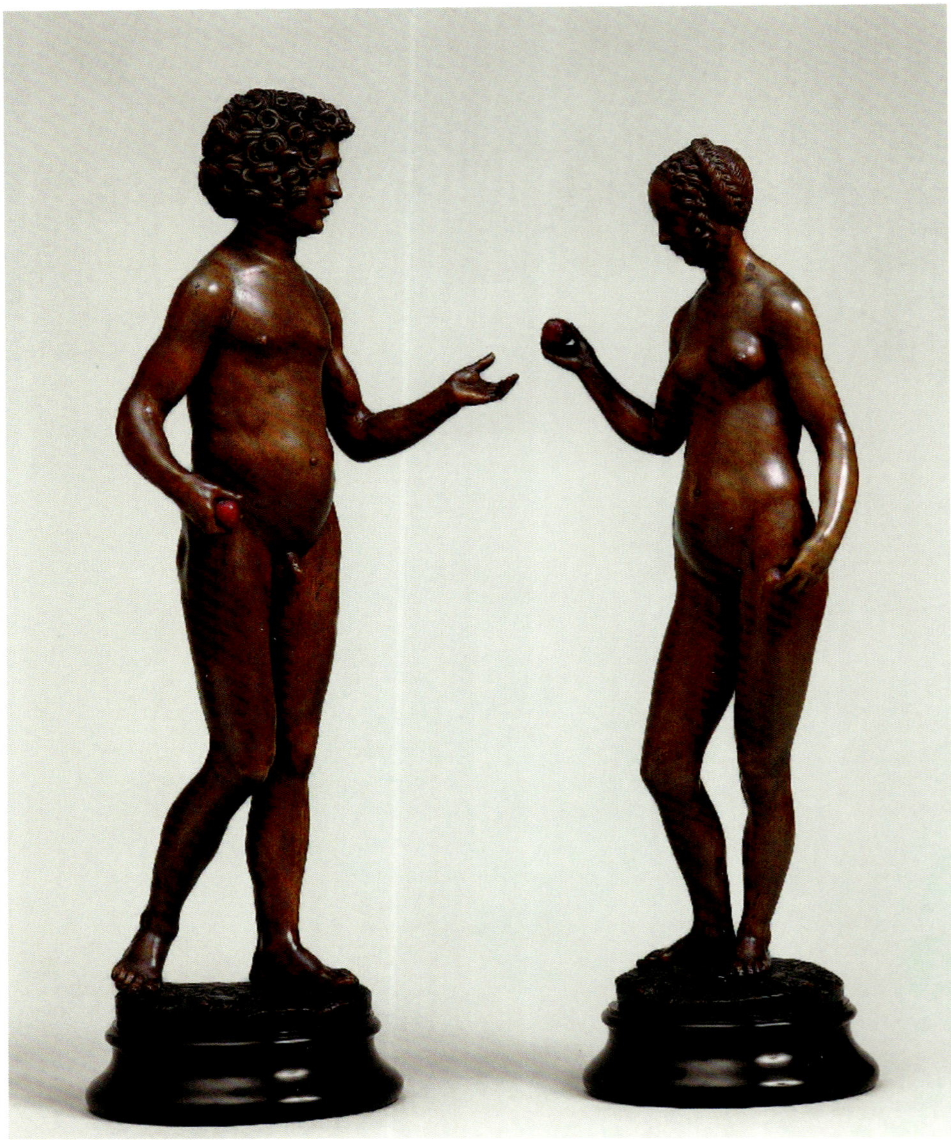

[fig. 31.1] Conrat Meit, *Adam and Eve*, *c.*1510, Gotha, Herzogliches Museum.

Netherlands – namely, with the late work of Jan Gossaert and Lucas van Leyden.[6] With their confident corporality, the heroic nudes of Van Leyden's *Fall* of 1530, which almost seem to dance, differ considerably from the childlike figures in his earlier versions of this theme.[7] And in Jan Gossaert's works depicting this subject from around 1525, we no longer find the coy figures of his earlier versions, but rather nudes that, with their expressive, mobile poses and almost exaggerated three-dimensionality, take up space and give the impression that the Fall is no longer a religious subject.[8]

That Adam in the Vienna figure group turns more toward Eve and the latter crosses her legs is owing to Dürer's *Adam* and *Eve* in Madrid, both from 1507, from which Meit has taken inspiration. The bodily concept is different, but the gestures of the two figures, the confident look Eve gives her companion, and Adam's questioning, doubting outstretched hand and slightly open mouth are very close to the two paintings. The psychology of the couple in Vienna is different to that of their counterparts in Gotha: the later work is more about the beauty of man and woman and the lust that drives them together.[9] In this respect Eve's pose is particularly significant. Crossed legs stood for closing off – barring the way to spirits and curses, hindering probing hands – and hence also for the opposite of sexual availability.[10] Lucas Cranach the Elder and Gossaert both painted nudes with this defensive attitude, but with Meit it is so foregrounded that it seems more like an alibi and therefore inviting rather than repelling.

The figures were not only meant to be looked at – the viewer was also meant to handle them. Grasping in order to understand, that was the intention: not only looking at the depiction of the Fall, but also literally coming into contact with the moral message of the images.

JLB

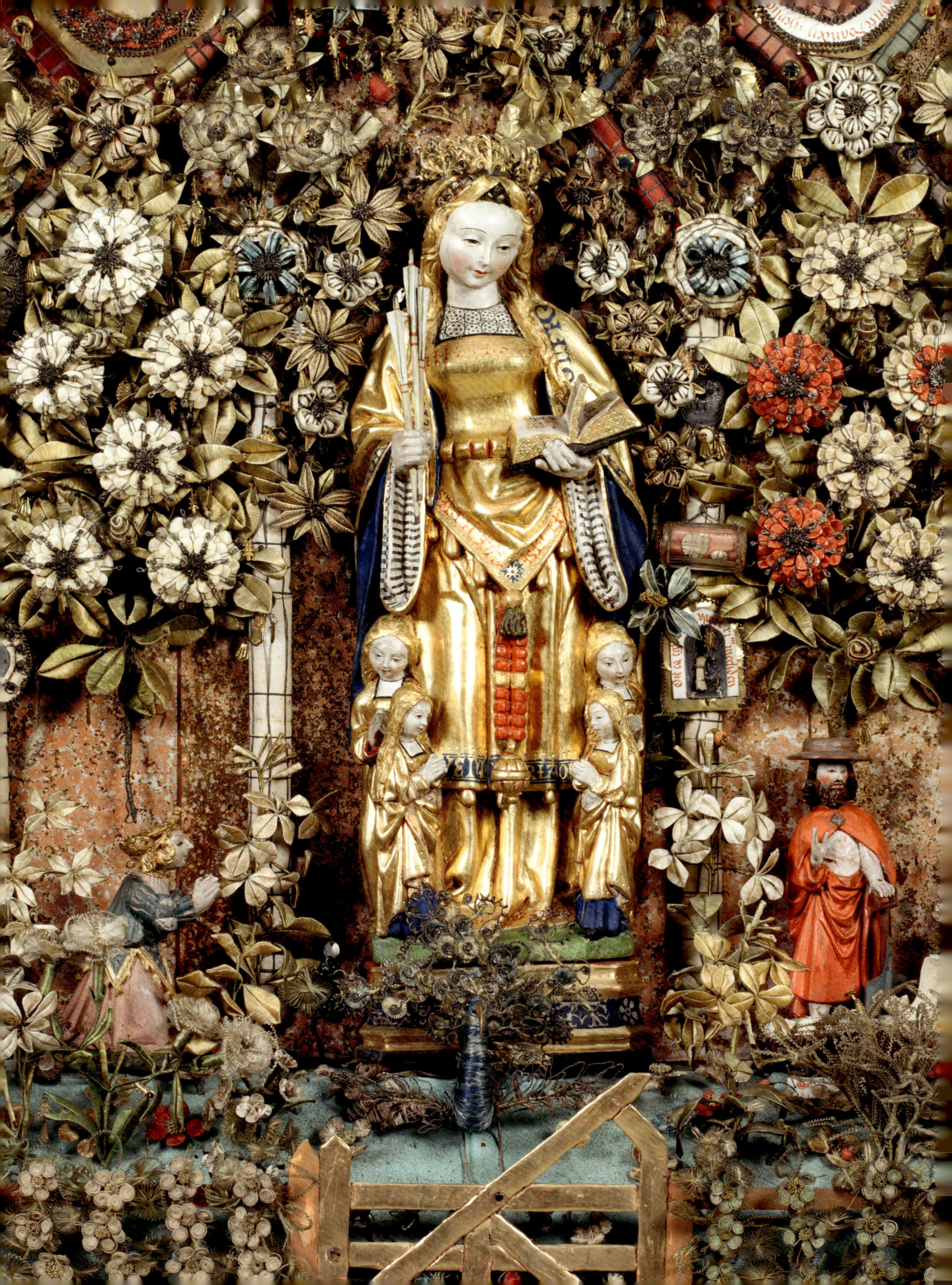

The Conservation of Enclosed Gardens from Mechelen
A Closer Look at Fragile Mixed Media

LIEVE WATTEEUW

DETAIL CAT. 33

Sometimes art is meant for getting as close as possible to some utopian paradise. This is essentially the meaning of a special group of retables from Mechelen (see the contribution by Barbara Baert in this publication). The 'Enclosed Gardens' or 'Besloten Hofjes' of Mechelen, as they are known, were made in the sixteenth and seventeenth centuries for and probably by Augustinian hospice nuns. No less than seven exemplars have survived the turbulent course of history. Three such gardens, all of which date from the first decades of the sixteenth century, were subjected to a complex conservation treatment specially for the exhibition.

The gardens are made of wood, silk, parchment, glass, semi-precious stones, crystal, sequins and metal thread. The wooden sculptures are delicately polychromed and some are stamped with Mechelen hallmarks. Between silk-encased stems, arranged in lozenge-shaped patterns in the background, countless relics and other devotional objects find their place in this miniature cosmos. These include metal pilgrimage insignia, wax seals, figurines in *papier mâché*, miniatures and inscriptions on parchment, small prints, and even a delicate spider of metal thread. All of these elements were mounted on a thin wooden panel at the beginning of the sixteenth century, and the whole was placed vertically in an oak retable cabinet. The sculptures (known as *poupées de Malines*) are secured with a wooden pin to a ground of turf covered with green taffeta. The garden is surrounded by an enclosure with a fine wooden gate decorated with devotional texts. Finally, the ceiling and walls of the retable are decorated with clusters of grapes made from cherry pits, covered with embroidery, and filled with miniature reliquaries, pilgrims' insignia and glittering sequins. Most Enclosed Gardens are equipped with painted wooden wings that protect the cabinet's fragile interior.

The Augustinian nuns used these gardens and preserved them with great care for hundreds of years. They were regularly restored from time to time: worn background paper was replaced, figurines were repainted and loose flower garlands re-attached. From the beginning of the nineteenth century, the wooden cabinets were sealed off with leaded glass in order to give additional protection to the disintegrating silk flowers.

In this way this exceptional collection of seven Enclosed Gardens was preserved within the cloister community until it was transferred to the Stedelijke Musea van Mechelen at the end of the 1990s. In 2011, the entire group – the only one of its kind in the world – was added to the official list of Flemish masterpieces.

The artefacts in the gardens, which amount to around 400 per cabinet, are made of various organic and inorganic materials. These have degenerated in different ways over the centuries, and have become particularly vulnerable through a combination of contamination and natural aging. Light has faded the natural colours of the silk flowers; fibres have broken, metal threads have corroded. The small objects come loose from their wooden support, they break, disintegrate or are crushed.

From 2014 onwards, the Enclosed Gardens were conserved by a team of eight conservators/restorers and scientists in a specially equipped, air-conditioned workshop in the Hof van Busleyden in Mechelen. New scientific imaging techniques revealed the 'skeleton' of the gardens. Through X-ray research, the fine metal structures in the stems of the flowers were made visible, thus enabling a highly precise conservation treatment (fig. p. 217). The hundreds of components were treated one by one. Respect for patina, timbre and finish played an essential role in their conservation. New research techniques and conservation methods were applied in order to preserve the unusual aura of the miniature utopian gardens for the future.

The three Enclosed Gardens featured here were conserved specially for the exhibition by Joke Vandermeersch, Hilde Weissenborn, Jean-Albert Glatigny, Anne-Sophie Augustyniak, Justine Marchal, Sarah Benrubi, Carola Van den Wijngaert, Ann Lievens, Ingeborg Tamsin, Derek Biront, Marjan Jacobs, Susan Verhagen and Lieve Watteeuw. The Musea & Erfgoed Mechelen received financial support for the conservation project from the Topstukkenfonds van de Vlaamse Overheid, the Fonds Baillet-Latour of the King Baudouin Foundation (2014), the Vrienden van het Hof van Busleyden, Mechelen, and the crowdfunding campaign 'Red de Besloten Hofjes'.

32

Enclosed Garden with Calvary and Hunt for the Unicorn

Mechelen
c.1500–1530
Mixed media, 124 x 158,5 x 33 cm
Musea en Erfgoed Mechelen – Collection Gasthuiszusters, inv. GHZ BH 0001

PROVENANCE

Onze-Lieve-Vrouw-Gasthuiszusters Augustinessen, Mechelen.

BIBLIOGRAPHY

Weale 1864: 16, no. 149; Neeffs 1891: 274-275, no. 215; Van Doorslaer 1912: 31-32, no. 131; Poupeye 1912: 61-62, 81-82; Godenne 1957: 120-125, no. XXXVIII-XLVI; De Borghgrave d'Altena 1959: 66; Crab et al. 1971: 451-454, no. MB/36; Vanderwiele 1981: 187; Vandenbroeck 1994: 120-125, 251, no. 175; Hoflack 1998: 55, no. 51; De Nijn 2000: 125-126, no. 4; Antoine 2002: 52-53, no. 12; Eichberger 2002: 379-399.

Enclosed Gardens bear witness to the longing to be allowed back into Paradise. The recreation of the divine garden is a virtuous and spiritual labour. In this Enclosed Garden, religious scenes, inscriptions, relics, seals and silken flora and fauna symbolise incarnation, virginity and salvation.

In the centre of the garden stands an altar on which the Crucifixion is depicted, surrounded by ten candlesticks. Above the cross, God looks down upon his suffering son. To the left of the altar is John the Evangelist with a goblet of poison containing two writhing snakes. He is accompanied by Mary Magdalene. Next to her, white lilies bloom. Before the altar we see an angel with a pack of dogs, the Virgin Mary, and a unicorn. Together this group represents the 'hunt for the unicorn'. This Christian theme alludes to both the Immaculate Conception and salvation from original sin. The unicorn is identified as Christ, who descended into the Virgin Mary and was subsequently crucified in order to save the world from its sins. This is confirmed by the inscriptions at the bottom of the Enclosed Garden: 'The unicorn, breaking out of a strong kingdom in paradise, becomes tame again in the lap of a virgin, thus cleansing us from a sinful poison.'[1] The theme of the hunt for the unicorn has strong sexual connotations. This erotic visual language seems to contrast with the chaste lifestyle of the sisters. Here, however, it is rather a *spiritual* longing for unity on the part of the nuns.[2] The scene depicting the hunt for the unicorn is supplemented according to visual tradition with various Old Testament scenes that prefigure the Immaculate Conception of Mary. Moses and the burning bush, the *urna aurea,* the *porta Ezechiel,* the priest Aaron and the warrior Gideon are all present and accounted for.[3] Moreover, out of the *Fons signatus* grows a tree trunk, atop which a small, female figure holds the coat of arms of the city of Mechelen, confirming the Garden's origins. Various figurines bear the hallmarks of the city as well.[4]

The heavenly garden is enclosed by a gate with a Latin inscription: 'You are the garden, entirely sweet and overflowing with various virtues, untouched and

1
The original inscription reads: *Reynosceron forti imperio Egressus de celi palatio Virginis mansuescit in gremio nos veneni purgans a vicio.* See Vandenbroeck 1994: 91–117.

2
Muir 2012: 2–12, 151–158.

3
This small pail has been moved several times. In 1912 the object still hung on one of the arms of the cross to catch the blood of Christ. As such it functioned not as a prefiguration of the Immaculate Conception, but as a reference to the Eucharist and the origin of the Holy Blood. Poupeye 1912, 81–82.

4
A gothic M has been drawn to the lower right of Mary Magdalene's dress. The back of the image is marked by three bars. Gideon bears a gothic M below on his armour and is likewise hallmarked with three bars. The same two hallmarks can be found on the clothing of Aaron, the Virgin Mary with the unicorn, and Moses. See Crab et al. 1971: 451–454; Kruip 2006: 27–39; Geelen & Steyaert 2009: 19–49.

5
Originally: *Tu es ortus cunctis deliciis affluens multisque divitiis umquam tactus spurriciis (sic) gignens florem refertum gratiie.* See Vandenbroeck 1994: 91–117. On the relationship between scent and virtue, see Baert 2014; Baert 2016c; Iterbeke 2015: 36–48.

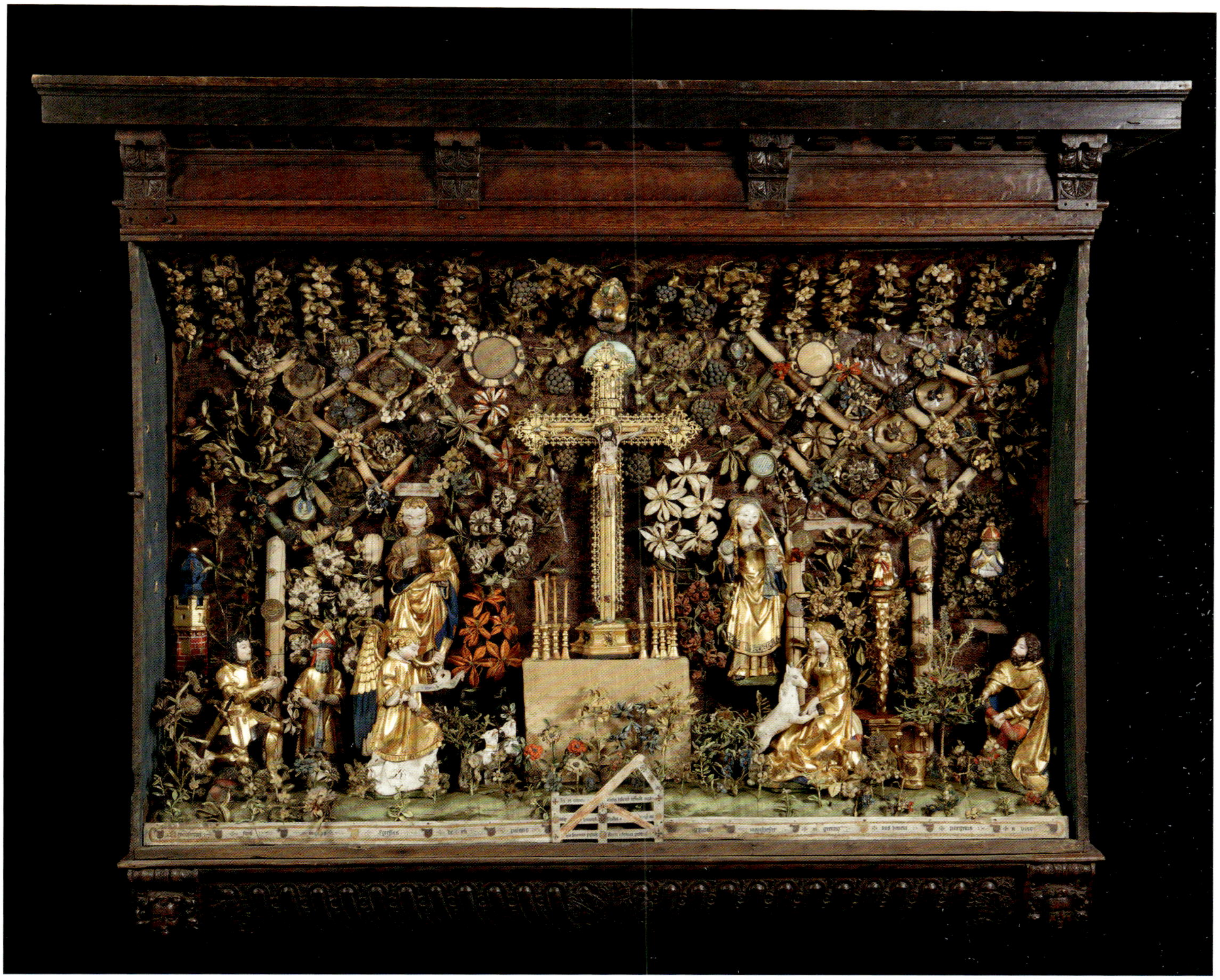

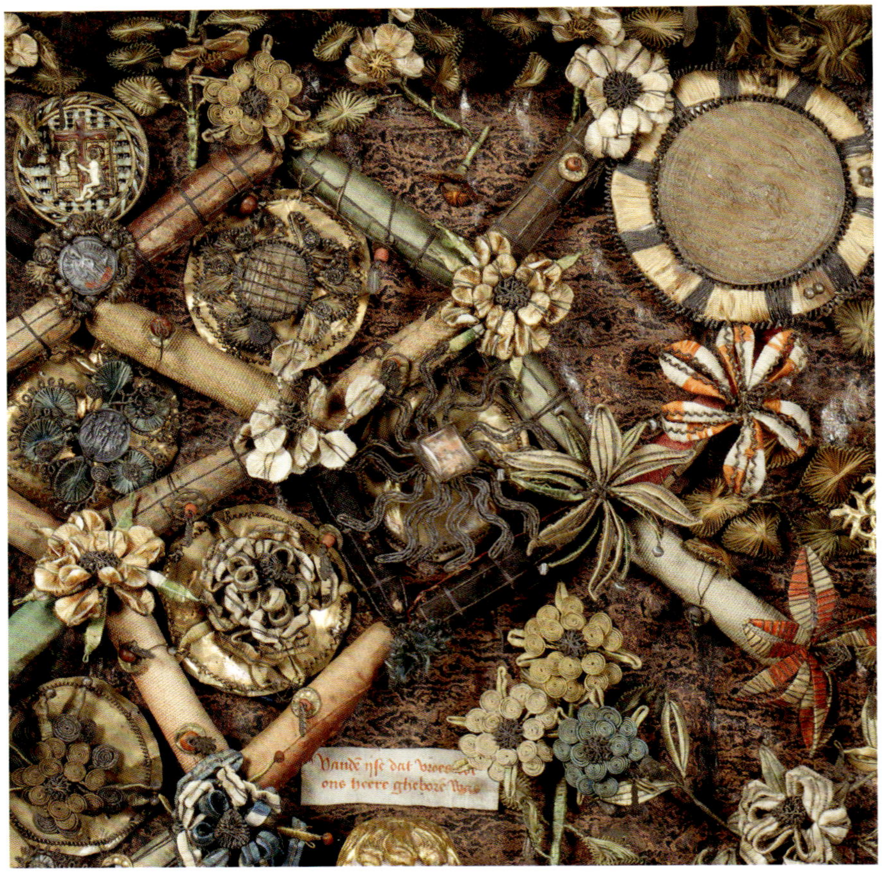

[cat. 32] detail

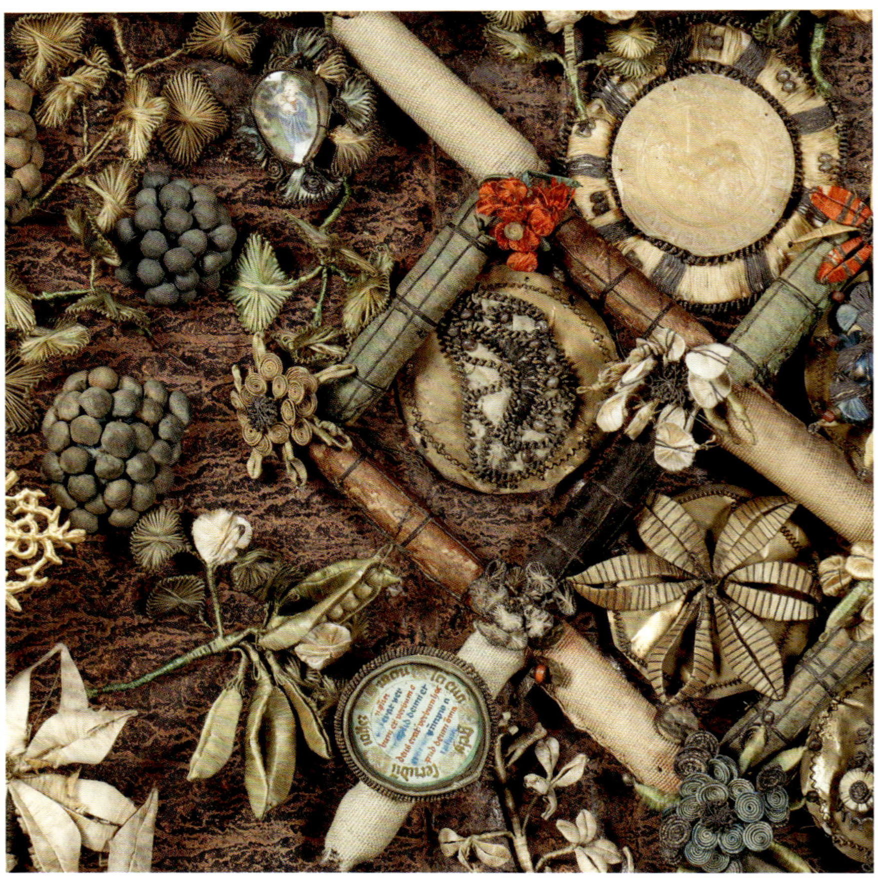

[cat. 32] detail

filled with flowers and grace.'[5] The inscription is both a panegyric on the Virgin Mary and a personal appeal to the viewer to lead a virtuous life. The gate marks the division between the immaterial sanctuary of the Enclosed Garden and the material world of the viewer. The gate symbolically blocks access to the heavenly garden and in this way protects it from 'sinful' influences.

HI

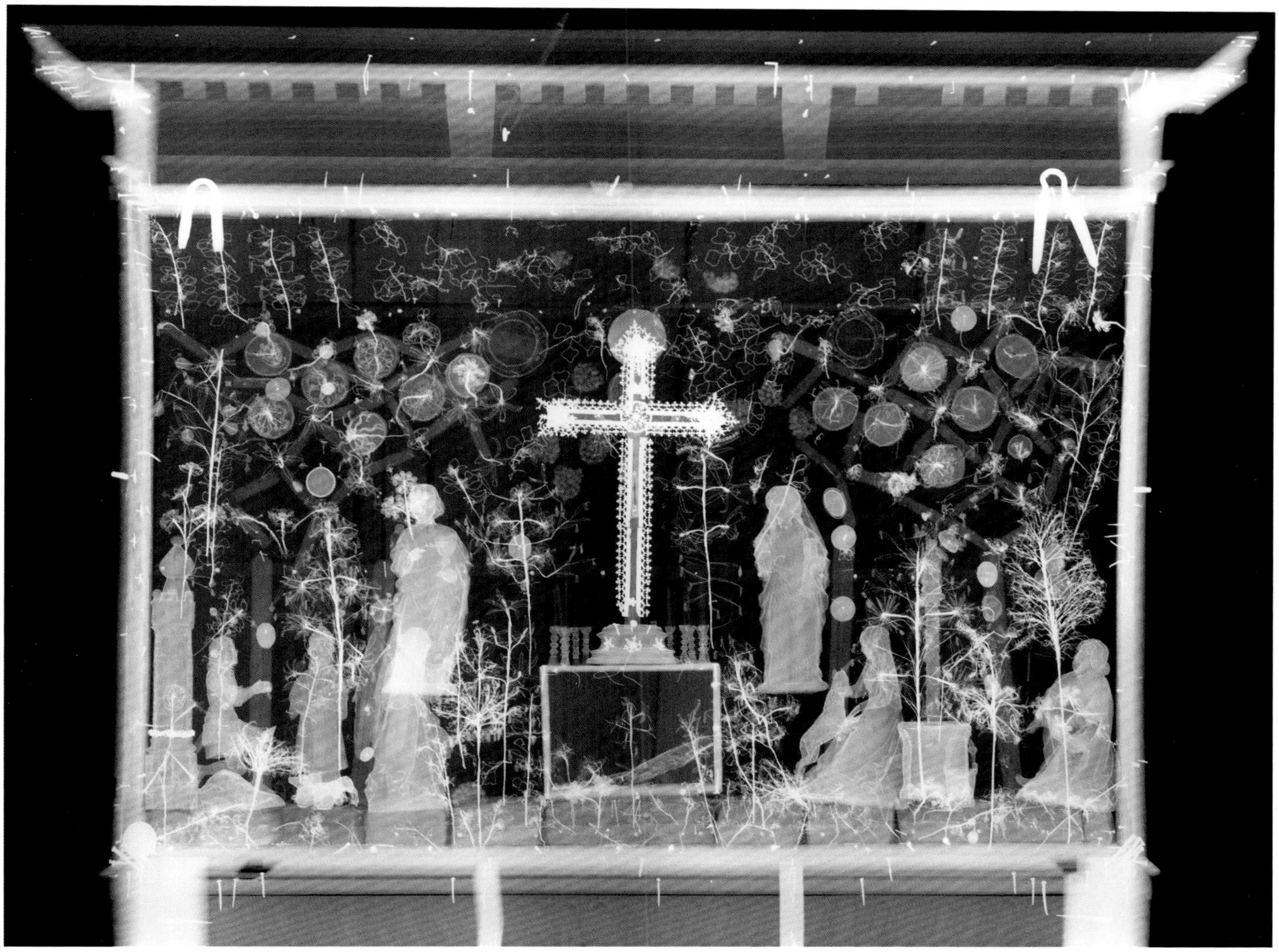

[cat. 32] X-radiograph

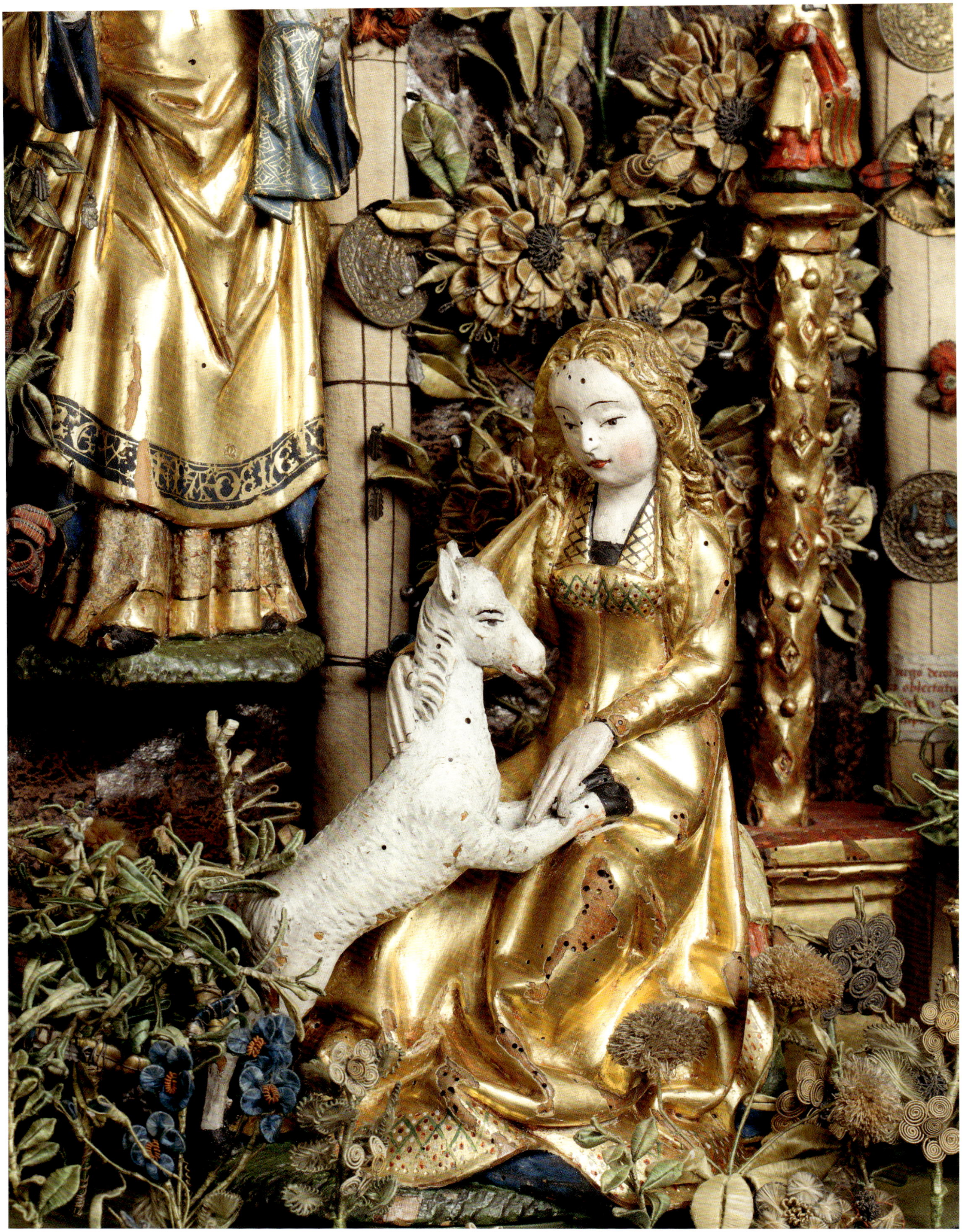

[cat. 32] detail

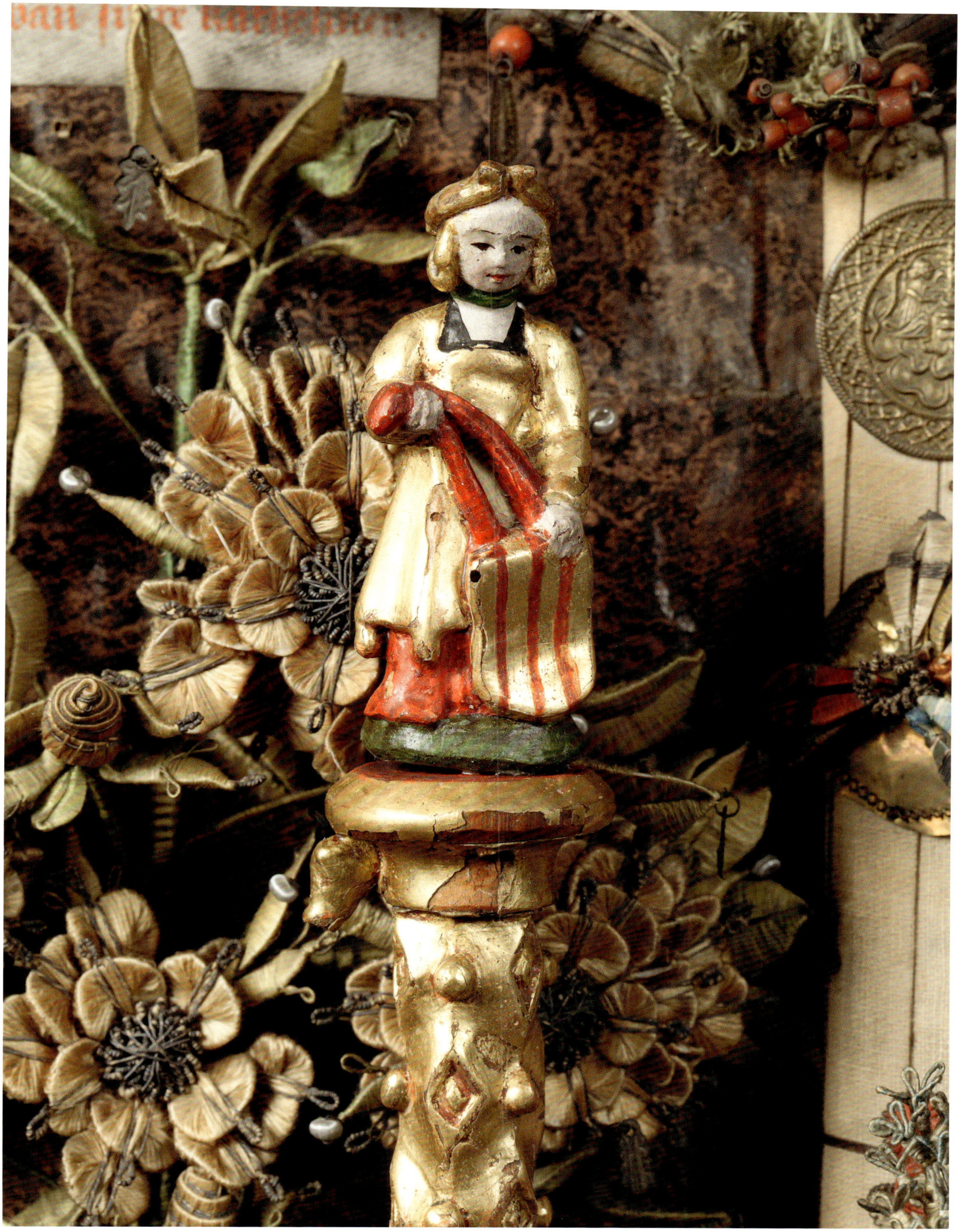

[cat. 32] detail

33

Enclosed Garden with St Elizabeth, St Ursula and St Catherine

Mechelen
c.1520–1530
Mixed media, 134 x 190 x 25 cm
Musea en Erfgoed Mechelen – Collection Gasthuiszusters, inv. GHZ BH 0002

PROVENANCE

Onze-Lieve-Vrouw-Gasthuiszusters Augustinessen, Mechelen.

BIBLIOGRAPHY

Weale 1864: 15, no. 147; Neeffs 1891: 272-273, no. 213; Van Doorslaer 1912: 20-22, no. 78; Poupeye 1912: 78-79; Antwerp 1948: 50, no. 352; Godenne 1957: 47-127; De Borchgrave d'Altena 1959: 67; Stockholm 1970: no. 26; Crab et al. 1971: 421-423, no. MB/37; Brokken 1990: 189; Vandenbroeck 1994: 239, no. 114; Hoflack 1998: 54, no. 49; Koldeweij 2000: 222-252; De Nijn 2000: 124-125, no. 3; Antoine 2002; Eichberger 2005: 271-272, no. 116; Koldeweij 2006: 247-256; Kruip 2006a: 55-56; Kruip 2006b: 231-234; Van Asperen 2009: 102-108; Rudy 2011: 111-118; Baert 2013: 111-152; Fraeters 2013: 65-94; Baert 2014: 9-45; Iterbeke 2015: 44-47.

Enclosed Gardens represent much more than just their iconographic programme. They are the incarnation of specific norms and values. Represented by three female saints, the notion of virtue dominates this Enclosed Garden. The heavenly surroundings contain not only the figurines of St Ursula, St Catherine and St Elizabeth of Hungary, but also a *noli me tangere* scene.[1] In the centre of the garden stands the figure of St Ursula, one of the eleven thousand martyred virgins. The legend of the eleven thousand virgins was particularly beloved of sixteenth-century nunneries. Due to a misinterpretation of an earlier text, in which the 'XIM' stood for 'eleven martyrs' instead of the Roman numeral 'eleven thousand', a very large number of relics of these three female saints were in circulation. The relics 'vanden XI Maechden' – of the eleven Virgins – as they are called in the inscriptions decorating this reliquary shrine, are also included as such in many other Enclosed Gardens.[2]

Ursula is accompanied on the left and right by St Catherine and St Elizabeth of Hungary. The pious lifestyle and engagement of the latter was considered a great example for the nuns who had taken her as their patron saint.

Finally, there are two more small figurines arranged in this garden. The small figure with the red mantle is Christ, depicted as a gardener. At the right, Mary Magdalene adores him. This group illustrates the *noli me tangere,* the first encounter of Christ with another person after his resurrection.[3] The story of the resurrection is moreover depicted within the Enclosed Garden on a wax seal above St Ursula.

This Enclosed Garden is dotted with silk flowers, plants, pilgrims' insignia and relics.[4] They offer the viewer the possibility of making a 'virtual' pilgrimage. This imaginary pilgrimage to the Holy Land or the Kingdom of Heaven takes place in the imagination of the faithful. By looking at the relics and insignia, he or she travels a spiritual path that leads to greater insight.

1 Vandenbroeck 1994: 239.

2 Rudy 2011: 117.

3 Baert 2013: 142–145.

4 Koldeweij 2000: 222–252; Koldeweij 2006: 247–256; Kruip 2006b: 231–234; Van Asperen 2009: 102–108.

5 In the middle ages, a distinction was made between the physical senses and the inner, spiritual senses. A sensual stimulus could be discerned or experienced by the physical senses, but interpreted by the inner senses. In this way, the soul could detect the sweet odour of God without its being physically present. Fraeters 2013: 65; Iterbeke 2015: 44.

6 Rudy 2011: 111–118; Baert 2014; Baert 2015; Iterbeke 2015: 45–47.

7 I would like to thank Andrea Pearson for sharing the results of her research. See Pearson 2017.

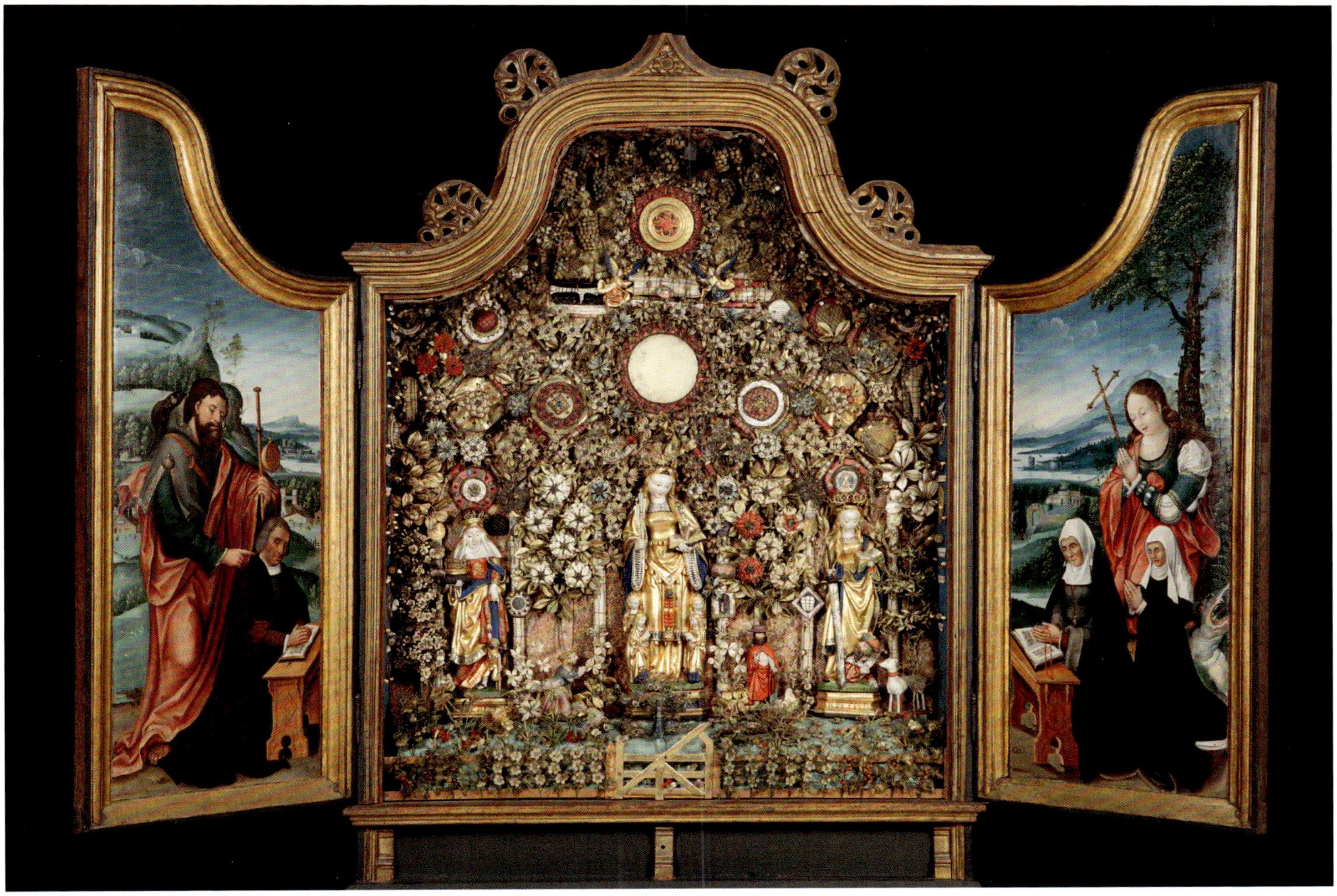

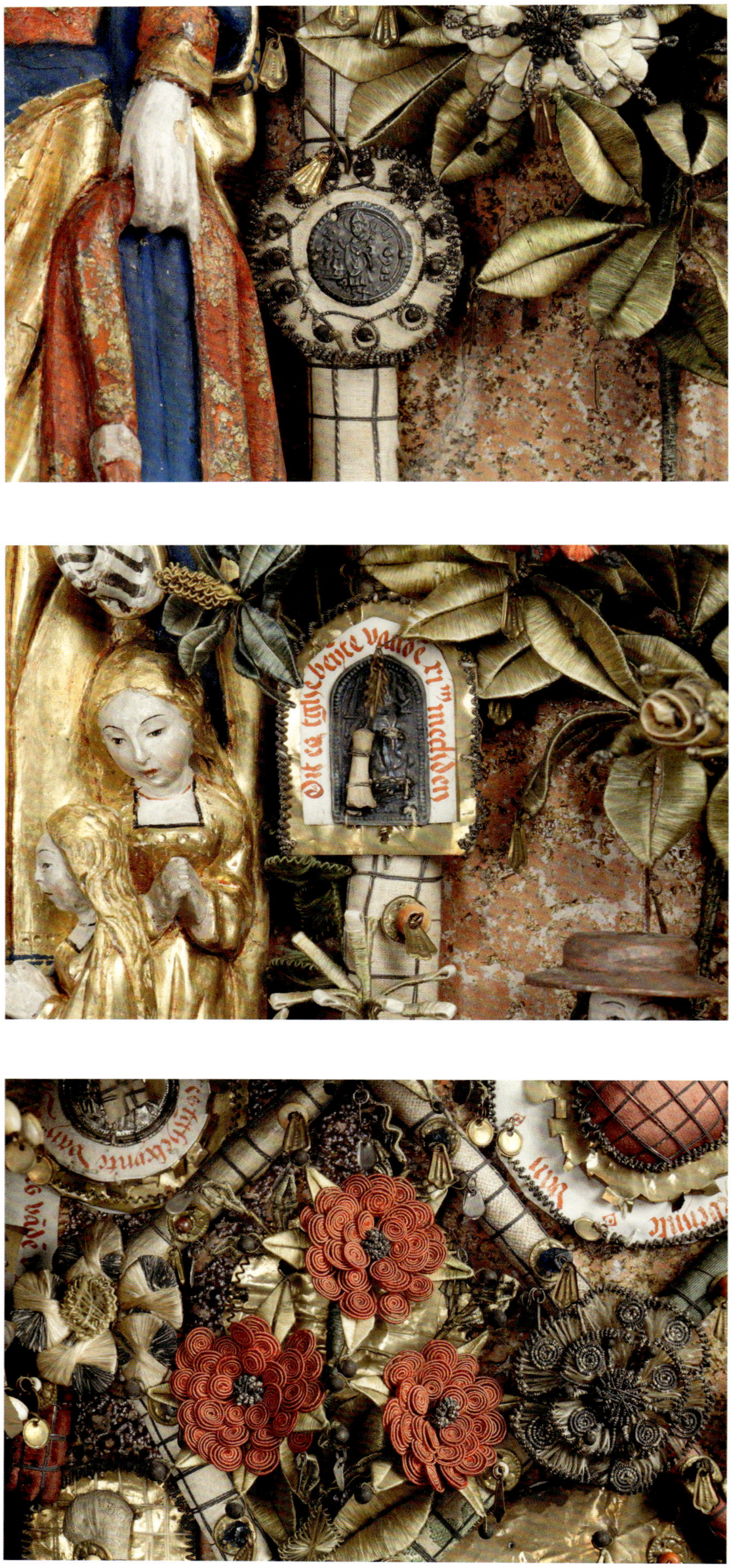

[cat. 33] details

In addition to the relics and pilgrims' insignia, the silk flowers likewise contribute to the spiritual journey. The flowers in the reliquary shrine are on the one hand a reference to Heaven and the symbolism of flowers based on the Song of Songs; on the other, they represent the mystical 'sweet scent' of God.[5] In this way they ask the viewer to see beyond the materiality of the Enclosed Garden and to open him- or herself up to what is not visibly present, but can only be experienced spiritually.[6]

The Garden can be closed by the two painted wings. Recent research has identified the figures on the wings as Jacob vanden Putte, his wife Margaretha Svos and their daughter Maria vanden Putte, who probably commissioned the Enclosed Garden. Maria vanden Putte was a hospice nun in Mechelen and took her vows around 1524.[7]

HI

34

Enclosed Garden with St Anne, the Virgin and Child, St Augustine and St Elizabeth

Mechelen

c.1520–1550

Mixed media, 150 x 240 x 38 cm

Musea en Erfgoed Mechelen – Collection Gasthuiszusters, inv. GHZ BH 0006

PROVENANCE

Onze-Lieve-Vrouw-Gasthuiszusters Augustinessen, Mechelen.

BIBLIOGRAPHY

Weale 1864: 17, no. 152; Neeffs 1891: 276; Van Doorslaer 1912: no. 82; Poupeye 1912: 79-80; Brussels/Mechelen 1954: 24, no. 81; Godenne 1957: 116-120, no. XXXVIII-XLVI; De Borghgrave d'Altena 1959: 66; Crab et al. 1971: 459-462, no. MB/43; Godenne 1972: 38; Brokken 1990: 188-189, no. 78; Van der Stock 1991: 447, no. 212; Vandenbroeck 1994: 240, no. 115; Hoflack 1998: 52-53, no. 48; De Nijn 2000: 125-126, no. 4; Antoine 2002: 55-56, fig. 14a.

In this reliquary shrine, paradise lost is recreated. Behind the golden gate in the centre of the Enclosed Garden stands St Anne. In her right arm she carries the Virgin Mary, who holds Jesus on her lap. Saint Anne with the Virgin and Child is not only a representation of three successive generations, but also the female genealogy of the Christ Child.[1] This doctrine of faith implies that Mary was born into the world without original sin, which in turn enabled the Christ Child to be born free of sin. At the very top of the cabinet is yet another image of the Virgin Mary with her son, this time in an aureole. The image is a representation of the *Lucerna Clarissima* (cat. 26), a motif that has its origins in the apocalyptic visions of John. The ensemble is surrounded by four cherubs and a grapevine.[2] A small image of God the Father watches from directly above.

This central group is flanked by figurines depicting St Elizabeth of Hungary and St Augustine. This church father is depicted as a bishop with mitre and staff. In his right hand he offers his hart, a gift that goes back to his *Confessiones*.

In this Enclosed Garden there is also an exceptional medallion in *papier pressé* (fig. p. 225).[3] Although the origins of these medallions are still unclear, the *patacons* in pipe clay are a product of Mechelen. Similar reliefs in *papier mâché* and pipe clay were made in cloisters.[4] The medallion depicts the hunt for the unicorn. In addition, reference is also made to various other Marian symbols such as the *fons signatus, virga Aaron, vellus Gedeonis, urna aurea, rubus Moysis, porta Ezechiel* and *hortus conclusus*. All these Old Testament elements are prefigurations of the Immaculate Conception. The seal is surrounded by a wreath of silk flowers carried by two polychrome cherubs.

The background of the Enclosed Garden is covered by a lattice of paper rolls (*paperollen*), silk flowers and small ornaments.[5] The pattern of the lattice creates the impression of an orchard, a frequently occurring backdrop in the devotional literature of the sixteenth century. The garden that is depicted here is the orchard

1
Bynum 1992: 79–85.

2
This Mary in an aureole corresponds to the Marianums of the Onze-Lieve-Vrouwen-in-de-zon (Our Lady in the Sun), which decorated many church buildings in the fifteenth century. There was also a Marianum in the Onze-Lieve-Vrouw Gasthuis in Mechelen. Vandenberghe 1983: 137–149.

3
The theme of the hunt for the unicorn is also discussed in cat. 32.

4
This is evident among other things from excavations and an inventory of the Bethania cloister in Mechelen. Cordemans de Bruyne 1895: 20–111; Van der Stock 1991: 448–450, no. 213 (with earlier literature).

5
In this Enclosed Garden there are neither pilgrims' insignia nor relics with inscriptions.

6
Falkenburg 1994; Baert 2014; Baert 2016c; Iterbeke 2015: 29–35.

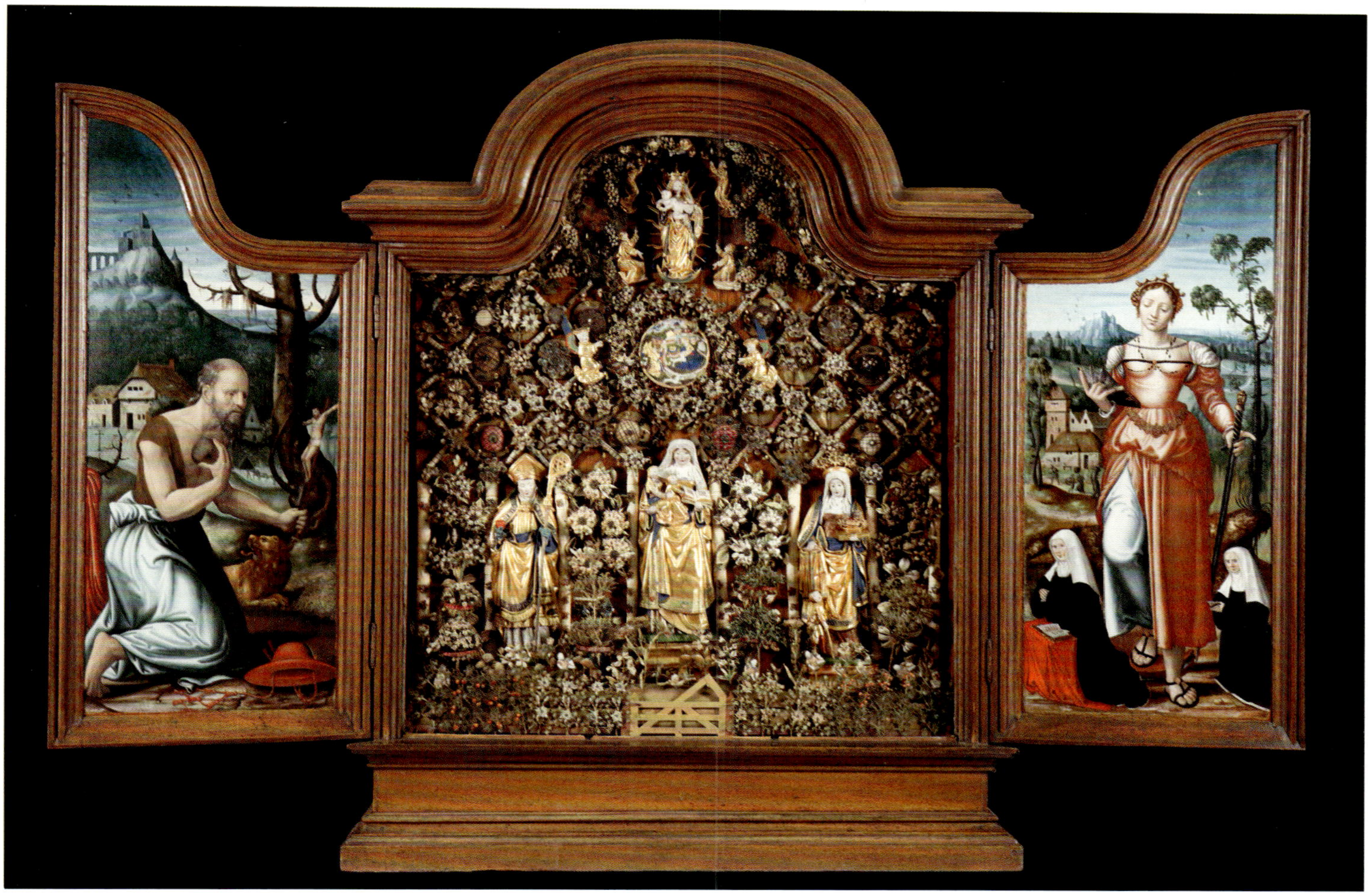

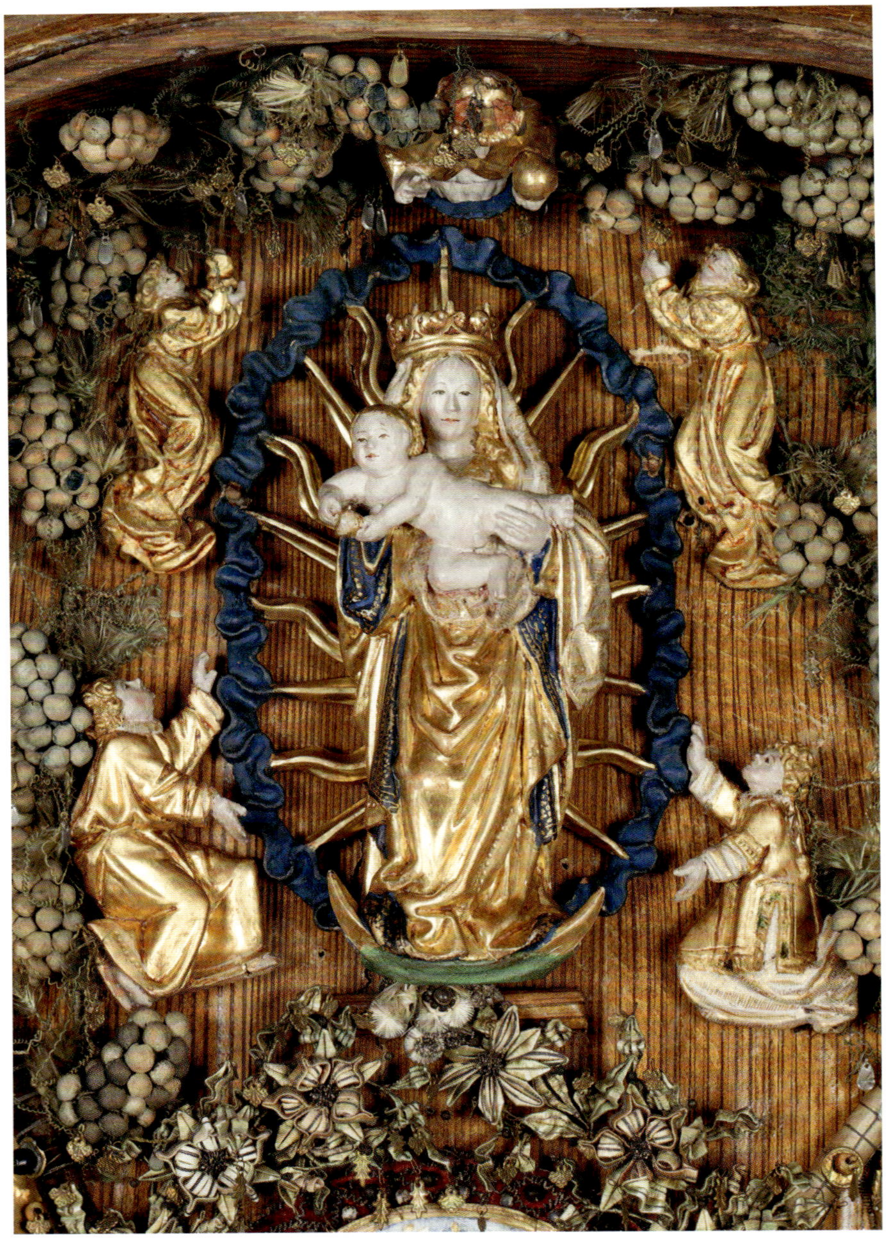

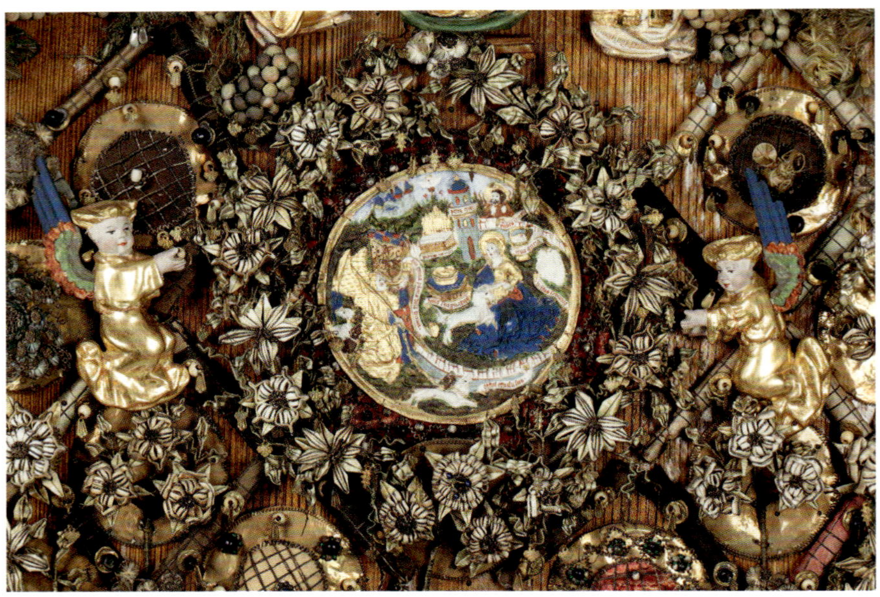

[cat. 34] details

of the cloister, the Enclosed Garden from the Song of Solomon and the Garden of Eden, as well as the loving soul of the believer. It is in the earthly, tangible objects that the spiritual or higher counterpart is represented. The garden allegory permits the devout viewer to explore the unknown while in the familiar environment of the nuns. The viewer is encouraged to (re-) create a virtuous, heavenly environment in which his or her encounter with the Lord can take place.[6]

This Enclosed Garden can be closed with two painted wings. On the left panel St Hieronymus is depicted; on the right, two unidentified nuns kneel in prayer, attended by St Catherine.

HI

35

Maître François
The Heavenly and the Earthly City

In: St Augustin, *La cité de Dieu,* Paris, *c.*1475–1480, fol. 6
Translated from Latin into French by Raoul de Presles (*c.*1364–1380)
Parchment, 270 x 185 mm
The Hague, Museum Meermanno, House of the Book, Ms. 10 A 11

PROVENANCE

Made for Jacques d'Armagnac, duke of Nemours (1433-1477), after his imprisonment, finished for Philippe de Commines (1447-1511), lord of Argenton (coat of arms in the margin and below); sale, J.-A. Crozat de Tugny (1696-1751), Paris (Thiboust), 1751, acquired there by L.-J. Gaignat, Paris; his sale, Paris (G.F. De Bure le Jeune), 1 February 1796, lot 242), acquired there for Gerard Meerman (1722-1771), Rotterdam and later The Hague; by inheritance to his son Johan Meerman (1753-1815); between 1816 and 1824 acquired by Willem H. J. van Westreenen van Tiellandt (1783-1848), The Hague, who bequeathed it to the Kingdom of the Netherlands.

BIBLIOGRAPHY

Boeren 1979: 14-15; Blöcker 1993: 297-299, no. 45, fig. 12a-i; Avril & Reynaud 1993: 52; Bartelink 1994: 2, 34; Le Guay 1998: 34-35; Wilkins 1999: 28, fig 16; Van Bueren 1999: 27-28, fig. 19; Wieser & Giorgi 2000: I, 52 & II, 244; Korteweg 2002: 17-18, 129, 132-134, 210, no. 56, fig. 103-104; Brayer & Korteweg 2003: microfiche 15; Reidemeister 2006: 76; Rehnig 2006: fig. 2, 52; Rabel 2007: 268; Shoham-Steiner 2008: fig. 2.

In his *Civitate Dei,* or *City of God,* the church father Augustine (354–430) describes as world divided into two cities: the City of God and the City of the Devil. The City of God is found in heaven as well as on earth. On the one hand, it is inhabited in Heaven by God himself, the angels and saints; on the other, people who lead a virtuous life on earth also belong. The City of the Devil, inhabited by the wicked, is only located on earth. Here, these two different worlds intermingle.

The *Civitas Dei* was Thomas More's preferred reading material. Of all the church fathers, Augustine is the one More cites most frequently and there are reasons to assume that he already knew the text quite well as a young man.[1] The *Cité de Dieu* was also one of his sources of inspiration for *Utopia*.

The *Civitas Dei* was written between 413 and 426 and consists of twenty-two 'books' that cover a wide variety of subjects. Greek philosophy is discussed, as is the origin of the Divine and the Worldly States and the relationship between them. Augustine also goes into politics and social and economic problems. The author's significance for Christian thought is immense. He was the first to formulate the concept of the 'doctrine of two kingdoms,' with a heavenly city (*civitas caelestis*) and an earthly city (*civitas terrena*). Life on earth is not bad in itself, as long as one focuses on the heavenly afterlife by striving to cultivate the virtues depicted and by turning away from wickedness.

1 Trapp & Schulte Herbrüggen 1977: 28, no. 20.

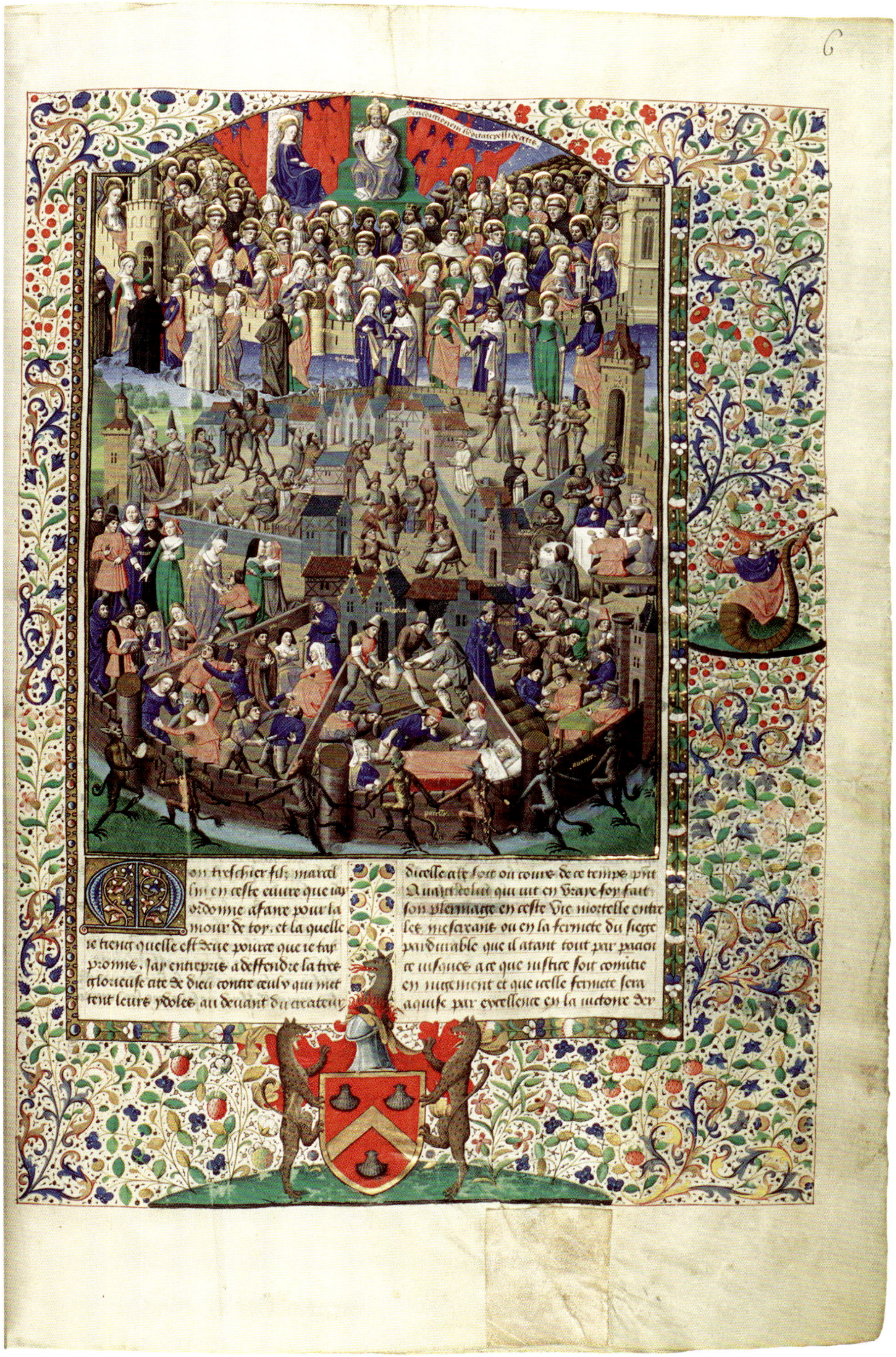
6
Mon treschier filz marcel
lin en ceste euure que iay
ordonne afaire pour la
mour de toy. et la quelle
ie tiens quelle est deue pource que ie lay
promis. Iay entrepris a deffendre la tres
glorieuse cite de dieu contre ceulx qui met
tent leurs ydoles au deuant du createur
d'icelle cite soit ou cours de ce temps pnt
Quant celui qui vit en vraye foy fait
son pelerinage en ceste vie mortelle entre
les mescreans ou en la fermete du siege
pardurable que il atant tout par pacien
ce iusques a ce que iustice soit conuertie
en iugement et que icelle fermete sera
aquise par excellence en la victoire der

The French translation shown here dates to around 1475–1480. It was made in Paris and it is one of the most richly decorated versions to survive. The coat of arms of the original owner have been cut away and replaced by the arms of Philippe de Commines (1447–1511), chronicler and advisor to King Louis XI. The existence of the two cities is magnificently depicted. On a miniature that takes up almost an entire page, the world is depicted as a walled city. In front of it seven devils dance in the foreground as representatives of the seven deadly sins. Within the walls, the city is divided into seven compartments, each of which contains a depiction of one of the seven virtues and its corresponding vice. In the foreground is Sloth, represented by several sleeping men and an unemployed spinster, opposite several people diligently working. Then follows Contentiousness, opposed to Peacefulness, represented by several praying figures. Jealousy stands opposite Friendship and Pride opposite Modesty or Humility. Lust opposes Chastity, represented by several monks and laypeople that turn away from its seductions. Gluttony is placed opposite Sobriety, and finally greed opposite Generosity. Above, seven virtuous men are conducted to heaven by female saints. Heaven is depicted as a walled castle in which a large number of saints are milling around the throne of God the Father, with Mary nearby.

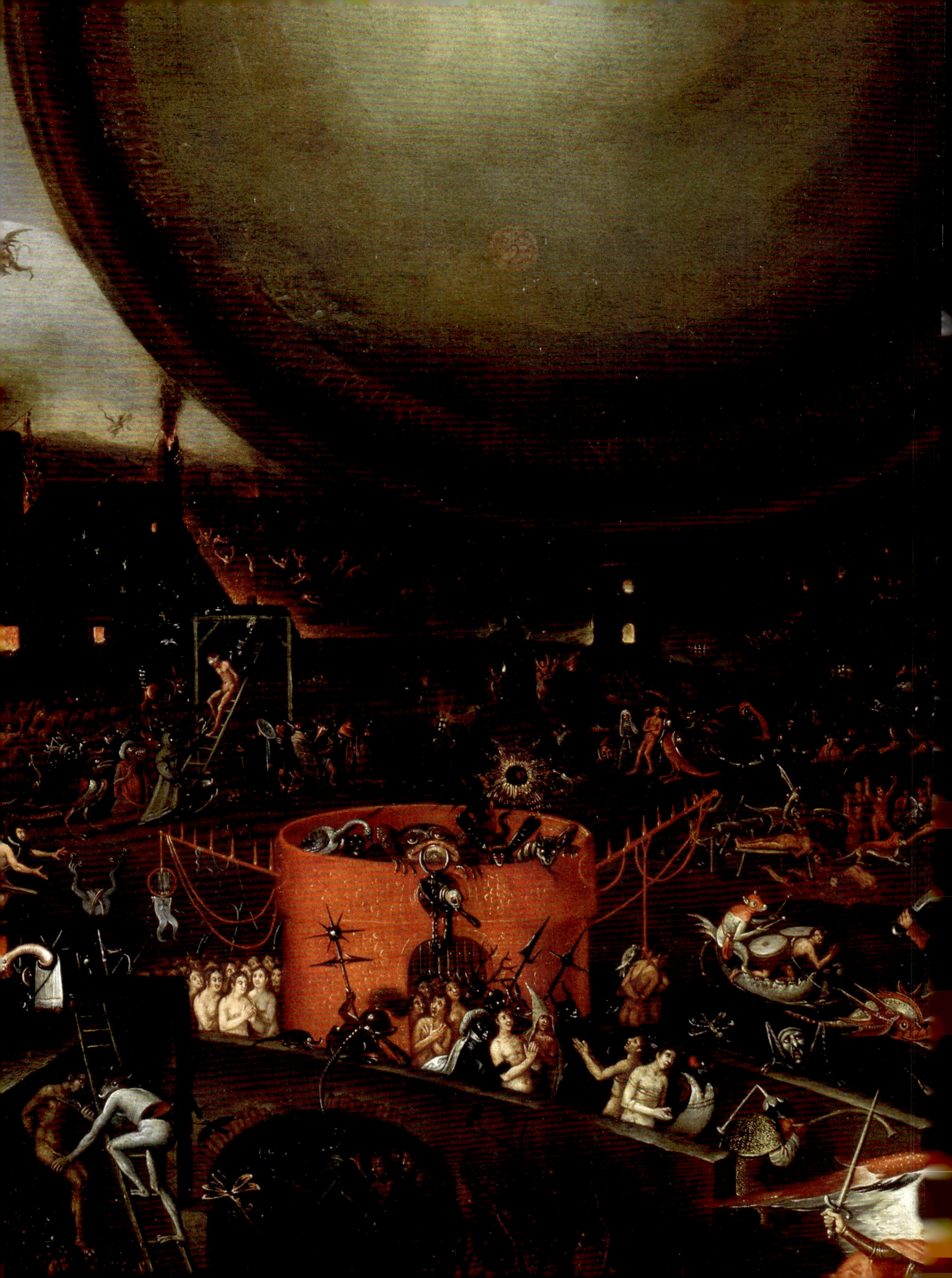

Dystopia

DAAN VAN HEESCH

[CAT. 39] DETAIL

In dystopia – a combination of the Greek words 'dys' (bad, abnormal, difficult) and 'topos' (place) – Paradise on earth makes way for a gruesome place where poverty, crime, chaos and inequality rule the day. The notion of an ideal city or state is turned around and replaced by a political or social bugbear, a negative inversion of what is actually desired. The term *dystopia*, a derivative of Thomas More's *Utopia*, first turned up in Victorian England, after which it evolved into an entire topos within the genre of twentieth-century science fiction. Nevertheless, the roots of utopia's dark side are centuries old. Artists gave not only positive but also negative form to a feverish imagination, driven by political, religious and social uncertainties. Works of art that depict divine retribution, eternal damnation and other fearsome images often refer not only to sinful human nature, but also to contemporary social problems. Themes such as the destruction of Sodom and Gomorra (cat. 38) were particularly popular in the visual arts the year *Utopia* was published. A godless and perverted city like Sodom was a true counterpart of the island Utopia and can thus be viewed as an archetype of the dystopian myth. In addition, artists like Hieronymous Bosch and his followers use the pictorial space of the underworld to give form to all manner of objectionable aspects of daily life (cat. 39). In their Hell scenes it is as if the damned are still on earth (cat. 41). In this way, aspects of daily life that are tied to a specific time and place are embedded in conventional Christian themes such as the Last Judgment and Hell. The terrible tortures that sinners must endure in Hell warn the viewer against the reprehensible behaviour that is characteristic of the anti-utopia described above. In the Last Judgment (cat. 40) and the popular text *The Vision of Tondal* (cat. 36-37), man is confronted by two extremes. The knight Tondal experiences not only the gruesome punishments of sinners in Hell, but also tastes the nectar of happiness in Heaven. In the end, utopia and dystopia (in fact two sides of the same coin) are inextricably bound to these ancient human fantasies concerning the eternal fire of Hell and the divine glory of Paradise.

36

Hieronymus Bosch, follower
The Vision of Tondal

c.1530
Oil on panel, 54 x 72 cm
Madrid, Fundación Lázaro Galdiano, inv. 2892

PROVENANCE

Nicolaas Beets, Amsterdam; before 1913 acquired by José Lazaro, Madrid.

BIBLIOGRAPHY

De Tolnay 1965: 384, no. 51; Boon, Bruyn, Lemmens et al. 1967: 151, no. 43; Corwin 1976: 215; Zehnpfennig 1979: 43; Unverfehrt 1980: 222, no. 83; Klein 2001: 125; Vandenbroeck 2002: 333, no. 64b; Tapié 2012: 182, no. 29; Cecchi, Hersant & Rabbi-Bernard 2013: 128-129, no. 58.

The painting of *The Vision of Tondal* from Madrid is entirely in the style of the diabolical creations of Hieronymus Bosch. It is not clear where and when the work was created. Various authors attribute it to an anonymous Antwerp master who was active in the years 1520–1530, but concrete indications of its origins are lacking.[1] Compositionally and stylistically, the work is closely related to two paintings with the same theme in the Denver Art Museum (fig. 36.1) and an unknown private collection. In all likelihood these three works were produced in the same workshop.[2]

The religious text *Visio Tnugdali* was written around 1149 by an Irish monk named Marcus. The main character in this vision of the afterlife is the licentious knight Tondal, who becomes unwell and falls into a state resembling death for three days. During this period an angel conducts his soul through Purgatory, Hell and Heaven. The terrible punishments meted out to sinners in the afterlife confront him with his own sinful existence. After Tondal awakens, he decides to lead a compassionate life of preaching.

The Visions of the Knight Tondal was a widespread text in the Low Countries in the sixteenth century. A printed edition appeared with Bosch's fellow citizen Gerardus Leempt in 1484. Several editions were also published in Antwerp, among which one with Mathijs van der Goes in 1472 and another with Henrick Eckert in 1515.[3] Several authors have tried to associate the text with works by Bosch and his followers (usually from Antwerp). In most cases, the hellish pictorial motifs in these paintings are too general to be able to speak of direct kinship with the literary text.[4]

In contrast to what the inscription at the lower left of the panel claims (*Visio Tondaly*), this painting is also not a direct visual translation of Marcus's text. It is instead a phantasmagorical catalogue of various punishments that reflect the missteps of the sinners. The knight Tondal lies sleeping at the lower left. As if in a dream, he experiences the horrors of Hell.[5] Behind him stands the angel who leads his soul through the afterlife. The sinners are tormented by devils that take on hybrid and at times anthropomorphic guises: an executioner with a pig's head, a monster with a trumpet-nose, a rat-like creature with orchid-butterfly wings, etc.[6] In the centre stands a frightening, mask-like face with glassy eyes and ears that have been pierced

1
Dendrochronological research has shown that the earliest possible felling date for the tree from which the panel is made is 1488. Klein 2001: 125.

2
Corwin 1976: 213–216; Unverfehrt 1980: 222 and Cecchi, Hersant & Rabbi-Bernard 2013: 128. The dimensions of the three panels are practically identical (Madrid: 54 x 72 cm; Denver: 55.5 x 71.8 cm; private collection: 52 x 70.5 cm). Each panel bears the inscription *Visio Tondaly(s)*; for the work from Denver see Bach 1949 and Unverfehrt 1980: no. 82, fig. 225; the work in a private collection is illustrated in Unverfehrt 1980: no. 84, fig. 224; according to De Tolnay 1965: 384 the work in Madrid may be by the same hand as the *Last Judgment* from the former Kadjar collection (sale Christie's, London, 9 July 2003, lot 25).

3
Palmer 1982.

4
Bax 1949: 275–277; Bax 1983: 379; Palmer 1982: 219–220 and Kren 1992: 19–20.

5
On the association of Bosch's visual language with dream images in the early sixteenth century, see Gibson 1992; see in addition Zehnpfennig 1979: 13–14.

6
It is not possible here to discuss each motif separately. For a cultural-historical interpretation of Bosch's formal language, see among others Bax 1949; Bax 1983 and Vandenbroeck 2002.

7
Among others see Vandenbroeck 1985.

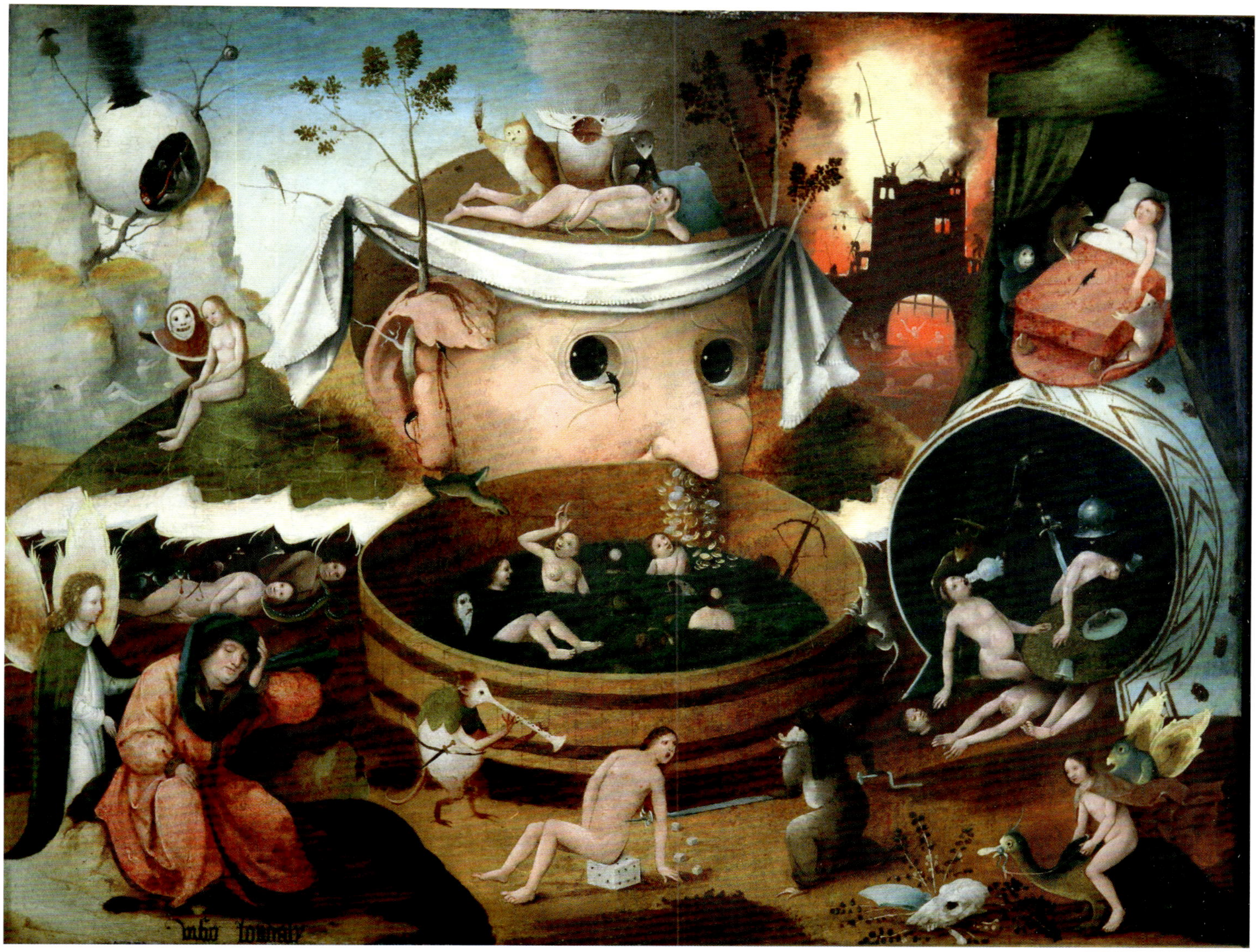

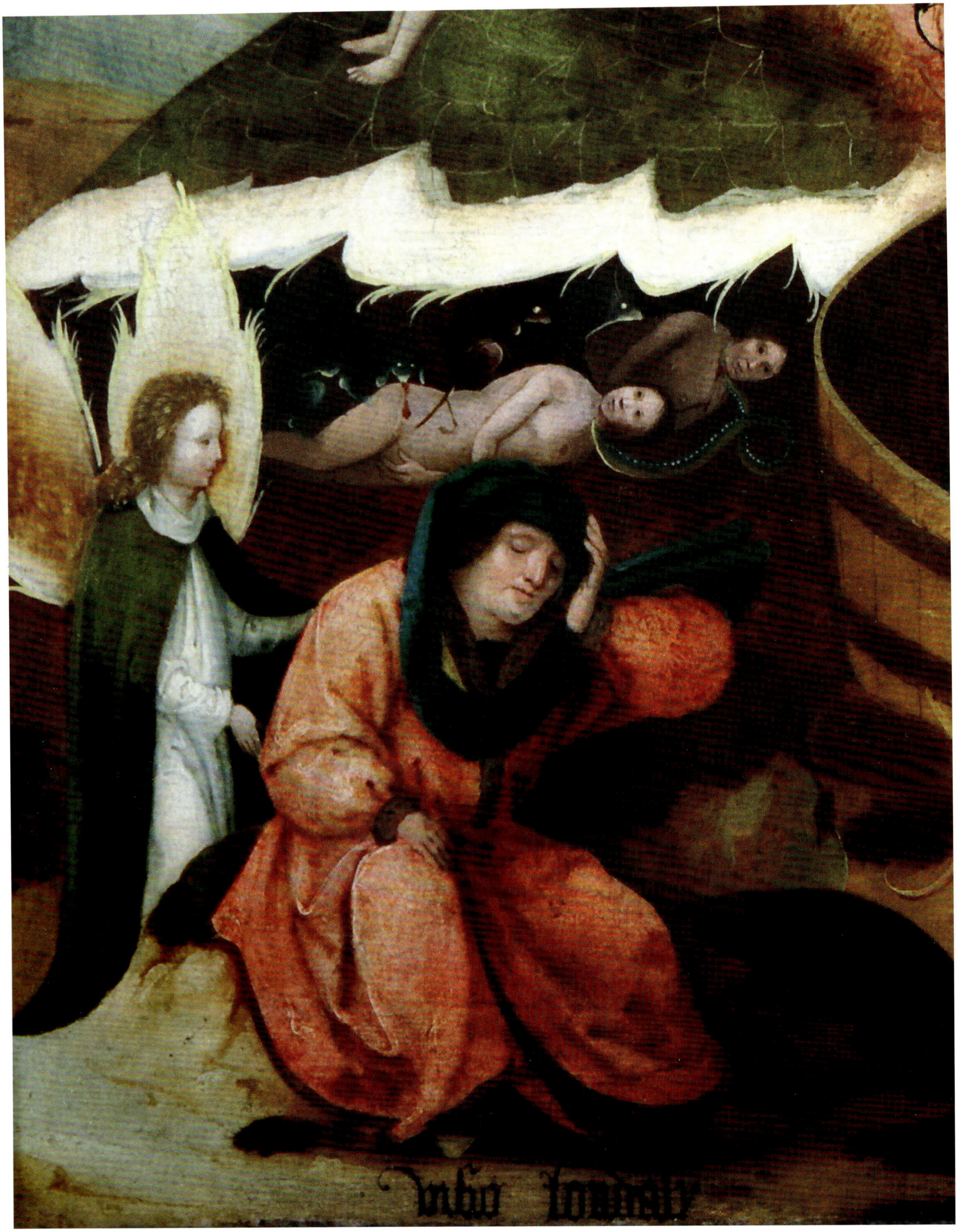

[cat. 36] detail

by trees. From the nose of this grotesque head, innumerable pieces of money stream into a large washtub in which souls are bathing. Various scenes refer to the deadly sins, such as pride (the demon with the mirror at the upper left) and gluttony (the diabolical tavern scene right of centre). The owl, which observes this terrifying scene from atop the hell-head, was a known symbol of folly and deceptive evil.[7]

DVH

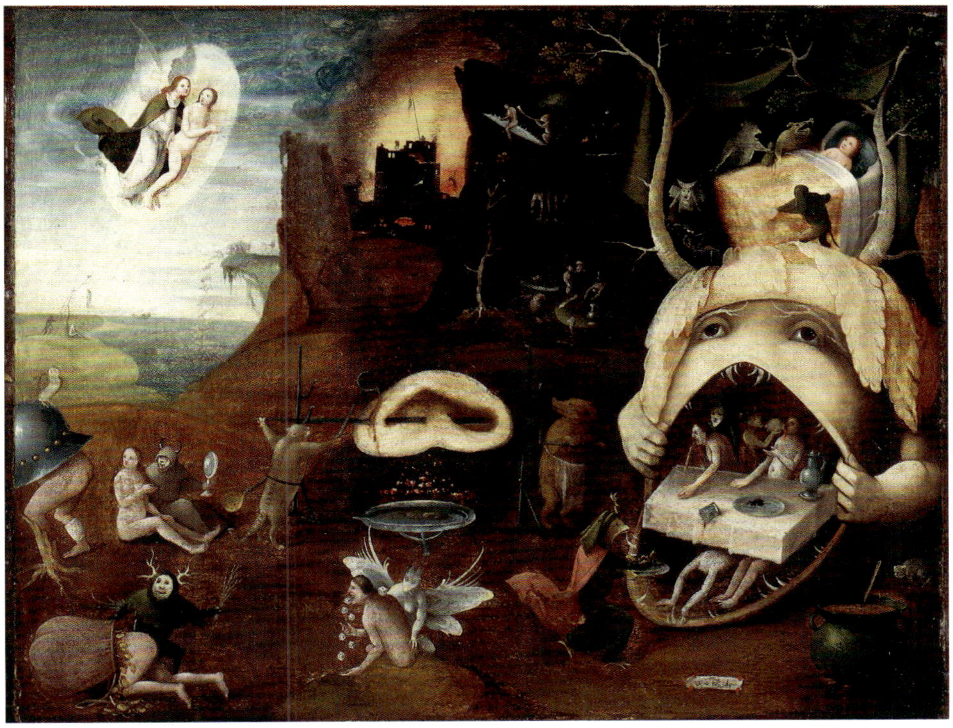

[fig. 36.1] Follower of Hieronymus Bosch, *The Vision of Tondal*, c.1520-1530, Denver Art Museum.

37

Simon Marmion
The Torment Reserved for Murderers

In: David Aubert, *Les Visions du chevalier Tondal*, Valenciennes and Ghent, 1475, fol. 13v–14
Ink, tempera colours, and gold on parchment, 363 x 262 mm
Los Angeles, J. Paul Getty Museum, Ms. 30

PROVENANCE

Margaret of York, duchess of Burgundy (1446-1503); Le Marquis de Ganay, in 1853 [his sale, Paris, Hôtel Drouot, May 12-14, 1881, lot 39]; Le Comte de Lignerolles [his sale, Paris, Charles Porquet, 1894, lot 17]; Baron Vitta; Baron de Brouwer, Manoir du Relais, Pommeroeul (Hainault); [to libraire Fl. Tulkens, *c*.1944]; [to H. P. Kraus, New York]; to Philip Hofer (1898-1984), Cambridge, Mass., in 1951 (his Ms. Typ 234 H); private collection, United States; acquired by the J. Paul Getty Museum, 1987.

BIBLIOGRAPHY

Cambridge 1955: no. 88; Miner, Carlson & Filby 1965: 53 no. 35; Kren & Wieck 1990: 3-36, 43, pl. 6; Eichberger 1992: 129-140; Nascimento & Miranda 2000: no. 119; Greenblatt 2001: 62-64, 73-74, 110-111, fig. 8-9; Dinzelbacher 2002: 104-105; Kren & McKendrick 2003: no. 14, and passim; Heck 2003: 227, 234, fig. 256; Kren & Ainsworth 2003: 46-47; Blanc 2004: 30-31, 51, fig. 22 and 40; Michalove 2004: 77; Giorgi 2005: 36, 43; Legaré 2005: 210, 211, fig. 4; Eichberger 2005: 210-211, 245-246 no. 84; Selliers 2006: 354-355, 368; Silver 2006a: 92-93, 348, 350; Turner 2006: 61-63, fig. 5.8; Morrison 2007: 91, 96-97, fig. 77; Klein 2007: 223, no. 82; Woods, Richardson & Lymberopoulou 2007: 116-118; Laneyrie-Dagen 2008: 169, fig. 3; Brown 2008: 241, 448; Cavagna 2008: 7-9, 16-18, 121-239; Marti, Borchert & Keck 2009: 251, fig. 103; De Pontfarcy 2010: XIV-XXII, passim; Kren 2010: 6, 21-23, 35, 26, 81-84; De Kesel 2010: 100-101, fig. 22-23; Mandabach 2010: 42-47, fig. 13-17; Göttler 2010: 228-29, fig. 100; Sciacca 2010: 9, 52, fig. 7 and 49; Wolfthal 2010: 188, fig. 136; Bousmanne & Delcourt 2011: 396-397; Carroll 2011: 253, 255; Van Buren & Wieck 2011: 224, 227, 374; Scott 2011: 41, fig. 24; Black 2012: 122, fig. 53; Schwarz 2012: 36-39, fig. 2-5; Kren 2013: 4-11; 18-35; Dubois 2013: 36-37; Guernelli 2014: 26, 31; Schinkel & Schmidt-Rutsch 2014: 128-131, no. 2.66, 2.66.1-2.66.13.

The twenty illuminations of the Getty's *Les Visions du chevalier Tondal* depict the varied horrors of Hell as well as the serene delights of Heaven.[1] The only known illuminated copy of this text, the Getty manuscript was illuminated by Simon Marmion at the behest of Margaret of York (1446–1503), duchess of Burgundy and wife of Charles the Bold (1433–1477).[2] It is perhaps the most original and visceral of all Marmion's creations.

The story of Tondal, originally written Latin but translated into a multitude of vernacular languages, became one of the most popular medieval accounts of the afterlife before the advent of Dante.[3] The text describes the spiritual journey of a knight who, after being led by an angel on an extended tour of Hell and then Heaven, vows to live a better life.[4]

1
This entry is indebted to the research of Thomas Kren and Roger Wieck. See especially Kren & Wieck 1990 and Kren 1992.

2
This text was originally bound together with a copy of *La Vision de l'ame de Guy de Thurno* (J. Paul Getty Museum, Ms. 31), and the *Vie de Sainte Catherine* (Bibliothèque nationale de France, Ms. n. a. fr. 28650): Kren 2013.

3
Kren & Wieck 1990: 3–7.

4
For a partial English translation, see Kren & Wieck 1990: 37–59; for editions of the French text found in the Getty manuscript, see Cavagna 2008 and De Pontfarcy 2010.

5
Kren & Wieck 1990: 41.

6
Kren & McKendrick 2003: 116.

7
Wieck in Kren 1992: 119–128.

8
Although Kren (1992: 19–20) concludes that Bosch probably did not see this manuscript, Carroll (2011, 253) points out that Margaret of Austria, who was named for her step-grandmother Margaret of York and may have known her manuscript collection, also owned a painting by Bosch, while Silver (2006a: 92–93) seems to leave the possibility open (see also bibliography cited in Kren 1992: 26 n. 58–61).

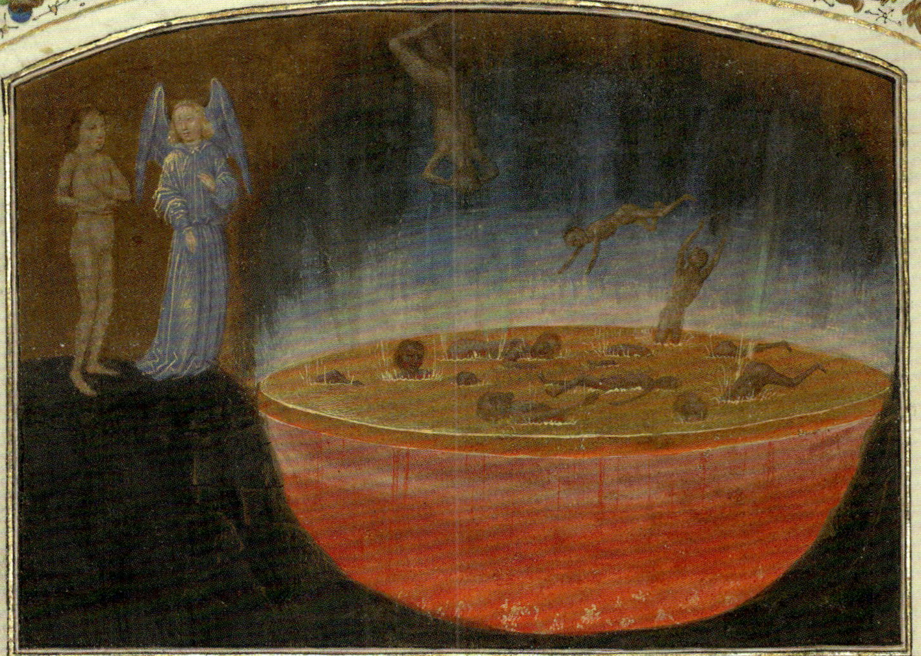

ample ouuerte et moult
obscure. Ceste vallee estoit
tres parfonde plaine de
charbons ardans. Et sur
icelle vallee avoit ung
couvercle de fer en rondeur
tout ardant moult grant
et massif a merveilles.
De celle horrible vallee par
tout tres obscure la puan
teur qui a sentir surmon
toit toutes les puantes
sta reure que oncques ame
sentist. Sur cest couvercle
descendoit lors une grant
multitude de ames desmp

nees qui estoient illecques
bruslees et arses et furent
couslees et passees parmi
cestui ardant couvercle en
la maniere que len passe
une faulse parmi lestam
ne et ainsi de la cheoient
ou feu diceulx charbons
ardans. Et la estoient les
tourmens a toute heure
renouvellez. Quant
lame de tondal apperchut
la grant horribleté diceulz
tourmens elle en fut tou
te espouentee si dist en tres
grant paour a langele

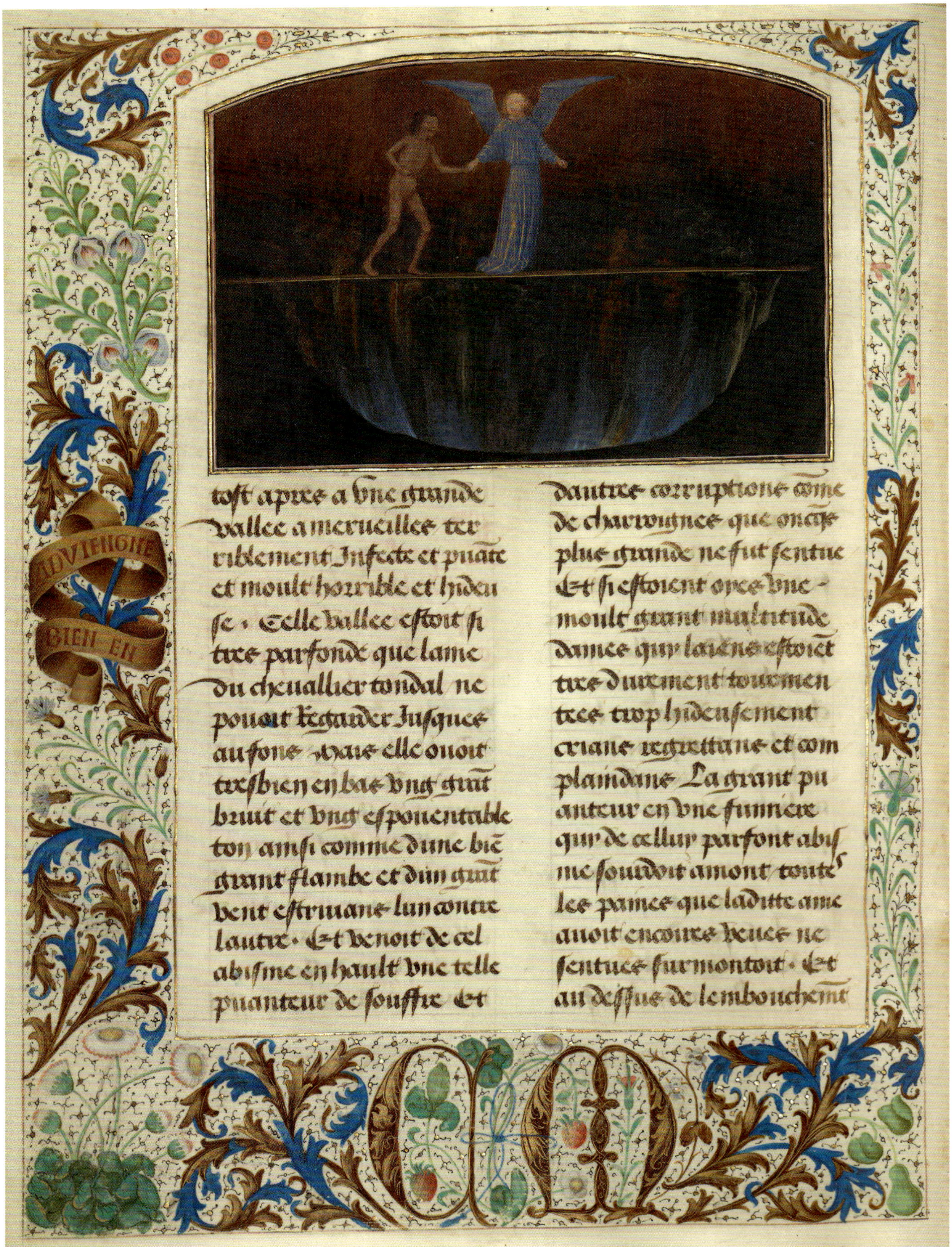
tost apres a vne grande
vallee a merueillee ter
riblement Infecte et puate
et moult horrible et hiden
se . Celle vallee estoit si
tres parfonde que lame
du cheuallier tondal ne
pouoit regarder Jusques
au fons . mais elle ouoit
tresbien en bas vng grant
bruit et vng espouentable
ton ainsi comme dune bie
grant flambe et dun grant
vent estriuans lun contre
lautre . Et venoit de cel
abisme en hault vne telle
puanteur de souffre et
dautres corruptions comme
de charoingnes que oncques
plus grande ne fut sentue
Et si estoient ores vne
moult grant multitude
dames qui laiens estoient
tres durement tourmen
tees trop hideusement
crians regrettans et com
plaindans La grant pu
anteur en vne fumiere
qui de celluy parfont abis
me sourdoit amont toutes
les paines que ladite ame
auoit encoures veues ne
sentues surmontoit . Et
au dessus de lemboucheme[n]t

[cat. 37] fol. 16

In Hell, a specific set of bodily tortures awaits those guilty of a particular vice or sin, such as the illumination on fol. 13v showing the torment reserved for murderers. The text describes a dark valley, over which is suspended a round lid of iron on which souls are roasted and then liquefied, falling into a searing fire below.[5] Marmion was a master at drawing out the physical terrors of Hell, making a point of evoking all five senses. One can almost hear the sizzle of the bodies as they hit the surface, feel the red-hot pain of Hellfire, and smell the odour of despair.

By comparison, it has often been noted that the scenes of Heaven that follow seem almost dull in their pastel peacefulness.[6] For an artist, it is no doubt much easier to envision the painful varieties of bodily torment than the evenness of spiritual bliss. In addition, in Hell, Tondal is often called upon to sample himself the particular torment awaiting those accountable for a particular sin. In the text as well as the illuminations of the Heaven scenes, it is clear that Tondal is constantly being separated from any personal experience of the delights of Paradise, often by walls or other barriers.[7] The difference between body and soul, Hell and Heaven, is thus underscored in multiple ways.

Although it is uncertain whether Bosch could have ever personally seen these visualizations of the Tondal story, the same nightmarish and almost claustrophobic concept of Hell permeates his works (cat. 36).[8] But while Marmion was intent on creating a convincingly naturalistic vision of Hell, Bosch's work is infused by a characteristic chimeric fantasy and frenzy that is absent from Marmion's more physically sensual Hell.

EM

38

Joachim Patinir
Landscape with the Destruction of Sodom and Gomorra

Antwerp
c.1521
Oil on panel, 22.5 x 29.7 cm
Rotterdam, Museum Boijmans Van Beuningen, inv. 2312, on loan from the Instituut Collectie Nederland, Amsterdam

PROVENANCE

In 1925 sold by P. de Boer to D.A. Hoogendijk (both Amsterdam); by 1927 in the possession of Ch.A. de Burlet in Berlin; before 1929 sold to F.W. Koenigs in Haarlem, who lent it to the Museum Boijmans Van Beuningen between 1935 and 1940; in 1940 sold to Goudstikker in Amsterdam; via the gallery Goudstikker/Miedl, the panel entered the possession of Hermann Göring; after the Second World War, the panel was regained by the Dutch state, which once again gave it on loan to the Museum Boijmans Van Beuningen in 1948.[1]

BIBLIOGRAPHY

Koch 1968: 24, 71; Corwin 1976: 220-221; Unverfehrt 1980: 178-179; Bätschmann 1981: 164, 174; Falkenburg 1988: 6, 55, 65, 106; Gibson 1989: 5-7; Lammertse 1994: 220-223; Silver 2001: 626-627; Silver 2006a: 367-369; Vergara 2007: 164-169, no. 2 (with exhaustive bibliography); Borchert 2010: 199, no. 63; Fricke 2011: 271-300; Vogelaar, Filedt Kok, Leeflang et al. 2011: 221-222, no. 16; Tapié 2012: 194-195, no. 35.

One of the earliest dystopian myths tells of the destruction of Sodom and Gomorra, two cities in moral decline that are described in the Old Testament book of Genesis (18–19). These immoral cities, destroyed by God with 'fire and brimstone', were the paragon of a corrupt society guilty of violence, cruelty and xenophobia.[2]

In this small panel, the landscape painter Joachim Patinir (1475/80–1524) situates the story in a panoramic, rocky landscape flooded with fire. The overwhelming inferno towers over the miniscule figures at the right side of the painting. The stark contrast between the low-lying cities around the Jordan River and the mountainous refuge is an image that is directly borrowed from the Bible (Genesis 19).[3] Lot and his daughters, the only ones spared by God, are led away by angels at the lower right. In the centre of the composition stands Lot's wife, who has been changed into a pillar of salt because she did not heed God's command not to look back; At the upper right, Lot is seduced in a tent by his two daughters, who hope in this way to ensure they bring forth descendants.

The panel in Rotterdam is regularly identified as the *Lot and his daughters* by 'Master Joachim', which the Antwerp pensionary Adriaen Herbouts gave to Albrecht Dürer in 1521. Although the painting is of higher quality than the versions in Oxford (Ashmolean Museum) and Antwerp (Koninklijk Museum voor Schone Kunsten), this statement remains hypothetical.[4] What is certain is that it is by the hand of the artist, who Dürer described as 'der gut Landschaftsmaler'.[5] On the basis of dendrochronological research, the work can be situated late in Patinir's oeuvre, or c.1521.[6]

1 For more details, see Vergara 2007: 164, 168–169.
2 Almog 2014: 157; Greene 2001: 294.
3 Silver 2001: 626.
4 Vergara 2007: 164–169, no. 2.
5 For a technical analysis of the painting, see Vergara 2007: 167–169.
6 Vergara 2007: 168.
7 Vandenbroeck 2004: 36–41; Vandenbroeck 2012: 107.
8 Falkenburg 2007: 69–79. See also Koch 1968: 24; Corwin 1976: 220–221; Unverfehrt 1980: 177–180; Silver 2001: 626–628 and Silver 2006a: 365–372.
9 Vandenbroeck 2002: 149, 310 and Vandenbroeck 2012: 105–109; on the relationship between natural phenomena and evil in the work of Bosch and Patinir, see also Falkenburg 2007: 69–73 and Falkenburg 2011: 102–116.
10 Gibson 1989; Silver 2006b: 26–52; Weemans 2013.

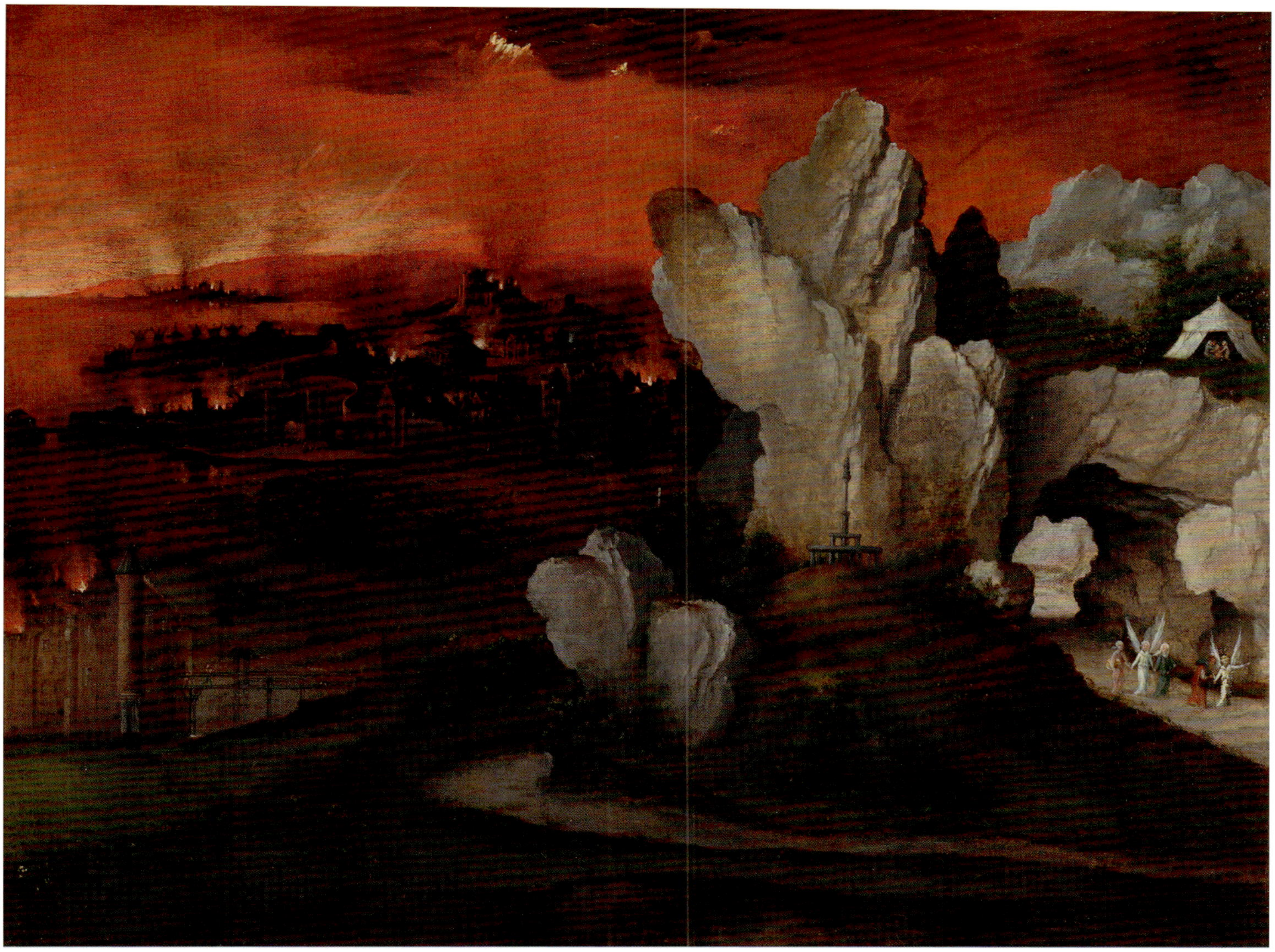

[cat. 38] detail

Themes such as the Destruction of Sodom and Gomorra and Lot and his Daughters were particularly popular in the visual arts from the early sixteenth century onwards. Such works reflected the prevailing idea that a fundamental disruption of the social and divine order led to universal disaster.[7] Patinir drew inspiration for his fiery landscapes from the phantasmal oeuvre of Hieronymus Bosch (*c.* 1450–1516).[8] The artistic kinship between the two masters is not limited to the strictly formal: in the works of Bosch, the idea that natural disasters and cataclysms are inextricably bound with moral wrongdoing also manifests itself.[9] Patinir's 'world landscapes' were widely copied in the sixteenth century and lay at the basis of landscape painting as an independent artistic genre.[10]

DVH

39

Hieronymus Bosch, follower
Apocalyptic Vision

*c.*1575–1600 (1595?)
Oil on panel, 120 x 170 cm
Venice, Palazzo Ducale (on loan from the Soprintendenza PSAE e per il Polo Museale della città di Venezia e dei comuni della gronda lagunare), inv. 269

PROVENANCE

Possibly identical to the large painting 'opposite the tribunal' in the chamber of the Council of Ten in the Doge's Palace in Venice, together with paintings of 'Il Civetta' (Herri met de Bles) described as 'various narrative and imaginary representations'.[1]

BIBLIOGRAPHY

Corwin 1976: 418–419, no. 157; Collobi Ragghianti 1990: 56, no. 95; Limentani Virdis 1992: 122–124; Danesi Squarzina 1995: 85–87; Bellavitis 2010: 191–195, no. 5; Squellati Brizio 2013: 133, no. 60; Meganck 2014: 181, n. 10.

The struggle of the divine order against diabolical evil is central to this apocalyptic vision.[2] In the sixteenth century, this cosmic battle was usually depicted as the battle between the archangel Michael and the seven-headed dragon, referring to the seven deadly sins. Michael, here in golden armour, does not fight the dragon of the apocalypse but a legion of infernal beings that swarm through a diabolical labyrinth.[3]

A gigantic star or globe appears to descend into the hellish chaos. This highly exceptional pictorial element has in the past been associated with two passages from the Apocalypse in which stars from Heaven fall to earth (6: 12–13 and 9: 1–3).[4] The concentric circles of the cosmic globe, however, suggest that these are descending heavenly spheres, among which the Empyrean, or highest Heaven, where God dwells with the angels and saints.[5] The Apocalypse relates that after the victory over Satan (20: 7–10) and the Last Judgment (20: 11–15), the old order will be replaced by a new, glorious Heaven (21: 1).[6] At the upper left, various souls are being borne up to Heaven, which suggests that the Last Judgment has already taken place. The painting in the Doge's Palace in Venice is probably a simultaneous depiction of the imminent conclusion of the Apocalypse, in which sin has been vanquished and the New Jerusalem established on earth.

Earlier authors have suggested that the painting functioned as a justice scene in one of the courtrooms of the Doge's Palace, but there is no concrete evidence to suggest it.[7] The anonymous artist clearly took inspiration from the hell scenes of Hieronymus Bosch, whose works could be found in Venice from 1521 onwards.[8] The overall composition and choice of pictorial motifs, however, are more closely related to the work of Bosch's Antwerp followers. The devilish horn-player at the lower right, for example, also appears in several works by Pieter Huys (*c.*1520–before April 1586) and his immediate circle.[9] The tormented soul with the swollen belly in the central foreground is moreover bor from a torture scene from

1 '[...] de più grande sta di rimpetto ad esso tribunale; tutto con varie istoriche e chimeriche rappresentazioni'. Zanetti 1771: 492. See also Bellavitis 2010: 192.

2 Apocalypse 12: 7–9: 'And war broke out in Heaven; Michael and his angels fought against the dragon. The dragon and his angels fought back, but they were defeated, and there was no longer any place for them in Heaven. The great dragon was thrown down, and that ancient serpent, who is called the Devil and Satan, the deceiver of the whole world – he was thrown down to the earth, and his angels were thrown down with him.'

3 The Musée des Beaux-Arts et d'Archéologie in Châlons-en-Champagne has a Boschian painting with similar iconography (inv. 861.1.104). This work is also based on the engraving by Hieronymus Cock (see below).

4 Squellati Brizio 2013: 133.

5 Compare with Limentani Virdis 1992: 122; Squellati Brizio 2013: 133 and Meganck 2014: 104, 181, n. 10.

6 The motif is highly unusual in both Boschian and apocalyptic iconography. A fifteenth-century manuscript in the Huntington Library (HM 83), which tries to map the geographic transformation of the world during the apocalypse and after the Last Judgment, shows a similarly unformed, crystallised 'new' heaven/earth. Van Duzer & Dines 2016: 185, 191, fig. 5.10.

7 See n. 1, Bellavitis 2010: 191–192 and Squellati Brizio 2013: 133.

8 Aikema 2001: 25–26; see also Aikema 2014: 30–43.

9 See cat. 40 [Pieter Huys] in this catalogue and Corwin 1976: 419.

10 Van Grieken 2013: 248–249, no. 62 and Bass 2015: 136–139, no. 14.

11 Compare with Limentani Virdis 1992: 122.

12 Squellati Brizio dates the panel to 1595 on the basis of an only partially (?) legible inscription '[15?]95', but it is as yet unclear whether it is actually a date, or the number 95. Squellati Brizio 2013: 133.

13 Limentani Virdis 1992: 122 (as 'JS') and Squellati Brizio 2013: 133 (as 'IS').

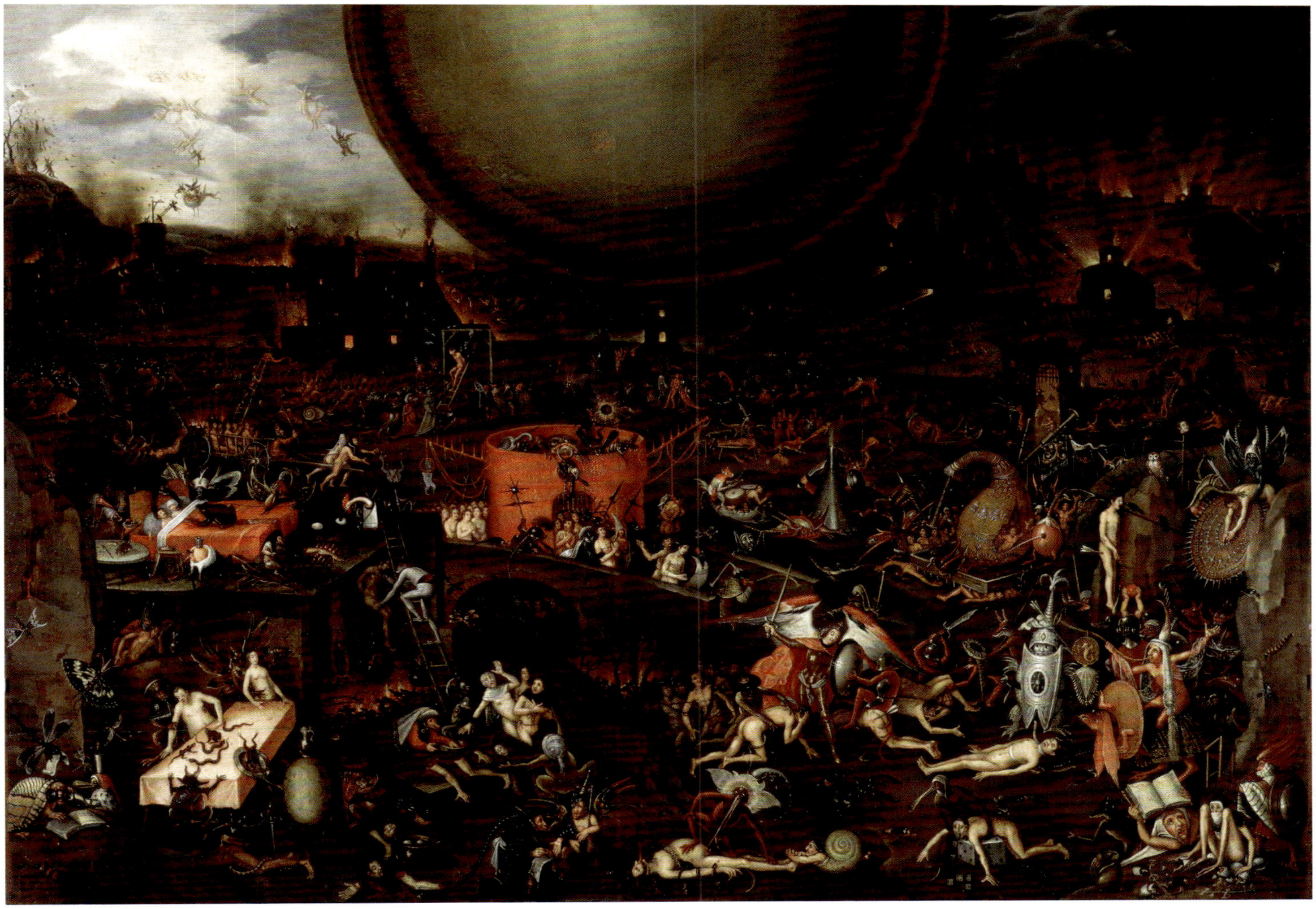

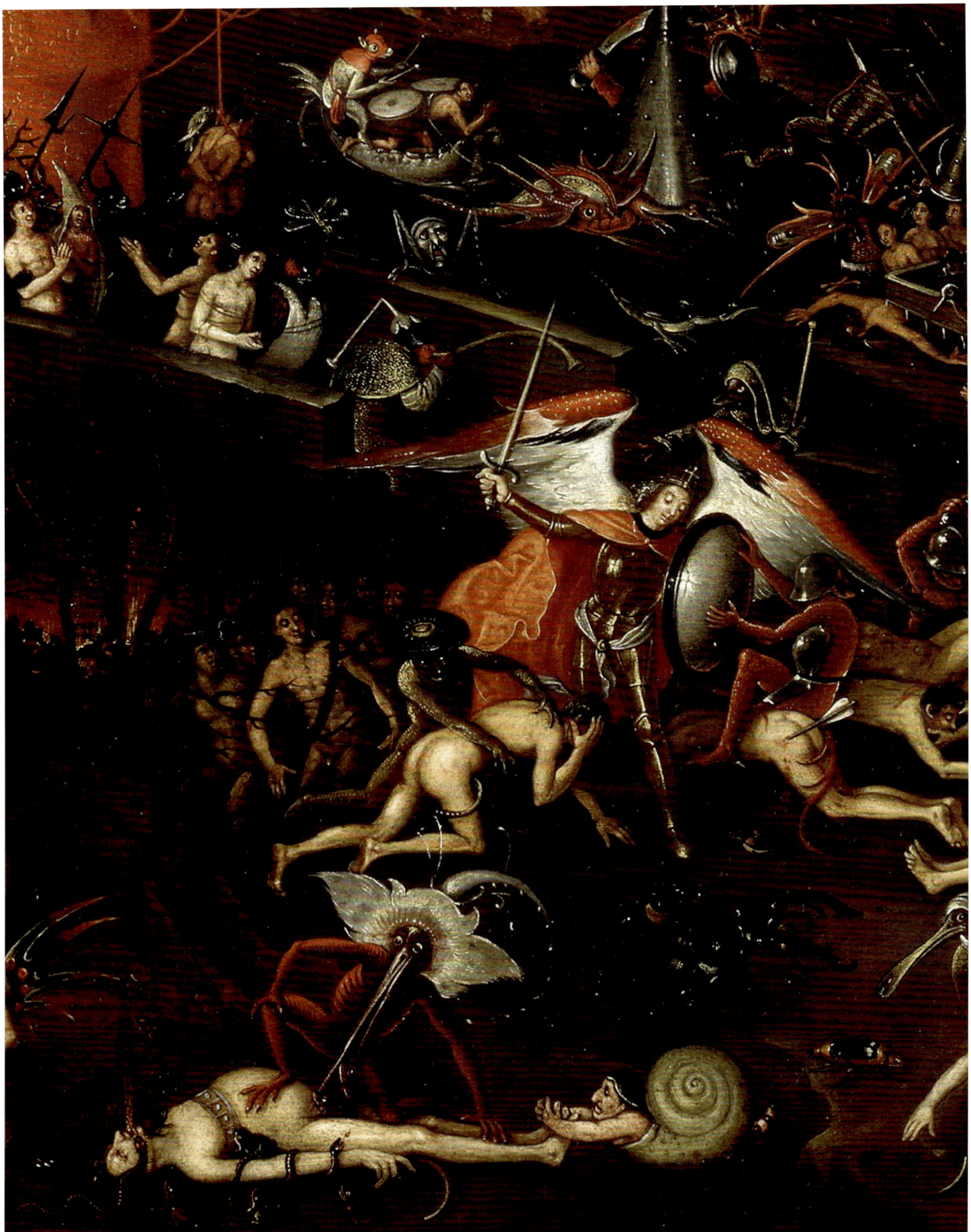

[cat. 39] detail

the Boschian engraving *Triptych with the End of Time, Heaven and Hell,* published by Hieronymus Cock (1517/18–1570) in the 1560s.[10] Finally, the carefully painted insects in the foreground bring to mind the artistic naturalism of artists such as Joris Hoefnagel (1542–1600).[11] Hence, the painting can probably be dated to the last decades of the sixteenth century.[12]

The work was previously attributed to the 'Monogrammist IS' on the basis of a supposed letter combination on the shield of a demon at the lower left of the panel.[13] It is unclear, however, whether this pictogram can be identified as a monogram. It is more likely a rod of Aesculapius, a staff entwined by a serpent that since time immemorial has been associated with medicine. In this way, a sadistic and absurd drama unfolds in which a damned soul is forced to eat poisonous reptiles and amphibians under the approving eye of a diabolical physician.

DVH

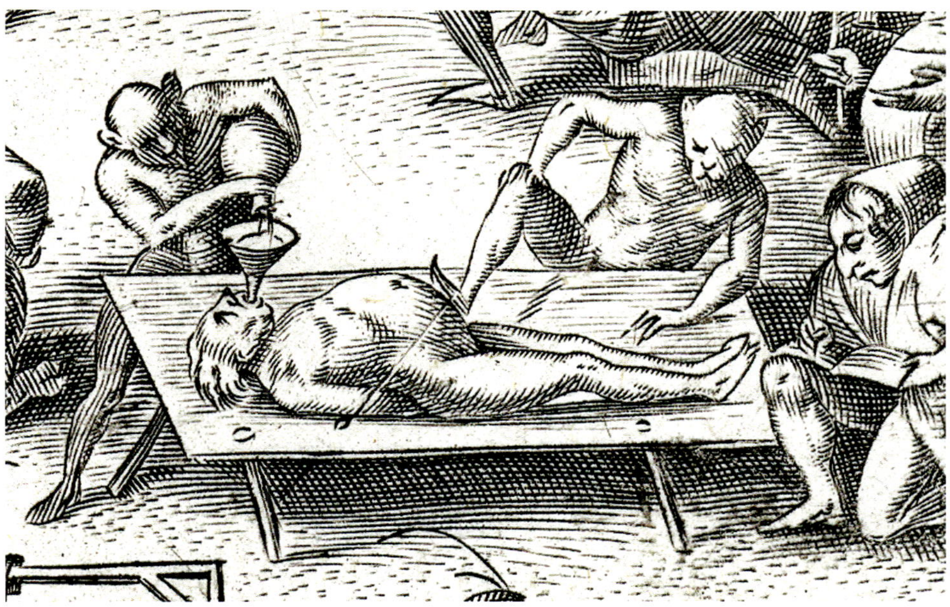

[fig. 39.1] Cornelis Cort after a follower of Hieronymus Bosch, detail from the right wing cf the *Triptych with the End of Time, the Heaven and the Hell*, Brussels, Royal Library of Belgium, Print Room.

40

Pieter Huys (?)
The Last Judgement

1554

Oil on panel, 133 x 100 cm

Brussels, Royal Museums of Fine Arts of Belgium, inv. 3900

PROVENANCE

Acquired by the museum via E. De Coninck in 1908.

The painting has been specially conserved for this exhibition by Etienne Costa.

BIBLIOGRAPHY

Ninane 1963: 123, no. 139; Boon, Bruyn, Lemmens et al. 1967: 230, no. 106; Corwin 1976: 325-326, no. 97; Mund 1977: 84-85, 153-154, no. 24; Unverfehrt 1980: 281, no. 129; Davies 1993: 86-89; Silver 1999: 37, 54 n. 18; Silver 2006a: 137, 280 n. 15; De Vrij 2006: 77-78.

In this painting Christ, as ultimate judge, passes his judgment on the blessed and the damned.[1] Above, Mary and John the Baptist assist him, each with six apostles at their side. The heavenly court is drowned out, however, by a swarming parade of multifarious and agonizing infernal beings. The possibility of salvation seems distant, but if one looks closely, one sees a tunnel of light opening up in the dark mass of clouds at the left to lead a few of the chosen to heavenly bliss. The figure on the slope in an antique style with a sickle probably refers to Revelation (14: 14–18) in which angels descend to earth in order to harvest the 'overripe grapes' – a known symbol of the Last Judgment.

The grotesque and playful demons embody various vices, such as pride, gluttony and ire, but refer above all to the phantasmagorical visual world of Hieronymus Bosch (*c.*1450–1516). The work is indeed punctuated by visual citations of this extraordinarily popular master. The compositional scheme is related to Bosch's *Last Judgment* in Vienna, and the divine tunnel of light seems borrowed from the *Ascent into Heaven* in Venice.

The painting is probably a relatively faithful copy of a *Last Judgment* by Jan Mandijn (*c.*1500–*c.*1559) in Springfield.[2] The Brussels *Last Judgment* is traditionally attributed to the painter and engraver Pieter Huys (*c.*1520–before April 1586), one of the many Antwerp artists who adapted Bosch's pictorial language to a more contemporary stylistic idiom.[3] There is no consensus as to the attribution, but what is certain is that Huys must have been familiar with this composition or one of its variants. The figure group with the trumpeting devil on the right, for example, recurs in Huys's *Hell* in Madrid (1570) and again in reverse in his *Temptation of St Anthony* in Antwerp (1577).[4]

The original frame with painted devils in grisaille (of which two figures below hold a banner bearing the date 1554) is quite striking and extremely rare. It is tempting to connect Huys's activities as a designer of printed margin decorations with this pictorial translation of what was then primarily a graphic

1
The lily and the sword above his head represent compassion and justice, respectively.

2
As a copy after the panel in Springfield in Corwin 1976: 325–326, no. 97; Davies 1993: 86 and De Vrij 2006: 77. By contrast Silver 1999: 37 and 2006: 137 considers the Brussels panel the prototype.

3
Attributed to Huys in Ninane 1963: 123, no. 139; Boon, Bruyn, Lemmens et al. 1967: 230, no. 106; Corwin 1976: 325–326, no. 97; Mund 1977: 84–85, 153–154, no. 24; Unverfehrt 1980: 281, no. 129; De Vrij 2006, 77–78; attributed to Mandijn and/or workshop in Corwin 1976: 325–326; for a recent study of the signed oeuvre of Huys, see Van Heesch 2014.

4
Corwin 1976: 324; see also cat. 39.

5
Already suggested in Ninane 1963: 123, no. 139; on Huys as an engraver and designer of book illustrations, see Bowen & Imhof 2008: 40, 47–49, 342–345.

6
Compare with Unverfehrt 1980: no. 21a, 33, 55a–b, fig. 31–33, 53–54, 56; see also the margin decorations of Bernard Salomon in *La Métamorphose d'Ovide figurée* (1557) in Elsig 2004: 183–184, fig. 2.

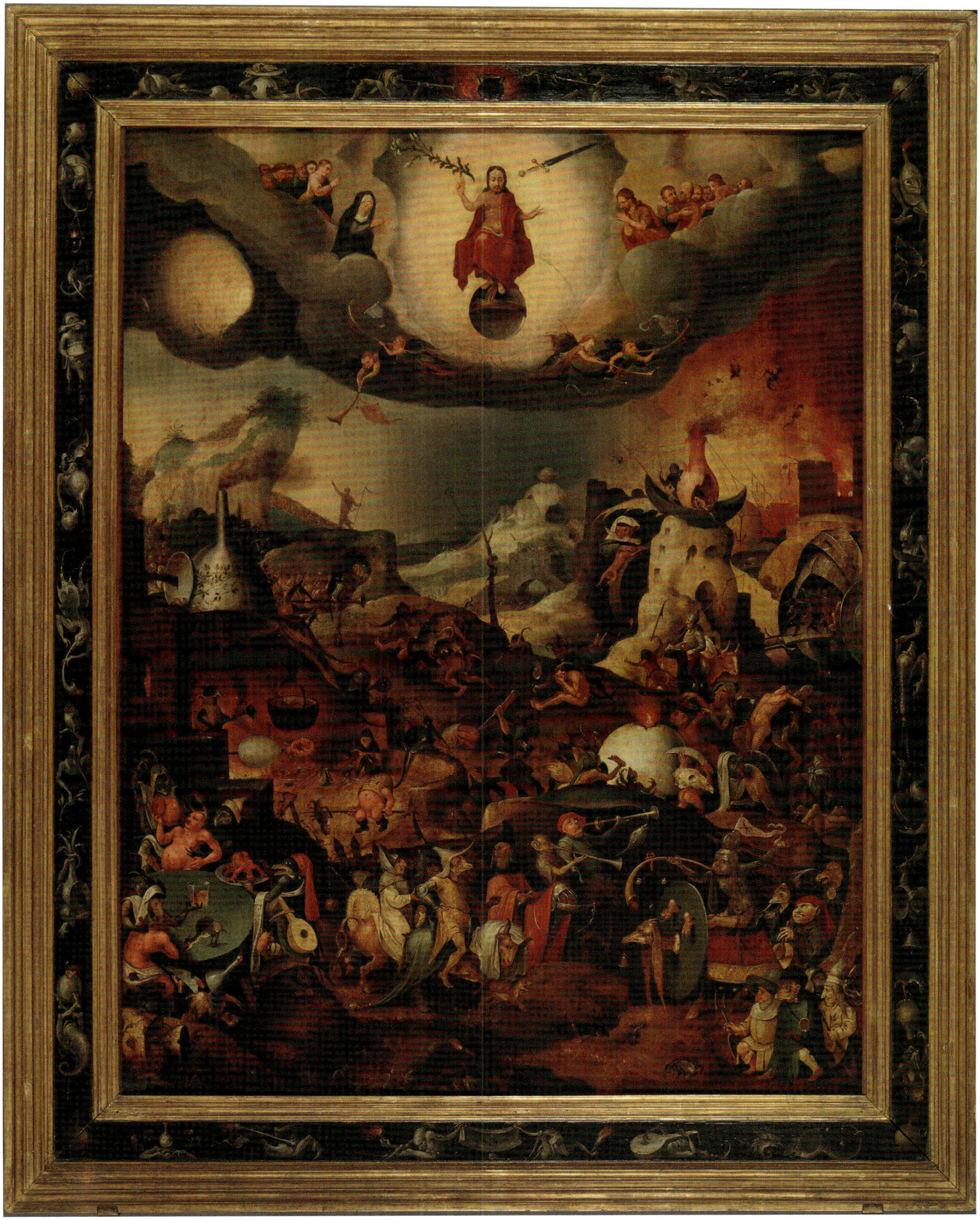

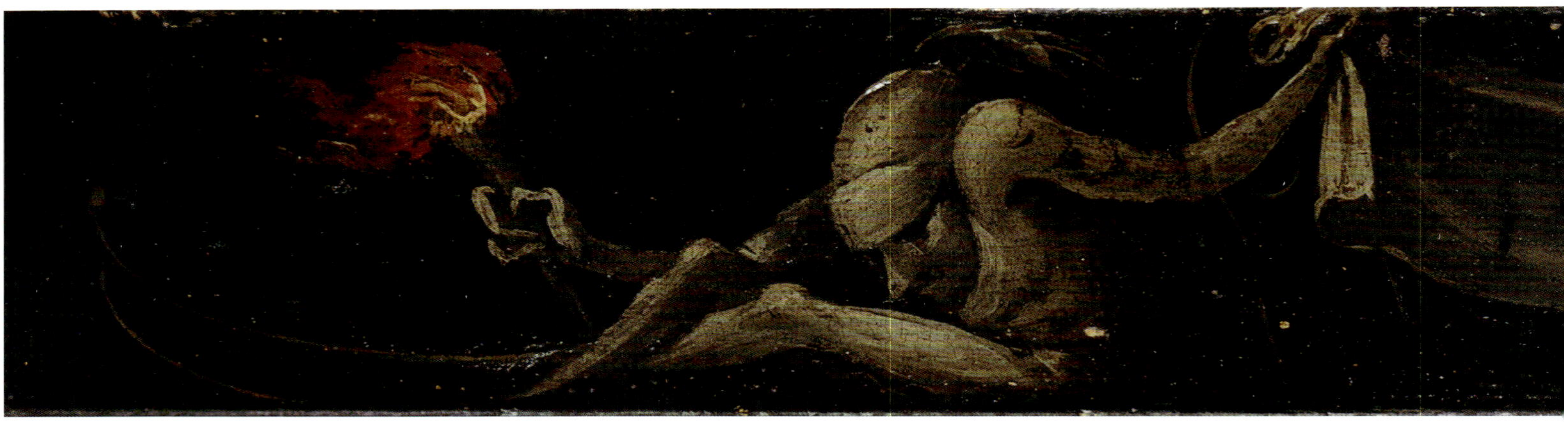

[cat. 40] detail

art form.[5] By the middle of the sixteenth century, Boschian demons were more often used as independent pictorial motifs in a decorative context.[6] Bosch's bizarre creatures were indeed recreated here in an ornamental context and thereby elevated to diabolical *mirabilia*, as it were.

DVH

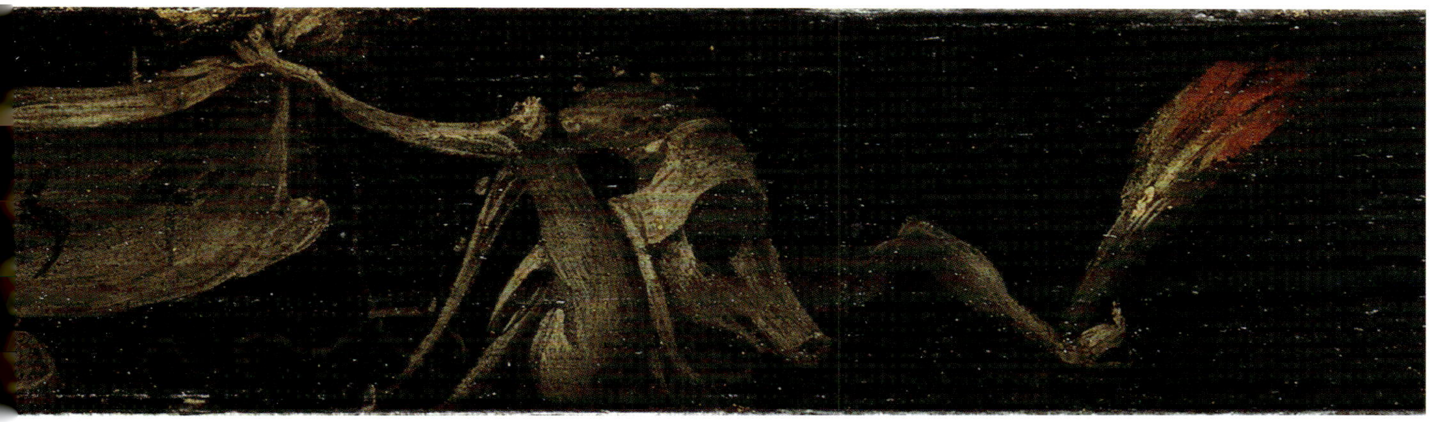

41

Hieronymus Bosch, follower
Christ's Descent into Hell

c.1525–1550

Oil on panel, 53.3 x 116.8 cm

New York, Harris Brisbane Dick Fund, The Metropolitan Museum of Art, inv. ACC 26.244

PROVENANCE

Philipp, count of Saint Genois, Vienna (before 1895); [Kominík, Prague, after 1895]; František Kominík, Prague (until 1926); [Hugo Feigl, Prague, 1926; sold to The Metroplitan Museum of Art].

BIBLIOGRAPHY

Donath 1926: 378-379; Friedländer 1927: 149, no. 88; Burroughs 1927: 272-274; De Tolnay 1937: 105, no. 58; Lewisohn 1937: 185; Baldass 1938: 21; Dallas 1947: 17; Wehle & Salinger 1947: 123-124; Larsen 1960: 98-99; Bergmans 1963: 136-137; Macagy & Sourian 1964: no. 2; De Tolnay 1966: 385, no. 58; Friedländer 1969b: 85, no. 88; Corwin 1976: 406-408, no. 144; Unverfehrt 1980: 289, no. 158; Sintobin 1998: 36, 254-256, no. 64; Larsen 1998: 136, no. 41; Stroo & Goetghebeur 2003: 401; De Vrij 2012: 568, no. E.24; Weemans 2013: 286-287; Büttner 2014: 31; Müller 2016.

This vividly populated Boschian panorama of the underworld presents Christ's descent into the first circle of Hell (limbo) between his death on the cross and his resurrection in order to redeem the souls of the Just of the Ancient Law, as outlined in the Gospel of Nicodemus in the *Apocryphal New Testament* (16: 1–13) and Jacobus de Voragine's *Legenda aurea*.[1] While armed demons try to prevent Christ's arrival, Old Testament patriarchs and prophets in anticipation of the Saviour populate a massive ruin, evoking the Tower of Babel. Besides Adam and Eve, the scene features Abraham and Isaac with the expiatory ram, Noah with a model of the Ark, Moses with the Tables of the Law, David with his harp, the Repentant Thief with his cross, and presumably Lot accompanied by his daughters. The figure at the rear carrying a lantern has been related to both Diogenes and Zephaniah.[2]

Besides the main narrative, a tiny St Christopher, carrying the Christ Child on his shoulders, can be identified in the tumult of the Antichrist's legion behind the tower. As the *Golden Legend* teaches us, St Christopher served the devil's army in his search for the world's greatest ruler before finding his lord in Christ.[3] Although unconventional, the appearance of the saint in this context might be read as an allegory for the human pilgrimage to Christ, challenged by trials of mischief and temptation.[4] In contrast, the painting confronts its beholder with Judas, prominently holding his moneybag, at the base of the right rock. Planning to slip off with the Old Testament figures, he is stopped by devils and prevented from seeking redemption, for he committed suicide by hanging, a deed strongly repudiated by the Church in the sixteenth century.[5]

1
James 1926: 83–85; Voragine 1969: 221–223. Unverfehrt 1980: 202 mentions two other versions of the scene; one, a faithful copy, is in a private collection in Milan; the second, the whereabouts of which are presently unknown, was property of the art dealership Matthiesen (Berlin) in 1928.

2
For the identification of the figure with Diogenes, see Bergmans 1963: 138–139; for its identification with Zephanias, one of the twelve minor Old Testament prophets, see the correspondence between Felix Thürlemann and the author, 2015–2016, New York, The Metropolitan Museum of Art, European Paintings Department, departmental file 26.244.

3
Voragine 1969: 377–382.

4
Müller 2016; for the cult of Saint Christopher during the sixteenth century and his different connotations, see Unverfehrt 1980: 187–201.

5
Büttner 2014: 31. The figure of Judas in such a pictorial context was emphasised by Karel van Mander in his description of a *Hell* that he had seen in Amsterdam and attributed to Bosch. This painting might have been an original panel by Bosch, which is lost today, or The Met's painting, as Büttner suggested; see Van Mander 1994: 124–125.

6
Donath 1926: 378; Müller 2016; for Sisyphus's early modern reception see Simon 2007: 136–192.

7
For the reception of Brant's book by Bosch and his circle see Silver 2006a: 243–252.

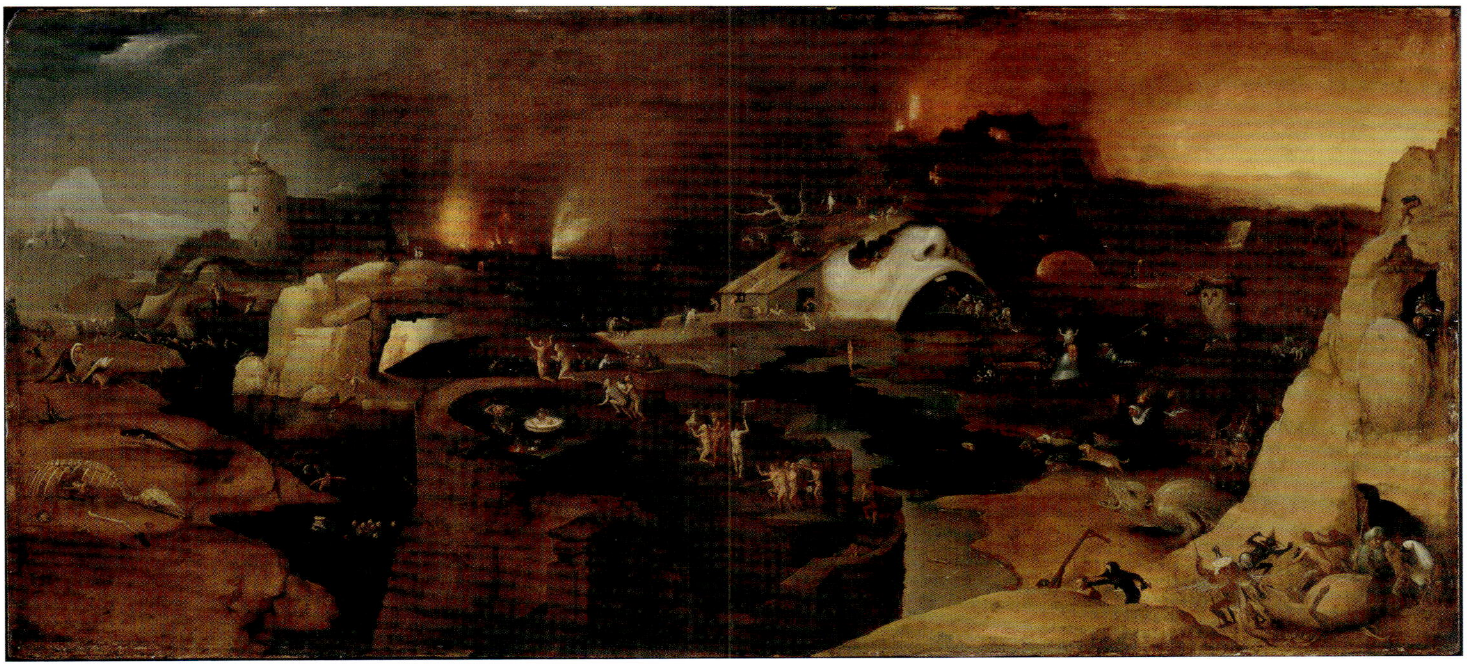

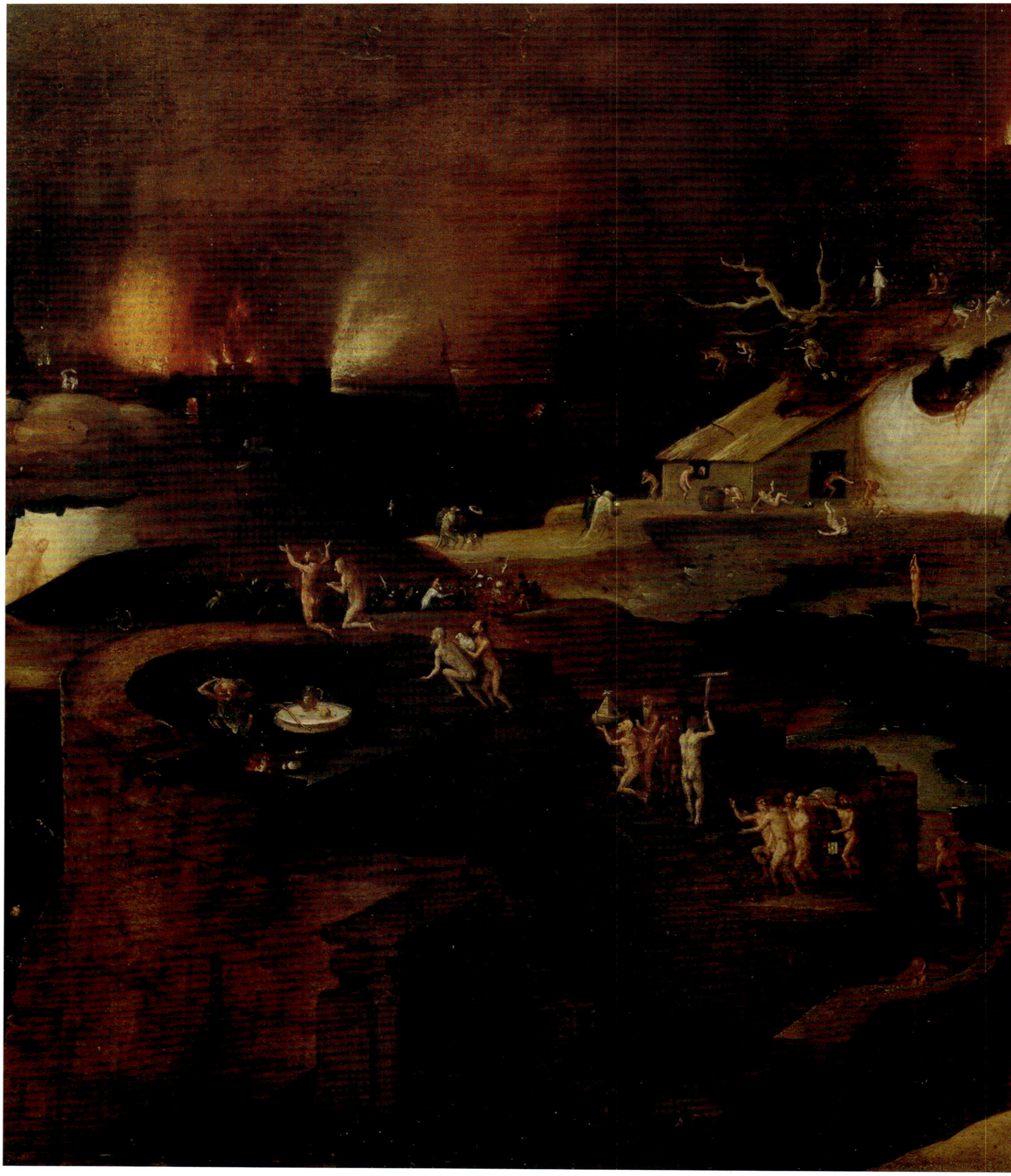

[cat. 41] detail

Moreover, Sisyphus, who climbs to the top of the right rock, plays a pivotal role for the image's subtext. In Greek mythology, Sisyphus was punished for his self-aggrandizing deceitfulness by being forced to roll an immense boulder up a hill in the underworld, only to watch it roll back down, repeating his action for eternity. In Christian thought and in contradiction to such saints as St Christopher, Sisyphus became the epitome of vain effort and senseless life. On the other hand, his endless aspiration to reach the rock's summit was received in humanist literature since Petrarch as an encouraging, moral metaphor for the human struggle to overcome a sinful lifestyle.[6] An allusion to immoral living is also ascribed to the burning ship and shipwreck in the painting's background. This recalls the image of the Christian community as a ship, inverted to serve as a satire on folly and vice in Sebastian Brant's *Ship of Fools* (1494), in whose conception the latter is ultimately an image of the followers of the Antichrist.[7]

In closely adopting Hieronymus Bosch's visual vocabulary, the painting's interwoven narratives offer a deep eschatological subtext for its audience, providing them with an encouragement to devote their lives to Christ and renounce immoral living in anticipation of the Last Judgment.

LMM

BEYOND THE HORIZON

Imaging the Unknown

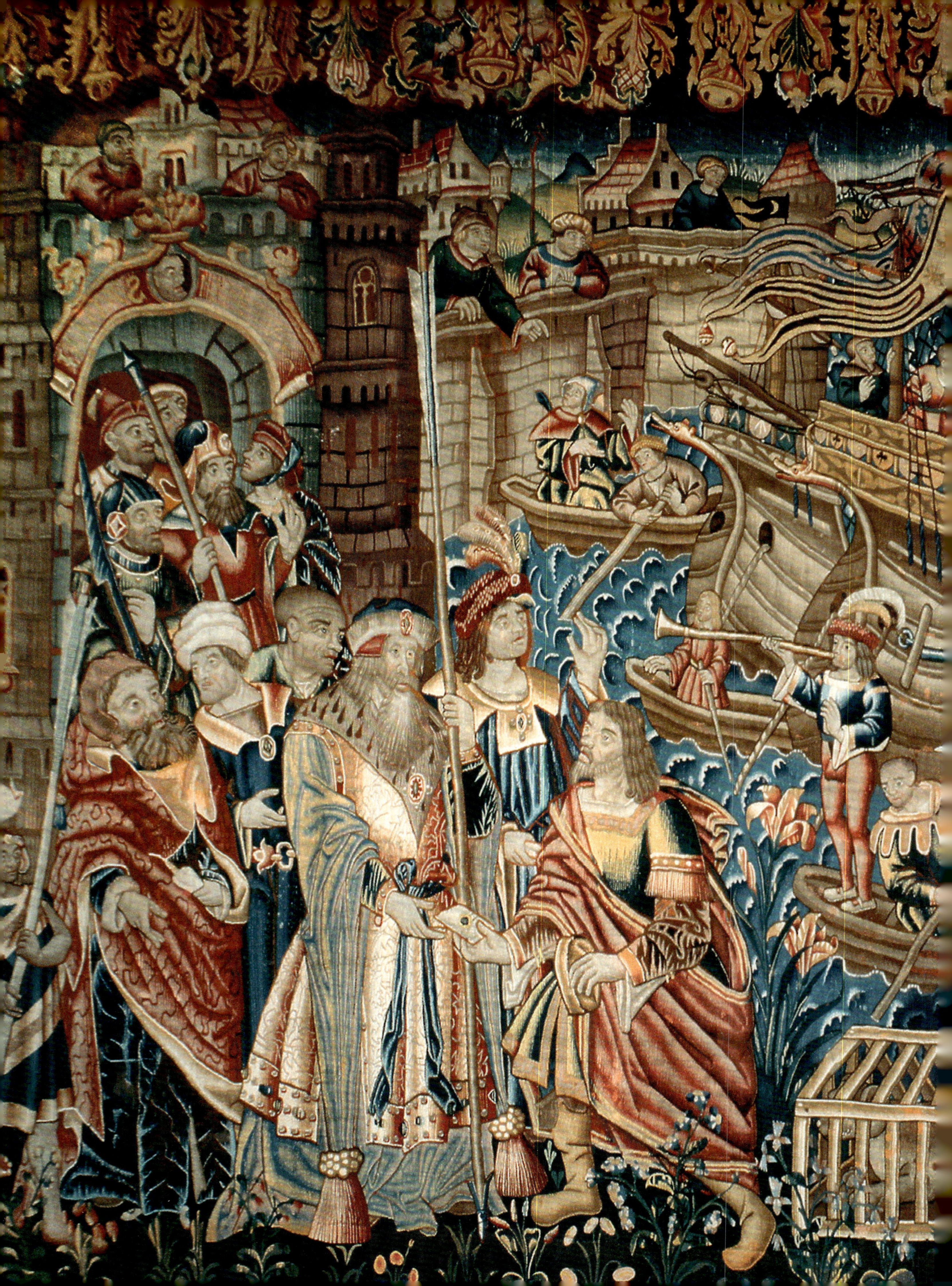

Terra Incognita

EMMANUELLE VAGNON

[CAT. 46] DETAIL

People living during the Middle Ages did not need to set out on faraway travels to come up against the boundaries of the known world. The way in which Europe, Africa and Asia are configured around the Mediterranean Sea in the schematic circular world maps often seen in medieval manuscripts was also a way of marking the boundaries that world. Often, these maps also depicted the history of mankind, from its origins in the East to man's recent spread across West-Europe. Looking at these maps gave people some idea of how the land and the sea related, allowing them to wonder as to what extent the world was inhafbitable and how the *terrae incognitae* could be reached: those uncharted regions that lay beyond the familiar circle of territories that had already been described in Antiquity, regions that were undoubtedly populated by monstrous races and mythical creatures. Not all maps from before the sixteenth century are this schematic. Some do a good job of accurately describing the regions. Yet, in their apparent simplicity, more than anything else they served as visual aids that stirred the imagination and caused people to reflect on the world in which they lived: the possible materiality of the biblical Garden of Eden, the climate around the equator, whether or not a 'world of the antipodes' existed, etc. The myriad hypotheses put forward by medieval scholars in this respect came to an end when, during the Renaissance, seafarers were able to establish *with their own eyes* whether or not any of these assumptions were true, sailing on to discover territories 'beyond the horizon' which no one had ever heard of.

The world of the Europeans had already expanded during the thirteenth century, with the travels of Marco Polo (1254-1324) and other merchants and missionaries. During the fifteenth century, fearless mariners allowed themselves to be tempted by hopes of discoveries and untold riches to enter the service of the kings of Spain and Portugal, and to take up employment with the major trading companies. The pope ensured the expansion of Christianity and divided the world among the European powers. Seafarers sailed to Africa and the Indian Ocean, remote territories of which little was known but which, all the same, had been part of the known world since Antiquity, and in doing so, they opened the path to completely new and wholly unsuspected *terrae incognitae*: the American continent and later the islands of the Pacific Ocean. All that is human was projected onto the new worlds: the dream of paradisiacal territories with unspoilt nature and humans unaffected by the sins of civilisation, the desire for quick riches and power, and the ability to start from scratch to build a perfect Christian society.

The cartographers of the Renaissance combined the legacy of Antiquity - Ptolemy's *Geographia* - with the tales of medieval travellers, the contributions of Arabic cartographers and of portulan charts and bundles with islands or townscapes. Until fairly recently, there were gaps in every representation made of the world. New worlds existed in people's imaginations before they were discovered and explored: fervent efforts were first made to locate *Terra Australis* (literally: South Land) and it was not until the seventeenth century that man started to explore the continent to which it owes its name, Australia. Home to uncharted territories and a wealth of islands packed with natives and exotic animals, the European maps of the sixteenth century were a visual geographical encyclopaedia that allowed for what was possible, what was probable, and for the past, present and future state of the world. There was every reason to believe that Thomas More's imaginary island Utopia would find its place in that world.

42

Simon Marmion
World Map (Mappamundi)

In: Jean Mansel, *La Fleur des Histoires*, Valenciennes, 1459–1463, fol. 281v
Parchment, 437 x 300 mm (folio), 300 x 220 mm (miniature)
Brussels, Royal Library of Belgium, inv. Ms. 9231

BIBLIOGRAPHY

Durrieu 1921: no. 6; Delaissé 1959: no. 058; Dogaer & Debae 1967: no. 175; Hottois 1982: no. 068; Lyna & Pantens 1984-1989: no. 301.1; Dogaer 1987: 14, 43-47, 51-53; Van den Bergen-Pantens 1989: no. 52; Van den Bergen-Pantens 1989: no. 52; Lemaire & Henry 1991: no. 04; Woodward 1991: 83; Avril & Reynaud 1993: 73-74; Zahlten 1995: 659; Cockshaw & Van den Bergen-Pantens 1996: 11, 92; Cardon 1996: 86; Bachfischer 1998: 118; Prevenier 1998: 101, 111, 189, 229, 337, fig. 90v, 176, 74, 120v, 116; Pelletier 1998: 31, 33-35, fig. 18; Orloff-Govaerts 1998; Smeyers 1999: 311, 338, 342, 345, 352, no. 38, 92, 102; Alexander, Marrow & Sandler 2005: no. 86; Van Hoorebeeck 2007: vol. II, 219-220; Wijsman 2010a: 233, 332, 362; Bousmanne & Delcourt 2011: 110, 389, 391, 393, no. 109; Dubois 2016.

Jean Mansel (*c.*1400–1473/74) served as an official in the employ of the Duke of Burgundy, Philip the Good. In addition to an *Histoire romaine* after Titus Livius, he also wrote *La Fleur des Histoires*, a compilation – in French – of the history of the world, from the time of Creation up until the reign of King Charles VI of France. This world map is the handiwork of Flemish artist Simon Marmion (1425–1489), who garnered considerable acclaim as a miniaturist, working at the Court of Burgundy (cat. 27 and 37).

The map has all the makings of a genuine painting. It has the time-honoured configuration of circulars maps, with an inscribed T. The inhabited world is divided into three parts, with all four elements duly represented: the earth is successively surrounded by the ocean (water), a sky filled with clouds (air), a circle of fire, and the stars. Asia takes in the top half of the circle. Separated by the Mediterranean Sea, Europe and Africa each take up a quarter of the circle at the bottom. The text underneath the image indicates that it is based on the *Etymologiae* by archbishop Isidore of Seville (560–636). This seventh century encyclopaedic work puts forward a description of the civilised world that is imbued with a Christian interpretation. Each continent is assigned to one of Noah's sons: Asia goes to Ham, Europe to Japheth and Africa is assigned to Chem. Jafet is dressed in Roman style (a red mantle is seen draped across his shoulder), whilst Sem and Cham are wearing an exotic tunic and hat.

This world map delivers a summary picture of the geography and history of the world. It tells of how, in the wake of the Flood – with Noah's Ark left stranded on a mountain in Asia – the Earth rose up again from the water and how this acted to separate the continents. In the same breath, the map shows the genealogy of the patriarch, who – according to legend – was the driving force that brought forth all the nations populating the inhabited world. Painted in the background of each continent are characters and delightful perspective landscapes. In the distance, 'ideal' yet unidentifiable fortified towns with towers and copulas are depicted in a blueish mist. The toponyms do not offer any clues that would enable them to be identified.

EV

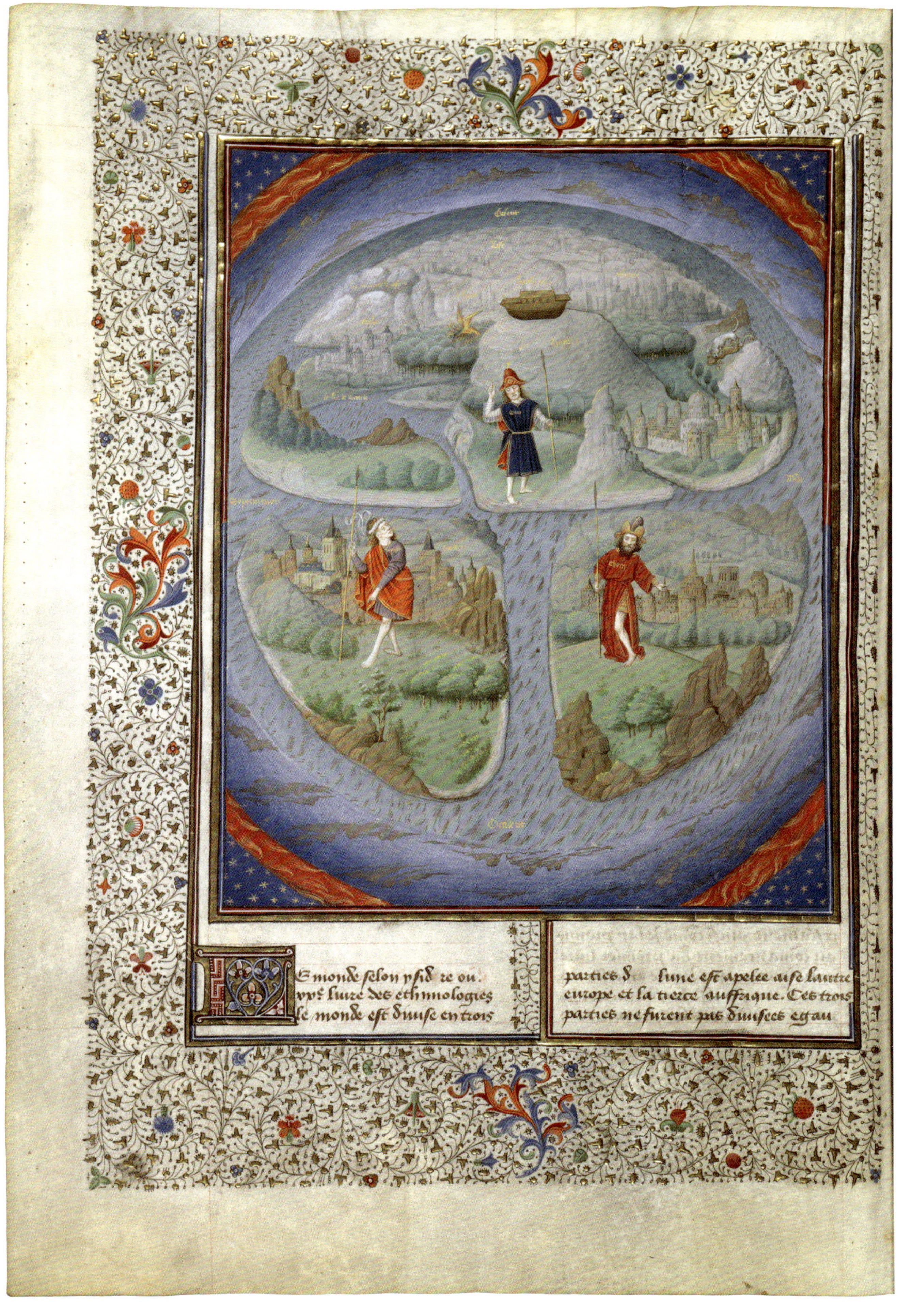
Le monde selon ysid re ou
xv.e livre des ethimologies
le monde est divise en trois
parties d line est apelee ase lautre
europe et la tierce auffrique. Ces trois
parties ne furent pas divisees egau

43

World Map (Mappamundi)

In: *Rudimentum novitiorum*, published by Lucas Brandis de Schass, Lübeck, 1475, fol. 74v–75
Coloured round woodcut, 375 mm (diameter)
Private collection

BIBLIOGRAPHY

Winter 1952: 102; Brown 1952: no. 42, pl. XV; Nordenskiold 1973: 3, fig. 2; Shirley 1984: XXI, pl. 3 & 1-2, no. 2; Campbell 1987: 144-151, no. 26; Nebenzahl 1995: 61-62, pl. 20; Brown 2000; Scafi 2006: 204, fig. 8.3, pl. 11a; Edson 2007: 169-172, fig. 7.2; Whitfield 2010: 34-35.

These *First principles* (*Rudimentum novitiorum*), first printed in Lübeck in 1475, make up a kind of history textbook aimed at young monks or students. In 1488–1491, the book was published by Pierre le Rouge and Guillaume in a French translation: *La Mer des Hystoires*. Before 1555, the book saw at least seven reissues. The work is illustrated with genealogical tables and maps, providing a summary of the history of the entire world according to medieval views which were Christianised concepts from Antiquity.

The schematic and didactic codes of the world atlas adopted in this book take little or no account of the right scales of topographical accuracy. As such, the map is in keeping with the type of world maps produced during the Middle Ages. So much so that, for a map that was created before the end of the fifteenth century, it looks quite archaic. The presentation of the local regions and towns looks more like a list than a spatial representation. The hills lined up one behind the other in different planes are used to create the illusion of natural scenery. Each hill represents a region. Brown ink was used to colour the water that separates the regions. The map observes the traditional scheme, involving a circle with an inscribed T (a cross in this particular case); the East is situated on top, with the names of the directions of the wind written at the four outer ends of the cruciform. The top half is taken up by Asia, the bottom half by Europe (left) and Africa (right) (cat. 42). Because of the fold in the book, the Mediterranean Sea – which separates Europe from Africa – is barely visible. The world extended from a walled garden (widely believed to be the Garden of Eden) at the top to the Pillars of Hercules (the Strait of Gibraltar) at the bottom. Right in the middle are Judea and Palestine and a depiction of Jerusalem. In Europe, viewers' attention is drawn to Rome (walled city, papal figure) and Venice. There is no sign of Constantinople. The two characters in the Garden of Eden are not Adam and Eve, but two men bearing olive branches. Historians have identified these men as prophets, representing the dialogue between the Old and the New Testament or a master and his pupil.

The map incorporates the usual locations seen in the accepted geography of the Bible and Classical Antiquity. For the towns and cities, simple building indications are used, whereas crowned figures are adopted for the kingdoms. In Asia, the Queen of Sheba is depicted on the Kingdom of Ophir and the Red Sea. Just outside the Garden of Eden is the Tree of the Sun and Moon, which is said to

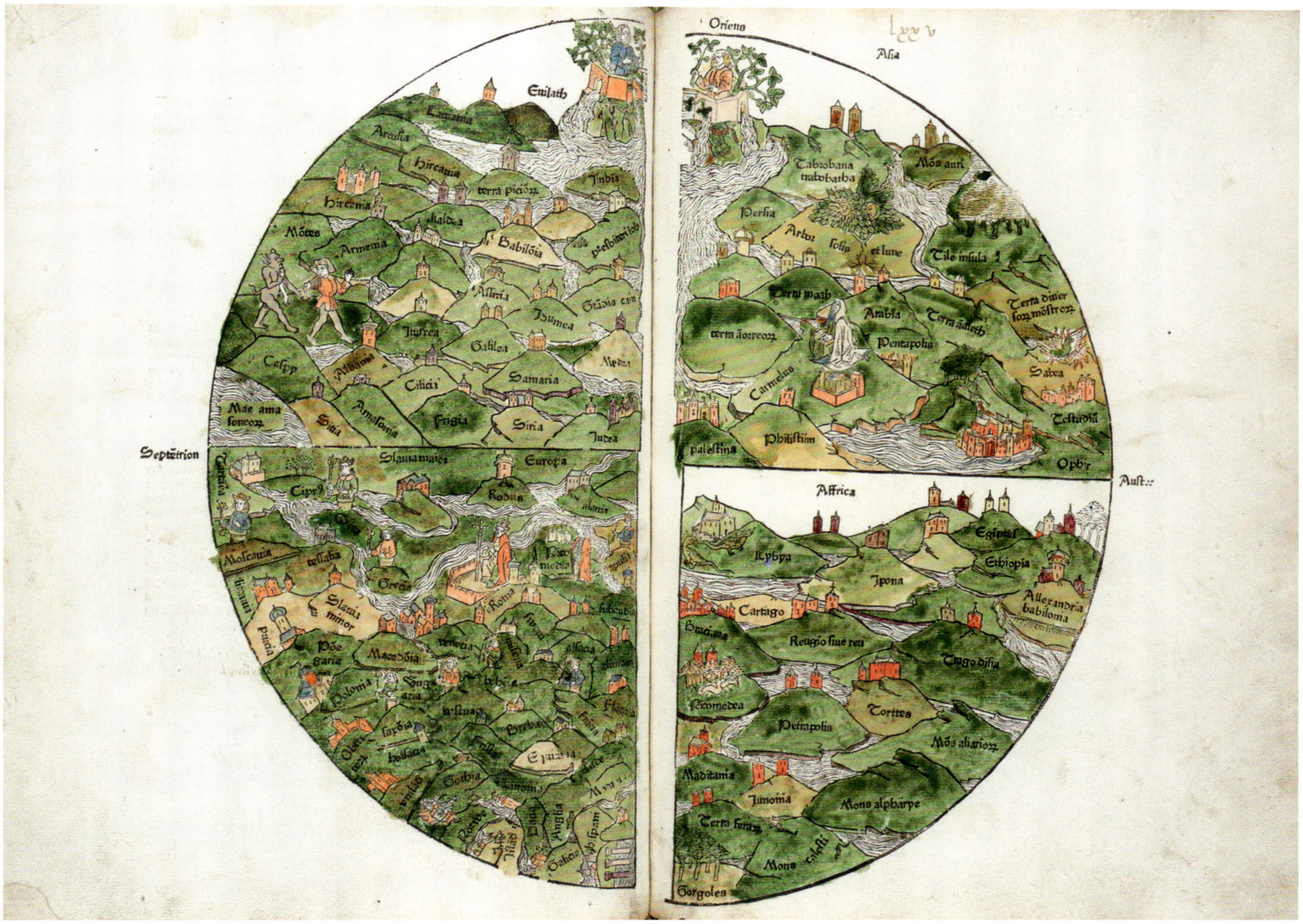
Oriens
Asia
Septētrion
Affrica
Persia
Arabia
Pentapolis
Sabea
Carmelus
Philistiim
palestina
Ophir
Egiptus
Ethiopia
Cartago
Alexandria babilonia
Mons alpharpe
Gorgoles
Armenia
Babilonia
Assiria
Idumea
Galilea
Samaria
Cilicia
Frigia
Siria
Judea
India
Europa
Rodus
Grecia
Roma
Macedonia
Anglia

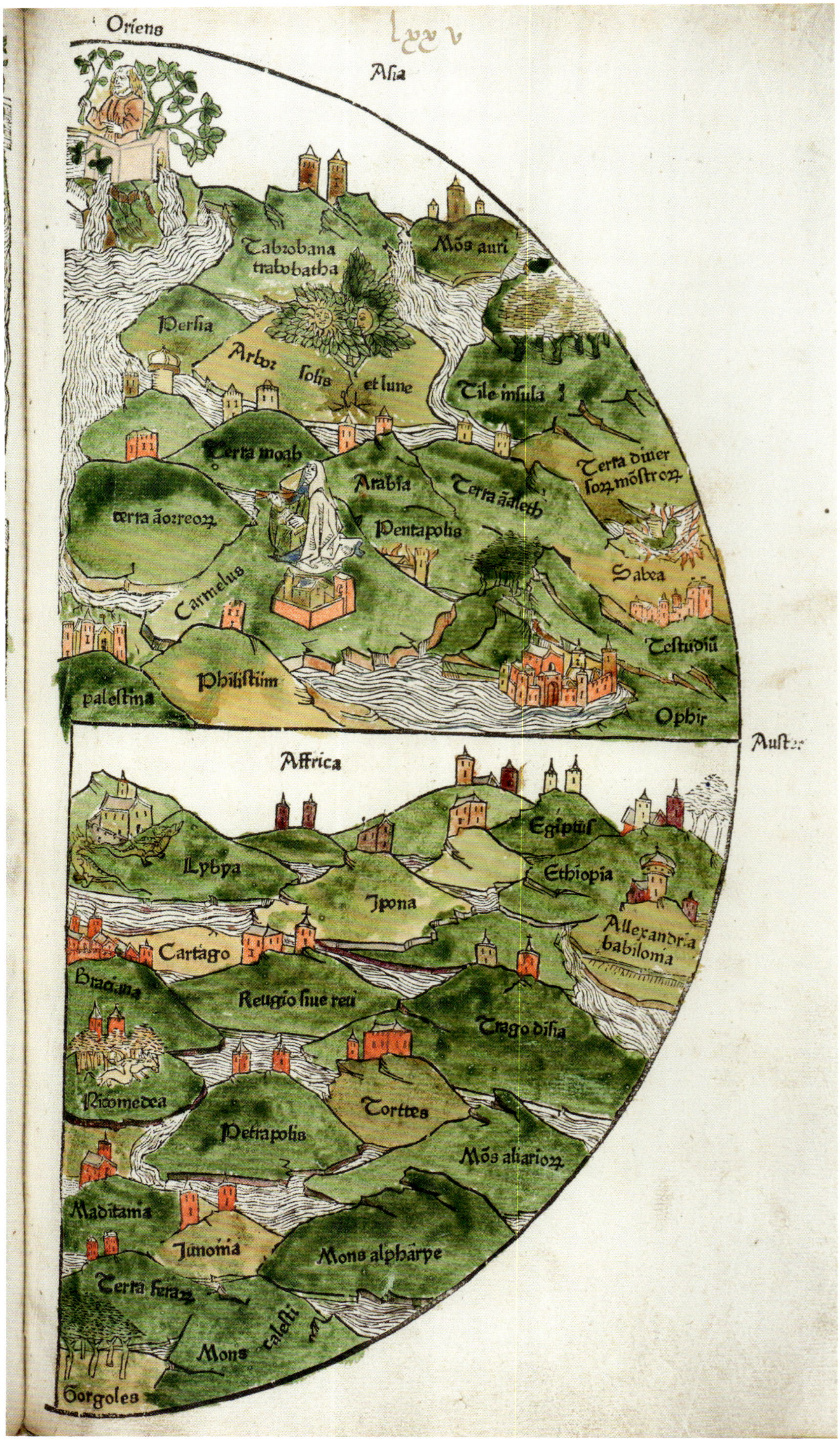

[cat. 43] detail

have been known to Alexander the Great. On the edge of the world, dragons and monsters are seen to loom large (*terra diversorum monstrorum,* a caption explains). The hair-covered character with the devil's tail in the north, close to the Caspian mountains (*Caspii*), could possibly be the King of Gog and Magog. The way in which the geographical objects are configured is a little confusing at times. On the right, by the Indian Ocean, is the island of Taprobana (modern-day Sri Lanka), as well as 'island of Tile' from Classic Antiquity, which is commonly thought to be Iceland.

EV

44

Map of the River Nile and Unknown Territories in North Africa

In: Claudius Ptolemy, *Geographia*, Ghent 1482–1485, fol. 79v–80
Latin translation of the original Greek text by Jacopo d'Angelo [Jacobus Angelus]
Scriptor: Jean de Kriekenborch (studio)
Patron: Raphael de Mercatellis
Parchment, 535 x 400 mm
Brussels, Royal Library of Belgium, inv. Ms.14887

BIBLIOGRAPHY

Stevens 1908; Van den Gheyn 1927: no. 7350; Bagrow 1945: 318-387; Brussel 1973: no. 24; Derolez 1979: no. 6; Elkhadem, Heerbrant, Calcoen & Wellens-De Donder 1994: 19; Elkhadem 2005: 86-87, no. 30.

The *Geographia* by Claudius Ptolemy (c. 90–168), written in Alexandria in the middle of the second century, sometimes also referred to in the humanist tradition as the *Cosmographia*, is one of the main geographical treatises that was studied during the Renaissance. In the Middle Ages, the work was simply unavailable in the Western world. It was not until around 1409, when Jacopo d'Angelo translated a copy from Greek into Latin, which Byzantine scholar Manuel Chrysoloras had brought with him when he travelled to Florence. In the fifteenth and the sixteenth century, the study of the *Cosmographia* – as the work was referred to at the time – resulted in a thorough overhaul of geographical insights and contributed to the expansion of Europe. The work is a description of the world as it was known at the time of the Roman Empire, based on mathematical and astronomical foundations. The first parts sets out the general principles of the structure of the world and explains how a map should be made. This is followed by a lengthy list of place names with matching coordinates, a map of the inhabited world quadrilled by parallels and meridians (*mappa mundi*) and, depending on the manuscript, twenty-six or sixty-four detailed regional maps (the oldest remaining Greek specimen of which date back to the thirteenth century) with the principal 'cities of interest' and their latitude and longitude. The work met with great favour among humanists. Numerous luxury versions were made for scholars and sovereigns.

Manuscript 14887, shown here, once belonged to an illegitimate son of Philip the Good, Raphael de Mercatellis (1437–1508), who subsequently served as abbot at Saint Peter's Abbey in Oudenburg and Saint Bavo's Abbey in Ghent. The decorations are much more sumptuous than seen in other manuscripts. Ptolemy's pursuit of scientific rigour is combined with an iconography rooted in the universe of imagination of Classic Antiquity and the Middle Ages and in the decor of portolan charts. As such, the map of Africa has been embellished with botanical, zoological and ethnographic elements in which the peculiar and the wondrous rule supreme: exotic characters, dromedaries, predators, snakes, dragons, etc. There can be no doubt: there is no comparison between the civilised world, with its many clearly identified towns along the Mediterranean coastline, and the Nile delta or the unexplored territories full of indigenous peoples, wild animals and monsters in deepest darkest Africa.

EV

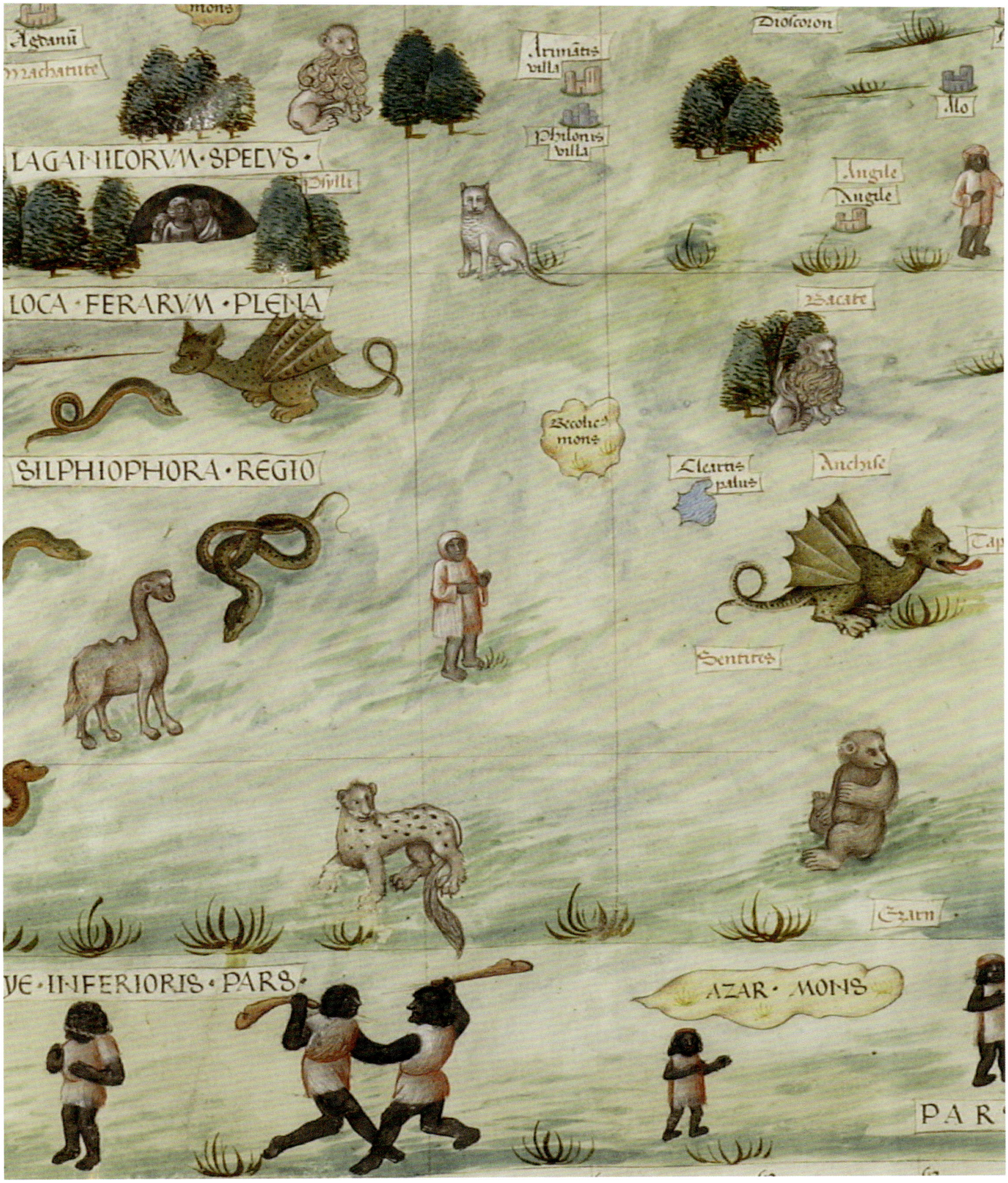

[cat. 44] detail

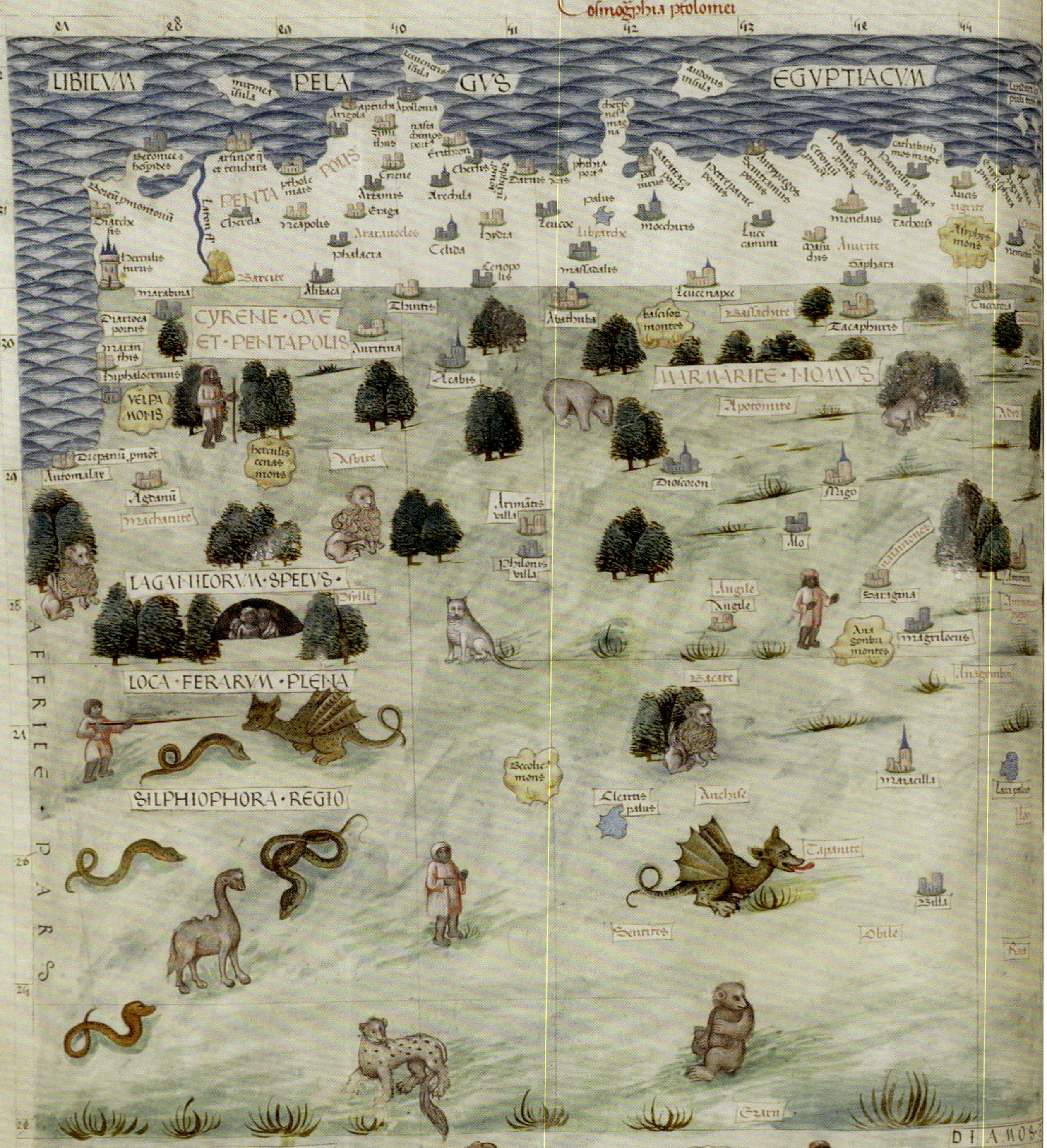

Cosmographia ptolomei
LIBICUM
PELA
GUS
EGUPTIACUM
PENTA POLIS
CYRENE · QUE ET · PENTAPOLIS
MARMARICE · NOMUS
LAGANICORUM · SPECUS ·
LOCA · FERARUM · PLENA
SILPHIOPHORA · REGIO
AFFRICE · PARS
LIBYE · INFERIORIS · PARS ·
AZAR · MONS
PARS · ETHIOPIE · QUE S

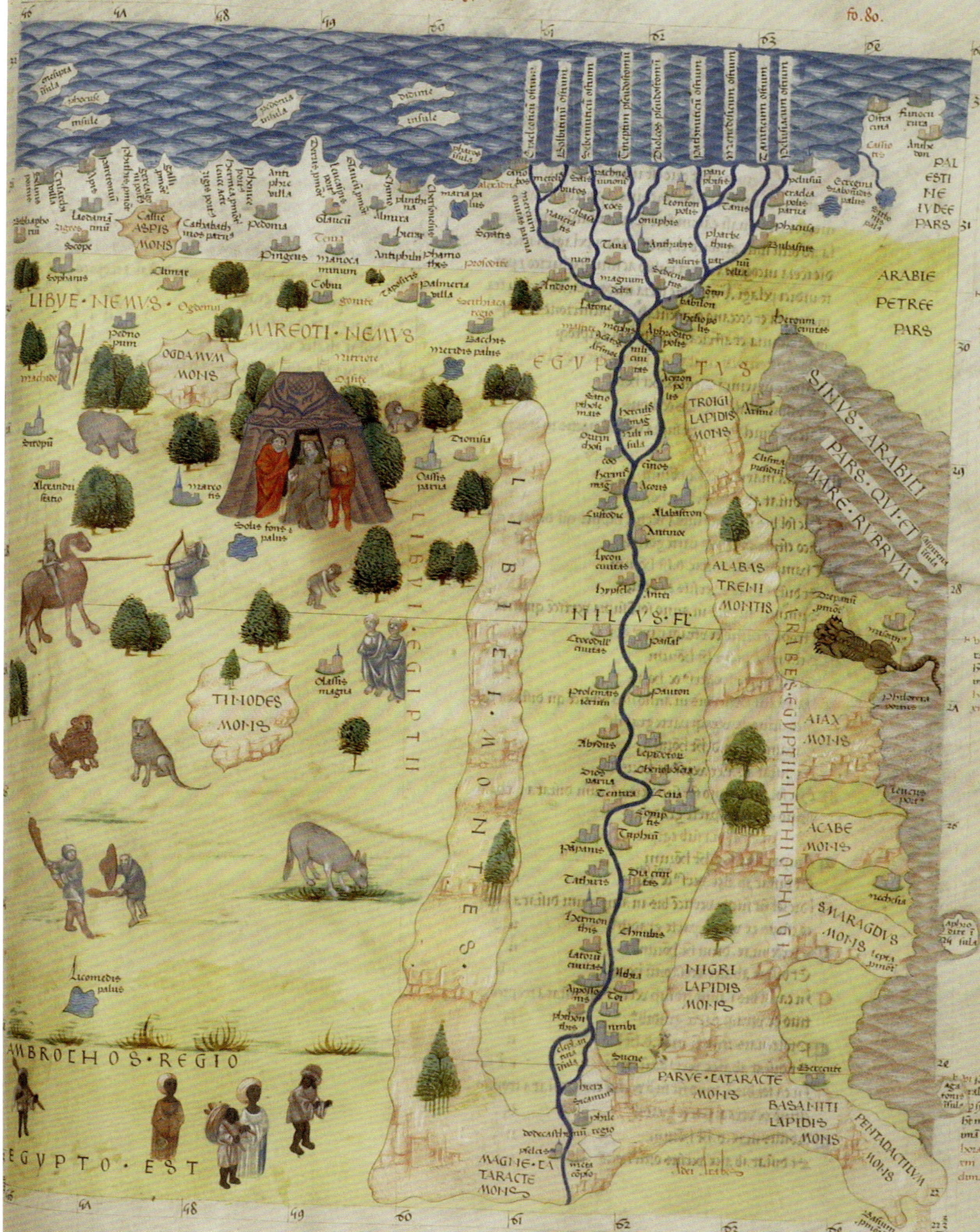
LIBYE · NEMVS
MAREOTI · NEMVS
OGDAMVM MONS
CALLIS ASPIS MONS
TINODES MONS
LIBICI MONTES
LIBVI · EGIPTII
EGYPTVS
NILVS · FL
TROIGI LAPIDIS MONS
ALABASTRENI MONTIS
ARABES · EGYPTII · ICHTHIOPHAGI
SINVS · ARABICI PARS · QVI · ET MARE · RVBRVM
ARABIE PETREE PARS
PALESTINE IVDEE PARS
AIAX MONS
ACABE MONS
SMARAGDVS MONS
NIGRI LAPIDIS MONS
PARVE · CATARACTE MONS
BASANITI LAPIDIS MONS
PENTADACTILVM MONS
MAGNE · CATARACTE MONS
AMBROCHOS · REGIO
EGYPTO · EST

45

Word Map (Mappa mundi)

In: Claudius Ptolemy, *Geographia (Cosmographia)*, Ulm, 1486
Paper, coloured woodcut, 420 x 565 mm
Antwerp, Museum Plantin-Moretus, inv. R 9.11

BIBLIOGRAPHY

Campbell 1987: 122-138, no. 39; Edgerton Jr. 1987: 13; Ptolemaei 1990: 1; Pühlhorn & Laub 1993: 660-661, no. 2.16; Schneider 2004: 10-11; Fiorani 2005: 84; Hiatt 2008: 150, fig. 20; Whitfield 2010: 8-9; Lehmann 2010: 181; Horst 2011: 19.

The *Geographia*, also known as the *Cosmographia* by Claudius Ptolemy (*c*.90–168), was translated into Latin in the early part of the fifteenth century and thereafter copied, adapted, commented and expanded with maps in numerous manuscripts. In 1477, the first printed edition was published in Bologna, soon followed by further impressions in Rome (1478) and Ulm (1482 and 1486). The coloured maps in the famous Ulm editions are woodcuts, inspired by the works of Nicolaus Germanus, a German geographer who worked in Italy during the second half of the fifteenth century. In the preamble, Germanus explains that he used Ptolemy's first projection for the maps, as set out in the latter's text, as this best enabled him to render the spherical shape of the Earth. In addition, he supplemented Ptolemy's work with 'modern' regional maps with an updated shape and toponymy: i.e., a map of Scandinavia that had been detailed by Claudius Clavus in the early part of the fifteenth century, and a medieval map of the Holy Land. In due course, some of his innovations would be adopted and disseminated by fellow mapmakers such as Martin Waldseemüller, Gerardus Mercator and Abraham Ortelius. In this way, territories that were discovered during in the Renaissance were given a place in a world view that was rooted in Antiquity.

The map shown in this tome is a *mappa mundi*: all land that rose from the sea surrounded by allegorical depictions of the winds. The known and unknown land is divided into latitudes and simultaneously transected by meridians with circular arcs in consideration of the duration of the day at the time of the summer solstice. However, before the precision instruments of the eighteenth century became available, calculating longitudes was particularly awkward. The measurements that had been used since Antiquity were largely based on estimates of distances between the towns and cities of the Roman Empire and on the observation of eclipses.

In the fifteenth century, the readers of Ptolemy's *Geographia* became aware that his world map was not entirely accurate. During the Greek scholar's lifetime, Scandinavia was unknown. Nicolaus Germanus rendered this territory by adding a 'protrusion' to the northern part of Europe. And whilst Ptolemy considered the Indian Ocean a sea that was sealed off on its southern edge by a continent that linked Africa to Asia, this assumption was already contradicted by medieval *mappa mundi*: Africa was depicted as a continent that could be circumnavigated. In 1488, the Portuguese explorer Bartolomeu Diaz was the first to sail around what is now known as the Cape of Good Hope. In the 1486 *Geographia*, this world map was the

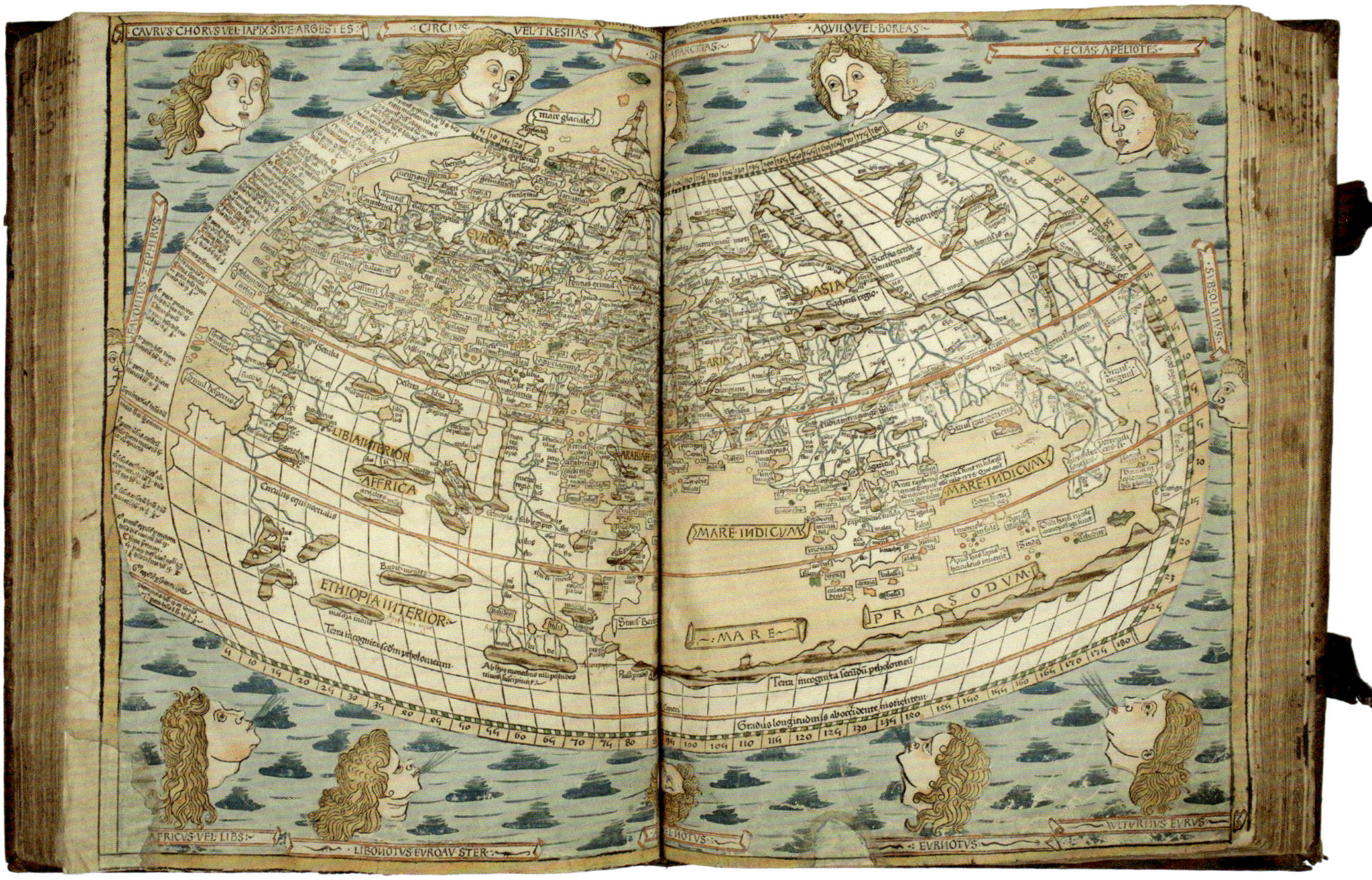
CAVRVS CHORVS VEL IAPIX SIVE ARGESTES
CIRCIVS VEL TRESIIAS
AQVILO VEL BOREAS
CECIAS APELIOTES
SVBSOLANVS
mare glaciale
EVROPA
ASIA
LIBIA INTERIOR
AFRICA
ETHIOPIA INTERIOR
MARE INDICVM
MARE
PRASODVM
Terra incognita secundum ptholomeum
AFRICVS VEL LIBS
LIBONOTVS EVROAVSTER
EVRNOTVS

first in which cartographers ventured to cast aspersions on the authority of a Greek scholar, adding the words *Terra incognita secundum Ptholomeum*, meaning 'unknown territory according to Ptolemy' beneath Africa and the Indian Ocean.

In spite of Bartolomeu Dias's feat, which was repeated in 1498 by Vasco de Gama, who sailed as far as Calicut (Kozhikode on the south-western coast of India), Ptolemy's hypothesis was not discarded. In fact, it gained fresh impetus when the Europeans set out to find the southern continent in the sixteenth century (cat. 55). The strip of land from Antique cartography might just point to the existence of a continental land mass on the other side of the Indian Ocean, in the southern hemisphere. This hypothesis prompted the exploration of the Pacific Ocean and the discovery of Australia and Oceania.

EV

46

The so-called 'Landing of Vasco da Gama in India'

Tournai

Early sixteenth century

Tapestry, wool and silk, 400 x 760 cm

Lisbon, Caixa Geral de Dépositos, inv. 3721. On loan to the Museu Nacional de Arte Antiga, Lisbon

BIBLIOGRAPHY

Gentil Quina 1998: 57-77; Campbell 2002: 140; Vandenbroeck 2007: 121-137; Levenson 2007: 100-101, no. P-30.

The tapestry depicts the port of the *Indas Novae* – the New Indias – as appears from the inscription above the gates of the city at the left.[1] The ruler of the city, with a full beard and sumptuous robes, stands before the gates. Soldiers and turbaned advisors follow on his heels. He receives a letter from a visitor who has removed his cap and greets him with a small bow from the knee. This man is the leader of the centrally represented fleet, on the verge of embarking for the West. Sailors load a handful of ships with exotic animals. In the foreground are a panther and parrots in cages. A unicorn is hoisted onto a ship. Two camels are already standing on the deck. Five ostriches are being rowed to another ship in a sloop. In the foreground at the right, we find a group of magistrates and courtiers who comment on the scene unfolding before them.

Like most tapestries, this piece is also part of an edition of a series comprising several scenes.[2] This series turns up for the first time in an archival document of 1504. It was then that Philip the Fair (1478–1506) purchased a number of tapestries *faicte à la maniere de Portugal et de Indie* (made in the style of Portugal and India) from Jean Grenier, a tapestry dealer from Tournai.[3] Other sixteenth-century documents mention seven other editions of what is presumably the same series: *histoire de calcou* (story of the turkey), *ystoire de gens et de bestes sauvaiges à la manière de calcut* (story of people and wild animals in the style of Calicut) and *histoire indienne à ollifans et jeraffes* (Indian story of elephants and giraffes).[4] Of these editions, some twenty tapestries have been preserved in public and private collections.[5] This tapestry may have belonged to one of these editions.

The scene known as *The Landing van Vasco da Gama* illustrates both the fascination for and ignorance of the New World, as well as the vivid imagination of the artists who turned all manner of fragmentary and unclear written and pictorial sources into a pseudo-realistic but essentially mythical-romantic vision of an unknown land. The tapestry dealer, who took the financial risk of ordering the cartoons and having the tapestries woven for the market, wanted a generic series of scenes that played on the admiration for – and latent fear of – the foreign and the Other. It is therefore *not* an attempt to document the discovery of Calicut by Vasco da Gama; nor is it an illustration of the Portuguese passion for discovery around 1500.[6] As Vandenbroeck has already emphasised in his convincing analysis of this tapestry and the series to which it belongs, there are three central themes:

1 Vandenbroeck 2007: 121–137 is the best analysis of this and related tapestries. For detail photos and references to earlier literature, see Gentil Quina 1998.

2 For a concise introduction to the complex problems of late medieval tapestries, see Cavallo 1993: 17–77. For the tapestries of Tournai, see Soil de Moriamé 1892.

3 De Farcy 1913: 107.

4 Cavallo 1993: 591.

5 Vandenbroeck 2007: 121–123.

6 Also demonstrated by Gentil Quina 1998: 144–165.

7 Vandenbroeck 2007: 137.

the peaceful encounter between cultures and peoples, the non-sedentary nature of exotic peoples, and finally the struggle between nature and culture.[7] Hence, this tapestry – now shown in isolation, but meant as part of a tapestry environment – shows an imaginative evocation of a *terra incognita* that, in expectation of further exploration and documentation, was already projected on the walls of the Old World in monumental form.

K B

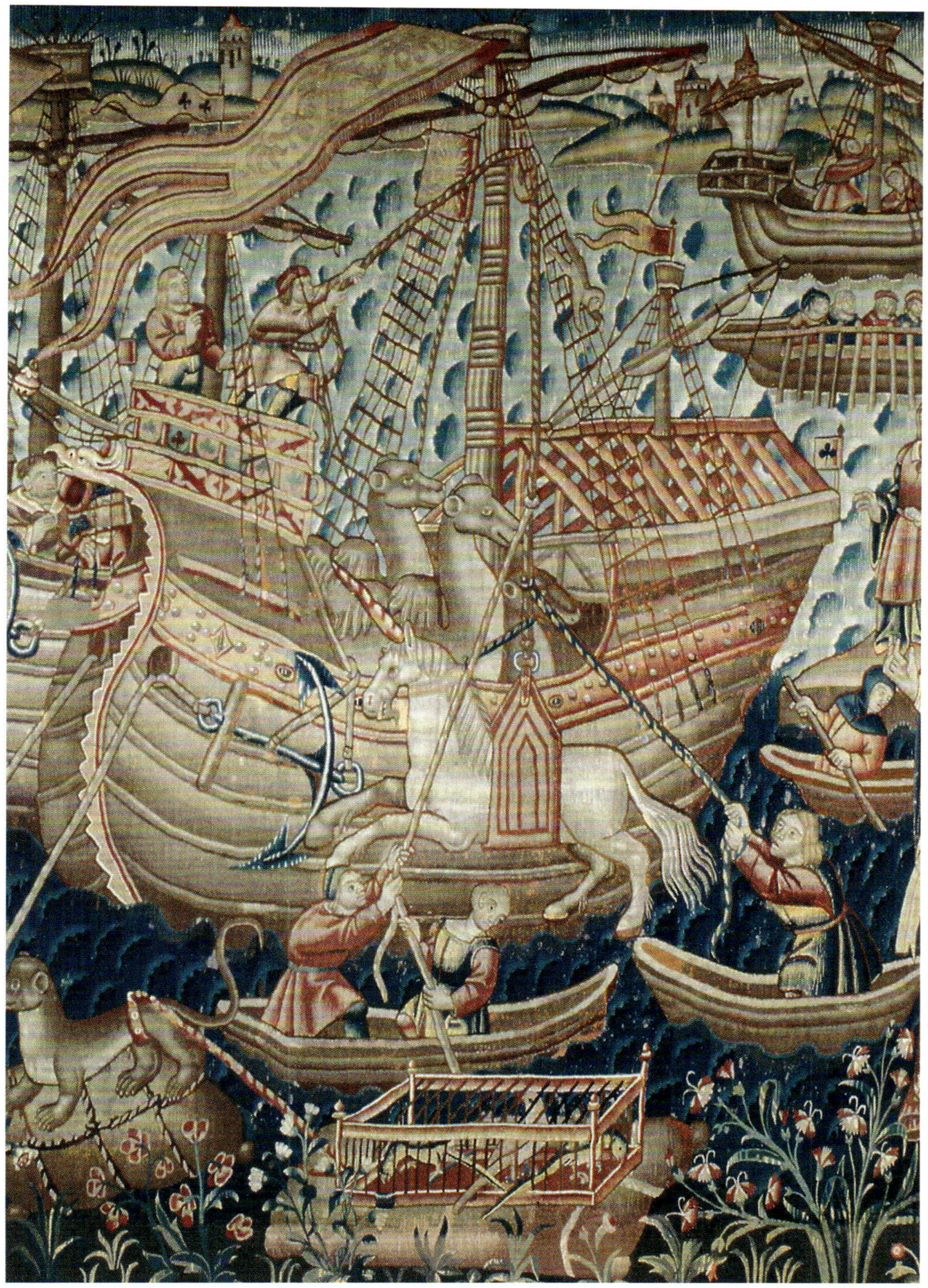

[cat. 46] detail

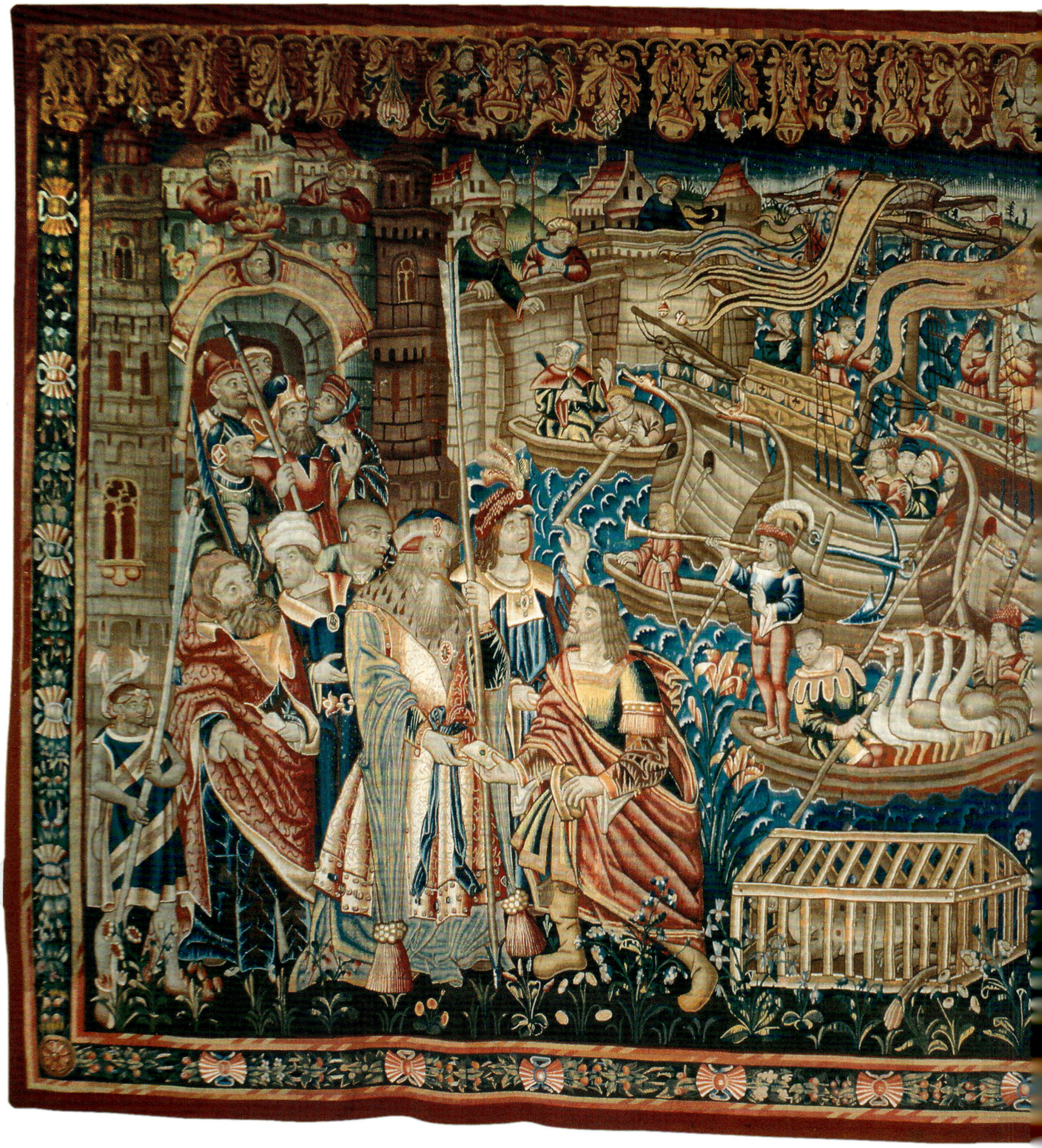

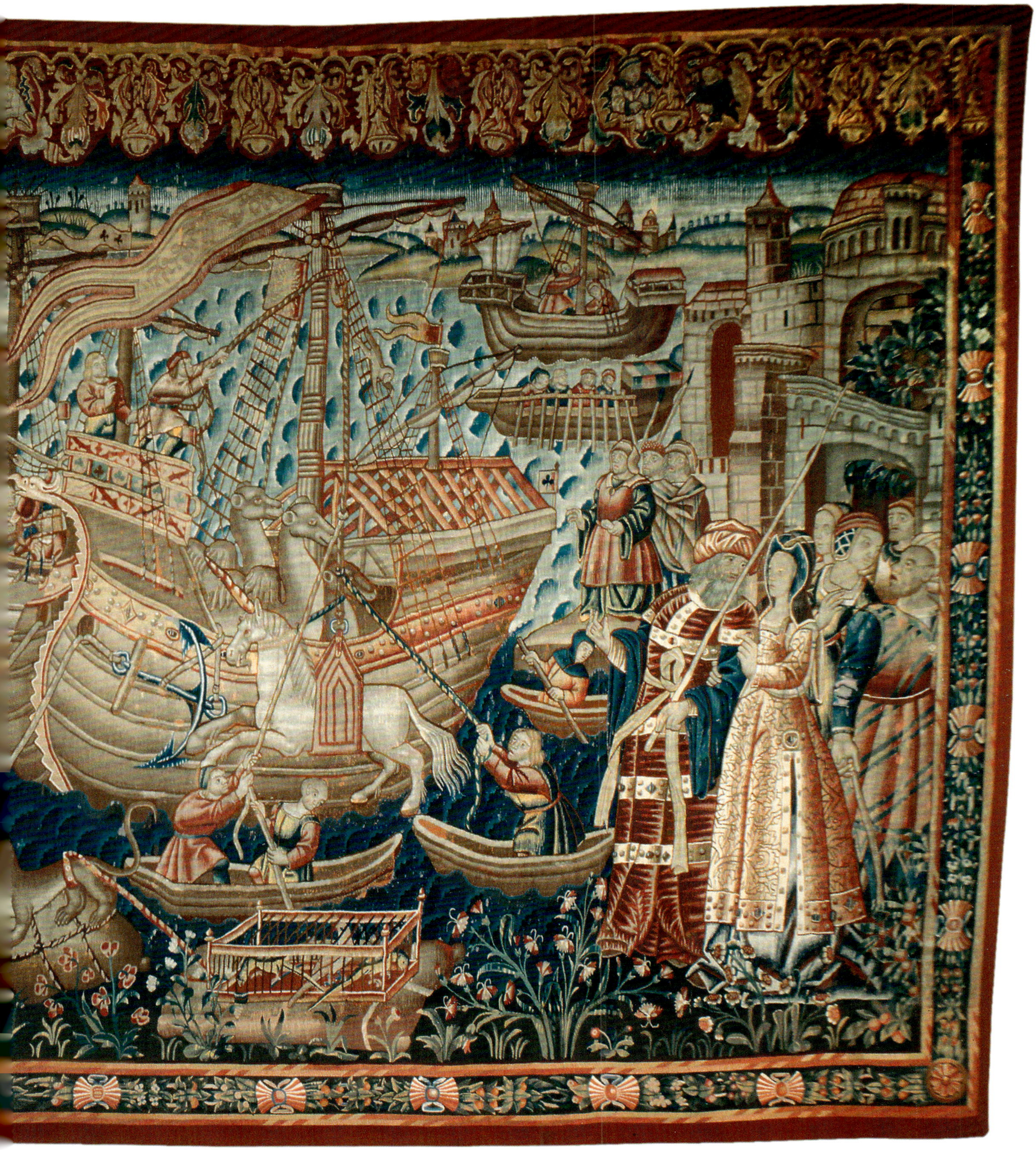

47

Edo/Bini people
Afro-Portuguese Luxury Saltcellar

Kingdom of Benin (in present-day Nigeria)
First half of the sixteenth century
Ivory, 19.2 x 7.2 x 8 cm
Antwerp, Museum aan de Stroom (MAS), inv. AE.1974.0025.0001

PROVENANCE

Collection of Frans Maria Olbrechts (1899–1958), Belgium; given to the Ethnographic Museum Antwerp in 1974 by his widow, Margriet Olbrechts-Maurissens.

BIBLIOGRAPHY

Claerhout 1976: 28; Anzelewsky, Mende & Eeckhout 1977: 95-111, no. 184; Lisbon 1983: 99, no. 31; Bassani & Fagg 1988: 161-186, no. 249; Claerhout 1989: 4; Herreman, Holsbeke & Van Alphen 1991: 18; Bott 1992: 853, no. 5.25; Ambrosio & Brusa 1992: 226, no. 147; Bassani 1997: 261; De Palmenaer 1999: 304, no. 213; Bassani 2000: 257-258, no. 798; Bassani 2003: 160, no. 2.24; Eichberger 2005: 325, no. 136; Curnow 2007: 299, no. 11.

This ornamental Afro-Portuguese vessel for pepper and salt is a precious witness to African-European cultural exchange in the sixteenth century. From the beginning of the Renaissance onwards, ivories in the Afro-Portuguese style appeared as *curiosa* on the tables of the European elite. Ivory saltcellars,[1] carved in 'white gold', were perfect accompaniments to the luxurious silver and gilt tableware that was made by European silversmiths and sculptors.[2] They ended up in the cabinets of curiosities and on the tables of the European upper classes, who, in the spirit of the *terra incognita* and discoveries of the times, sought out all that was rare and exotic.

At the end of the fifteenth century, Portugal gradually succeeded in opening an alternative southern spice route to the East along the west coast of Africa (cat. 46). From then on, local chieftains of the African coastal regions built up trade relations with Portuguese seafarers and tropical products such as black pepper and salt reached European markets.[3]

This rare ivory artwork attests to the first trade contacts between Europe and the Sub-Saharan African kingdom of Benin.[4] Already in 1486, the Portuguese reached the kingdom of the Edo peoples, which were governed from the capital Benin.[5] Here they were confronted for the first time with the artists' guilds of the *Oba,* or king, who were specialised in high-quality ivory carving.[6] Although trading and working in ivory was strictly reserved for the regalia of the ruler of Benin and ritual objects for an affluent elite, the ivory carvers' guilds also made luxury tableware commissioned by the Portuguese for export to Europe.[7]

The Afro-Portuguese saltcellars are probably made after European models.[8] This exemplar is one of the oldest surviving 'double' ivory saltcellars in the Edo-Portuguese style.[9] The cellar comprises two parts stacked one on top of the other and sealed by a lid decorated with a caparisoned horse, but no rider. In the central part are two hollow bowls, probably meant to hold what were then costly 'spices': salt and pepper.

1
Saltcellars were known as *saleyros* in Portuguese, *salières* in French. See Bassani 2008.

2
From the sixteenth century onwards, luxurious cups were produced that incorporated exotic materials such as nautilus shells, ostrich eggs and 'Indian' nuts. One of the earliest surviving coconut cups from the Netherlands is mounted in a silver gilt setting (1543–1544) and is now preserved in DIVA. Antwerp Home of Diamonds (inv. S 63/5). See De Palmenaer 1999: 301, no. 207.

3
West-African regions were given names such as the 'Grains of Paradise Coast' (after the malaguetta pepper), 'Tooth/Ivory Coast', 'Gold Coast' and 'Slave Coast' in the profit-oriented commercial policy of the time. Meridith 2014: 117–118.

4
Afro-Portuguese ivory carving is a collective name describing artworks that were commissioned by the Portuguese (between 1490 and *c.*1600) and made by African craftsmen for export to Europe. Typical objects include luxury tableware such as saltcellars and flatware, but also hunting and powder horns, the hafts of arms. De Palmenaer 1999: 304; Blackmun 1992: 163.

5
Bassani 2000: XXII; Duchâteau & Van Damme 1990: 15; Meridith 2014: 119.

6
Inneh 2007: 107.

7
Bassani & Fagg 1988 distinguish between ivory carving in Edo-Portuguese style from Benin and the Sapi-Portuguese exemplars with a single recipient made in the same period by craftsmen of the Sapi, forefathers of the present-day Temne/Bullom peoples from Sierra Leone.

8
Probably engravings or sketches of covered boxes. Bassani 2008: 52.

9
There are fifteen known Edo-Portuguese saltcellars. Only three 'double' exemplars survive, including this exemplar from the MAS in Antwerp. The other examples are by the 'Master of the Merchant Ship' with standing Portuguese from the collections of the British Museum, London (inv. 78.11.1.48; fig. 47.1), and the Musée du Quai Branly, Paris (inv. 70.2008.14.1; fig. 35).

10
Curnow 2007: 299.

11
This ambiguous nude figure is also depicted on the Edo-Portuguese saltcellar in the Museum Volkenkunde Leiden (inv. RMV 1323-1). According to Curnow, the prototype was the angel on the exemplar in the Smithsonian Museum Washington (inv. 2005-6-36).

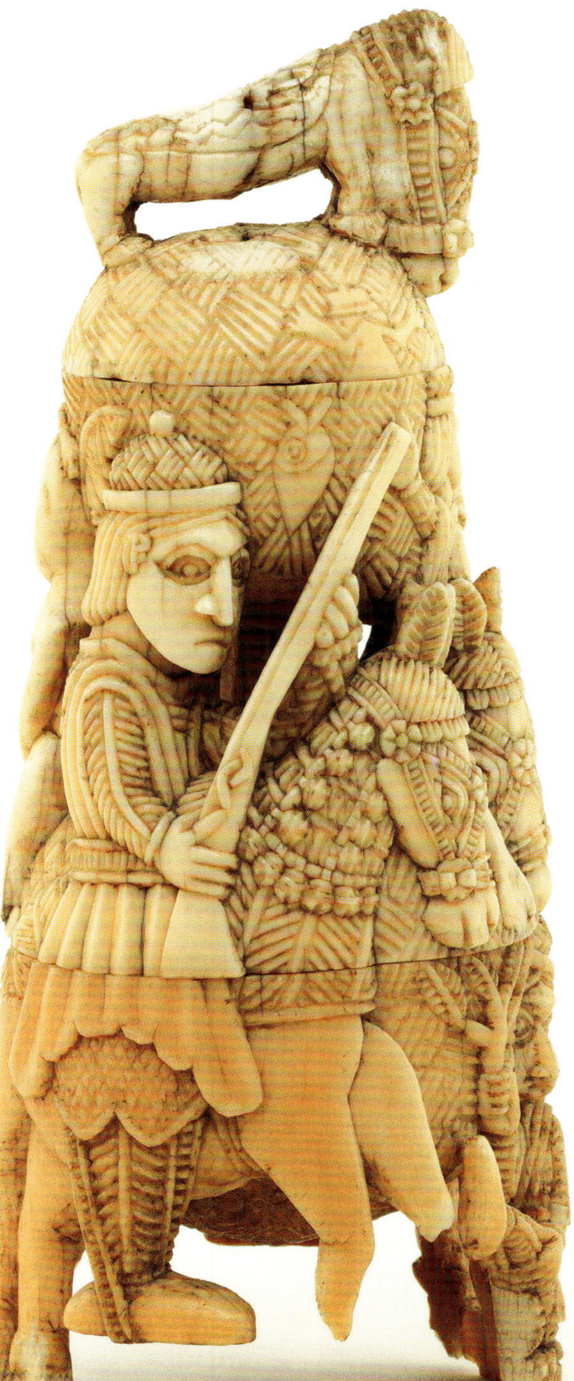

[cat. 47] Left view

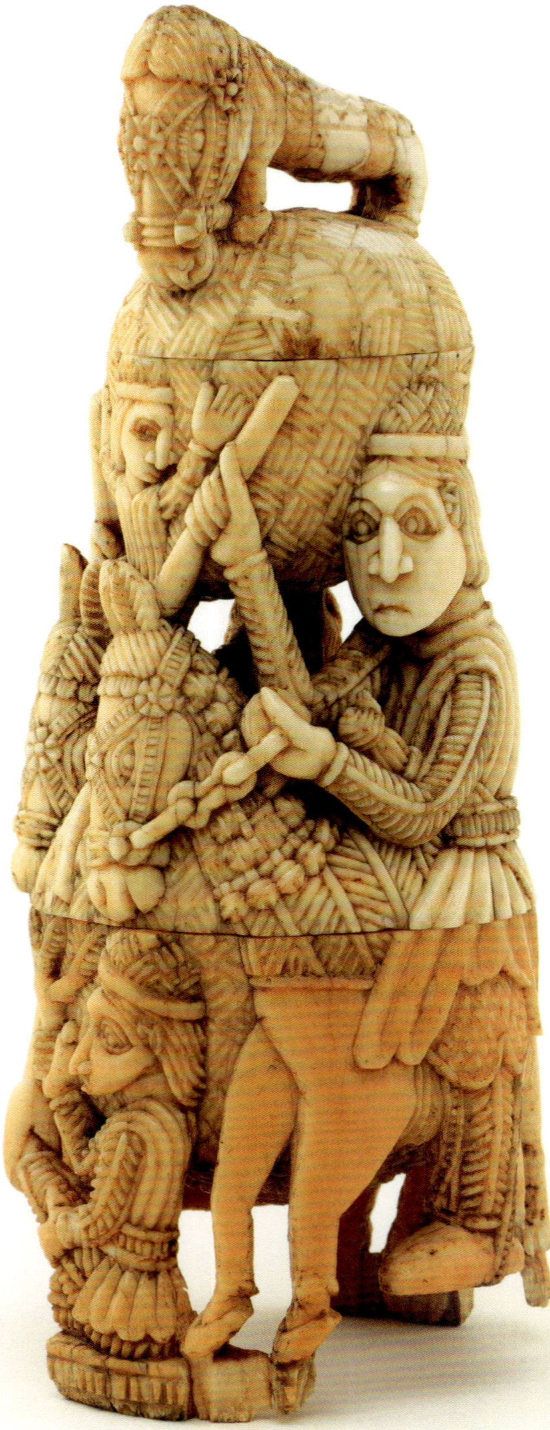

[cat. 47] Right view

The artwork is a synthesis of European-inspired formal language enriched with African decorative and iconographic motifs carved in relief. The European influence is expressed in the depiction of figures in western dress, two Portuguese horsemen depicted in profile. In front, between their horses, sits a smaller man in an unusual position.[10] The hatched decoration refers to African weaving motifs. The fish motif carved in relief symbolises *Olukun*, the god of the ocean and bringer of wealth. The fish motif alludes to the trade relations with the Portuguese seafarers, who brought the Edo people good fortune via the ocean. On the reverse stands a nude male figure with the same facial features as the Portuguese horsemen. Perhaps he is an angel, an un-African being, which the ivory carver has interpreted in his own way. The figure's wings are not attached to his shoulders, as in Christian iconography, but appear in vague relief on either side of his ears.[11]

EDP

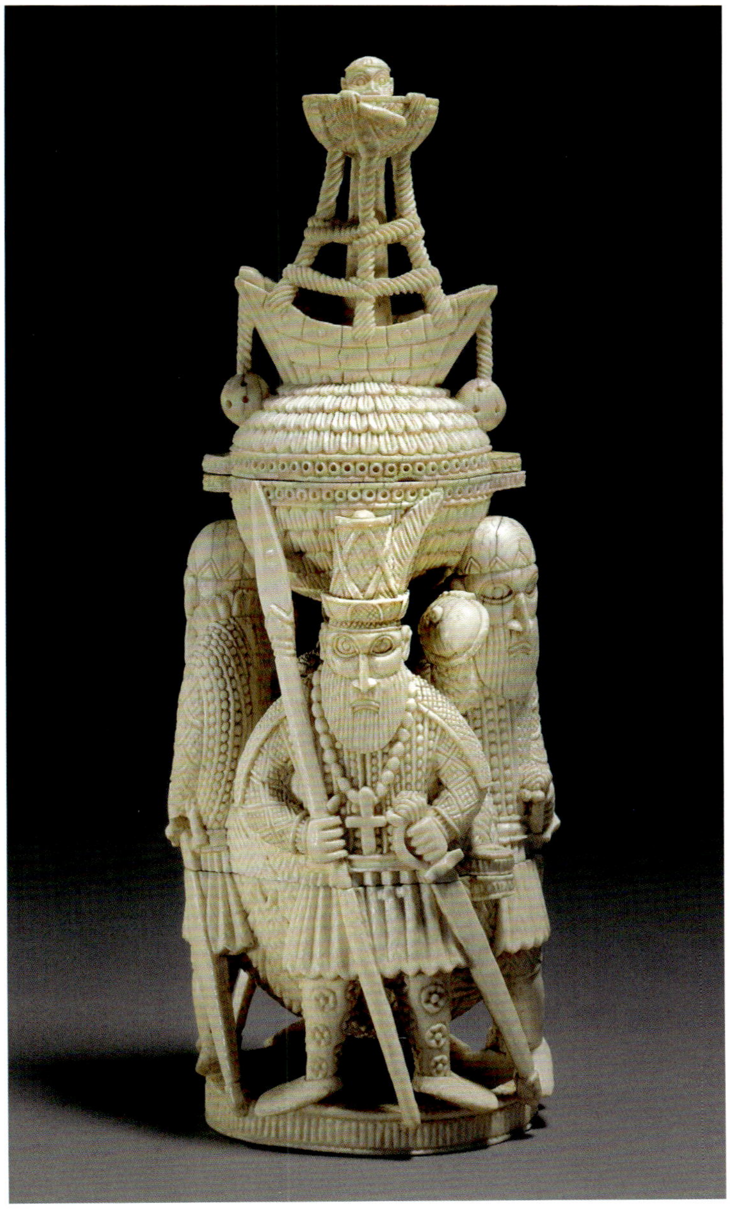

[fig. 47.1] Edo/Bini People, Kingdom of Benin (in present day Nigeria), *Afro-Portuguese salt cellar*, first half of the sixteenth century, London, British Museum.

48

New Spanish Artist after an Antwerp Prototype
Adoration of the Magi

Mexico

*c.*1540–1560

Feather mosaic on amate paper and panel (recto), oil and gouache on panel (verso), 43.5 x 31 cm (centre panel), 43.5 x 16.5 cm (wings)

Madrid, Museo de América, inv. 12.343

PROVENANCE

Possibly in possession of Cristóbal Vaca de Castro (*c.*1492-1566/71), governor of Peru under Charles V; after Vaca de Castro's death presumed to have been acquired by the Convento de Santa Isabel in Valladolid where the de Castros had their family tomb; acquired by Enrique Traumann in 1935 through the Convento de Santa Isabel in Avila; gifted by Traumann to the Museo de America in 1954.

BIBLIOGRAPHY

Feest 1987: 176; Vandenbroeck 1991b: 417-418; Vandamme 2003; Vandamme 2006: 93-135; Russo 2014: 30-31; Estrada de Gerlero 2015: 308-309.

In this New Spanish *Adoration of the Magi* of *c.*1540–1560, the Mesoamerican feather mosaic technique is seen to engage in dialogue with the traditional Western medium of the triptych.[1] The hybrid work of art is a prime example of the cross-pollination between Mesoamerican and European visual cultures right after the age of colonisation got under way. All the more so as feather art in modern-day Mexico was traditionally associated with the worship of various indigenous deities, such as the Aztec god Quetzalcoatl ('feathered serpent'). By depicting a conventionally Christian topic such as the Adoration of the Magi using a familiar, indigenous technique, links were established between the spiritual world of the time before and the time after the arrival of European missionaries. Traditionally, the origins and dissemination of this mixed style of art are traced back to the workshop of Netherlandish Franciscan fray Pedro de Gante (*c.*1480–1572), who served as a missionary in Mexico from 1524. [2]

The European prototype of the feather mosaic at the Museo de América was entirely in keeping with the triptychs depicting the *Adoration of the Magi*, which were highly popular at the time, and which were being produced for the open market by Antwerp workshops over the 1520–1540 time period.[3] This type of work by the so-called 'Late Gothic Mannerists' must have been exported via the Iberian Peninsula to the overseas colonies, after which they presumably went on to play an active part in the Christianisation of local populations. The exact example or prototype on which the New Spanish artist modelled his work is untraceable, but the indigenous work of art bears striking similarities with an anonymous triptych of an *Adoration* in the style of Pieter Coecke van Aelst (1502–1550) (fig. 48.1).[4] Particularly striking is the likeness in the crown worn by Balthazar, with points ascending at a right angle, which is rare in the Antwerp pictorial tradition. In

1
On this complex transformation process, see Russo 2014 and Russo, Wolf & Fane 2016.

2
The *Adoration* displayed at the Museo de América is attributed to Pedro de Gante's workshop in Vandenbroeck 1991b: 417–418 and Russo 2014: 31.

3
See Vandenbroeck 1991b: 417–418 and Vandamme 2006: 97–98; on the production and dissemination of such *Adorations* see Vermeylen 2003: 29–30 and Ewing 2006: 274–299.

4
Sale, Paris (Sotheby's), 21 June 2012, lot 2.

5
As such, an earlier assertion that this type of crown steps away from the Antwerp *Adoration* tradition needs to be revised. See Vandamme 2006: 128; on this topic, also see Vandenbroeck 1991b: 418.

6
See Vandamme 2006: 98–99.

7
Vandamme 2006: 24–26 and 128–130.

8
Vandamme 2006: 123–124.

9
See above under 'Provenance'. For further details, see Vandamme 2003: 31–33, 144 and Vandamme 2006: 119–120.

10
Respectively *imagen de los Tres Reyes hecha de tapicería a la manera de las Indias* and *imagen del ofrecimiento de los Reyes en tela de India*. See Russo 2014: 30–31, 261, n. 48–49.

11
See Vandamme 2006: 121–124.

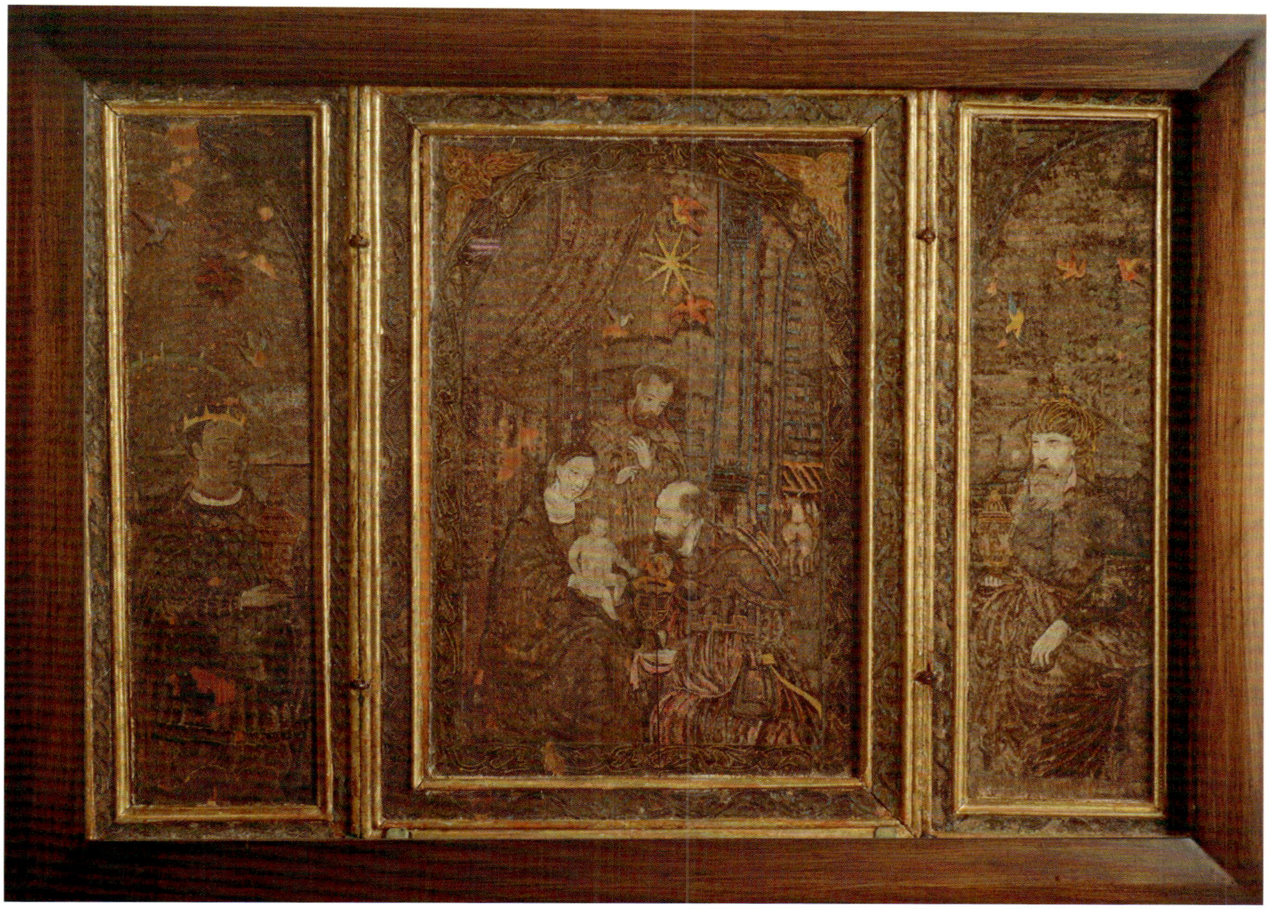

[cat. 48] Front

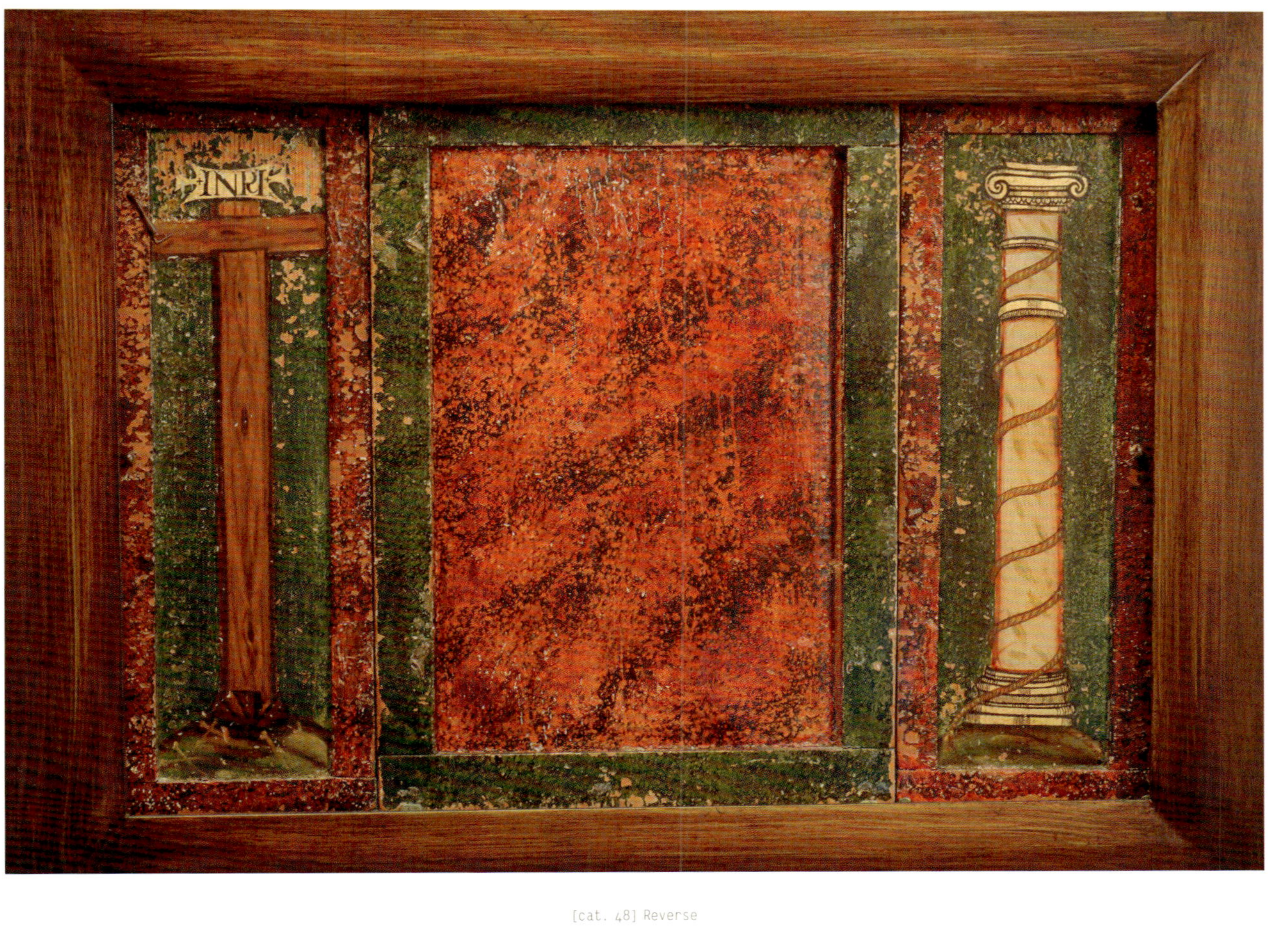

[cat. 48] Reverse

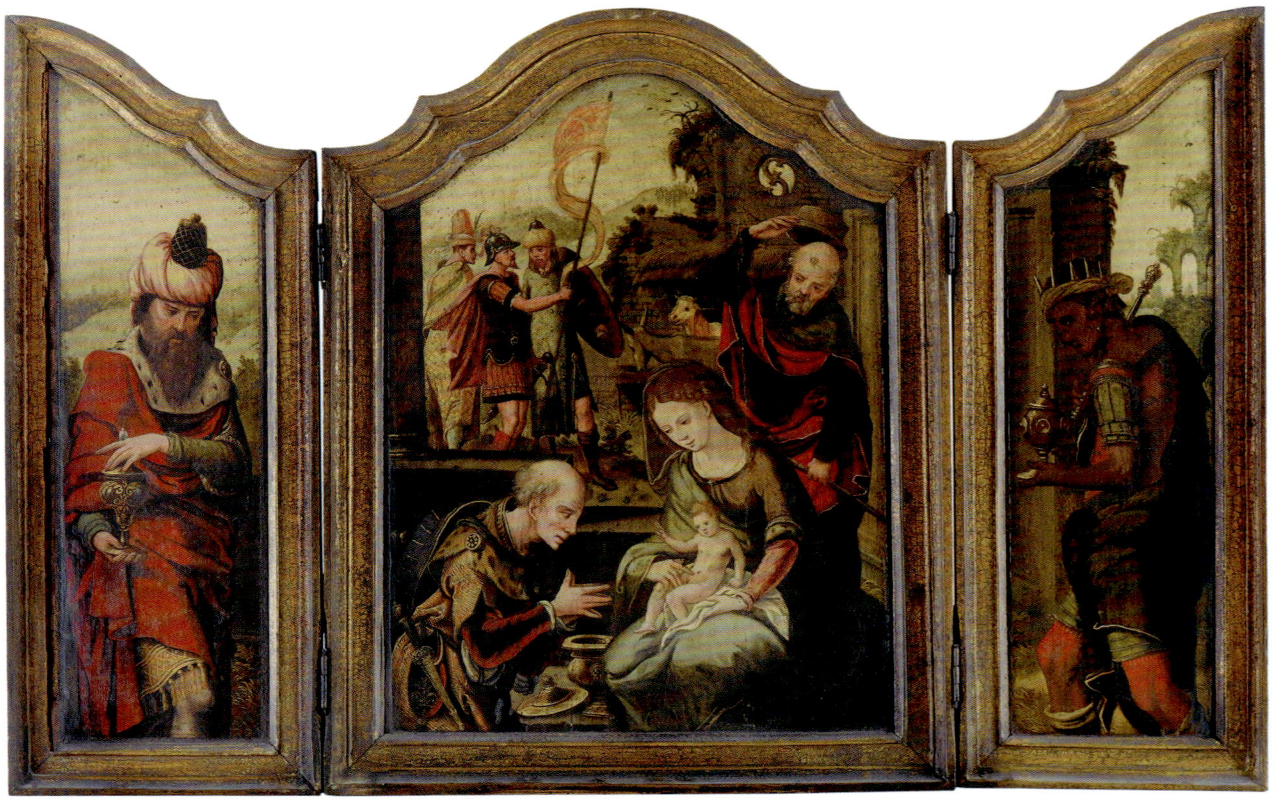

[fig. 48.1] Circle of Pieter Coecke Van Aelst, *The Adoration of the Magi*, 1520-1550, sold at Sotheby's, Paris, 21 June 2012.

the latter tradition, the black king is usually depicted bareheaded or wearing a fantastical turban or pointed hat.[5] The fact that the composition of the feather mosaic is the mirror image of what is presumed to be the Antwerp exemplar may be explained by the production process of colonial feather mosaics, in which the artist first traced the Western exemplar before transferring the composition onto *amate* paper.[6]

For all that, this feather triptych is not a straightforward transposition of a European model. For instance, the unusual star of Bethlehem in the centre was associated with the use of similar gilded stars in indigenous evangelisation theatre, and the plain back of the triptych differs substantially from those seen in the Antwerp *Adorations*.[7] The reverse of the *Adoration* shown here depicts the Cross of Golgotha and the pillar on which Jesus was tortured, rendered in red faux marble in the centre. In addition, the composition was also enlivened with varicoloured birds, which may refer to the origin of the medium used, and also underscores the exotic, non-native nature of the work. The latter, self-referential element therefore suggests that the feather triptych may have been produced for a European audience – for instance, as a diplomatic gift.[8] Indeed, the provenance of this *Adoration* points in this direction.[9] A similar work held as part of the collection of Charles v in Brussels was documented as a 'depiction of the Three Magi made of tapestry in the manner of the Indians' (1545), and another exemplar in Simancas was described as a 'depiction of the gifts of the Magi of fabric of India' (1559).[10] Very likely, these colonial *Adorations* served not only devotional purposes in the Christianisation process of New Spain, but also as part of the rising demand for *mirabilia* from the New World among European dignitaries of the time.[11]

DVH

49

Diverse Species of Monsters from Distant Lands

In: Thomas of Cantimpré, *Liber de naturis rerum creatarum*, Ghent, 1492, fol. 1v–2
Patron: Raphael de Mercatellis
Manuscript on parchment, 280 fols., 409 x 295 mm
Ghent, Sint-Baafskathedraal, inv. Ms. 15

BIBLIOGRAPHY

Wittkower 1942: 159-160; Versyp 1950: 32-33; Cantimpratensis 1973: 98-99; Smets 1974; Derolez 1979: 168-175; Derolez, De Kok & De Schryver 1984: 281-282, no. 166; Arnould 1991: 21, 185-186; Massing 1991b: 125, no. 6; Jackson 2006: 101, fig. 6.2.

This lavish compendium of medieval knowledge of the natural world was produced for Raphael de Marcatellis (1437–1508), abbot of St Bavo and a tireless bibliophile.[1] It is mostly derived from the *Liber de natura rerum* of Thomas of Cantimpré (1201–1275),[2] a testimony to the strong fascination that scholastic encyclopaedism could still exert on an erudite mind at the dawn of the modern era. Cantimpré's formidable work, in keeping with the much older encyclopaedic tradition of Isidore of Seville and Hrabanus Maurus, included as part of its sweeping panorama of created species an account of the marvellous peoples of Africa and Asia.[3] The repertoire rallies the familiar figures of the Cyclops, the headless Blemmie, the dog-headed Cynocephalus and the crane-fighting Pygmy; the Sciopod is here shown resting in the shade of his huge foot, while a mouthless man feeds off the scent of an apple. Some specimens are rendered wondrous on account of their behaviour, like the Amazons or the Indian tribes of the Gymnosophists and the Bragmanes (derived from old travellers' accounts of the Brahmin caste).[4]

These races enjoyed a venerable pedigree, for the most part traceable to Greek *periploi* like those of Ctesias, Megasthenes and the Pseudo-Scylax – whose descriptions reached the Middle Ages through the compilations of Pliny and Solinus.[5] The long tradition associated with the Alexander saga provided another rich hunting-ground for the prodigious, while pseudo-epistolary concoctions like the *Letter of Alexander to Aristotle*,[6] or the purported exchanges between Hadrian and an Eastern king about the wonders of Asia,[7] afforded a pretext for distilling narrative sources into tantalizing lists of *mirabilia* – some of which would blossom into finely-illustrated catalogues by the turn of the first millennium.[8]

The enthusiasm with which this material was consumed and circulated, spanning a variety of media and genres, yields precious glimpses into the inner workings of the medieval mind. Within the intellectual and imaginative modes proper to the fruition of these fantastical accounts, natural history walks hand in hand with antiquarianism, science with literature, experience with authority. A properly empirical curiosity about the outer world coexists with a bookish pleasure in the far-fetched and implausibility for its own sake: the monster disturbs and delights not because he is radically different, but because he is at once *other* and *same*, and so arrestingly ambiguous. When re-presented on the page, the spectacle of his contradictions is expressed as a cacophonous combination of familiar forms

1
For biographical notes, see Van Acker 1960 and Massing 1991b: 125. On this theme see, among others: Davies 2016.

2
Cantimpratensis 1973. The volume includes also a selection of texts of historical interest, ranging from the byzantine historian Jordanes' *De origine actibusque Getarum* to the more contemporary John of Thurocz's *Chronicle* and Aeneas Silvius Piccolomini's *Historia Bohemica* (Massing 1991b: 125).

3
The text given here corresponds to Cantimpratensis 1973: 98–99.

4
For a catalogue of monstrous races of the classical and medieval legendaries, including those shown here, see Friedman 1981: 9–21.

5
For an overview of Greek sources of the legendary of monstrous races and their transmission, see Wittkower 1942: 159–160.

6
On which see overview in Orchard 1997: 116–139, and translation of the Latin and Old English texts: 204–253.

7
On the *Epistle of Fermes*, alternatively Farasmanes, Feramen, Premo or Parmoenis, see James 1929: 1–59 and Gibb 1977: 15–30.

8
Most notably, the remarkable *Wonders of the East* manuscripts in England, on which see Gibb 1977: esp. pp. 28–29 and Friedman 1986; the three extant manuscripts are reproduced in James 1929.

Da inquiunt nobis immortalitatem quam super omnia desideramus. Alias autem divicias non curamus. Quibus alexander. Mortalis inquit cum sim immortalitatem quomodo vobis dare possum. Cui illi. Si ergo mortalem te cognoscis quare tot mala faciendo vagabundus in orbe discurris.

De bragmannis qui xpistum dei filium confitentur et colunt et antequam xpistus veniret in carne.

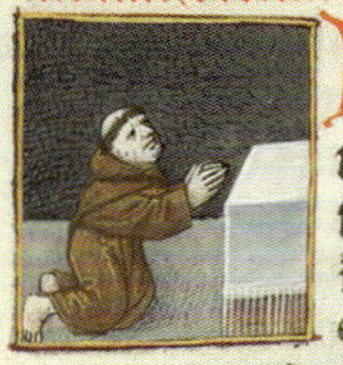

Homines preterea alii sunt valde mirabiles ultra gangem fluvium habitantes quos bragmannos appellant. Quos mirabiles religio mores et vita decorant. Hii etiam antequam xpistus veniret in carne de eius coeternitate cum patre aperte scripserunt. Nam quidam didimus dydascalus ipsorum bragmannorum rogatus ab Alexandro mirabilem epistolam scripsit ad eundem alexandrum de vita et sanctis moribus bragmannorum et de cultu unius dei. et de coeternitate filii cum patre. Dicit enim inter cetera. Deus verbum est et verbum illud mundum creavit et per hoc vivunt omnia. Nos autem hoc verbum colimus hoc adoramus. Deus spiritus est et mens est et ideo non amat aliud nisi mundam mentem.

De aliis hominibus et moribus eorum.

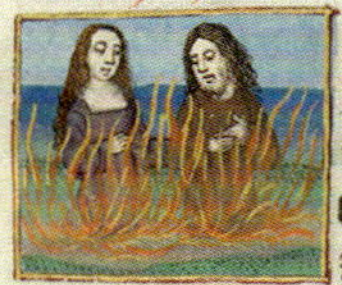

Homines alii sunt in partibus supradictis qui amore alterius vite in ignem se mittere non formidant.

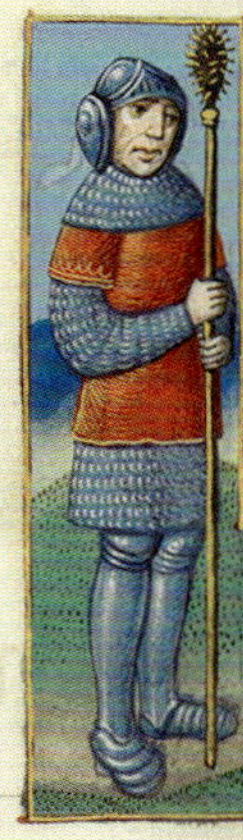

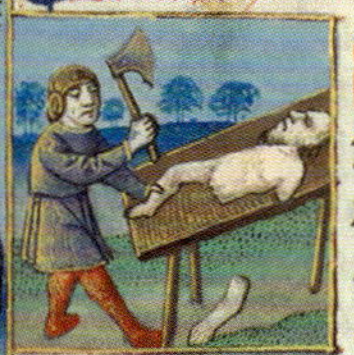

De aliis hominibus et moribus eorum.

Sunt homines alii qui parentes proprios nimio confectos senio mactare et carnes eorum ad epulas preparare immensam putant pietatem et magnam religionem. renuentes vero tamquam impios detestantur.

De aliis hominibus.

Homines etiam sunt ibi magni velud gygantes ut ele-

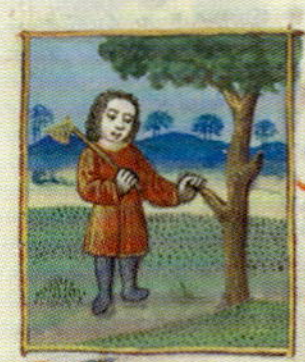

phantes facile transiliant.

De hominibus aliis.

Homines alii sunt ibi in tantum parvi ut cubitum longitudinis vix excedant.

Item de aliis.

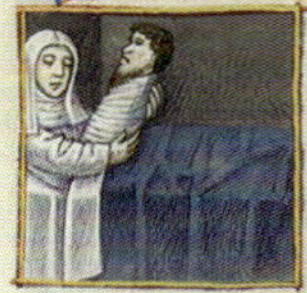

Sunt matres que semel parientes canos partus proferunt. Qui tamen diu viventes in senectute nigrescunt.

Item de aliis mulieribus.

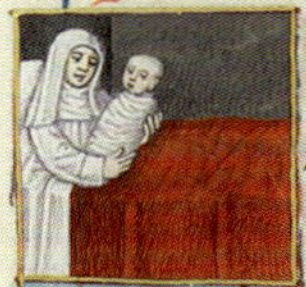

Sunt alie matres que quinquennes pariunt. Sed partus nisi octo annis vivere possunt.

Item de aliis hominibus pisces commedentibus.

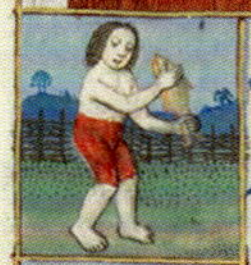

Homines alii sunt qui pisces crudos manducantes ipsum salsum mare bibunt.

De hominibus aliis

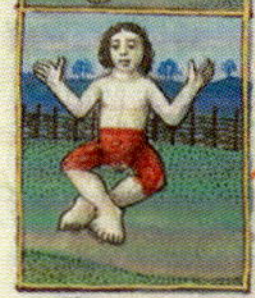

Homines alii similiter sunt in partibus orientis qui manus aversas habent et in pedibus octo digitos.

Item de hominibus aliis.

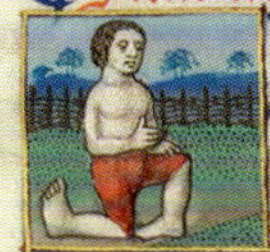

Homines similiter alii sunt in partibus orientis qui plantas habent transpositas.

Item de aliis hominibus.

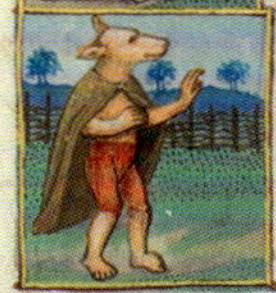

Homines alii sunt quos beatus Jeronimus cynocephalos appellat. Qui canina capita habentes ungues aduncos pellibus pecudum induti pro voce latratus canum proferunt.

Item de hominibus aliis.

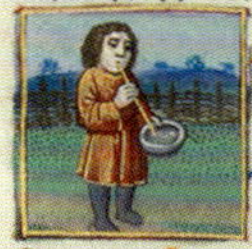

Homines alii sunt qui ita parvum os habent quod gracili calamo non nisi sorbilia comedunt.

Item de aliis hominibus.

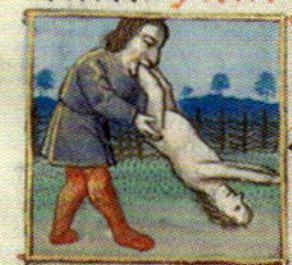

Homines alii ibidem sunt qui carnibus humanis vescuntur. Hii hominum vestigia tam diu olfactu percipiunt quibus

– unsatisfying analogies that draw us back to his indescribability. Such tensions remind us of these exercises' indebtedness to another ancient tradition, that of paradoxography, and to a dimension of formal and conceptual play extending far beyond the positive categories of true and false. The earth's manifold monstrosities are not, in this sense, mere reference points around the edges of the *mappa mundi* (cat. 50) or guardians of some external and unreachable frontier – they are pretexts for introspection and self-discovery, points where knowledge becomes reflexive and is made to meditate on its own limits.

MAC

50

Michael Wolgemut and Wilhelm Pleydenwurff
Mappa Mundi and Strange Monsters from Distant Lands

Paper, coloured woodcut
In: Hartmann Schedel, *Liber chronicarum*, Nuremberg, published by Anton Koberger, 1493, fol. 12v–13
Private collection

BIBLIOGRAPHY

Brown 1952: no. 44, pl. XII; Rucker 1973; Wilson 1976; Husband 1980: no. 4; Shirley 1984: 18-19, no. 19; Schoch 1986: 97; Campbell 1987; Woodward 1987: pl. 16; Massing 1991c: 124, no. 5; Puhlhorn & Laub 1993: 661-662, no. 2.18; Vogel 1994: 73-97; Landau & Parshall 1994; Fussel 1994: 7-30; Mund-Dopchie 1994: 25-26, 34; Bartrum 1995; Zahn 1996: 230-248; Padmos & Vanpaemel 2000: no. 105, 201-202; Reske 2000; Fussel 2001 (with exhaustive bibliography); Werfel 2001: 3; Dackerman 2002: 102-104, no. 7, Schneider 2004: 33 fig. 19, 126-127 fig. 73, Clobus & Hebert 2009: 35-56; Leitch 2009: 144; Silver 2010: 220; Helman 2013: 40-41.

The Nuremberg world-chronicle, elaborated by the humanist Hartmann Schedel (1440–1514), is conceived as a vast panorama of space and time, spanning the proverbial seven ages of the world as well as the full surface of the *mappa mundi*.[1] This monumental digest is illustrated with 1804 woodcuts from the workshop of Michael Wolgemut, which effectively turn it into a diorama of natural and human history.

While in every technical sense a fruit of its own age, its contents are in many respects traditional, looking back to the legacy of the great medieval encyclopaedias. Its conception of time, which provides the framework for its sweeping views of geographical space, relies on the scheme devised by Isidore in his *Etymologies*, in the seventh century; the world map on these pages, with its adjacent gallery of marvellous men from far-away lands, finds its place within an account of the Flood in the Book of Genesis,[2] an event marking the dawn of the earth's second age. The graphic composition here is important in ordering this body of knowledge, inserting it into its traditional biblical context. The figures of Japheth, Shem and Ham correlate geographical knowledge with chronological time, rooting an understanding of space in the historic moment when the continents were divided among Noah's sons (cf. cat. 42) – but they insert it also in the broader timeline of salvation history, and thus in the teleological framework proper to Christian accounts of the created world. The progeny of the reprobate Ham, cursed because he did not avert his eyes from his father's nakedness, was scattered across the wild and wonderful lands of Africa and Asia.[3] A second Cain, he becomes the patriarch of a degenerate humanity, represented by the fantastical peoples of the classical and medieval legendary.[4] They are illustrated here according to longstanding iconographic types: a six-armed man, a feral woman, a man with six fingers on each hand, a satyr, a hermaphrodite, a man with four eyes and a Ciconian (a member of the eponymous people from classical lore, often imagined literally as a bird-men). Wolgemut's reliance on older models is betrayed by these pictures' close resonance with those in the Mercatellis compendium (cat. 49) and in the related Bruges, Stadsbibliotheek Ms. 411.[5]

1
On this topic, see inter alia Friedman 1981: 37–58; Davies 2016.

2
Namely, Gen. 9: 22–27.

3
Gen. 9:25; an expressive commentary on this passage is provided by the anonymous author of a Middle Irish treatise on the Six Ages of the world (Oxford Bodleian Library Rawlinson Ms. 502b): 'His famous father cursed the son called Ham so that he – he excelled in perversity – is the Cain of the people after the Flood. From him with valour sprung horse-heads and giants, the line of maritime leprechauns, and every unshapely person; those of the two heads [...] and the two bodies in union, the dun-coloured, one-footed folk, the merry blue-beaked people [...]' (cited in Friedman 1981: 99).

4
On Cain's and Ham's association with the legendary of monstrous races, see Friedman 1981: 87–107; on Cain's curse and the history of its representations see Mellinkoff 1981.

5
Massing 1991c: 124.

6
Jacobsson 2002: 6. II. 29–30 (66–68).

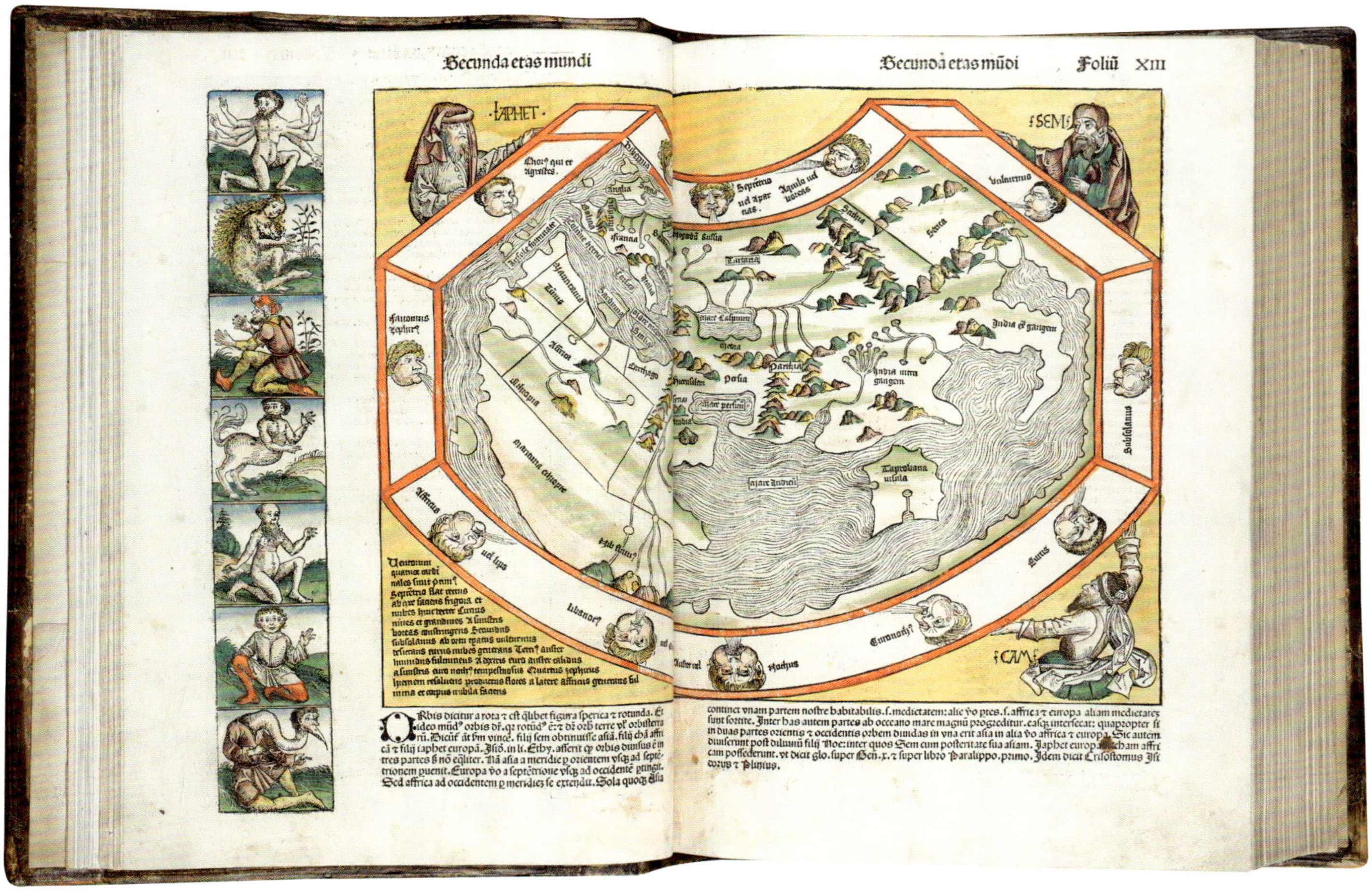

Secunda etas mundi
Secunda etas mūdi
Foliū XIII
IAPHET
SEM
CAM

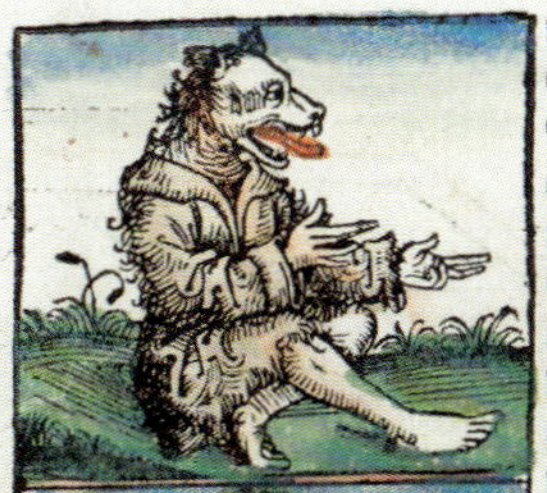

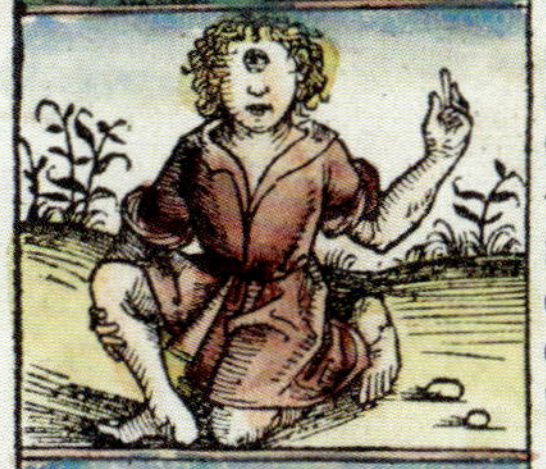

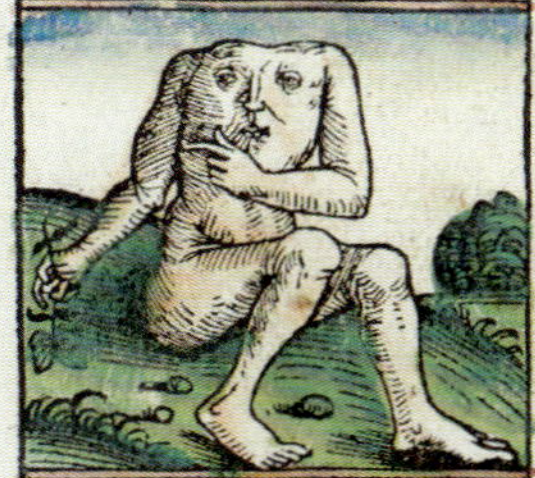

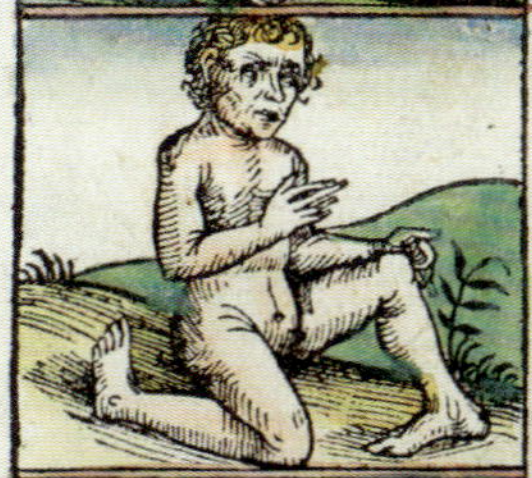

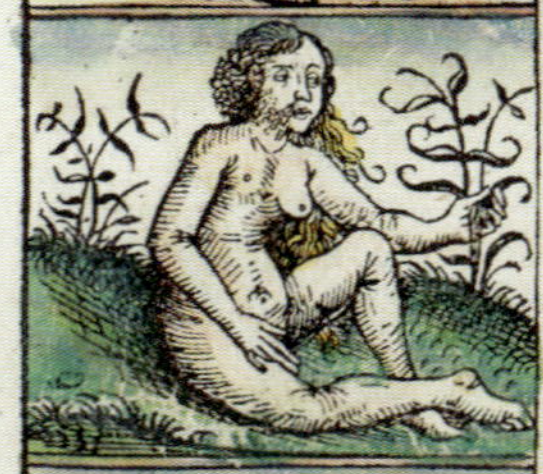

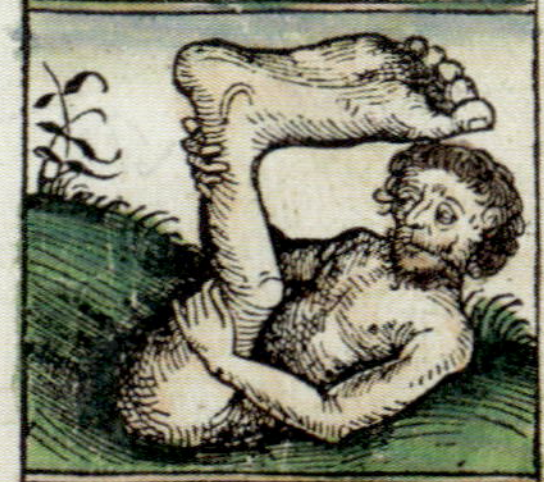

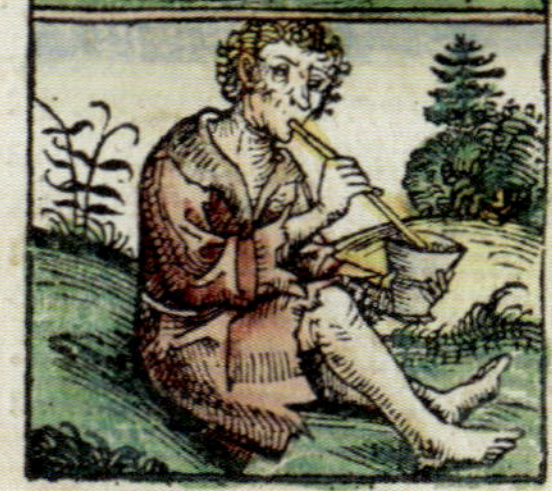

De homib⁹ diuersaꝝ formaꝝ dicit Pli.li.vij.
ca.ij. Et Aug.li.xvi.de ci.dei.ca.viij. Et Isi
dorus Ethi.li.xi.ca.iij. oīa q̄ sequūtur in in
dia. Cenocephali homines sunt canina capita habē
tes cū latratu loquūtur aucupio viuūt. vt dicit Pli.
qui omnes vesciūtur pellibus animaliū.
Cicoples in India vnū oculum hn̄t in fronte sup na
sum hij solas ferarū carnes comedūt. Ideo agriofa/
gite vocātur supra nasomonas confinesq3 illorū ho/
mines esse: vtriusq3 nature inter se vicibus coeūtes.
Calliphanes tradit Arestotiles adijcit dextram mā
mam ijs virilem leuam muliebrem esse quo hermo
froditas appellamus.
Ferunt certi ab oriētis pte intima esse homines sine
naribus: facie plana eq̄li totius corpis planicie. Ali
os supiore labro orbas. alios sine linguis ꝛ alijs cō
creta ora esse modico foramine calamis auenaꝝ po
tū haurie̅tes.
Item homines habentes labiū inferius. ita magnū
vt totam faciem contegant labio dormientes.
Item alij sine linguis nutu loq̄ntes siue motu vt mo
nachi.
Pannothi in scithia aures tam magnas hn̄t. vt con
tegant totum corpus.
Artabrite in ethiopia pni ambulāt vt pecora. ꝛ ali
qui viuūt p annos.xl. quē nullus supgreditur.
Satiri homūciones sunt aduncis naribus cornua i
frontibus hn̄t ꝛ capraꝝ pedibus similes qualē in so
litudine sanctus Antonius abbas vidit.
In ethiopia occidentali sunt vnipedes vno pede la/
tissimo tam veloces vt bestias insequantur.
In Scithia Ipopedes sunt humanā formaz eq̄nos
pedes habentes.
In affrica familias quasdā effascinātiū Isigonus ꝛ
Memphodorus tradūt quaꝝ laudatōne intereāt p/
bata. arescāt arbores: emoriātur infantes. esse eius/
dem generis in tribalis et illirijs adijcit Isogon⁹ q̄
visu quoq3 effastinent iratis p̄cipue oculis: quod eo
rū malū facilius sentire puberes notabili⁹ esse ꝙ pu
pillas binas in oculis singulis habeant.
Item hoīes.v.cubitoꝝ nūq3 infirmi vsq3 ad mortē
Hec oīa scribūt Pli. Aug. Isi. Preterea legit̄ i gest
Alexādri ꝙ i india sunt aliq̄ hoīes sex man⁹ hn̄tes.
Itē hoīes nudi ꝛ pilosi in flumine morātes.
Itē hoīes manib⁹ ꝛ pedib⁹ sex digitos habentes.
Itē apothami i aq̄s morantes medij hoīes ꝛ medij
caballi.
Item mulieres cū barbis vsq3 ad pect⁹ ꝭ capite pla/
no sine crinibus.
In ethiopia occidētali sūt ethiopes.iiij.ocl'os hn̄tes
In Eripia sunt hoīes formosi ꝛ collo gruino cū ro/
stris aīalium hoīmq3 effigies mōstriferas circa extre
mitates gigni mīme mirū. Artifici ad formanda cor
pora effigiesq3 celandas mobilitate ignea.
Antipodes āt eē.i.hoīes a ꝯria pte terre vbi sol orit̄
qn̄ occidit nob̄ aduersa pedib⁹ nr̄is calcare vestigia
nulla rōe credēdū ē vt ait Aug.16.de ci.dei.c.9. In/
gēs tn̄ h̄ pug̃ lr̃aꝝ ꝯtraq3 vulgi opīoē circūfundi ter
re hoīes vndiq3 cōuersisq3 iter se pedib⁹ stare et cūct
silem eē celi v̄ticē. Ac sili mō ex q̄cūq3 pte mediā cal
cari. Cur āt n̄ decidāt: miret̄ ꝛ illi nos n̄ decidere: nā
em̄ repugnāte: ꝛ quo cadāt negāte vt possint cadere.
Nā sic ignis sedes nō ē nisi i ignib⁹: aq̄rū nisi i aq̄s.
spūs nisi in spū. Ita terre arcentibus cūctis nisi in se
locus non est.

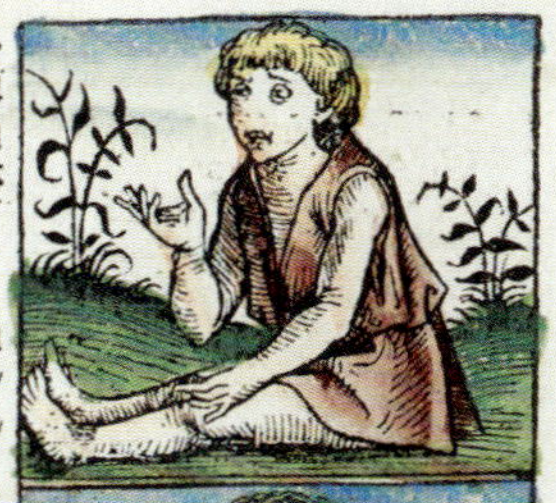

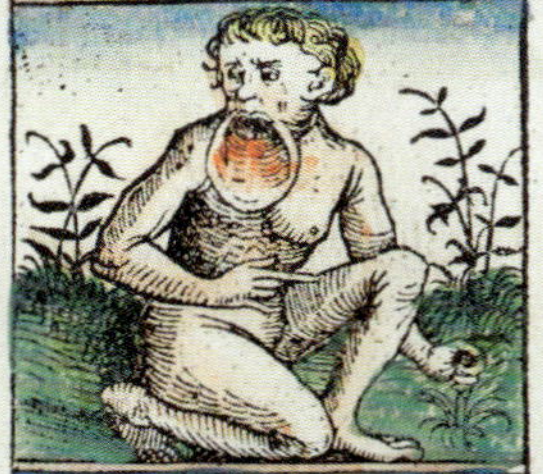

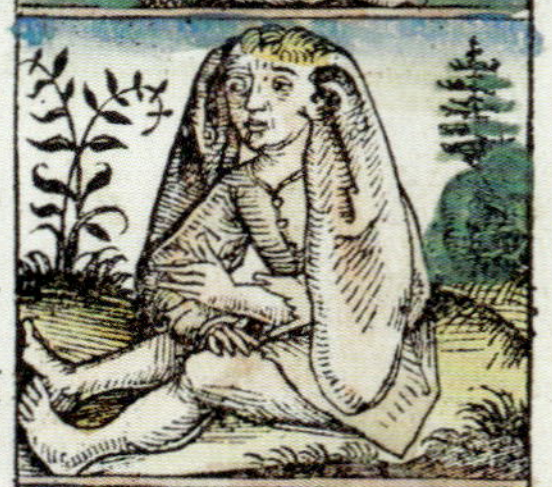

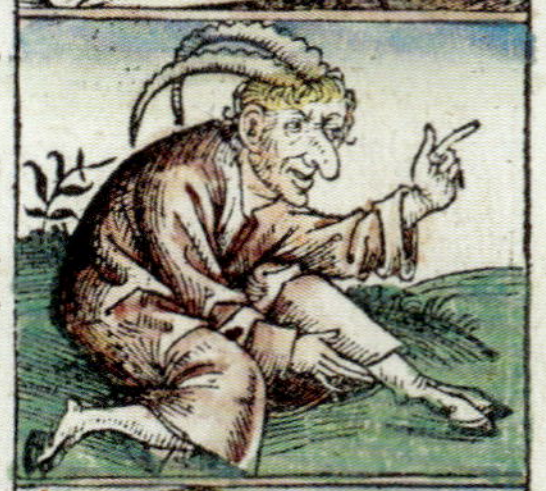

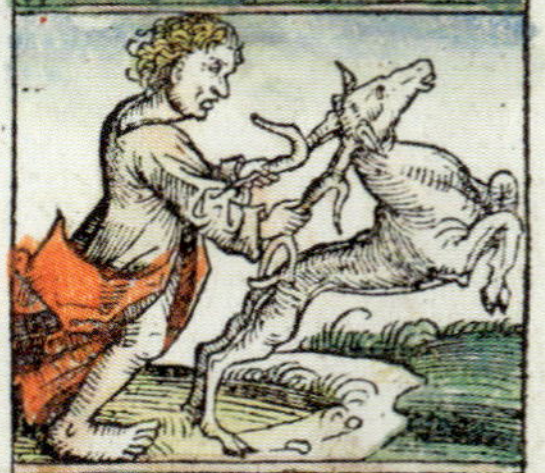

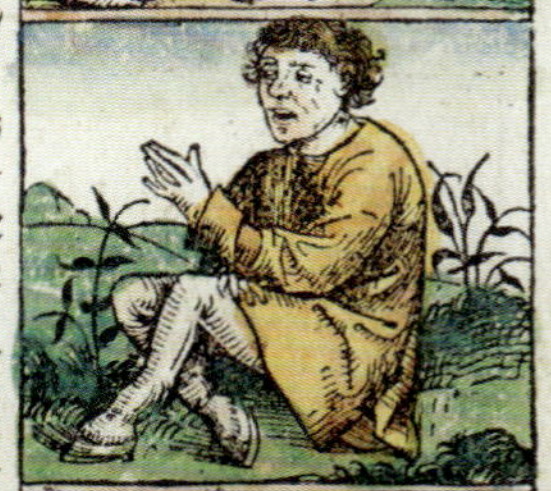

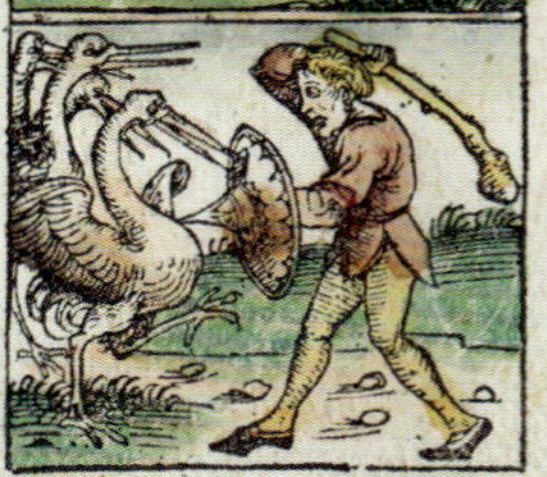

[cat. 50] fol. 12r

To interrogate the humanity of these fantastical beings, to assess their place within categories of *sameness* and *difference*, is also to seek their place in this divine scheme. Framed by these troubled genealogies, the *mappa mundi* becomes the object of a special understanding of the created world, one that is also axiological and whose overarching harmony does not preclude disquieting dissonances. In this, it reverberates with another longstanding idea, issued from Augustinian aesthetics: God's creation is rendered beautiful on account of an interplay of luminous and dark elements – a chiaroscuro lost on the mortal eye, which can only be captured by the vast, scanning gaze of the first maker.[6] This integrated vision of the varied individual threads that make the tapestry of nature is in some sense akin to the great visual artifice that the map, as a picture, represents: an inconceivable whole assembled out of endless conflicting parts.

MAC

51

Wild Man with Club

Southern Netherlands
Early sixteenth century
Wood, 179 x 45 x 42 cm
Private collection

In medieval imagination – which continued to live on unabated during the Age of Discovery – the 'wild man' represented all things alien to 'civilisation'.[1] As such, he was the perfect figure for inversive self-definition.

The late medieval yearning for the primeval state manifested itself in numerous literary works, in folklore entertainment at the courts of noble families, and in middle-class circles (where people would play-act the 'wild man' on festive occasions), as well as in countless works of art. In the Upper Rhine centres of the fifteenth and sixteenth century, for instance, a great many wall tapestries were woven that took the wild man and wild woman as their central motif. Here, wildness is often a metaphor for unbridled sexual desires, to be 'tamed' – that is to say sated – by one's lover. Here, the 'wild man/savage' epitomised the utopian dream of a blissful and satisfied 'natural' life.

In the wilderness, all things that are unthinkable in the 'civilised' world are possible, enabling people to give free rein to their uninhibited impulses. The 'wild men/savages' epitomise the absence of all constrictive norms and obligations of normal social intercourse. Their image is seen in countless forms and shapes in the courtly culture of the aristocracy and in middle-class culture. 'Wild/savage' encapsulated all things that lay beyond human culture, morals and norms, all things demonic and supernatural, all things unfaithful and amoral, all things rough and raw, all things wrong and muddled, all things odd, frightening and wondrous. Wildness was menacing and terrifying, but equally fascinating and exhilarating. The hairy savage equally embodied all things people desired, but which were forbidden under the conventions and norms of regular society: insouciant living in unison with nature, idyllic love and freedom from restraint by social obligations. The savage is 'Desire' (in the fifteenth century Spanish novel *Cárcel de amor* by Diego de San Pedro, he is called *Deseo*) or the 'Law of nature' (in Jean Germain's *Le chemin de paradis* from 1457; Brussels, Royal Library of Belgium, Ms. IV 823), according to which humankind had lived from the time of Adam up until the Flood.

But the wild primeval state of man was also a source of fascination for medieval folk in the religious aspirations of reformed Augustinism from the fourteenth century forward (*ut reddamur ad pristinum statum Adae*!), in subversive movements such as the 'Brethren of the Free Spirit', and in the Adamite cults from Bohemia to Picardy and from the Low Countries to the Basque Country (such as the fifteenth century heretics of Durango). These groups endeavoured to breathe new life into the supposed traits and behavioural patterns of primitive man in preparation of the Thousand Year Kingdom of the Holy Spirit. Which is why they sought to abolish all institutions and propagated free love (viewed from the male perspective). In doing so, 'primeval wildness' became the guiding principle that would usher in a utopian dream, which they were prepared to turn into a reality – by fire and sword if need be.

PVDB

1 On this subject, see inter alia Bernheimer 1952; Möller 1963: 69–70, no. 99, Husband 1980; Silver 1983; White 1986: 150–182; Vandenbroeck 1987: 7–21; Tchalenko 1990; Isaacson 1998; Moseley-Christian 2012; Pouvreau 2015; Stock 2016.

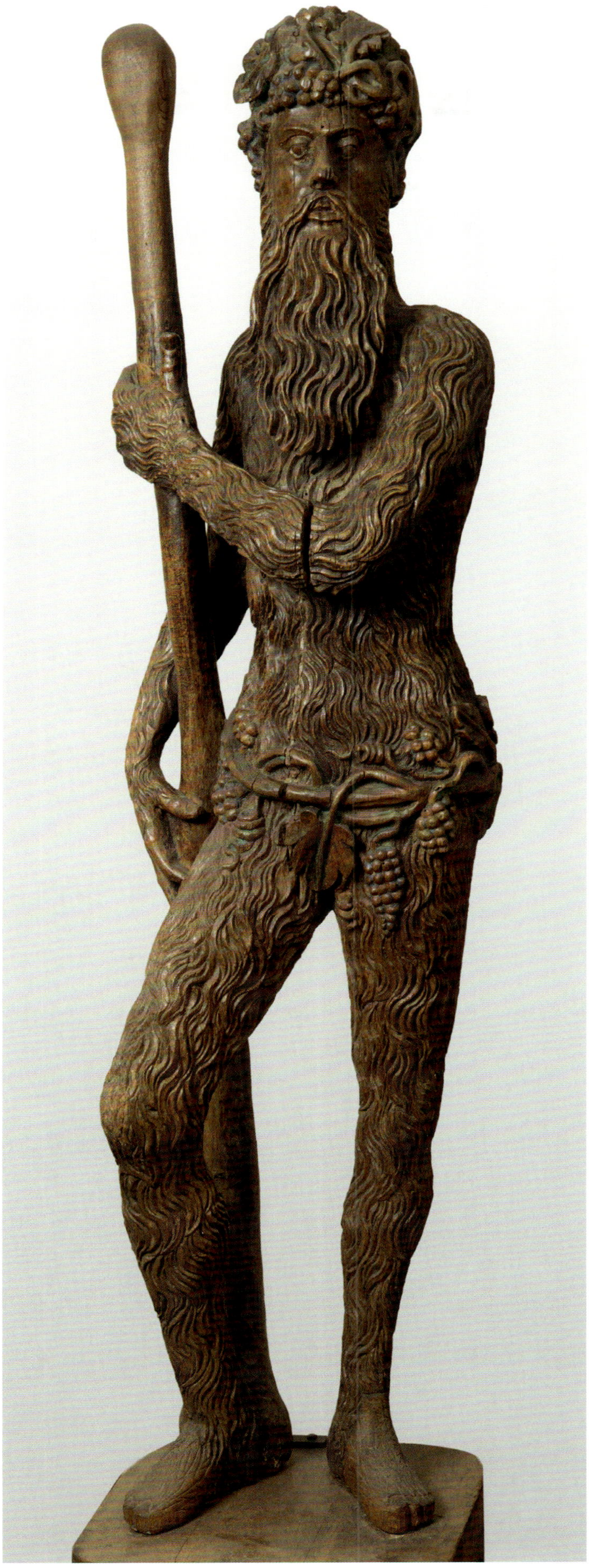

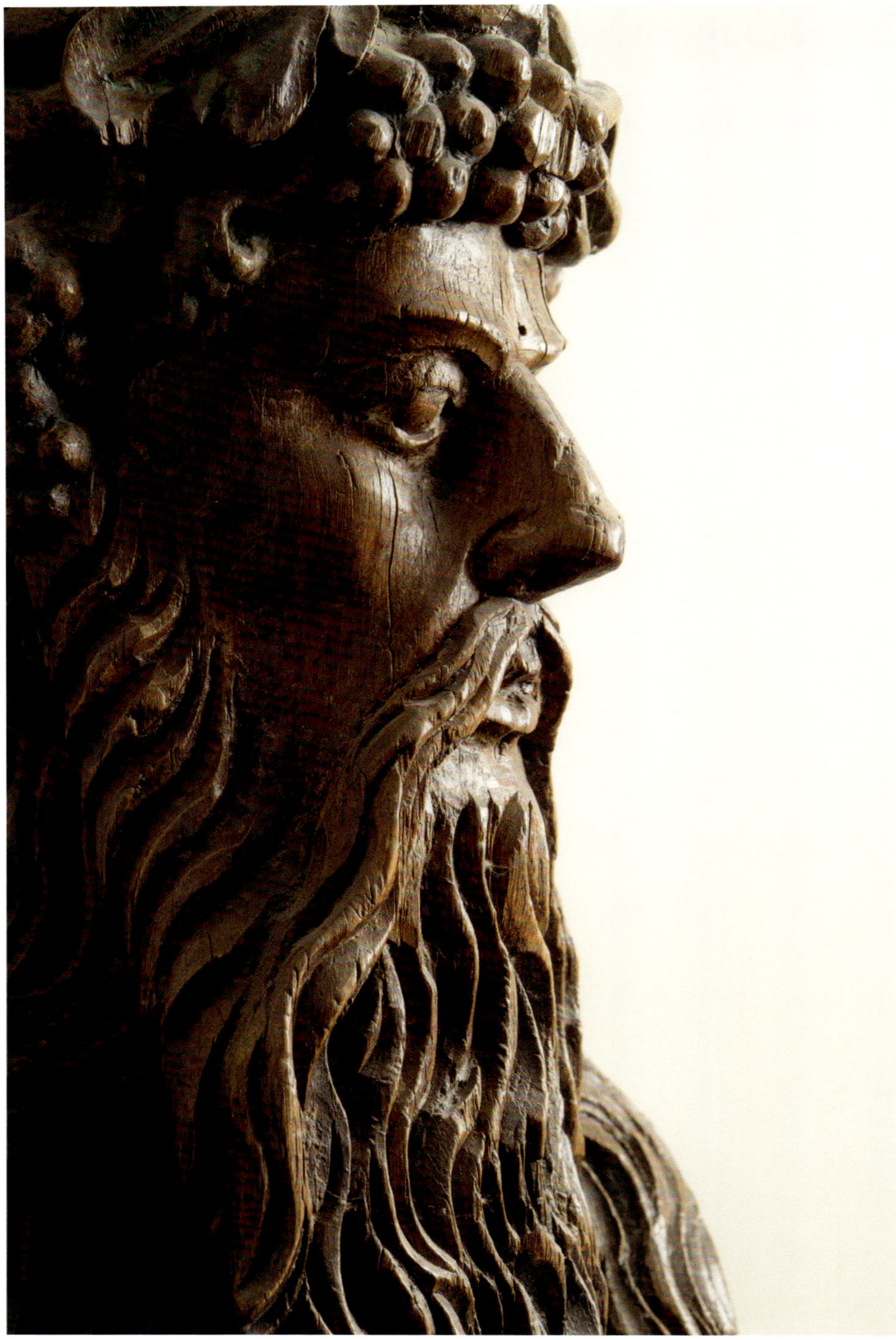

[cat. 51] detail

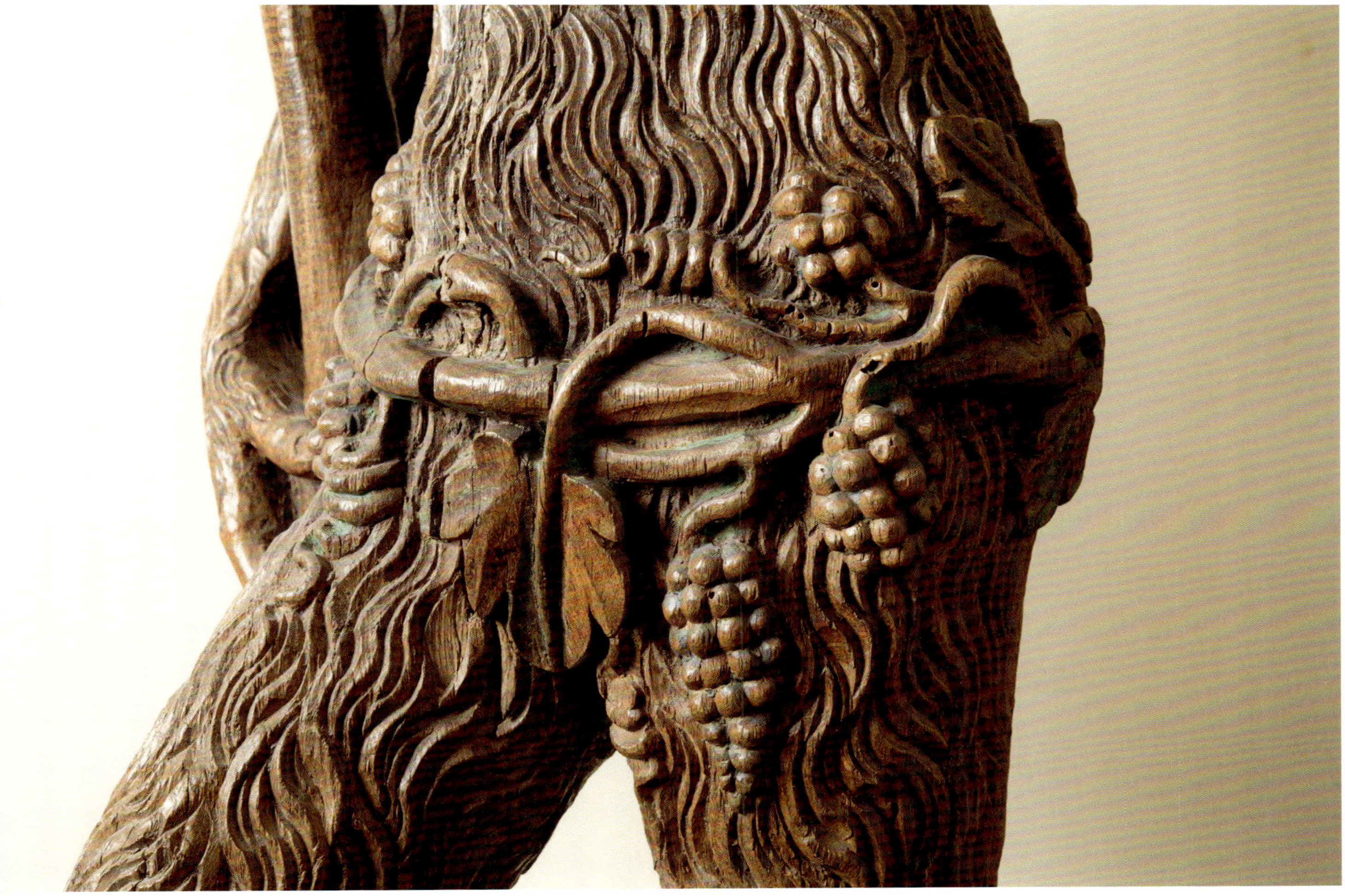

[cat. 51] detail

52

Kaart van Brazilië

In: *Atlas de Dauphin*, Dieppe, *c.*1538, fol. 15v–16
Parchment, 39 fol., 440 x 312 mm
The Hague, National Library of the Netherlands, inv. 129 A 24

PROVENANCE

Louis César de la Baume le Blanc de la Vallière (?-before 1783); Pierre Antoine Bolongaro Crevenna (1783); Joost Romswinckel (1790); Royal Library, The Hague (1807)

BIBLIOGRAPHY

Anthiaume 1916: 193-194; R.A.S. & Destombes 1964: 89; Cortesão & Teixeira da Mota 1987: 133-134, 141; Destombes & Gernez 1987: 144-145; Colin 1988: 138-140, 141, 304; Levenson 2007: 187; Raaflaub & Talbert 2010: 334; Stols, Piccoli, De Peuters & Petre 2011: 17-18; Davies 2012: 332; Palmenaer & Parmentier 2015: 66; Heremans 2015.

The *Atlas de Dauphin* contains fourteen illuminated nautical charts of Europe, Africa and America, with a particular focus on the depiction of the Portuguese colony in Brazil. The atlas was intended for the French dauphin and subsequent King Henry II of France, and is believed to have been a present from merchant seamen from the north of France.[1] The atlas was fashioned by the so-called *School of Dieppe*, a group of cartographers, illuminators and cosmologists working out of Normandy *c.*1540–1560.[2] From a cartographic perspective, their nautical atlases are in keeping with the Portuguese tradition, although they stand out by virtue of their own distinctive figurative details.

The cartographer and the illuminator(s) of the *Atlas de Dauphin* are unknown.[3] The atlas is dated to around 1538, based on the then geographical understanding and cartographic expertise and the overseas trade policy of King Francis I.[4] The Portuguese crown sought to secure its supremacy in overseas trade by concluding a friendship treaty (1536) with Francis I, which resulted in a royal ban on French ships sailing the seas to Brazil and Guinea.[5] However, this was against the will of the French merchant seamen, who maintained very frequent trade links with the indigenous populations of the Portuguese colony.[6] The *Atlas de Dauphin*, which came about shortly after, was presumably ordered and paid for by merchant seamen from the north of France in hopes of preserving transatlantic trade relations.[7] All the more so as there was considerable demand for Brazilian products, such as certain wood varieties and animal species, pigments and pearls.[8] In addition, traders were importing small monkeys and parrots, which were highly sought after as pets in the West. It is assumed that most illuminators never clapped eyes on these overseas territories for themselves, although it is likely that seamen were encouraged to gather information on the appearance of the indigenous tribes and the local fauna and flora.[9]

Folios 15v–16 in the *Atlas de Dauphin* comprise the most resplendent and most colourful depictions of recently discovered Brazil,[10] showing large numbers of small monkeys and parrots. In addition, this map shows the contacts that existed

1
Levenson 2007: 187; De Palmenaer & Parmentier 2015: 66.

2
Cortesão & Teixeira da Mota 1987: 133.

3
In all probability, the cartographer was a Portuguese who had emigrated to France. His dual nationality can be seen in the nomenclature. Salient place names, such as the names of islands, have been recorded in a French translation, whilst place names along the French coastline have been written down in Portuguese or in a hybrid of the two languages. Cortesão & Teixeira da Mota 1987: 133–134.

4
Destombes & Gernez 1987: 144–145; Cortesão & Teixeira da Mota 1987: 135–136; Colin 1988: 141.

5
Anthiaume 1916: 193–194; Colin 1988: 138–140.

6
In 1537 and 1538, Francis I ordained stricter rules: all who were caught flauting the ban on sailing to Brazil or Guinea and were captured by the Portuguese were now required to hand over their ships, colonial products and commodities. The Portuguese were also allowed to inflict corporal punishment, and those who failed to obey were liable to face the death penalty. Anthiaume 1916: 193–194; Colin 1988: 138–140.

7
Heremans 2015: 10.

8
Brazil-wood was used in the French textile industry; it was ground and dried to obtain the coveted red pigment. Raaflaub & Talbert 2010: 334; Stols, Piccoli, de Peuter & Petre 2011: 17–18 and also cat. 54A–B in this catalogue.

9
Davies 2012: 332.

10
This map shows the south coast of Brazil. South is situated on top and is indicated by a crescent moon, whereas north is situated at the bottom and is indicated by a *fleur-de-lys*.

11
These kinds of positive depictions of the Brazilians make for a stark contrast with the cruel scenes of cannibalism in the pictorial language of Portuguese nautical atlases.

12
Colin 1988: 304.

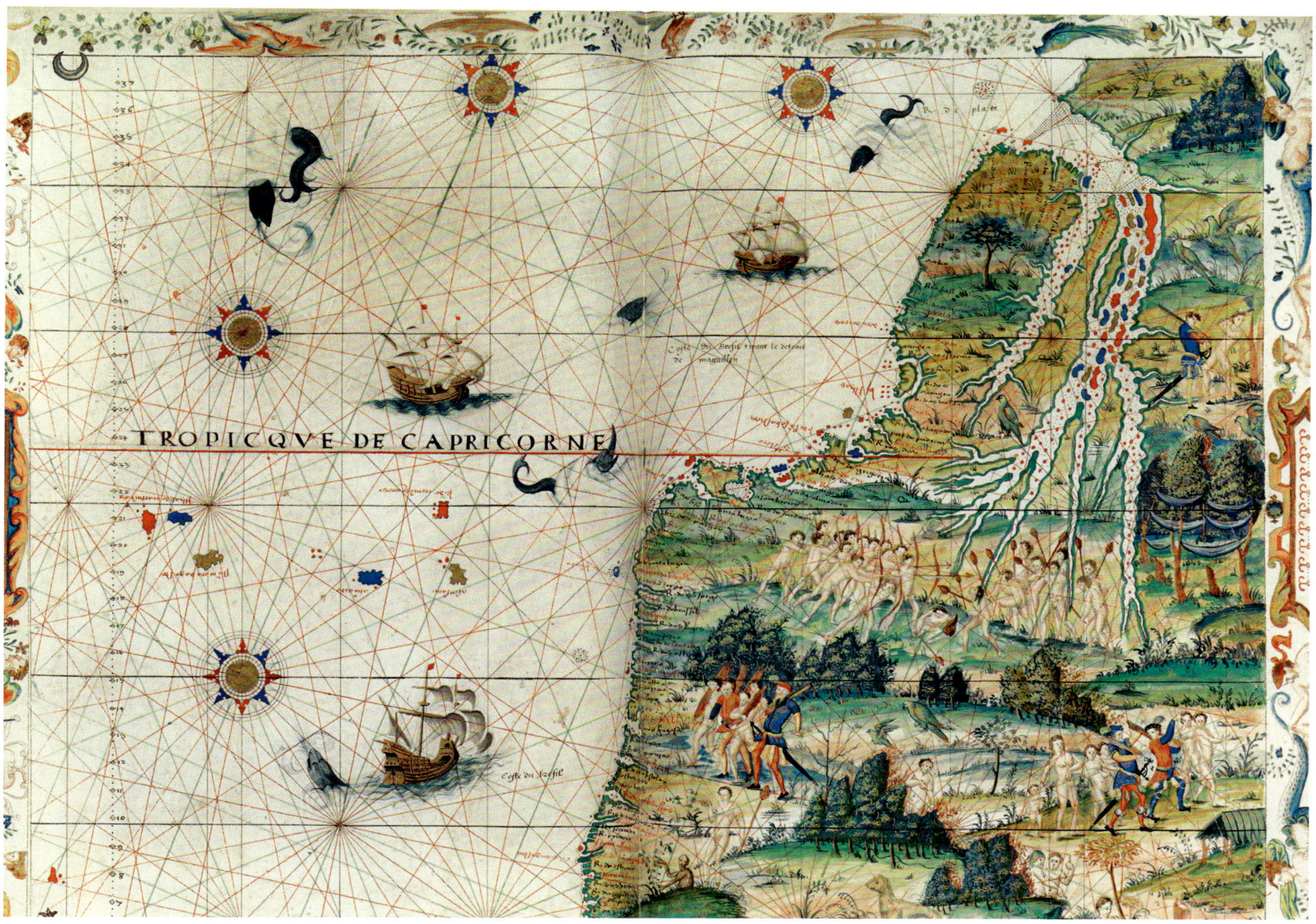
TROPICQVE DE CAPRICORNE

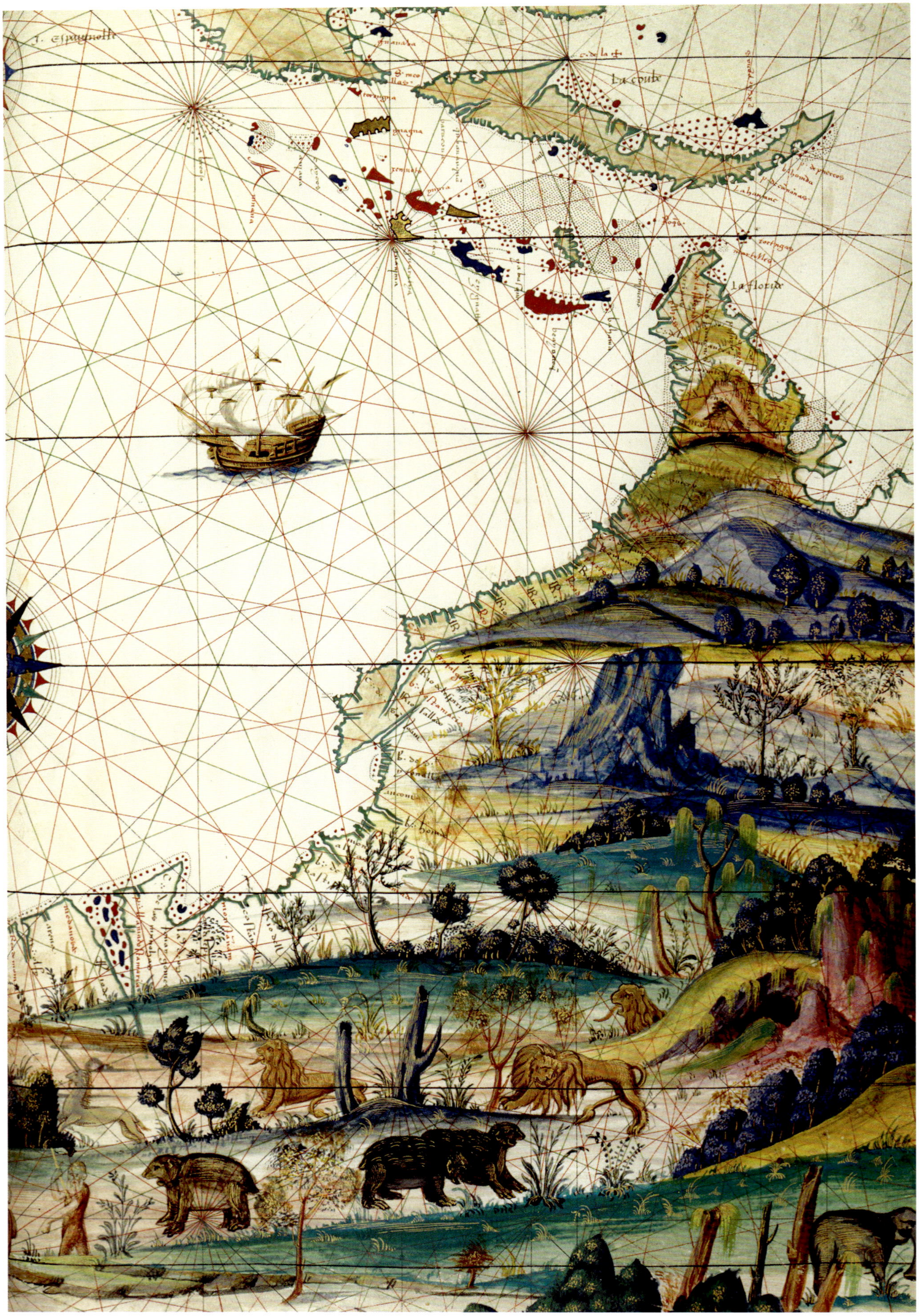

[cat. 52] detail of fol. 25v-26

between the indigenous populations amongst themselves and the relations they entertained with Westerners. The way in which Brazilians are seen to fell trees and lug them to French ships is indicative of a 'good' relationship that legitimates the mutual trade links (cat. 53 and 54).[11] Moreover, their cooperation enabled the French to trade with these native Brazilians without having to set up permanent infrastructures. Folios 15v–16 are also believed to incorporate a symbolic request for peaceful colonisation; Brazilians are seen to head inland with a handful of Frenchmen in tow, invitingly pointing to the soil.[12] Perhaps this pictorial programme was meant to encourage the crown prince to adopt a different course of foreign policy that stepped away from the policy pursued by his father. The *Atlas de Dauphin* exhibits an expanding cartographic understanding and a fascination with the unknown, tinged by political and economic interests.

SH

53

Pedro & Jorge Reinel, Lopo Homem & António da Hollanda
Map of Brazil

In: *Miller Atlas*, c.1519, fol. 5v

Parchment, 415 x 591 mm

Paris, Bibliothèque nationale de France, BnF Cartes et Plans, inv. GE-DD 683 (RÉS)

PROVENANCE

The maps were offered for sale as loose-leaf charters to the Viscount de Santarém in 1855 by the book dealer Jacques Charavay; later in the collection of the former's French friend, Emmanuel Miller, as debt settlement; sold to the Bibliothèque nationale de France by his widow in 1897.

The title page was only later connected with a group of maps in the Bibliothèque nationale de France. First mentioned in the collection of Major J. A. Morrison of Basildon Park; subsequent sale, London, Sotheby & Co, May 1930;[1] later in the collection of Marcel Destombes, who linked the sheet to the collection of maps in the Bibliothèque nationale, then known as the *Cartes Miller*;[2] in 1975 taken up in the collection of the Bibliothèque nationale.[3]

BIBLIOGRAPHY

Denucé 1908: 42; Heawood 1930: 159-160; Heawood 1931: 250-255; Destombes 1937: 460-464; Caraci & Destombes 1938: 263-266; Kammerer 1938: 450-453; Destombes 1939: 485-492; Adonias 1963: 23; Morison 1971: 231; Cortesão & Teixeira da Mota 1987: 55-60; Colin 1988: 298-300; Marques 1994: 53-57; Traub 2000: 63; Marques 2005; Alegria, Daveau, Garcia & Relaño 2007: 1029-1030; Marques 2009: 68-73; Hofmann, Richard & Vagnon 2012: 180; Davies 2016: 121-123.

The *Miller Atlas*, also known as the *Lopo Homem-Reineis Atlas*, is a richly illustrated manuscript that was made in 1519 by the Portuguese cartographer Lopo Homem. The geographic extent of the atlas is considerable. In addition to a *mappa mundi,* it contains maps of the North Atlantic Ocean, Northern Europe, the Azores, Madagascar, the Indian Ocean, the Sea of China, the Moluccas, Brazil, the Atlantic Ocean and the Mediterranean Sea.[4] Although the provenance of the atlas cannot be determined with any certainty, it is generally assumed that it was given by the Portuguese king Manuel I to Francis I of France (cf. cat. 52). Later it belonged to the collection of Catharine de' Medici (1519–1589), daughter-in-law of Francis I, as is evident among other things from the presence of her coat of arms on the title page.[5] An inscription on the same title page explains: 'This is a map of the entire globe of the universe such as it is now known, which I, Lopo Homem, cosmographer, comparing many other maps both old and knew, have drawn with great exertion and diligent work in the illustrious city of Lisbon, in the year of our Lord 1519, at the behest of Manuel, eminent king of Portugal'.[6] Although their names are not mentioned on the title page, these maps may have been made with the cooperation of cartographers Pedro and Jorge Reinel and the Dutch painter António de Holanda.[7]

1 Heawood 1930: 159–160; Heawood 1931: 250–255.

2 Destombes 1937: 460–464; Kammerer 1938: 450–453; Caraci & Destombes 1938: 263–266; Destombes 1939: 485–492.

3 Marques 1994: 57.

4 It is presumed that several maps, such as the map of the African continent, have been lost.

5 Cortesão & Teixeira da Mota 1987: 57. The coat of arms of Catharine de' Medici were only added to the title page at a later date.

6 The inscription reads: 'Hec est vniuersi orbis ad hanc usqe diem cogniti tabula Quam ego Lupus homo Cosmographus in clarissima Ulisipone ciuitate Anno domini nostri Millessimo quingentessimo decimo nono Jussu Emanuelis incliti lusitanie Regis collatis pluribus alijs tam vetustorum quam recentiorum tabulis magna industria et dilligenti labore depinxi.'

7 Marques 1994: 53–57.

8 Cortesão & Teixeira da Mota 1987: 57. The complete description is as follows: 'Tabula hec Regionis magni brasilis est ad partem occidentale$_D$ Antilias castelle regis obtinet. Gens uero eius nigrescentis coloris. fera immanissima carnibus humanis uescitur. Hec eadem gens arcu sagittis egregie utitur. Hic psytaci (sic) uersicolores alieq$_D$ innumere aues fereq$_D$ monstruose. et Scymuarum (sic) plura genera reperiuntur plurimaq$_D$ arbor nascitur que brasil nuncupata uestibus purpureo colore tingendis, opportuna censetur.'

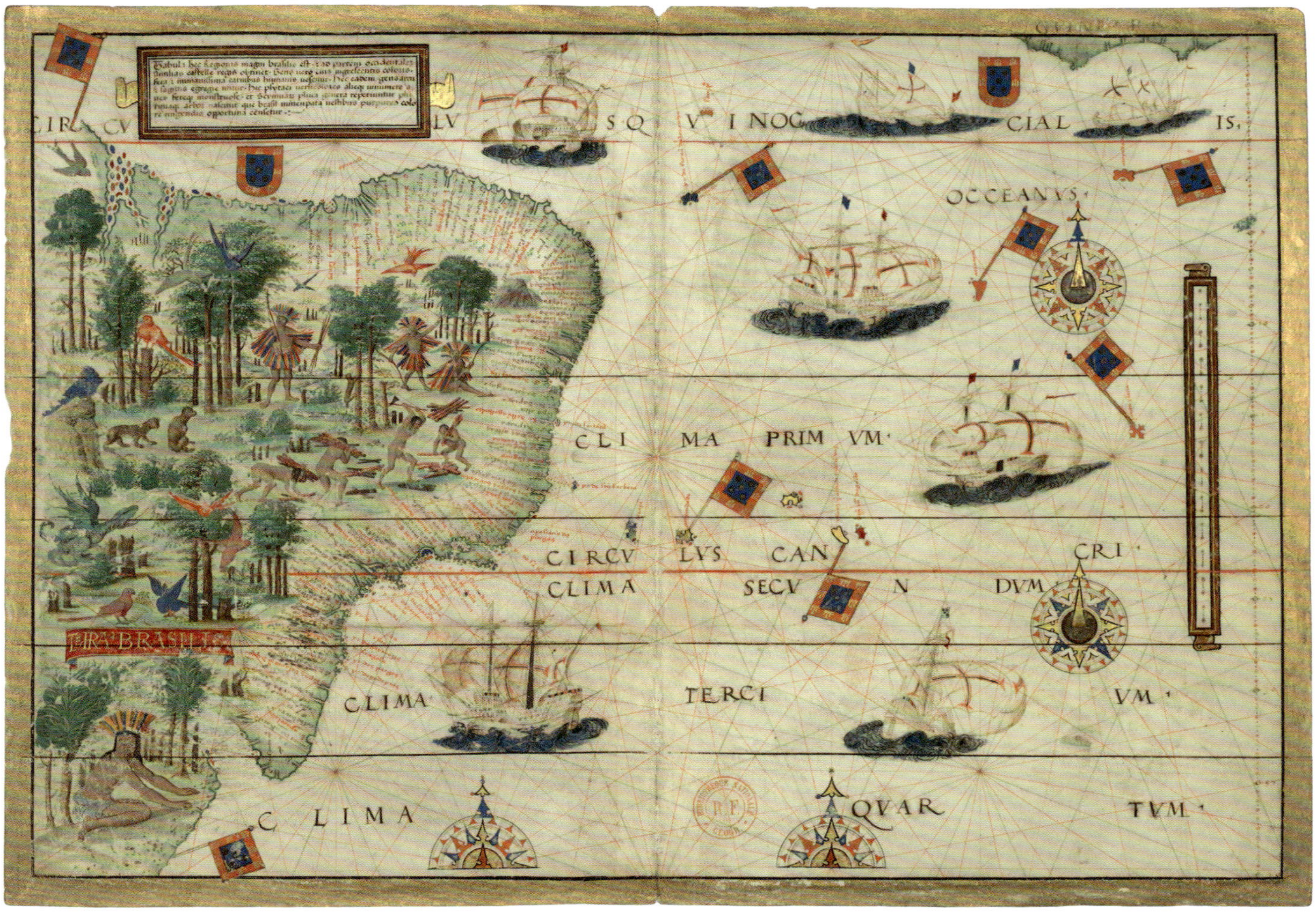
CIR CV
LV
S Q
V I NOC
CIAL
IS.
OCCEANVS
CLI
MA
PRIM VM
CIRCV
LVS
CAN
CRI
CLIMA
SECV
N
DVM
CLIMA
TERCI
VM
C LIMA
QVAR
TVM

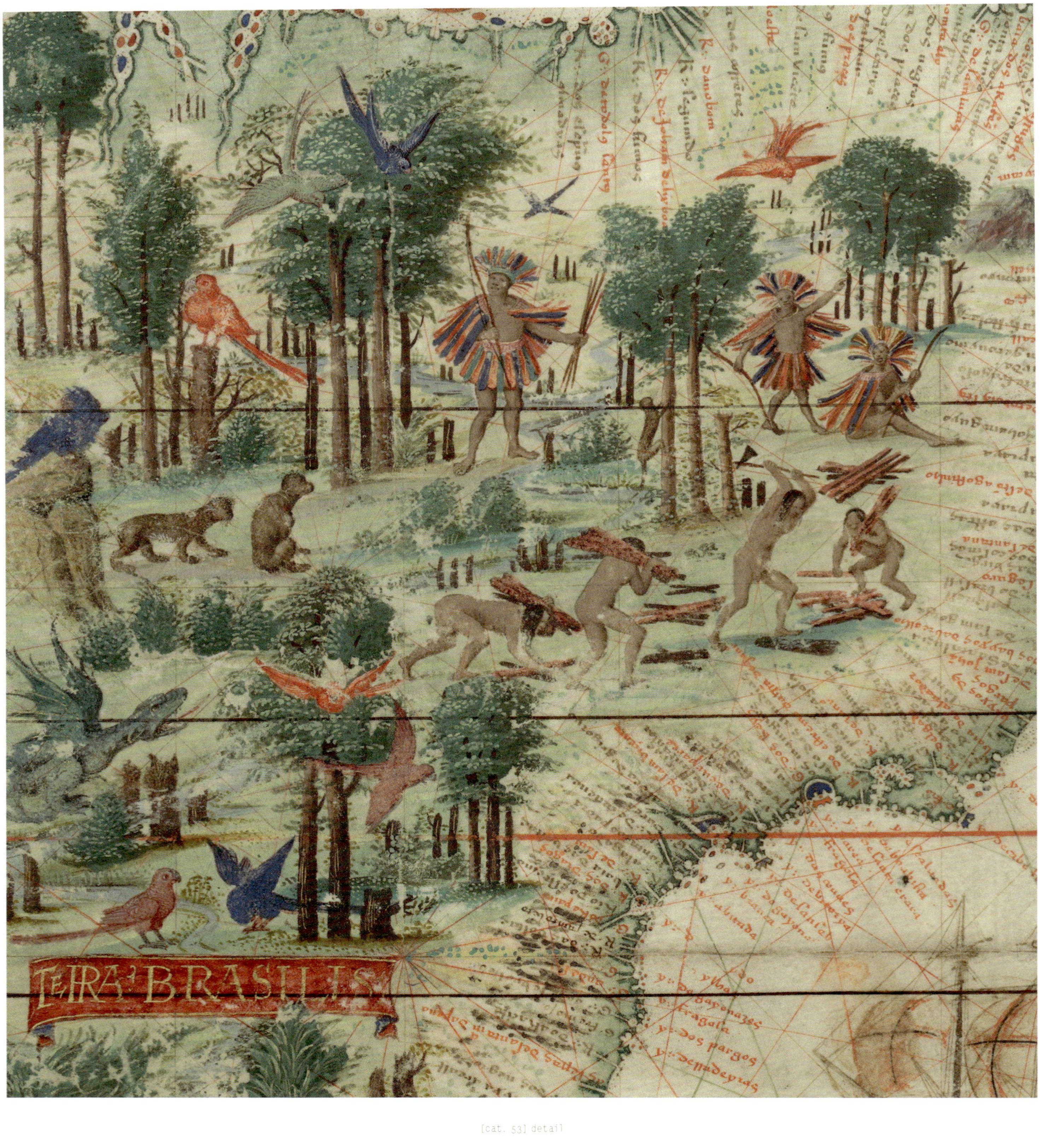

[cat. 53] detail

The map of Brazil, shown here, shows the Brazilian colony on the left and the South Atlantic on the right, dominated by Portuguese ships. Although the map is equipped with navigation guidelines, it is unlikely that this showpiece actually functioned as a maritime map. The luxuriant and exotic illustrations of overseas territories instead suggest that it had a purely cultural and ceremonial function within the trans-Atlantic trade policy of Portugal and France, on the one hand, and the promotion of geographical knowledge of recently discovered areas on the other (cat. 52 and 55). The inscription at the upper left, in addition to indicating the geographic position of the area depicted, provides some explanation of the customs, flora and fauna of the Portuguese colony. The inscription reads: 'This is the chart of the region of great Brazil and on the western side it reaches the Antilles of the King of Castile. Its people are somewhat dark in colour. Savage and very cruel, they feed on human flesh. These same people are most skilful in the use of bow and arrows. Here are multi-coloured parrots and innumerable other birds and monstrous wild beasts. And many kinds of monkeys are found and there grows in great quantity the tree called brazil, which is considered proper for dyeing clothes in purple' (cat. 54).[8] Thanks to its geographical depictions – accurate for the time – and minutely executed illuminations, the *Miller Atlas* is counted as one of the most important and earliest representations of a new, trans-Atlantic order.[9]

HI & DVH

54

L'Ile du Brésil

Frankrijk (Rouen?)
c.1530
Bas-relief in oak, two panels, 52 x 221 cm and 53 x 170 cm
Rouen, Musée départemental des Antiquités, inv. 1401-2 D

PROVENANCE

L'Hôtel du Brésil, the wooden house of a wealthy ship owner at Rue Malpalu 17 in Rouen; acquired by the museum in 1837.

BIBLIOGRAPHY

De la Quérière 1821: 150-153; Dubosc 1921; Rouen 1932: 134-135, no. 362; Paris 1955: no. 16; Costa 1964: 64, no. 286; Honour 1976: 15, no. 8; Mauduech 1976: 71, no. E7; Chirol 1980-1981: 25-26, no. 22-23; Chirol 1984: 230-231, no. 117; Collin 1988: 347-328, no. M.6 and M.17; Justo Guedes & Lombardi 1990: 230; Jaffe, Viola & Rovigatti 1991; De Moraes Belluzo 1994; Van den Eynde 2015; Buono 2016.[1]

These two bas-reliefs from the second quarter of the sixteenth century[2] were originally part of L'Hôtel du Brésil in Rouen,[3] the home of a wealthy ship owner who traded in Brazil-wood. Only a few decades after the discovery of the New World, the owner of this house provided lodgings to mariners who had come back from or were about to depart for Brazil.[4] The commercially important harvesting and transportation of Brazil-wood is the central theme of these two panels (cat. 53). This species of wood, from which Brazil gets is name, was used by the textile industry in Rouen to make scarlet dye.

The panels, which together form a whole, must be read from right to left. The first panel shows how the Tupinambá (an indigenous Brazilian people) penetrate deep into the interior in order to fell trees. After the trees have been stripped of their bark, the Indians carry them on their naked shoulders (fig. 54.1) – as Hercules bore the columns – all the way to the coast.[5] The second panel shows the trees being loaded onto a galleon,[6] after which the ship departs for France.

For Europeans of the time, America was not only the *new* world but also the *other* world. The anonymous sculptor of the reliefs has probably tried to explain this confusing reality by drawing parallels with familiar episodes from classical antiquity. The early colonisers still felt that the New World resembled a dream, the eternal spring of the Golden Age in Ovid's *Metamorphoses*.[7] For the sculptor, the cedar trees that appear throughout the two panels must have seemed the ideal means of referring to classical antiquity. In this way he could emphasise the evergreen character of nature in Brazil and transport the ideal – which until then had been situated far off in *time* – to a great distance in *space*.[8]

The manner in which the male Tupinambá are represented is unusual: with the exception of the Indian who strips a tree of its bark, all male members of the tribe have either a moustache or a moustache and beard, whereas in reality they removed their facial hair.[9] The Tupinambá are probably being compared here to

1
The study by Amy Buono, which was only completed after the catalogue was published, could not be consulted.

2
Chirol 1980–1981: 25; Dickason 1984: 135. The frieze is variously dated. De la Quérrière 1821: 151–153. The two panels are directly linked to the *fête brésilienne* that the city of Rouen organised in 1550 for the joyous entry of Henry II (1519–1559) and Catherine de' Medici (1519–1589). Chirol 1980–1981: 25. At the beginning of the nineteenth century, two panels were discovered under the shingle of the house at number 115 rue du Gros-Horloge that *L'Ile du Brésil* resembles considerably in terms of style. For this reason, Chirol proposes the decade 1527–1537. Van den Eynde 2015: 3–7. The most recent publications situate the work 'around 1530'.

3
De la Quérière 1821: 151.

4
Chirol 1980–1981: 25; Dickason 1984: 135.

5
Cloulas-Brousseau 1963: 108, 140, 173–175, 177; Chirol 1984: 230.

6
Chirol 1984: 231.

7
Book I: v. 107.

8
De Léry 1994: 147–148.

9
Hemming 1978: 64–65.

10
Cro 1994: 384–385; Davies 2012: 346

11
Book IV: 1–82; Van den Eynde 2015: 18–19, 32.

12
De Léry 1994: 15.

13
Hemming 1978: 199.

14
De Léry 1994: 432–433.

15
Van den Eynde 2015: 19–20.

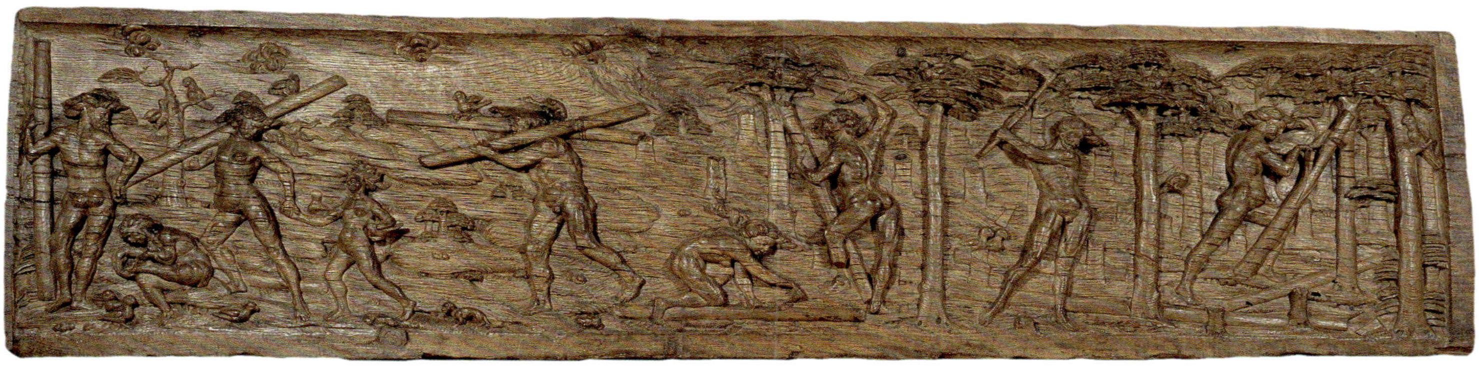

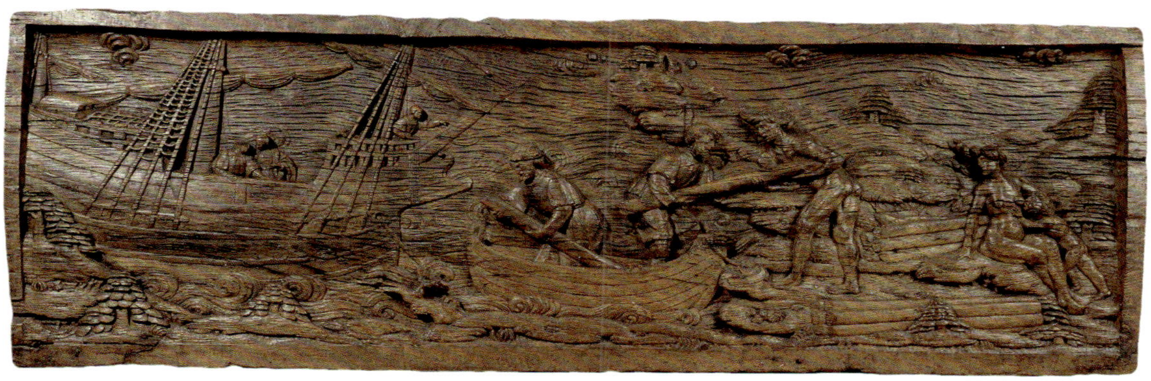

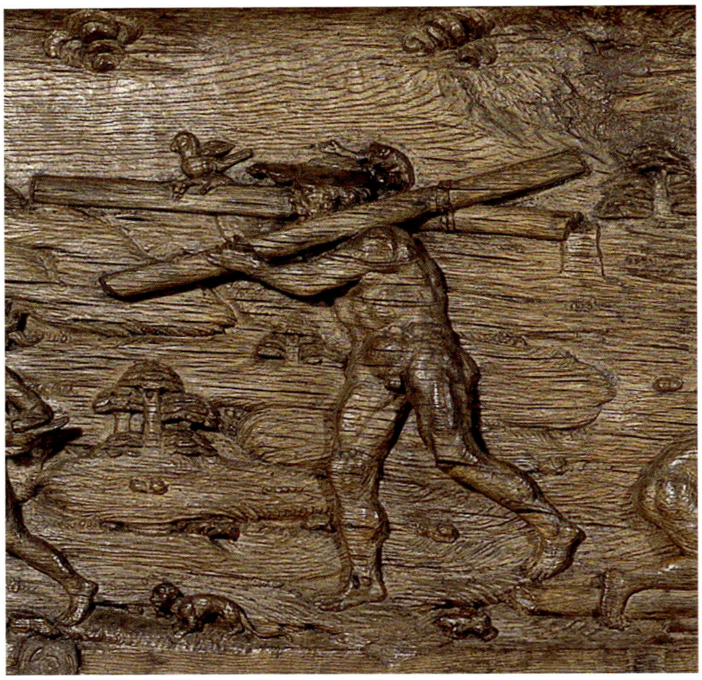

[cat. 54] detail

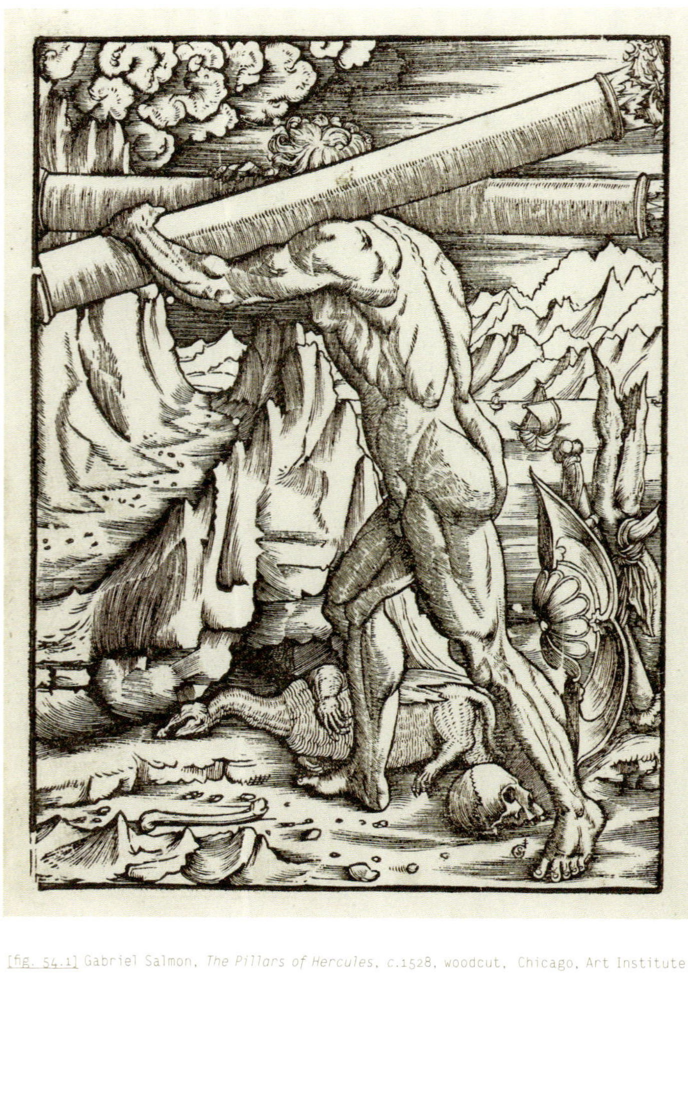

[fig. 54.1] Gabriel Salmon, *The Pillars of Hercules*, c.1528, woodcut. Chicago, Art Institute.

the primitive tribes familiar to the humanists from classical literature, such as the Germans in Tacitus's *Germania*[10] or the Scythians in Herodotos's *Histories*.[11]

The Tupinambá were fond of laughing.[12] In *L'Ile du Brésil*, the liveliness of the Indian men, women and children immediately catches the eye: indefatigable, they undertake even the hardest tasks for the French with a smile on their faces.[13] The fact that the women suckled their infants themselves – instead of giving them to a wet nurse, as was often customary in France – was a source of admiration for the early colonists, as attested by the prominent place given to the women nursing her child on the far right in the second panel.[14] The many depictions of animals are also striking: across the entire frieze we find parrots, and in the first panel we see a monkey. Both species were coveted items of trade.[15]

SVDE

55

Pierre Desceliers
World Map (Mappamundi)

Dieppe
1550
Parchment, 135 x 215 cm
London, British Library, inv. Add. Ms. 24065

PROVENANCE

Cristoforo Negri (1852); British Museum (1861).

BIBLIOGRAPHY

Collin 1988: 314, no. K. 20; Massing 2003: 231-248; Schneider 2004: 109-112; Massing 2005: 53, 57, 60; Toulouse 2007: 1550-1568; Hiatt 2008: pl. 7; Barber & Harper 2010: 21, 84-85; Whitfield 2010: 62-63; Baverel, Goutagny & Measson 2011; Baverel, Goutagny & Measson 2011: 95-115; Hofmann, Richard & Vagnon 2012: 136, 160, 171, no. 98; Van Duzer 2013: 399, 416, 425; Van Duzer 2015.

In 1550 the Norman cartographer Pierre Desceliers made this large world map which had been commissioned as a gift for the French king Henry II. The gift was intended to help rehabilitate the career of Claude d'Annebault, Admiral of France, one of Henry's ministers who had fallen out of favor, and no expense was spared in its decoration. In fact, as Henry already owned a large world map made by Desceliers in 1546, it is likely that a team of artists was recruited specifically to decorate the map at a higher artistic level than the earlier map. As a result, the 1550 map is one of the most beautiful to have survived from the Renaissance. It was designed to be consulted by being laid out on a large table: the images and texts in the northern half of the map are oriented so as to be right-side up when viewed from the north, and the images and texts in the southern half of the map are designed to be viewed from that edge of the map. It also includes 26 long descriptive texts set in cartouches that supply details about the geography and the inhabitants of various regions none of which texts appears on Desceliers' 1546 map or on his 1553 world map. These texts constitute another deluxe feature that sets the 1550 map apart from the 1546 map.

Desceliers was one of several cartographers who worked in Normandy from the 1540s to the 1560s, who are sometimes collectively referred to as the *Dieppe School* of mapmakers, and one thing that sets him apart from his peers is his greater interest in Asia. Atlases and maps made by other cartographers working in and around Dieppe at this period omit most of Asia. We do not know why the other Norman cartographers gave short shrift to Asia, but Desceliers created a rich and detailed image of the continent. He includes images of the Amazons, the Great Khan, and other Asian potentates; cynocephali (men with dogs' heads) and a man with no knees; elephants, a camel, bears, a baboon, and a flying serpent. And most of his descriptive texts relate to Asia.

A feature of the Norman maps that has attracted much attention is their inclusion of a southern continent, which some have argued represents a 'pre-discovery' of Australia, before its generally accepted discovery by Willem Janszoon in 1606. I believe that this continent is merely an elaboration of the purely hypothetical southern continents that had appeared on earlier maps.

CVD

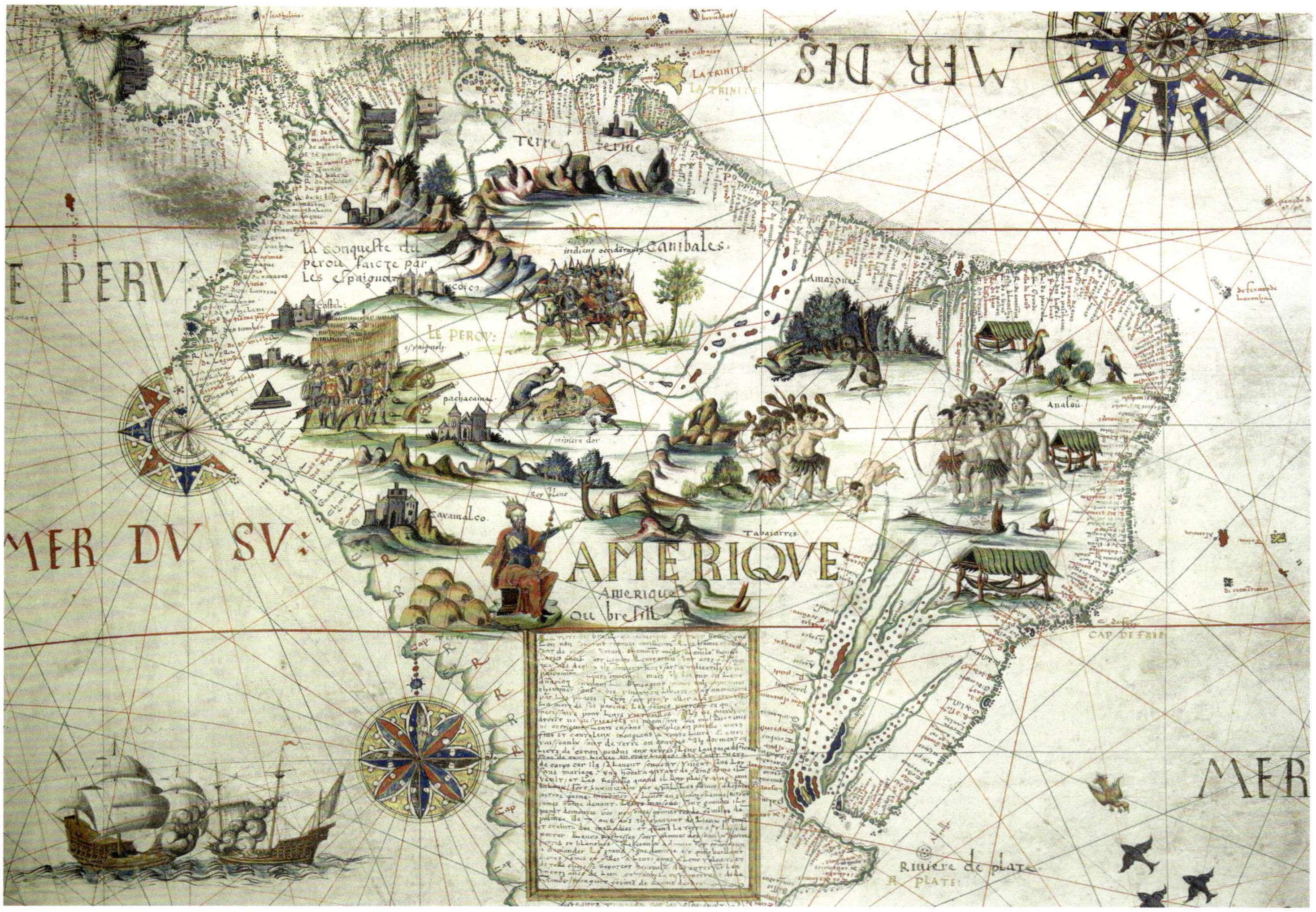

[cat. 55] detail

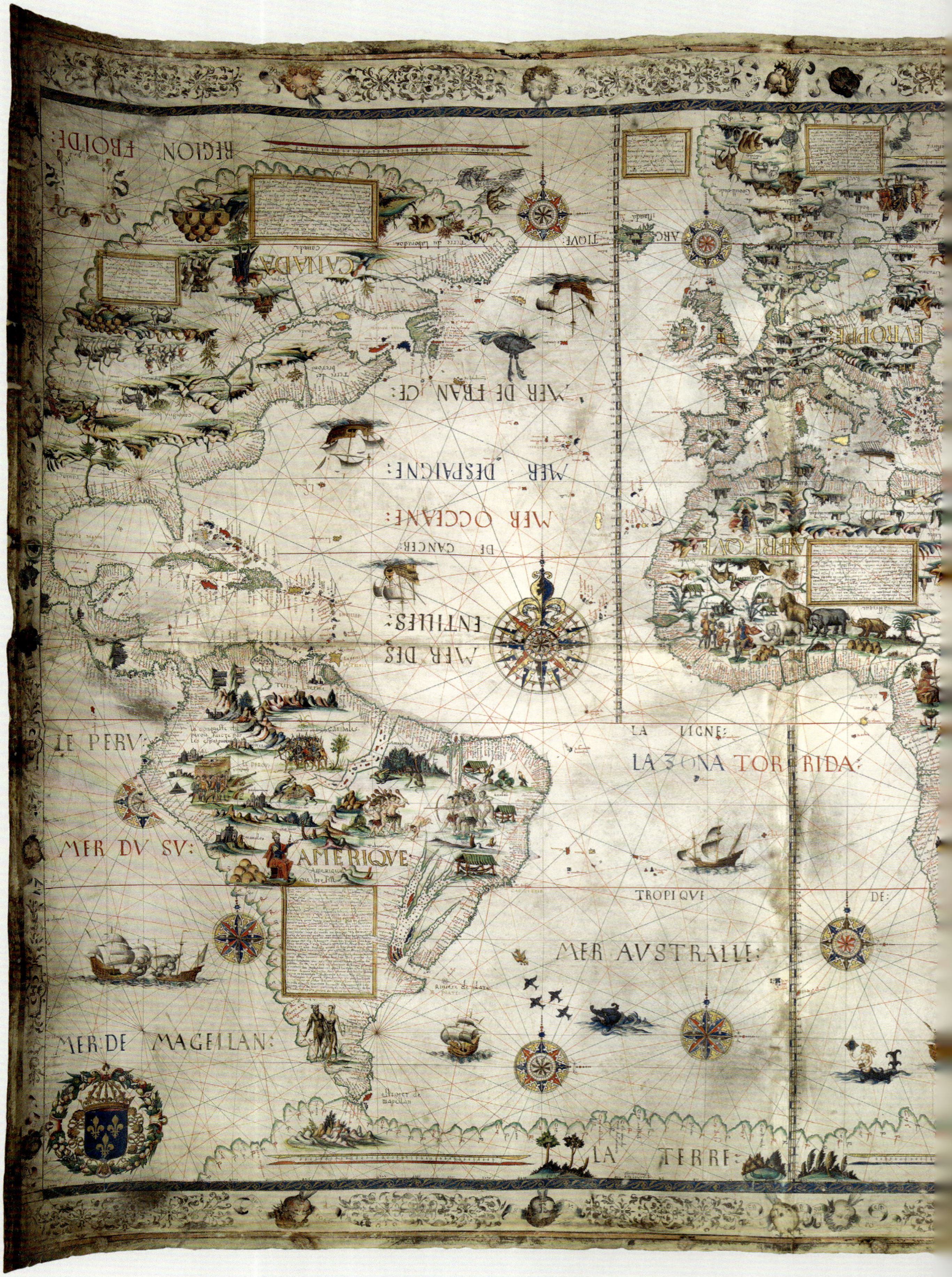
REGION FROIDE
CANADA
MER DE FRANCE
MER DESPAIGNE
MER OCCEANE
MER DES ENTILLES
EVROPE
AFRIQVE
LE PERV
LA LIGNE:
LA ZONA TORRIDA:
MER DV SV:
AMERIQVE
TROPIQVE DE:
MER AVSTRALLE:
MER DE MAGELLAN:
LA TERRE:

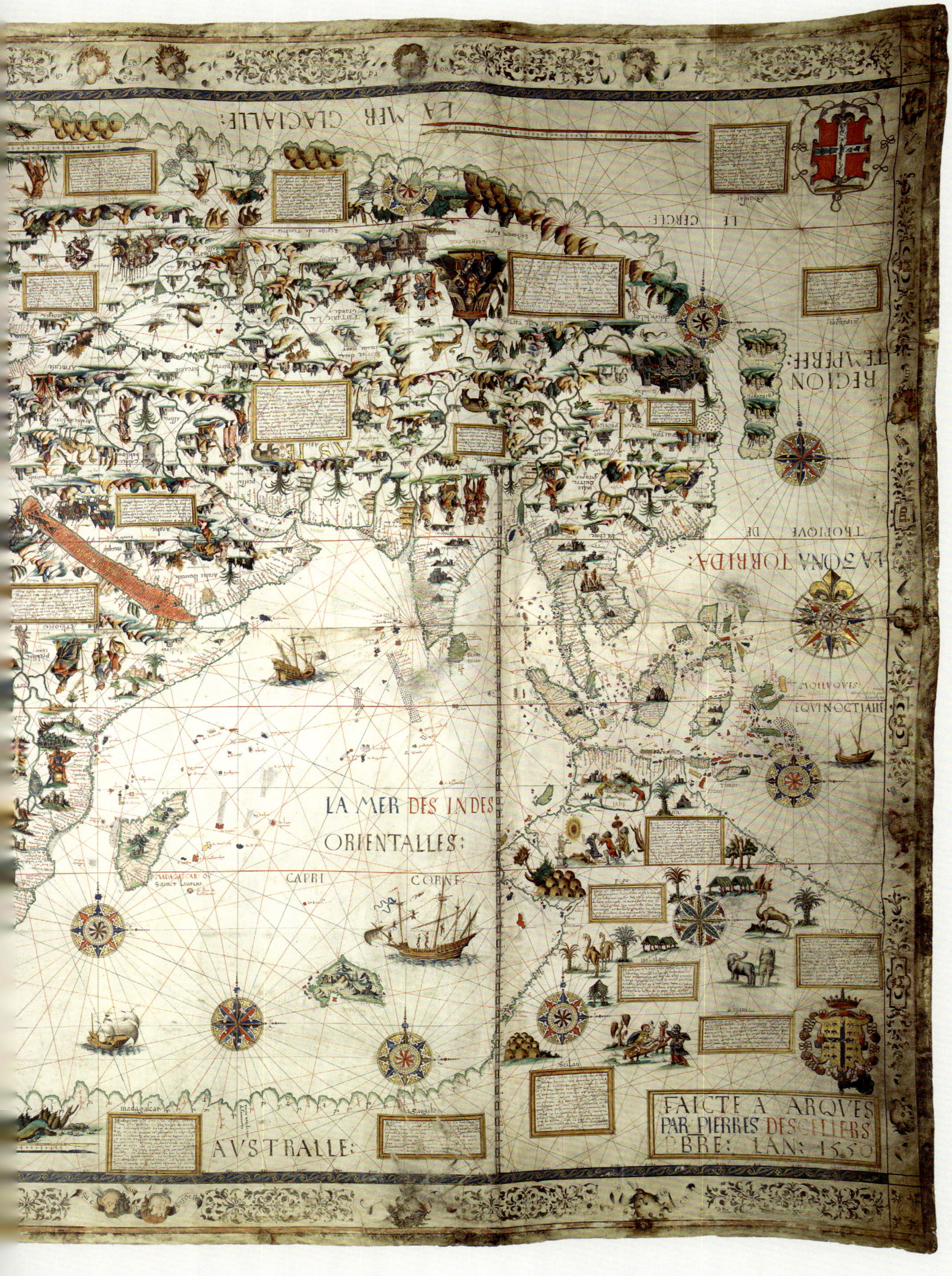
LA MER GLACIALLE:
LE CERCLE
REGION TEMPERE
TROPIQVE DE
LA ZONA TORRIDA:
MOLVQVES
EQVINOCTIALLE
LA MER DES INDES
ORIENTALLES:
MADAGASCAR OV SAINT LAVRENS
CAPRI CORNE
AVSTRALLE:
FAICTE A ARQVES
PAR PIERRES DESCELIERS
PBRE: LAN: 1550

56

Hans Burgkmair, after
Triumph of the King of Cochin

Netherlands
c.1540
Oil on panel, 45.5 x 149 cm
Paris, Galerie De Jonckheere, inv. AKG 966025

BIBLIOGRAPHY

Levenson 2007: 102-103 no. P-31, 335.

Technological, economic and ideological developments in the fifteenth and sixteenth century made it possible for the famed exploratory expeditions to be undertaken that went on to expose the West to new and unexpected worlds.[1] Printers were keen to seize on the exotic travel stories that emerged from these voyages and tried to sell them to the public at large by way of news reports, illustrated or otherwise. The frieze (1508) by Hans Burgkmair (1473–1531) published in Augsburg is to be viewed against this background (fig. 56.2).[2] Of the series of woodcuts depicting the peoples encountered along the coastlines of Africa all the way up to Cochin in the East Indies, as we know them today, the anonymous imitator of *c.*1540 exposes three prints, entitled 'DER KVNIG VON GVTZIN'.[3] The entire scene is based on the travel logbook of Balthasar Springer, a merchant who sailed on board of one of the German ships captained by Portuguese Francisco de Almeida, during the 1505–1506 expedition.[4] The frieze was ordered by one of their bankrollers, the powerful Welser family from Augsburg.[5]

For ethnographic accuracy, such prints depended on the artefacts the artist had to hand.[6] Due to the absence of living models, equivalent schemes from pictorial tradition, such as monsters (cat. 49–50) and wild men (cat. 51), were often recycled.[7] An early example of a work based on factual models is the Augsburg pamphlet from 1505 about Amerigo Vespucci's expedition. The anonymous woodcut devotes an unprecedented focus of attention to the rendition of the appearance and customs of the indigenous people (fig. 56.1).[8] The representation of the Brazilian Tupinambá even went on to become the new archetype for all recently discovered peoples, regardless of region, and the visual motif of the Indian dressed in a feathered skirt and a feathered crown spread like wildfire (cat. 58).[9] Building on the interest in more accurate depictions, Burgkmair embedded his figures in recognisable family structures, each with their own physiognomy, geographical context and cultural traditions.[10] The *Triumph of the King of Cochin* depicts a whole repertoire of activities, ages and classes. The cortege is headed by an elderly camel driver and an elephant trainer, followed by a group of exuberant musicians flocking around their maharaja.[11] Seated on his *palanquin*, the leader is escorted by manly looking armed warriors. The ostentatious cortege is rounded out by women and children.

To arrive at such a consistent rendition, it took more than Springer's self-aggrandising texts, which were packed with yarns of violence and driven by mercenary motives.[12] In all probability, alongside this literary information, the

1
Feest 2014: 288.

2
Hausberger & Biedermann 1973: no. 26; Kroll & Schade 1974: 39–42, 66–67 no. 9; Colin 1987: 5–6, 26–27; Bartrum 1995: 130–132 no. 130; McDonald 2003: 230; Leitch 2009: 136, 139–140; Silver 2010: 222–223; Feest 2014: 296.

3
Not a single full print of the frieze has survived. The frieze consisted of at least eight, possibly as much as thirteen woodcuts, Leitch 2009: 136 and 155 n. 16. See McDonald 2003 for a hypothetical reconstruction. In all probability, Cochin – modern-day Kochi – was the final stop of the scene. Cochin remained the capital city of Portuguese India up until 1530 and has continued to serve as a crucial seaport on India's west coast to this very day.

4
Springer 1509: 3. The text fragments on the frieze are derived from Springer's travel logbook. In literature, this ensemble as a whole is referred to as the 'short report'. The book version *Die Merfart un[d] erfarung nuwer Schiffung und Werge zu viin onerkanten Inseln und Künigreichen* – the 'long report', published in 1509 – contains Springer's full report, illustrated by woodcuts of a lesser quality by Wolf Traut. See Feest 2014: 296 for similarities and differences between Burgkmair's frieze and the woodcuts from this later report. Bartrum 1995: 131; Leitch 2009: 134, 140, 155 n. 6; Silver 2010: 224.

5
Leitch 2009: 136; Silver 2010: 222.

6
Oral sources were unreliable as they far too easily result in misinterpretations. Colin 1987: 5.

7
Colin 1987: 6; Leitch 2009: 142.

8
Feest qualifies Colin and Leitch's position which suggests this is the earliest known representation with ethnographic specification. Colin 1987: 26–27; Leitch 2009: 142–143; Feest 2014: 297–298.

9
Sturtevant 1988; Egmond & Mason 2000: 325; Leitch 2009: 134; Silver 2010: 222; Feest 2014: 293–294.

10
Leitch 2009: 134; Silver 2010: 222.

11
During the rule of the then maharaja – Unniraman Koil II or Rama Varma, there is some ambiguity over the matter – the region enjoyed rising prosperity; Sasikumar 2011: 32.

12
Leitch 2009: 145.

13
McDonald 2003: 230; Leitch 2009: 134–135; Feest 2014: 296.

14
Leitch 2009: 141, 143–144; Feest 2014: 301.

15
Leitch 2009: 143, 146; Feest 2014: 301.

16
Leitch 2009: 137.

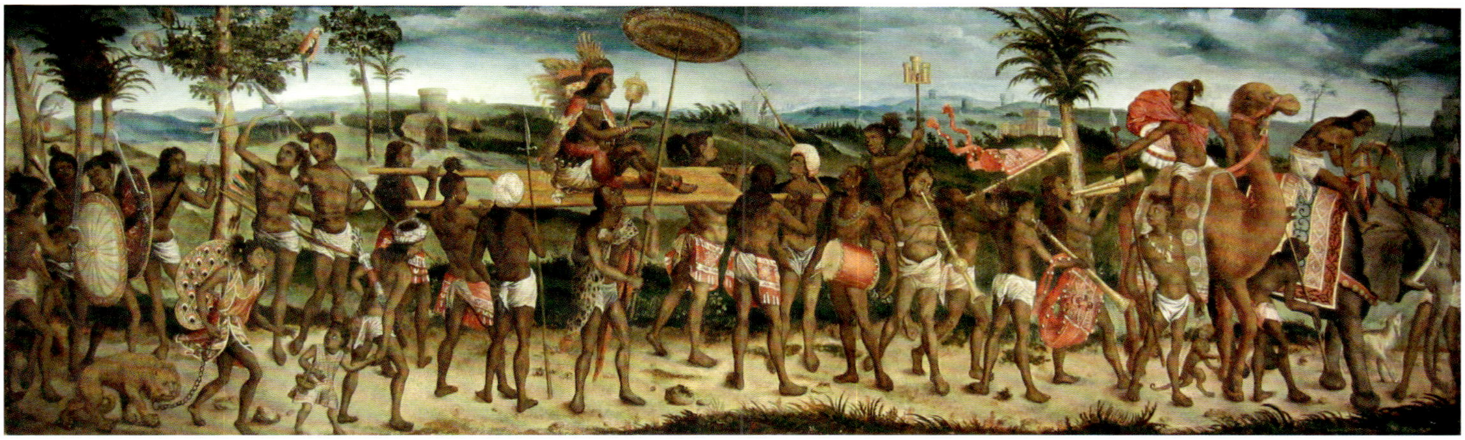

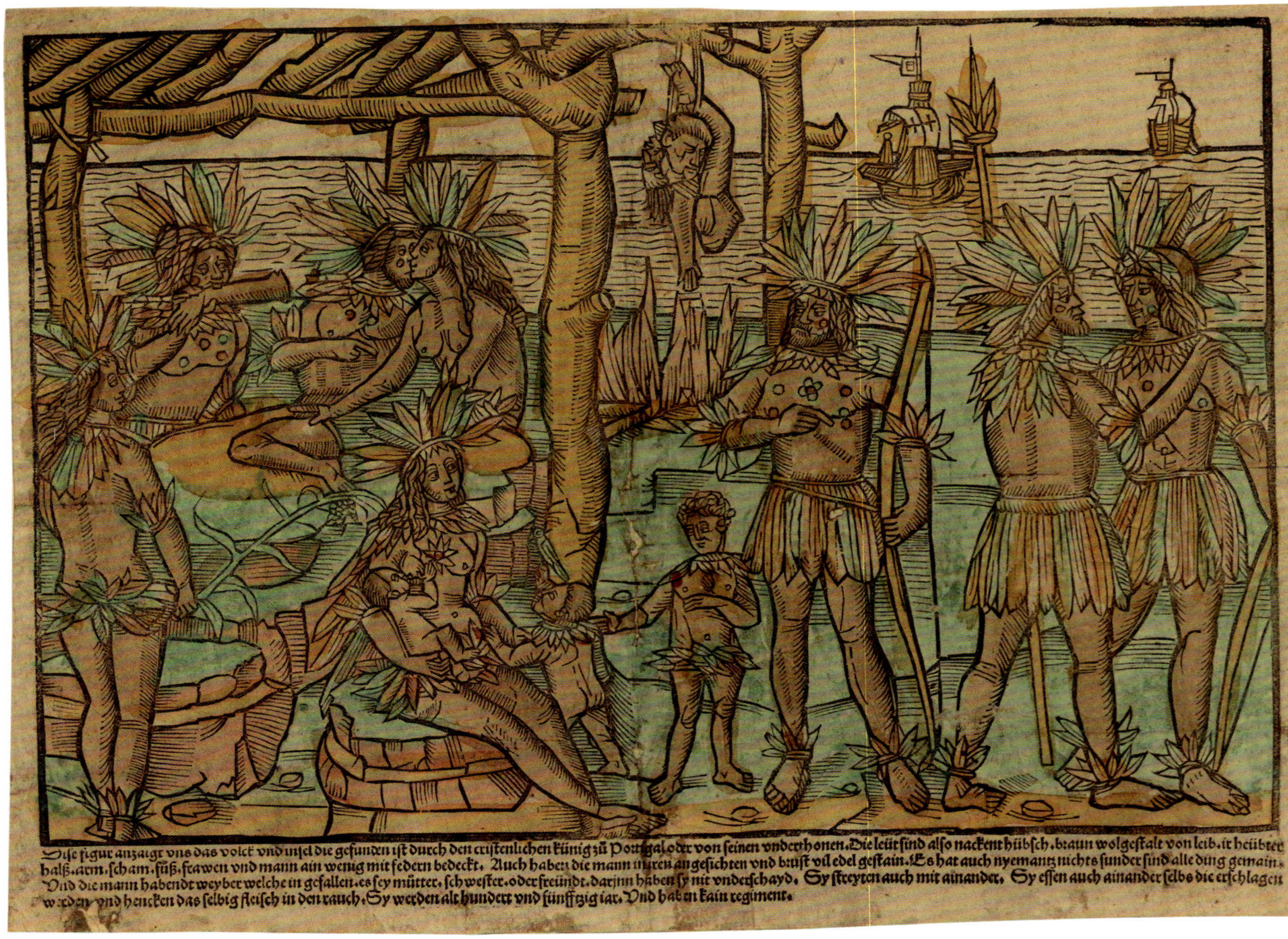

[fig. 56.1] Augsburg, *Dise figur anzaigt uns das volck…*, coloured woodcut, 1505. Munich, Bayerische Staatsbibliothek.

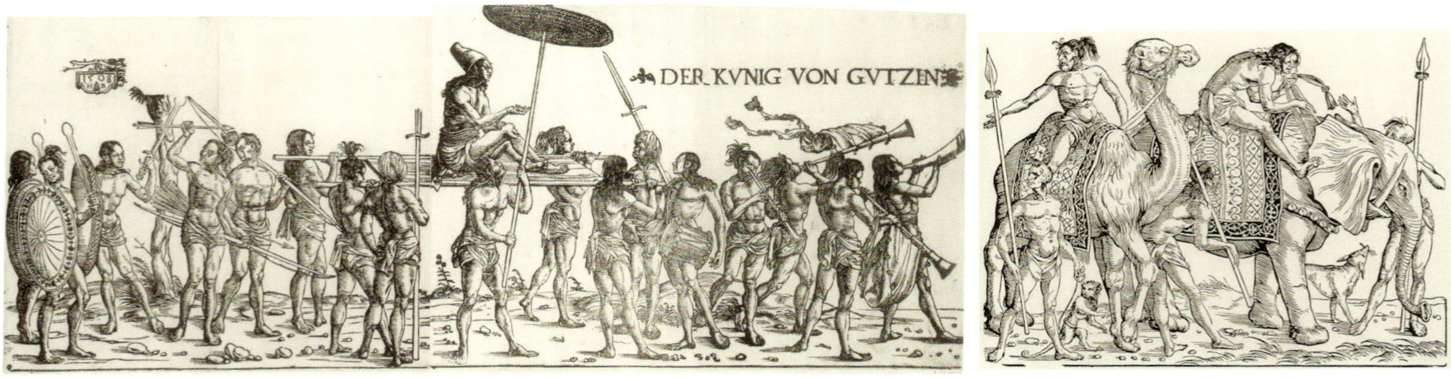

[fig. 56.2] Hans Burgkmair the Elder, *Natives of Africa and India*, detail: *Triumph of the King of Cochin*, 1508. London, British Museum (woodcut 1 and 2); Berlin, Staatliche Museen zu Berlin, Kupferstichkabinett (woodcut 3).

merchant also brought back visual source materials, such as sketches and local weaponry.[13] Moreover, Burgkmair was assisted by Konrad Peutinger, a humanist and cartographer with whom he was friendly. Both the materials from his library – including maps and manuscripts of recent German trade missions to India – and the humanist conceptual frame of reference served as important sources. In this milieu, new peoples were no longer viewed as exotic monsters, but as a major contribution to the on-going expansion of cosmographic awareness.[14] The schematic nature of the composition references techniques for organising time and space from the field of cartography. In due course, this ethnographic classification system went on to become a standard formula for depicting peoples of various description.[15] This places Burgkmair's frieze at the beginning of empirical data visualisation and the rendition of human diversity.[16]

WW

57

Jan Mostaert, attributed to
Scenes from the Life of Cain and Abel

c.1525

Falsified monogram of Albrecht Dürer, specifying the year as 1511

Oil on panel, 97.2 x 68 cm

Schorr Collection

PROVENANCE

Part of the collection of Johann Peter Weyer (1794–1864), Cologne, first recorded in his collection catalogue of 1859; sale, Cologne (Lempertz), 25 August 1862, lot 39, as a 'Dürer imitator'; purchased there by an art dealer by the name of Müller from Würzburg; at some point between 1862 and 1999 in possession of one Van der Bank, Frankfurt (according to a label on the rear of the panel); sale, New York (Sotheby's), 14 October 1999, lot 125, as the 'Circle of Jan van Scorel'; currently held as part of the Schorr collection.

BIBLIOGRAPHY

Weyer 1859: no. 40; Cologne 1862: 12–13, no. 39; Kier & Zehnder: 504; Wright 2011: 62–63; Wright 2014, vol. I: 180–181, no. 252 & vol. II: 56.

After they had been banished from the Garden of Eden, Adam and Eve ended up living in the wilderness, where an existence of hard toil and terrible ordeals awaited them. Here, the mother of humanity gave birth to Cain and Abel and – depending on the interpretation of the Bible and the apocryphal texts one goes by – later on gave birth to one or several daughters, and a son by the name of Seth.[1] This motif of the extended First Family, largely faded into oblivion in this day and age, was repeatedly picked up in the sixteenth century world of painting in the Low Countries by artists such as Quinten Metsys (1465/66–1530) and Jan Mostaert (*c.*1474–1552/53).

The work presented here shows different episodes in the narrative of the brothers who fell foul of one another. In the foreground, Eve is seen sitting in a hollowed-out tree with Cain and Abel on her lap. Next to her, the rest of her offspring are jostling each other around a piece of fruit, which some believe to be in reference to the ruinous Forbidden Fruit. In the background, Adam is seen ploughing the field. Behind him, half hidden on a hilltop, Cain and Abel's offering can be seen. The Lord's preference for Abel's offering is about to provoke Cain into a seething rage. The direct result of these events, the First Fratricide, is depicted up against the hill flank. Left in the background, in front of an open cave, Adam and Eve are seen lamenting over Abel's body. To the right of the cave entrance, Cain's dire fate is depicted: Lamech, accompanied by his son, unintentionally kills Cain whilst out hunting.

In spite of Albrecht Dürer's false signature, the painting was attributed to Jan Mostaert a few years ago.[2] Karel van Mander (1548–1606), a Flemish biographer of painters, described an unfinished landscape painting of the New World by Jan Mostaert (fig. 57.1).[3] Nowadays, Mostaert is credited with several panels that have

1
The general picture of Genesis 3 and 4 may be further supplemented with the details from apocryphal texts such as the *Testament of Adam*. For a more comprehensive discussion of the composition of the First Family, see Byron 2011: 23–29.

2
There is no question that Dürer's (1471–1528) monogram and the date 1511 were added later on to drive up the panel's value on the art market. It was Robrecht Janssen who recently attributed the panel to Mostaert. See Wright 2014: 180–181, no. 252.

3
'…een Landtschap, wesende een West-Indien, met veel naeckt volck, met een bootsighe Clip, en vreemt ghebouw van huysen en hutten' (… a West Indian landscape with many nude figures, a curious cliff, and exotic houses and huts). See Van Mander 1604: fol. 229v.

4
Williamstown, *The Sterling and Francine Clark Art Institute*, inv. 1955.946. See, amongst other authors: Berk 1982: 201–212; Moseley-Christian 2012: 437–438.

5
Zimerman 1885: CXIX–CXX. See below Vandenbroeck 1991a: 99–119; Feest 2014: 287–303.

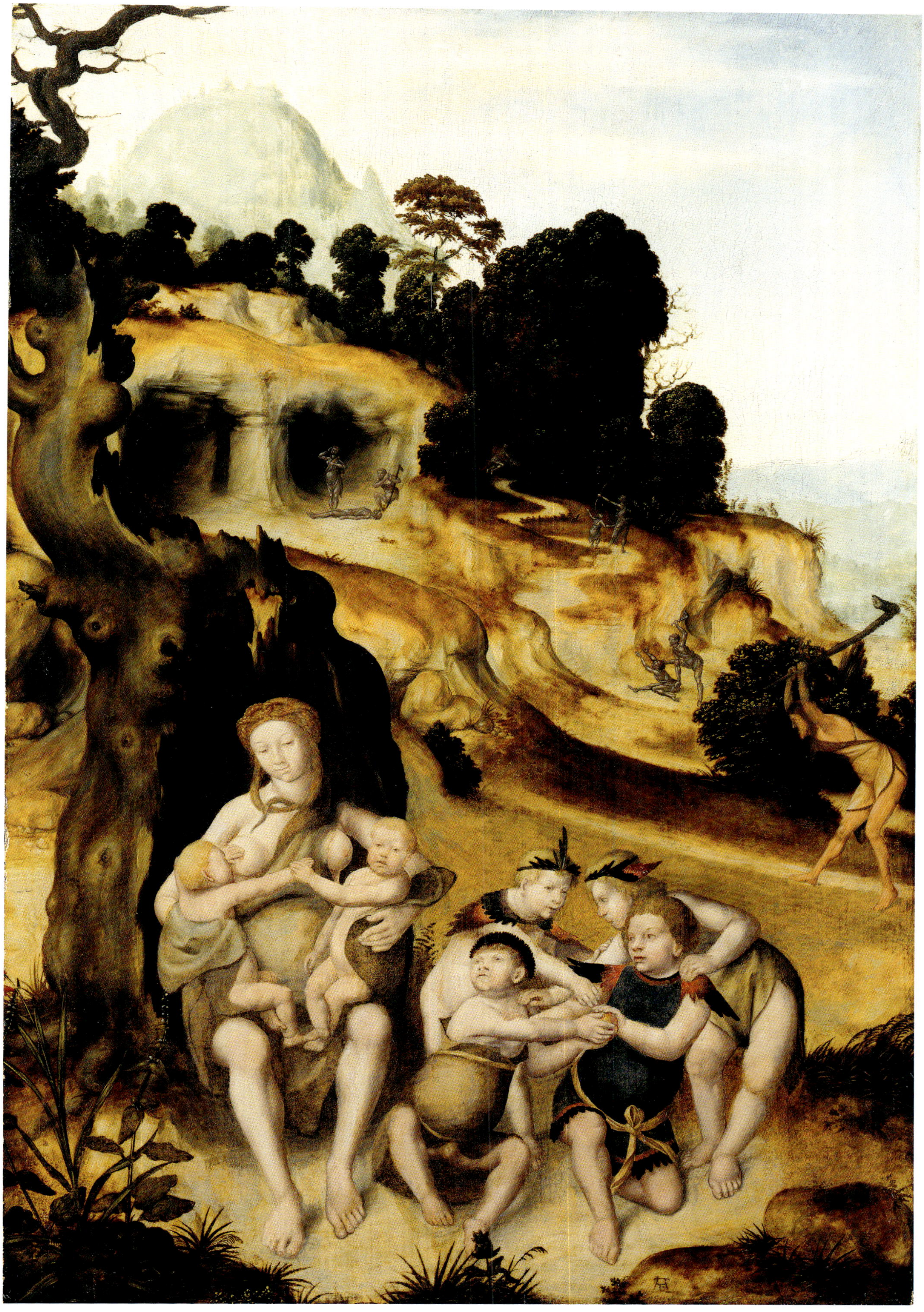

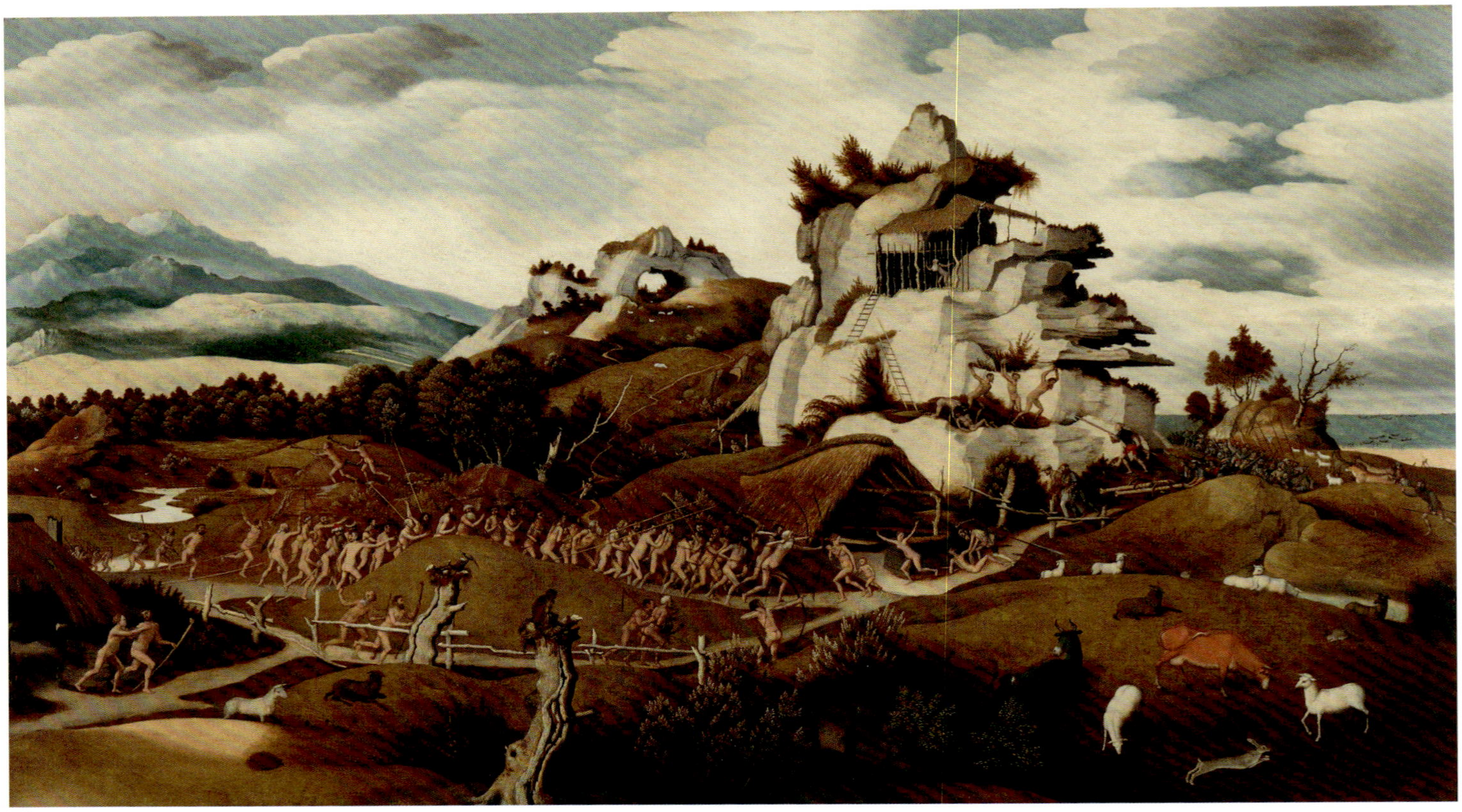

[fig. 57.1] Jan Mostaert, *Landscape with an Episode from the Conquest of America*, c.1535, Amsterdam, Rijksmuseum.

a similar setting, with or without biblical iconography.[4] As such, the connection between the First Family and the 'primitive life forms' encountered in the newly discovered territories would appear to speak for itself. The reference to the West Indies is rendered even more concrete in the way the children are dressed. In the painting, they are adorned with plumed crowns and feathered capes. There can be no doubt that the latter elements refer to the New World (cat. 56 and 58). For a long time, Mostaert – who lived and worked in Haarlem – was attached to the Mechelen court of the regent-governess Margaret of Austria (1480–1530). On 20 August 1523, the latter was sent a collection of *accoustremens de plumes, venuz des Indes* by her nephew, Holy Roman Emperor Charles V.[5] This may have been the source that inspired the artist's depiction.

RJ & MB

58

Calicut Resident

Germany (?)
c.1505–1520
Coloured woodcut, 316 x 181 mm
Berlin, Staatliche Museen zu Berlin, Kupferstichkabinett, inv. AM 90-1962

PROVENANCE

After 1920 in the collection Friedrich Quiring, Eberswalde (Lugt 1041c); in 1962 acquired by the Staatliche Museen zu Berlin.

BIBLIOGRAPHY

Osterwold & Pollig 1987: 304, no. 1.99; Egmond & Mason 2000: 324-326, 344-345; Feest 2014: 299.

This coloured woodcut of a so-called 'Indian' was made by an anonymous German artist. The standalone print depicts a bearded man in a subtle contrapposto pose, dressed in a frock and cape, his head adorned with feathers. The man is seen to carry a bow, quiver and staff, his body lavishly adorned with feathers and precious stones. The detailing of the feathered sections suggests that the woodcut was intended to be coloured immediately after printing. Even though a large run of this print is believed to have been printed, only one copy has survived. As recent as 1987, an unknown text sheet was attached to the rear of this copy, which has since been lost.[1]

The exploratory expeditions in the late fifteenth and early sixteenth century sparked a growing fascination with artefacts from faraway territories and foreign cultures (cat. 46–48). In the logbook of his journey through the Low Countries, German artist Albrecht Dürer describes how – filled with admiration – he was able to examine a collection of sumptuous artefacts from America on 27 August 1520 in Brussels.[2] This treasure trove had been dispatched from Mexico by Hernán Cortés to the future Emperor Charles v, although other sixteenth century collections likewise included exotic weapons and feathered clothing items.[3]

Nonetheless, this woodcut is not based on direct observation by an artist or a study of the actual objects. The print has strong similarities with the figures seen on two unrelated German woodcuts of around 1505 that report on the discoveries made by Amerigo Vespucci (fig. 56.1).[4] The woodcut shown here is more detailed, but not for that reason more lifelike. For instance, the bow is too small to serve any functional purpose and no artefact or iconographic model is known that accords with the special feather-covered staff.[5]

The woodcuts of German artists such as Hans Burgkmair (cat. 56) and Jörg Breu played a pivotal role in the dissemination of information on foreign cultures.[6] A particular example of the reception history of this anonymous woodcut is the *Visboeck* by Adriaen Coenen from 1577–1581 (fig. 58.1).[7] In the *Visboeck* manuscript, the Scheveningen beach comber, fisherman and fish auctioneer copied the woodcut and added a Dutch comment.[8] This text identifies the sketch as a portrait of the noblemen of Calicut (modern-day Kozhikode in India), and describes their attire

1 Osterwold & Pollig 1987: no. 1.99. The rare absence of a Kupferstichkabinett inventory stamp on the print would appear to confirm that another sheet had been attached to the rear. I would like to thank Georg J. Dietz for this information.

2 Rupprich 1956–1969: vol. 1, 155. On the relationship Dürer had with these objects, see Feest 2013.

3 On the history of these objects, see Feest 1990: 33–48. For other sixteenth century collections that include exotic feathered artefacts, see Fleischhauer 1976: 2–3; Vandenbroeck 1991a; Leitsch 2009: 157 n. 70.

4 Colin 1988: 186–188, 392–394; Leitsch 2009: 135, 145; Feest 2014 : 298–299.

5 Feest 2014: 299.

6 Publications to come out of Antwerp and elsewhere also contributed to this dissemination, see Colin 1988: 191–192, 437; Pisters 1991, no. 14–17.

7 The entire manuscript is available to be consulted online at https://galerij.kb.nl/kb.html#/nl/visboek/.

8 Egmond & Mason 2000: 324–326, 344–345; Egmond 2005: 156–159.

9 Calicut also features elsewhere in the manuscript; see n. 8.

10 Feest 2013: 368.

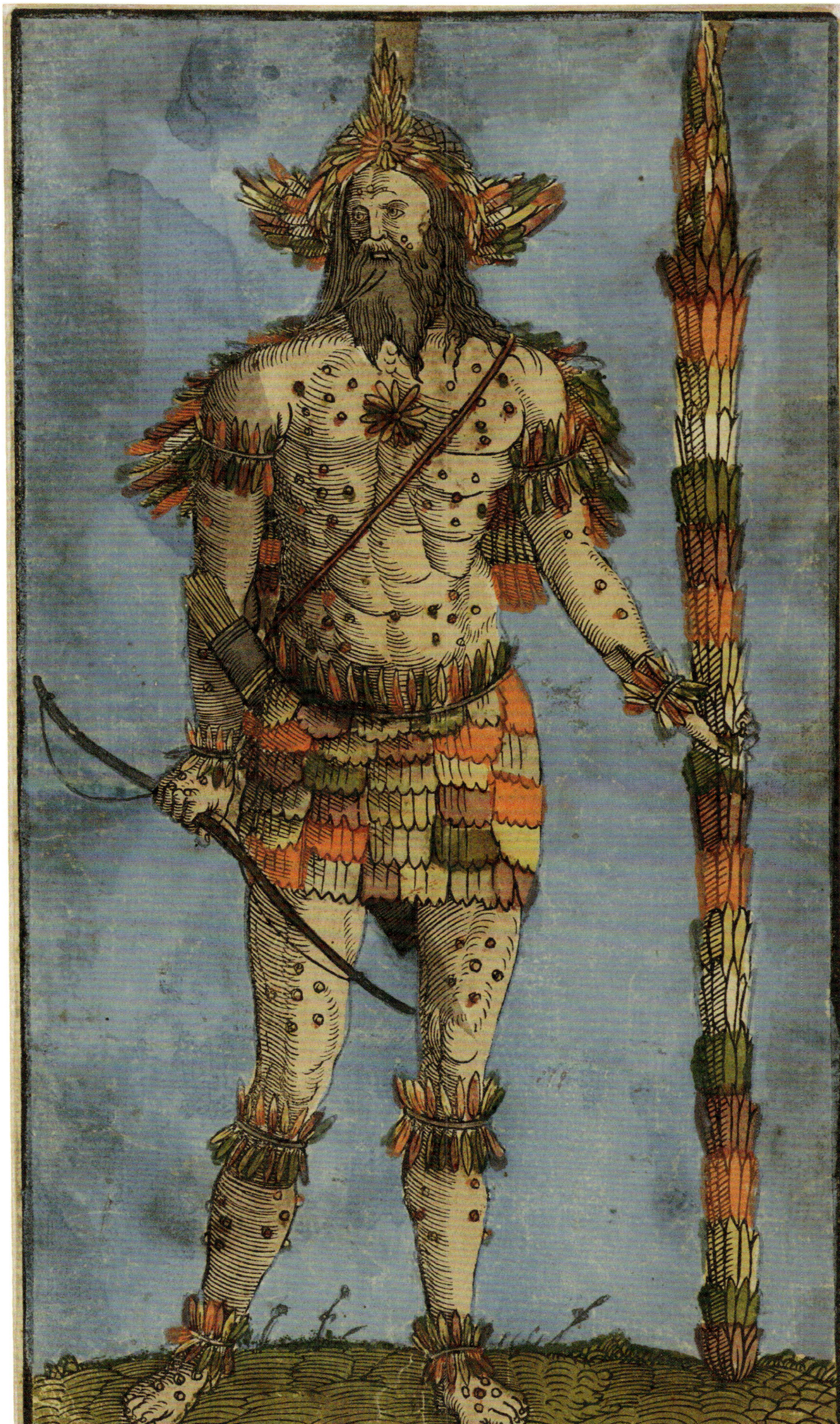

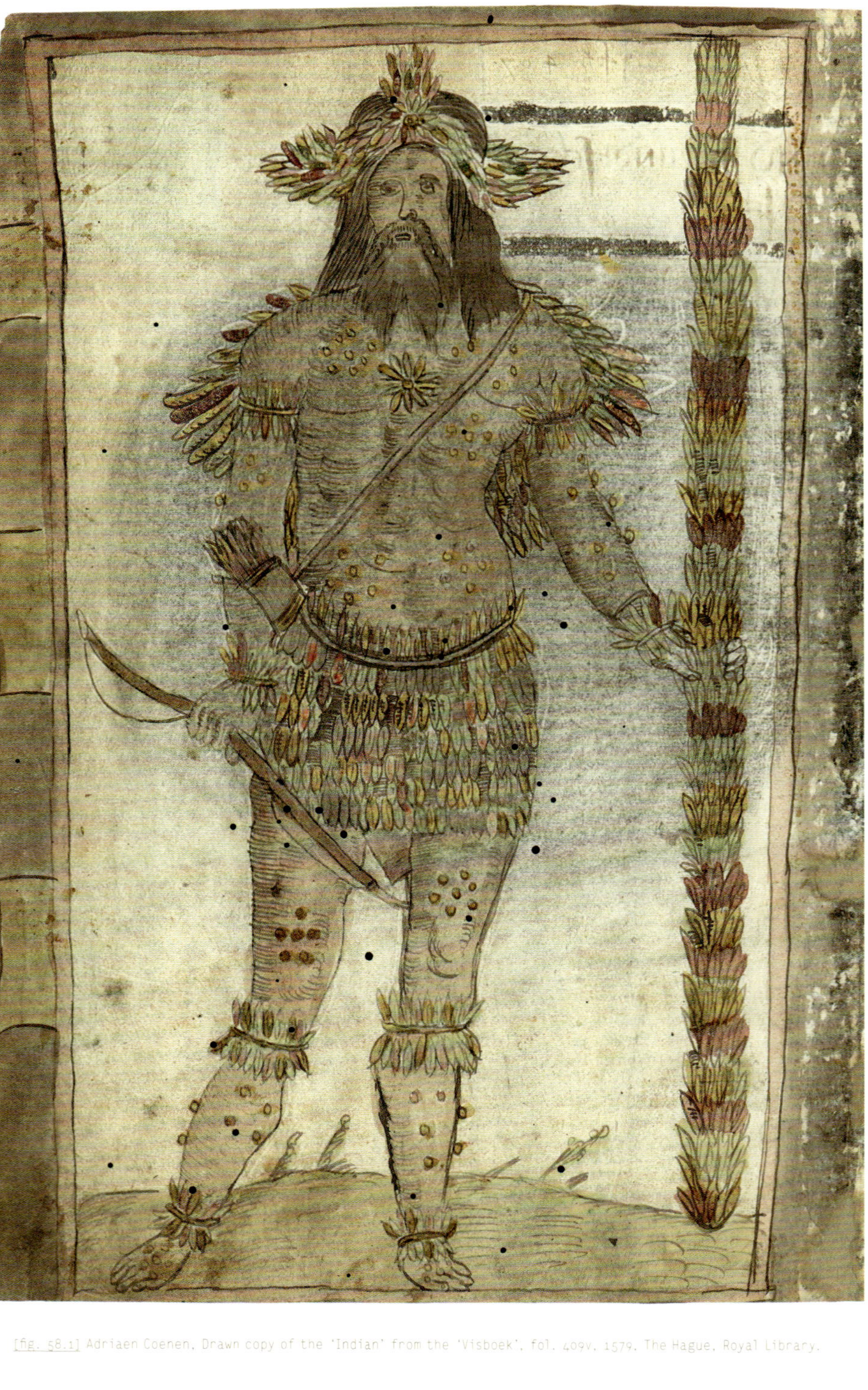

[fig. 58.1] Adriaen Coenen, Drawn copy of the 'Indian' from the 'Visboek', fol. 409v, 1579. The Hague, Royal Library.

and everyday lifestyle.[9] During the sixteenth century, the name 'Calicut' was used as a collective term to refer to Indian as well as American indigenous peoples, all the more so as America was not considered to be continent in its own right in those early days.[10]

JL

HOLDING THE UNIVERSE

Dreaming of Space and Time

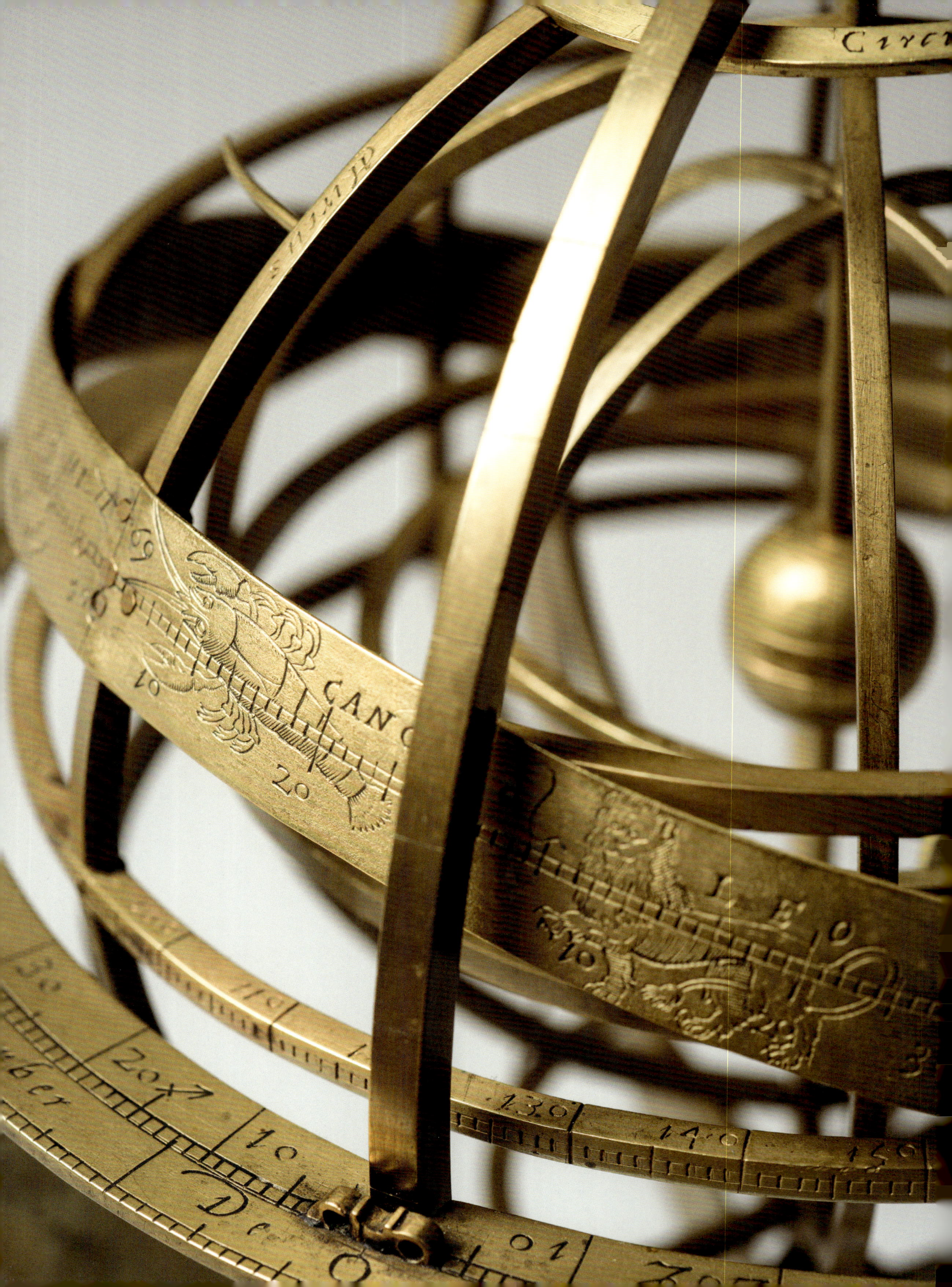

In Search of Utopian Perfection: Builders of Scientific Instruments – the 'Leuven School'

KOENRAAD VAN CLEEMPOEL

[CAT. 67] DETAIL

In the second half of the sixteenth century the scientific instruments made in Leuven gained international fame. Here, for the first time, were devices in which new cosmographic research, scientific accuracy, astronomical information and exceptional elegance of execution were perfectly balanced. This unique combination made them particularly attractive to the European market. The printer Christopher Plantin, whose printing office was in Antwerp, was the agent through whom the instruments were sold to an international clientele. They were bought not only by scholars and scientists who used them to take reliable measurements but also by princes who collected them to display in their libraries.

The most important names among the instrument makers are those of Gemma Frisius (cat. 59), Gerard Mercator (cat. 63, 71), Gualterus Arsenius (cat. 65, 68) and Adrianus Zeelst (cat. 72). Initially, Frisius and Mercator produced mainly globes and maps, but they also designed and refined other scientific instruments made in brass. Chief and most evocative of these are the armillary sphere (cat. 65-68) and the astrolabe (cat. 71, 72). Both are models of the universe that you can hold in your hand - in 3D by way of the armillary sphere's set of rings; in 2D by way of the astrolabe's stereographic projection of the heavens and the earth.

The armillary sphere consists of a series of rings placed parallel to each other: two polar circles, two tropics and in the middle the equator. Then, at an angle of 23.5° relative to the equator, is a ring representing the ecliptic, or apparent path of the sun throughout each year. The ecliptic touches the tropics at the solstices: the longest day on 21 June, and the longest night on 21 December. The equator is cut at right angles by two rings: the colures, which intersect at the poles. In the middle of the central axis that connects everything is the earth, which sometimes has additional rings around it for the sun, moon and other planets. This first system is mounted in a (vertical) meridian ring, which in turn sits in a (horizontal) horizon ring. The sphere can be tilted in the horizon ring: the degrees on the meridian ring denote latitude, so the sphere can be tilted to the user's latitude - 51° for Leuven, for example, or, as can be seen in Jan Gossaert's *Portrait of a Girl with an Armillary Sphere* (cat. 64), 55-56° for Copenhagen and Denmark.

An astrolabe consists of two essential parts, one fixed, one rotating, whereby heaven and earth come together in one plane: a set of plates representing different latitudes (the fixed 'earth' part) and the rete, on which various stars and the zodiac (the rotating 'Heaven' part) are projected. A central pin holds the plate and the rete in position in the mater - a brass disk hollowed out in the centre to hold the set of plates. The plates are often engraved on both sides with a projection for a particular latitude. To take an observation you insert into the mater the plate that corresponds to your location at the time - 51° for Leuven, for example. The rete shows the ecliptic and a number of 'fixed' stars that are represented by little pointers and engraved with a name (*ursa maior*, for example). The rete is pierced so that the plate beneath it is visible and the zodiac and the stars can move over the various lines. It evokes the rising sun and stars above the horizon, among other things.

Astrolabes were used for taking observations, as calculators and for demonstration. Importantly, given the widely-held belief in astrology, they could also be used to draw up horoscopes. You could use an astrolabe to tell the time during the day or night so long as the sun or a few recognizable stars - that were marked on the rete - were visible. Typical of the Leuven astrolabes were maritime applications, such as being able to set a course for a sea voyage.

59

Maarten van Heemskerck
Portrait of Gemma Frisius

c.1545

Latin inscription on tablet left of the sitter: *LVX TENEBRIS RVRSVS LVCI TENEBRE FVGIENTI SVCCEDVNT. STABILIS RES TIBI NVLLA MANET* (literally: 'After darkness comes light; after the fleeting light follows darkness; nothing that lasts remains for you')

Oil on panel, 104 x 83.5 cm

Rotterdam, Museum Boijmans Van Beuningen, inv. 1347

PROVENANCE

Between 1814 and 1935 in the collection of the Earl of Radnor, Longford Castle; offered for sale in 1935 at Thomas Agnew's, Ltd, London; purchased there in 1936 by Daniel George van Beuningen (1877-1955), Rotterdam, and transferred that same year to the museum with the support of the Rembrandtfonds.

BIBLIOGRAPHY

Passavant 1833: 136; Passavant 1836: 294; Waagen 1838: 265; Waagen 1854: 139; Waagen 1857: 360; Woltmann 1866: 201; Wornum 1867: 152; Scheibler & Bode 1881: 213; Matilda & Barclay 1909: no. 145; Winkler 1924a: 288; De Vries 1934: 27, 125; Friedländer 1936: 82, 160, no. 228, fig. XLIII; Rotterdam 1936: 51-52, no. 172; Hoogewerff 1941-1942: 356-358, fig. 168; Hoogewerff 1947: 159, no. 72; Van Schendel 1948: 25, no. 47, fig. 13; Van Luttervelt 1955: 67-68, no. 63; Haak 1967: 28-30, fig. 6; Vey & Von der Osten 1969: 186; Von der Osten 1973; Friedländer 1976: 45, 92, no. 228, pl. 116; Grosshans 1980: 153-154, no. 38, pl. IV, fig. 59; Bruyn 1987: 152, fig. 16; Bruyn 1988: 104, fig. 16; Harrison 1987: 427-440, no. 33; Van der Heijden & Roegiers 1994: 89-90, no. BB-28; Lammertse 1994: 356-359, no. 83; Borchert & Jonckheere 2015: 122-123, no. 22.

In 1814, when this painting was still part of the collection of the Earl of Radnor, it was described as a portrait of Maarten Luther (1483–1546) made by Hans Holbein (1497–1543). However, grave misgivings existed as to the identity of both the artist and the person portrayed, which made for debate in art-historical circles.[1] It was not until 1924, when Winkler credited Maarten van Heemskerck (1498–1574) as the rightful painter, that the right attribution appeared to have been made. Since then, no further reservations have been expressed regarding this attribution. Moreover, the painting is considered today to be one of the best-preserved works by the artist.[2] The issue of the model's identity was resolved seventy years later. It was professor Jan Roegiers, head librarian at Leuven University, who linked the painting from Rotterdam with a portrait engraving made in Leuven by Jan van Stalburch (active from 1555 to 1565) of sixteenth century scholar Gemma Frisius (1508–1555).[3] A second engraving, Gemma's effigy in the portrait series of famed European scholars by print publisher Philips Galle (who worked from 1555 to 1565; fig. 43), lent added weight to this hypothesis (fig. 59.1).[4]

In the autumn of 1525, Jemme 'Gemma' Reinerszoon relocated from Friesland to Leuven, to enrol at the town's university, where – after training to

1
For a more comprehensive discussion of the various credits, see Lammertse 1994: 357.

2
Friedländer 1976: 45.

3
Hollstein Dutch, XXVII: 43, no. 17.

4
Sellink 1997: vol. 2, 223–227, appendix 2B; Sellink & Leesberg 2001: 99, no. 626.

5
Sprinkled with anecdotes, Miraeus chronicles Frisius' friendship with the tall, obese and often red-faced scholar Hieronymus Triverius. When seen together on the street, the two friends were referred to as *par impar* by dint of their completely opposite physique. Miraeus 1609: 96.

6
The origin of the inscription is unclear. Gombrich associated the inscription with the writings of Ovid, whilst Grosshans saw a link with Catullus. Bruyn interpreted the inscription as a paraphrased quote from the Book of Job (17: 12), although there no indications to support this. This may be a poem authored by Frisius himself. See Grosshans 1980: 154, n. 1; Bruyn 1988: 104, n. 37; Van der Heijden & Roegiers 1994: 90; Lammertse 1994: 357.

7
Miraeus 1609: 96; Adam 1620: 75.

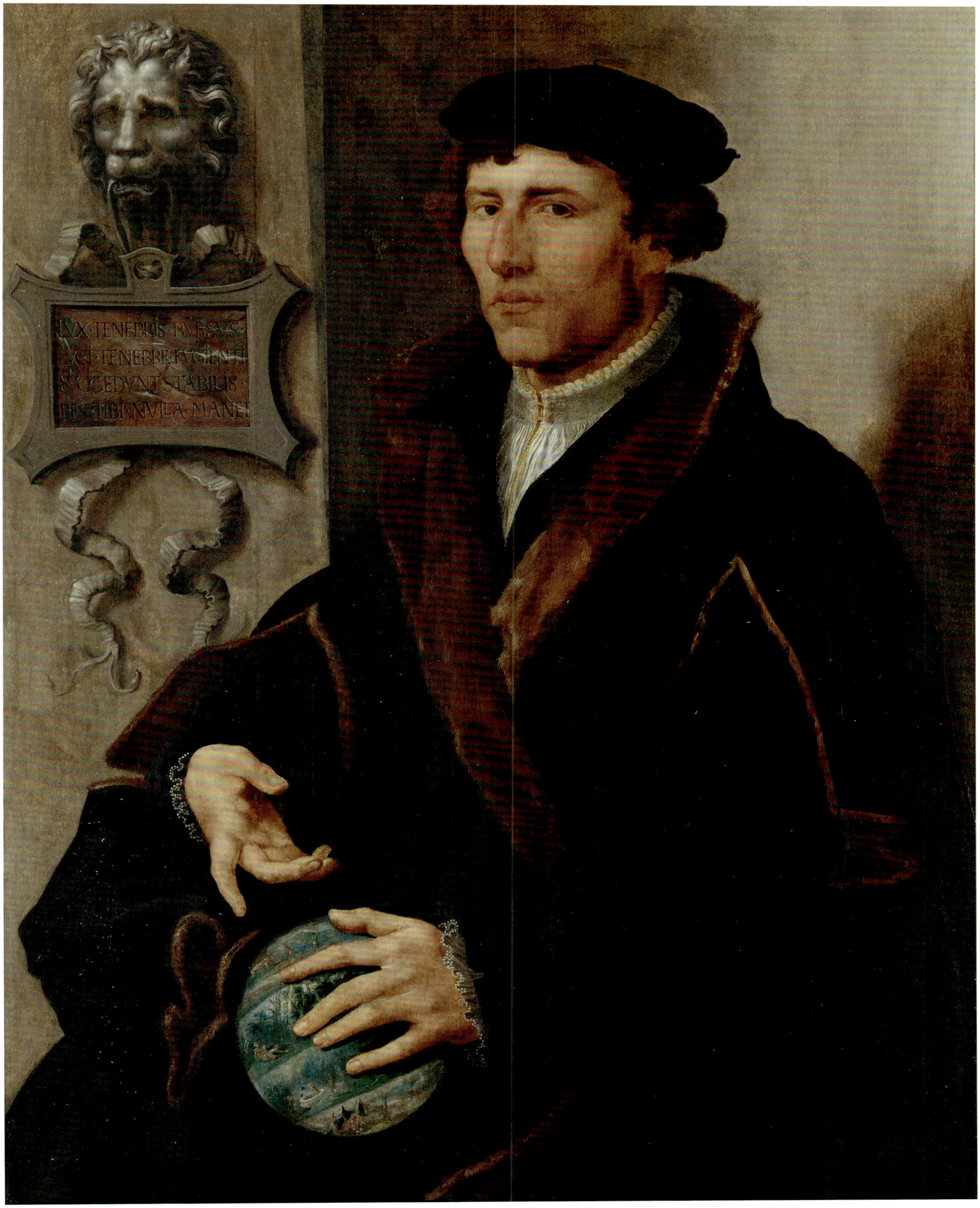
LVX·TENEBRIS·RVRSVS
LVCI·TENEBRE·FVGIENTI
SVCCEDVNT·STABILIS
RES·TIBI·NVLA·MANET

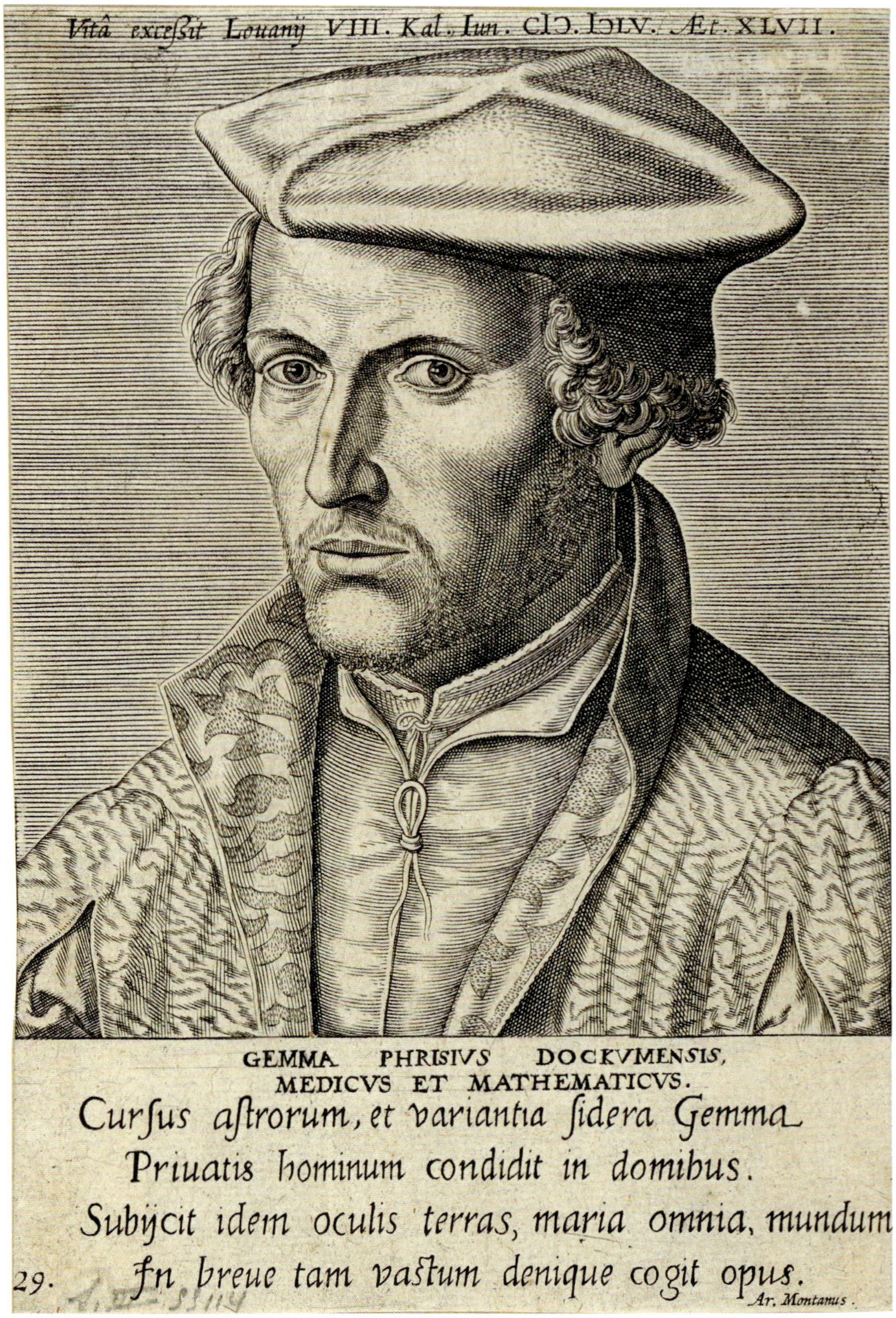

[fig. 59.1] Philips Galle, *Portrait of Gemma Frisius*, 1572, engraving, Brussels, Royal Library of Belgium, Print Room.

become a physician – he honed his skills in geography, mathematics and medicine. Some years later, the academic was offered a chair at his Alma Mater. Among his students were Andreas Vesalius, who would go on to acclaim as a leading physician and anatomist, botanist Rembert Dodoens and cartographer Gerardus Mercator, to name but a few. Even though the fame garnered by these pupils may transcend the renown of the master, when it comes to scientific accomplishments, the latter certainly was not to be outdone by any of them. Gemma's enhancements of contemporary geographical theories, his writings setting forth a surfeit of new findings, and the scientific instruments developed by him played a key role in the advancement of geography.

In spite of his 'small stature, slender build and pale countenance', Gemma Frisius is portrayed here in monumental fashion.[5] Gazing sternly at the beholder, the humanist is dressed in a damask gambeson beneath a fur-lined tabard. His hair is covered by a black cap. There is no sign of any scientific instruments that might refer to the scholar's academic accomplishments. His left hand holds a painted crystal ball representing the four seasons. From top to bottom, spring, summer, autumn and winter are rendered as an allegorical reference to the transience of life. This Vanitas motif is underpinned by the Latin inscription on the marble tablet on the left.[6]

Gemma died on 25 May 1555, just 46 years of age, as a result of kidney stones. His portrait was placed over his tomb at the Leuven Predikherenkerk (Dominican Church).[7] Whether or not this is the Rotterdam painting is difficult to establish. However, it would certainly explain the inscription and the globe with the allegorical reference to the finite nature of all living things.

M B

60

Albrecht Dürer after Conrad Heinfogel
The Celestial Maps: the Northern and Southern Skies
Published by Johannes Stabius

60A

Map of the Northern Sky
'Imagines coeli septentrionales cum duodecim imaginibus zodiaci'
1515
Woodcut, 433 x 432 mm
Brussels, Royal Library of Belgium, Print Room, inv. S.II 23031

60B

Map of the Southern Sky
'Imagines coeli meridionales'
1515
Woodcut, 430 x 435 mm
Berlin, Staatliche Museen zu Berlin, Kupferstichkabinett, inv. 778-10

BIBLIOGRAPHY

Bartsch 1808: no. 151; Weiss 1888: 209-213; Meder 1932: no. 260; Panofsky 1943: no. 365; Voss 1943: 89-150; Zink 1968: no. 99; Vincent & Chandler 1969: 382-383, fig. 16; Hamann 1971: no. 309; Talbot 1971: no. 199; De Solla Price 1975: 187, no. 203; Strauss 1980: no. 176, Appendix B, pl. 17; Strauss 1981: 431, no. 151; Chipps Smith 1983: no. 21; O'Neill & Schultz 1986: 315, no. 134; Levenson 1991: 220-221, no. 118-119; Dekker & Van der Krocht 1994: 264; Schoch, Mende & Scherbaum 2002: 430-435, no. 243-244; Dackerman 2011: 90-93, no. 16; Sumira 2014: 19.

Art and science meet in a particularly felicitous manner in the two celestial maps that Albrecht Dürer brought onto the market in 1515. The Nuremberg artist's direct source was a pair of hand-drawn star charts made in 1503 after a design by his fellow townsman, the astronomer Conrad Heinfogel (fig. 60.1 and 60.2).[1] These are the very first European printed star charts. Made in the form of woodcuts that could be endlessly reproduced on the printing press, they reached a wide international audience and were extremely influential (cat. 61).

These celestial maps build on the knowledge of classical Antiquity, added to and transmitted by way of medieval Arab astronomers.[2] Which is why, in the corners of his map of the northern sky, Dürer has portrayed four major classical and

1 Unknown artist after Conrad Heinfogel, *Maps of the Northern and Southern Skies*, drawings in brown ink on parchment, heightened with silver, gold and red, 674 x 672 mm and 669 x 670 mm, Nuremberg, Germanisches Nationalmuseum, inv. Hz 5576–5577, see Dackerman 2011: 86–88, no. 15; Voss 1943.

2 On this topic see inter alia the essay by Koenraad Van Cleempoel in this catalogue.

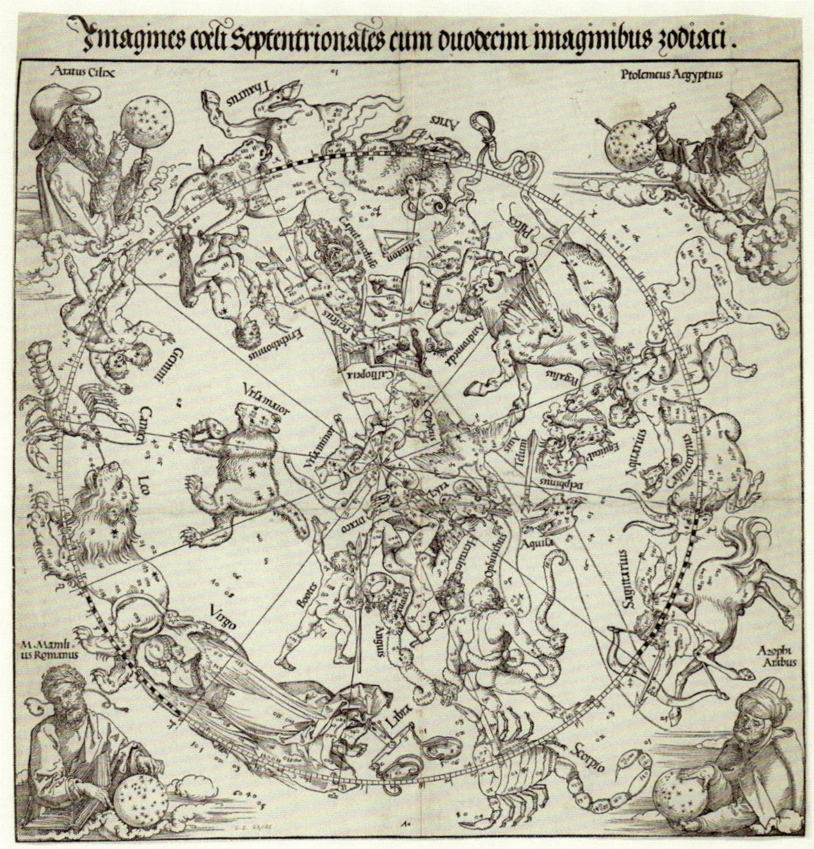

[cat. 60A]

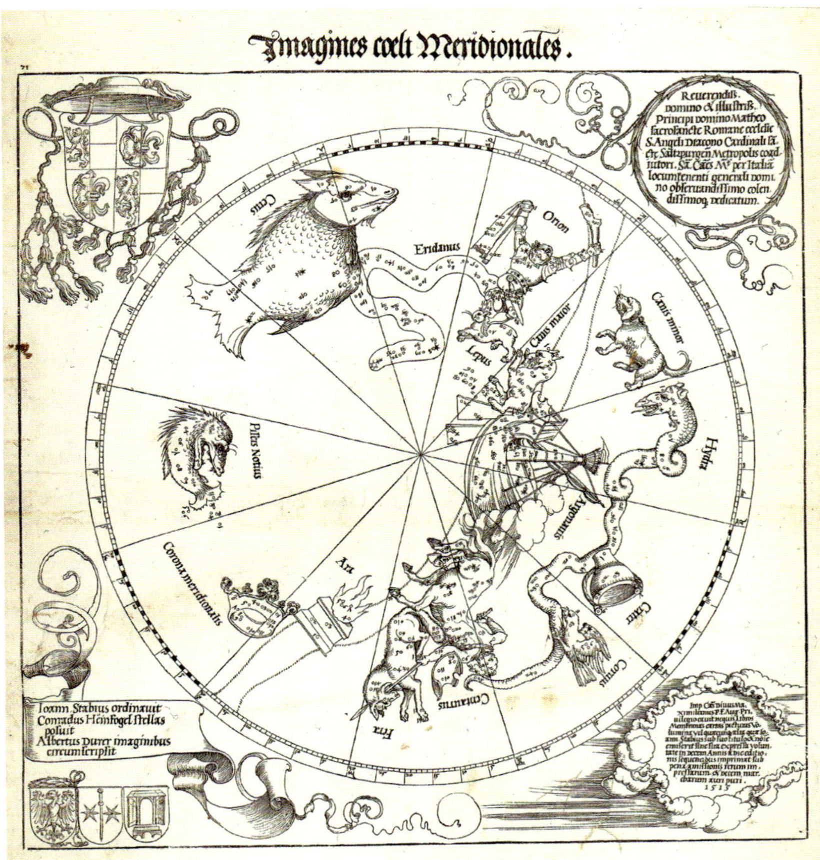

[cat. 60B]

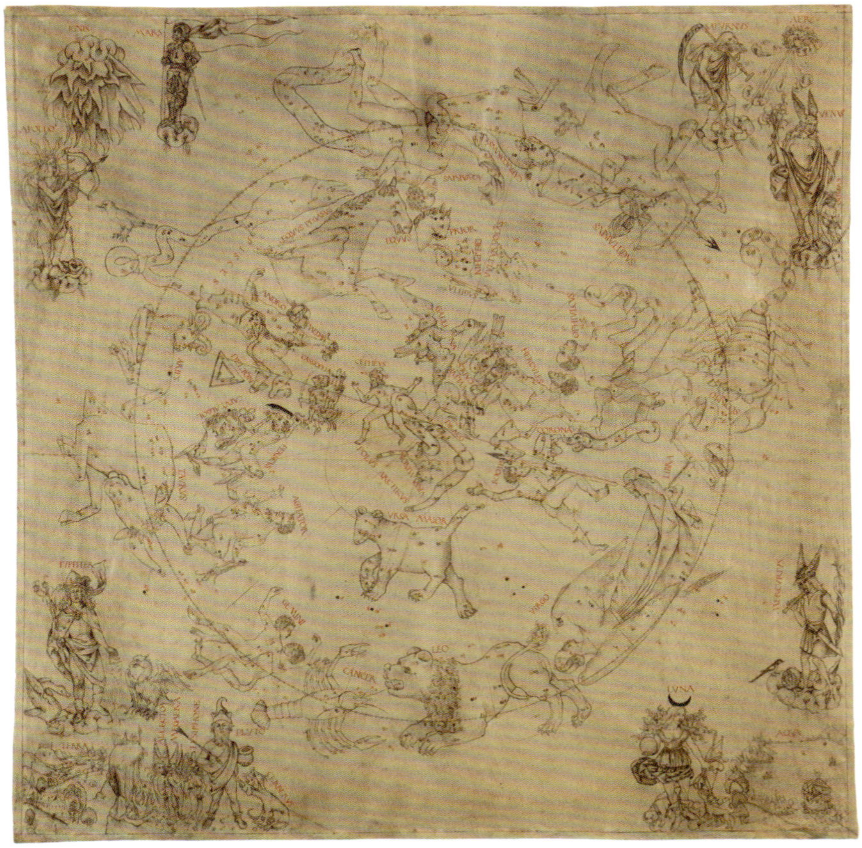

[fig. 60A.1] Conrad Heinfogel, *The Northern Firmament*, 1503, drawing, Nuremberg, Germanisches Nationalmuseum.

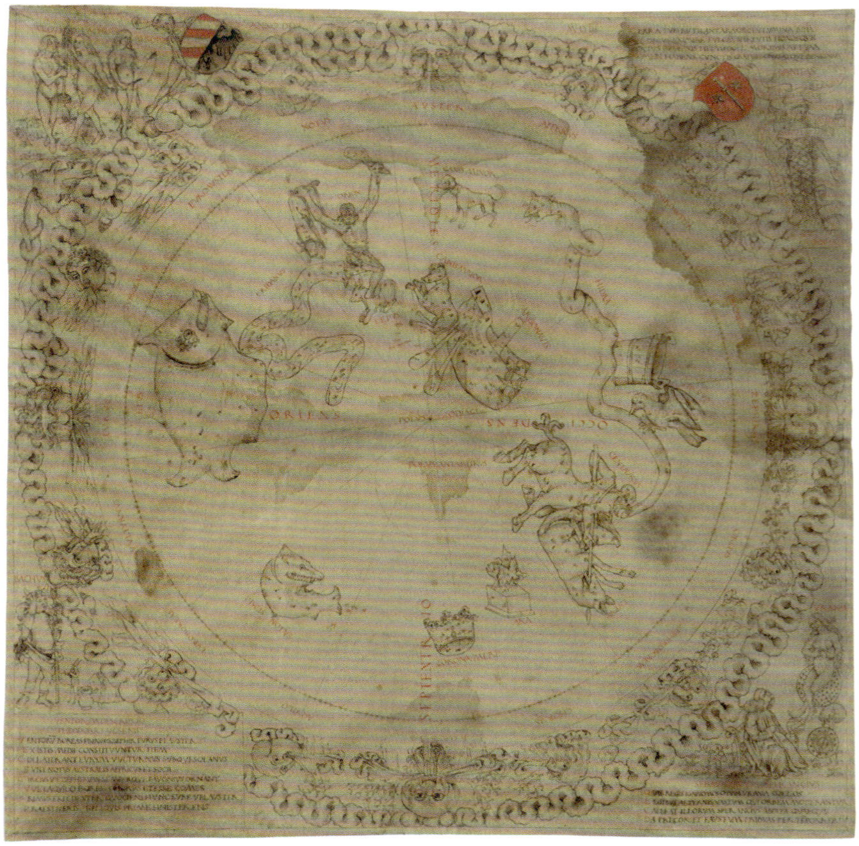

[fig. 60B.1] Conrad Heinfogel, *The Southern Firmament*, 1503, drawing, Nuremberg, Germanisches Nationalmuseum.

Arab astronomers or astronomical poets. Ptolemy (*c.*90–168) is included, of course, and is joined by Aratus Cilex (315–245 BCE), Marcus Manilius (first century) and the Arab astronomer Abd al-Rahman al-Sufi (903–985).

The positions of the stars are plotted and numbered according to the star catalogue in books eight and nine of Ptolemy's *Almagest*. They are represented by six-pointed stars or open or filled circles, depending on their magnitude. Here, Heinfogel reduced Ptolemy's classes of magnitude from six to just three.

These star charts are known as planispheres – maps formed by the projection of the two halves of the celestial sphere on a plane surface. For until the modern period it was thought that the fixed stars occupied an outer crystal sphere that moved round the universe. Celestial globes often represent that sphere as if seen from the outside. That external viewpoint is also adopted in these two celestial maps, so that we see the back of the constellations as it were.

The circle bounding each star map is the ecliptic, the sun's apparent annual path relative to the stars. This means that on the map of the northern sky the signs of the zodiac lie exactly on this circle, which as usual is correspondingly divided into twelve equal parts of thirty degrees. The central point is therefore the pole of the ecliptic, not the pole of the northern sky, which in reality is close to the pole star in the tail of *Ursa Minor* (Little Bear). In Antiquity comparatively little was known about the southern sky. New information recorded by explorers was not integrated into these celestial maps. The precession of the Earth's axis meant that the position of the stars had changed slightly. Heinfogel replotted the star field to show it as it was around 1500.

In the top left corner of the map of the southern sky is the coat of arms of Cardinal Matthäus Lang von Wellenburg, a leading church authority, diplomat and chief counsellor to Emperor Maximilian I; a dedication to the cardinal occupies the top right corner. In the lower left corner an inscription in a banderole informs us that Johannes Stabius (*c.*1468–1525) was the publisher and initiator of the maps: *Johannes Stabius Ordinavit*. Stabius was a humanist, geographer and scientific advisor to Emperor Maximilian. He was also the emperor's court historiographer and as such was closely involved in the commission for the spectacularly monumental woodcut image of a triumphal arch glorifying Maximilian's achievements, on which Dürer worked in the same period.

Dürer did not sign these celestial maps with his famous monogram but instead added his coat of arms – a pair of open doors or 'Türen', a play on his name Dürer or 'Thürer' – next to the arms of Stabius and Heinfogel. That the spiritual authors of these star charts should sign with their full names and arms is characteristic of the emerging humanist culture. Scholars and artists increasingly sought to be seen as self-aware individuals, proud of the achievements with which they claimed their place among the eternal stars.

JVG

61

Petrus Apianus (Peter Apian)
Astronomicum Caesareum

Ingolstadt

1540

Typographic printed book illustrated with hand-coloured woodcuts by Michael Ostendorfer and volvelles with silk strings, 480 x 340 mm

Brussels, Royal Library of Belgium, Department of Rare Books, inv. VB 5123 C

BIBLIOGRAPHY

Van Ortroy 1902: 112-114 no. 112, 115 no. 113; Dodgson 1911: 242 no. 4, 250 no. 5, 286 no. 7; Hollstein, German, XXX: 206, no. 54 (1-41); Muris & Saarmann 1961: 77, fig. 21; Zinner 1964: no. 1476; Stillwell 1970: vol. I, 19-22, vol. II, 135-138, vol. VI, 809-814; Gingerich 1971: 168-177; Snyder 1984: 57, fig. 25; Van Gijsen 1994: 232; Gingerich 1994: 293-296; Gingerich 1995: 113-122; Ernst 1995: 35-38; Wolfschmidt 1995: 96-98; Schmeidler 1995: 107-112; Röttel 1995: 309-313, no. 7; Kruse 2000: 283-284, no. 286; Horst 2011: 68; Karr Schmidt 2011: 104-109, no. 19; Kanas 2014: 89-91.

This book by the German humanist and astronomer Petrus Apianus is without doubt the most sumptuous scientific publication of the Renaissance, as indeed might be expected, given that, as the title indicates, it was dedicated to Emperor Charles V, the most powerful monarch of his time, and to his younger brother Ferdinand. Charles thought so highly of Apianus's work that he raised him to noble rank. The newly ennobled astronomer's coat of arms adorns the last page of the book.

Apianus (1495–1552) was professor of Astronomy at the university of Ingolstadt, where he also ran a small private printing shop with his brother. There they not only looked after of the design and printing of books but also the hand-colouring and assembly of the various printed instruments with movable parts that they contained. Apianus had already included several of the ingenious rotating paper devices, known as 'volvelles', in his *Cosmographia* of 1524.[1]

The *Astronomicum Caesareum* contains a remarkable twenty-one beautifully designed volvelles with which diverse astronomical calculations could be made. They include, for instance, a set of planetary *equatoria*, which are used to work out the position of a planet in the zodiac system. The *equatorium* for determining the longitude of Mercury has no fewer than nine visible printed parts with a hidden mechanism in paper and silk string that allows the parts to rotate around several axes. The moon's orbit can be calculated with help of two complex volvelles on facing pages. Apianus also included an instrument for computing the date of Easter and other holy days. How the printed instruments should be used is explained by means of appropriate examples, such as ascertaining the astronomical conditions in effect at the birth of the emperor and his brother. He also discusses the calculation of lunar and solar eclipses and illustrates his methods by reconstructing the precise astronomical conditions that prevailed at historical eclipses.

The painter Michael Ostendorfer (*c*.1490–1559) designed the magnificent woodcuts. The paper instruments are represented in trompe l'oeil, with charmingly

1 Volvelles had already been used in manuscripts and they were also used later in printed astronomical handbooks. For a survey, see Gingerich 1994.

2 In letters patent published in 1532, which protect Apianus's future publication for thirty years with an imperial privilege, there is already reference to an *Astronomicum Imperatorum*. This suggests that the book had been worked on for at least eight years. Moreover, Apianus published nothing else between 1534 and 1540. See Karr Schmidt 2011: 19 n. 3.

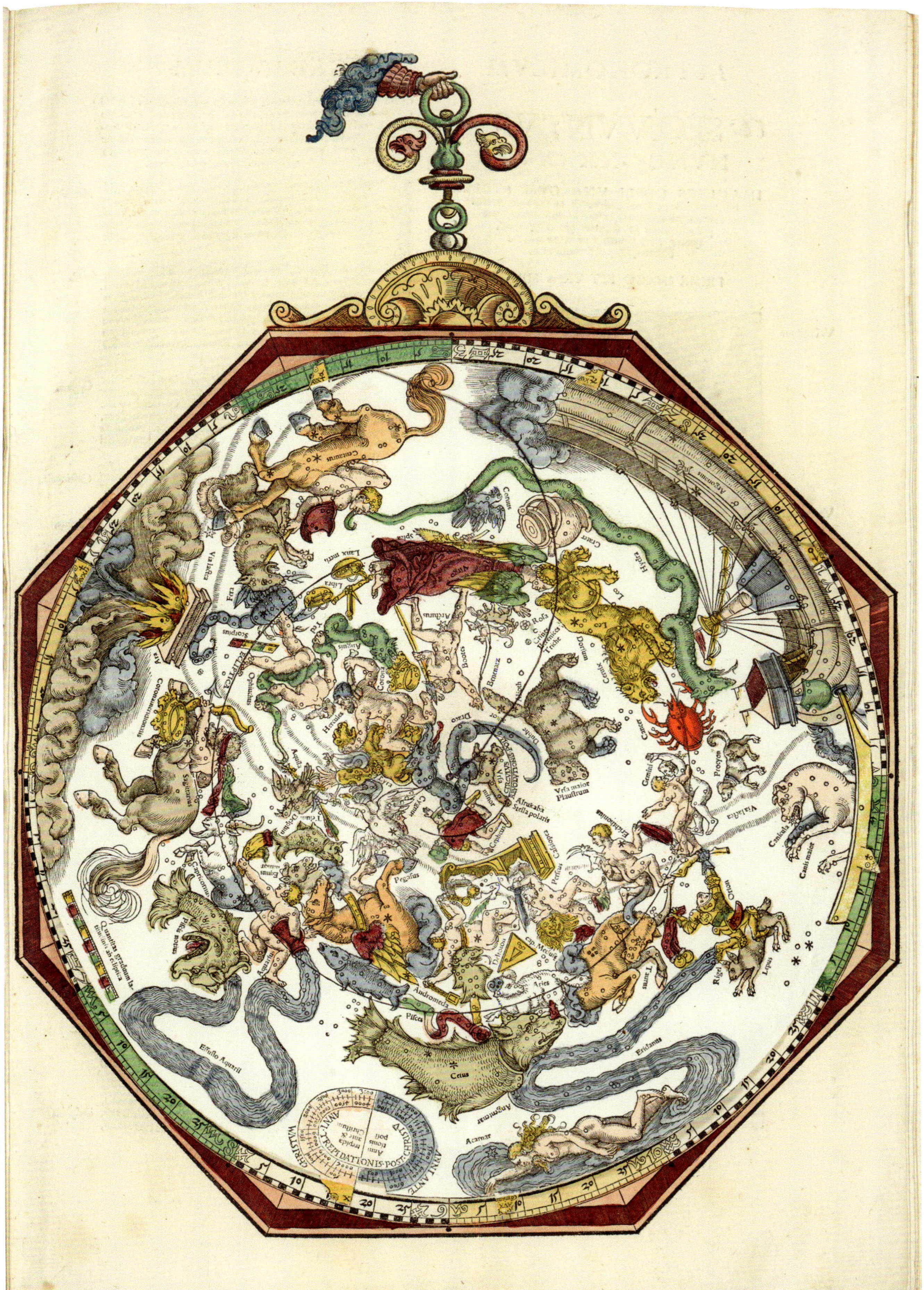

[cat. 61] Hemelkaart

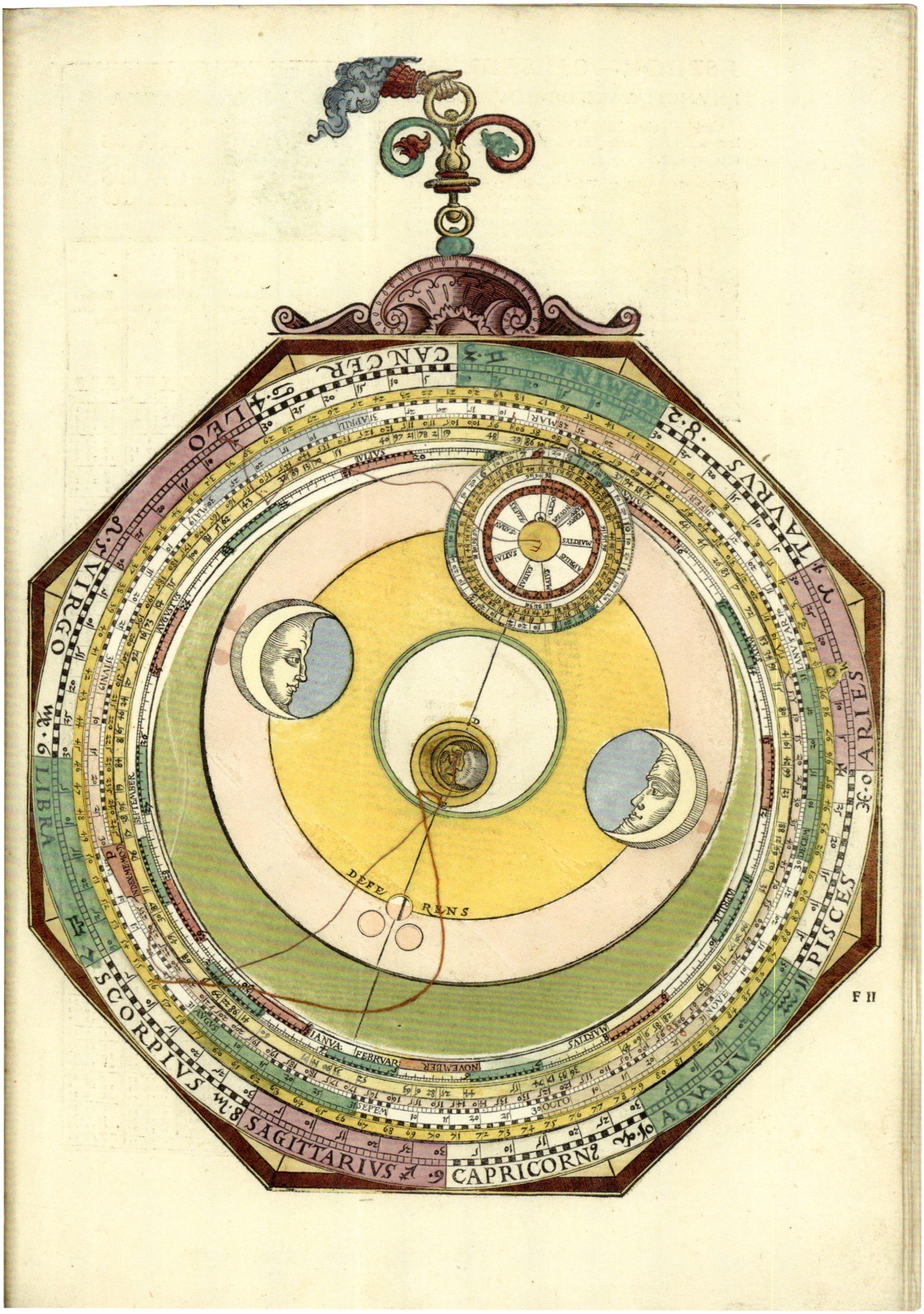

[cat. 61] p.46 - *Volvelle* showing the phases of the moon before and after Christ's birth by means of four discs and two epicycles

designed hooks from which to 'hang' them, thus alluding directly to their three-dimensional counterparts. Ostendorfer derived the volvelle on which the zodiac system is depicted from Albrecht Dürer's celestial maps (cat. 60A–B). Here the northern and southern skies are combined in a single plane, with the northern sky reduced in size and placed on the inside and the southern sky turned inside out and projected on the outside of the circle.

The second part of the book contains no movable instruments but it does have numerous illustrations. It focuses on methods of observation and on making the instruments that are needed for that purpose. Here Apianus describes the paths of several comets, including Halley's, and notes – the first to do so – that the tails of comets always point away from the sun.

The *Astronomicum Caesareum* is the ultimate synthesis of astronomical knowledge according to the Ptolemaic system, which assumes that the sun and the planets revolve around the earth. It appeared three years before the publication of Copernicus's new heliocentric doctrines that place the sun in the centre of the universe. It is generally thought that Apianus was already familiar with the new heliocentric theory, but the investments he had already made and his relationship with the geocentrically-minded imperial and ecclesiastical authorities deterred him from making such a momentous change in his publication.[2]

JVG

62

Celestial Globe

Italy or Central Europe, 1502

Dated on the 85th line of latitude: MCCCCCII

Diameter 69 cm

Ecouen, Musée national de la Renaissance Château d'Ecouen, inv. E.CL. 3218

BIBLIOGRAPHY

Du Sommerard 1883: 563, no. 7022; Paris 1936: no. 1184; Destombes 1968; Paris 1973: no. 204; Chapiro, Meslin-Perrier & Turner 1989: 118-121, no. 88; Daynes-Diallo 2009: 88.

Boasting a diameter of close to 70 cm, this globe is the largest known celestial globe from the early part of the sixteenth century.[1] What makes this globe unique is not just its size, but also the substance of the cartographic information; the globe shows over 750 stars, divided across 85 star signs. The engraving style is particularly refined and the sheer accuracy of the various scales and coordinates is quite high. The construction consists of two hemispheres that have been conjoined at the equator.[2] Due to the exceptional scale, both spheres have been braced from the inside. In addition, each pole has a hole to accommodate a central axis.

The *ecliptic*, or the apparent path of the Sun, is graduated down into 360°, numbered 6, 12, 18, 24 and 30. The name of each sign is engraved in Roman numerals. Running in parallel with the Zodiac is a net of circles: one for each degree up to 83°. The band between the second and the fifth circle North and South of the Zodiac is not engraved. At right angles to this band are the circles of the meridians, which transect at the poles, one every 6°. In addition, there are circles for the polar circles, the tropical circles and the equator (perfectly divided into 360° and numbered for every 5°).

This spectacular web of parallels and meridians acts as a frame of reference or system of coordinates for an unprecedented number of 750 stars, clustered in 85 constellations. What is more, the 'magnitude' or brightness of each star is specified on a scale of one to six. The position of the stars concurs with the catalogue of Alfonso of Cordoba, published in Venice in 1503: *Tabulae astronomiae Elisabeth Regina*. Also of note here is the position of the human figures: they are seen from behind, as it were, with their faces in profile, as though the beholder is inside the celestial sphere (see also cat. 60A-B, 61 and 63).

Establishing the location of origin with any degree of certainty remains an arduous task. The celestial globe displays a unique combination of classical know-how and Arabic astronomy. Central Europe is one possibility by dint of its kinship with the work of Hans Dorn from Vienna and Nuremberg. Dorn built a similar albeit smaller brass globe in 1480, which is on display at the Museum Jagellon in Cracow. Almost all early celestial globes were built in Central Europe, in workshops in Vienna, Prague and Cracow to be more exact. Yet the engraved letters have a more Italian 'feel'. They also show a degree of affinity with an edition of Ptolemy's *Geographia* of 1478, published by Arnold Brecknick in Rome in 1490, 1507 and 1508.

KVC

1 The description is based on the examination by Destombes 1968; on the development of astronomy from Antiquity up to the fifteenth century, see Dekker 2012.

2 On this topic, see Van Raemdonck 1873–1875; Stevenson 1921; Duprat 1973: no. 210; Dekker 1995; Bertsch 1997: in particular 149–150; Kugel, Van Cleempoel & Sabrier 2002: 19–41; Meskens 2012: 41.

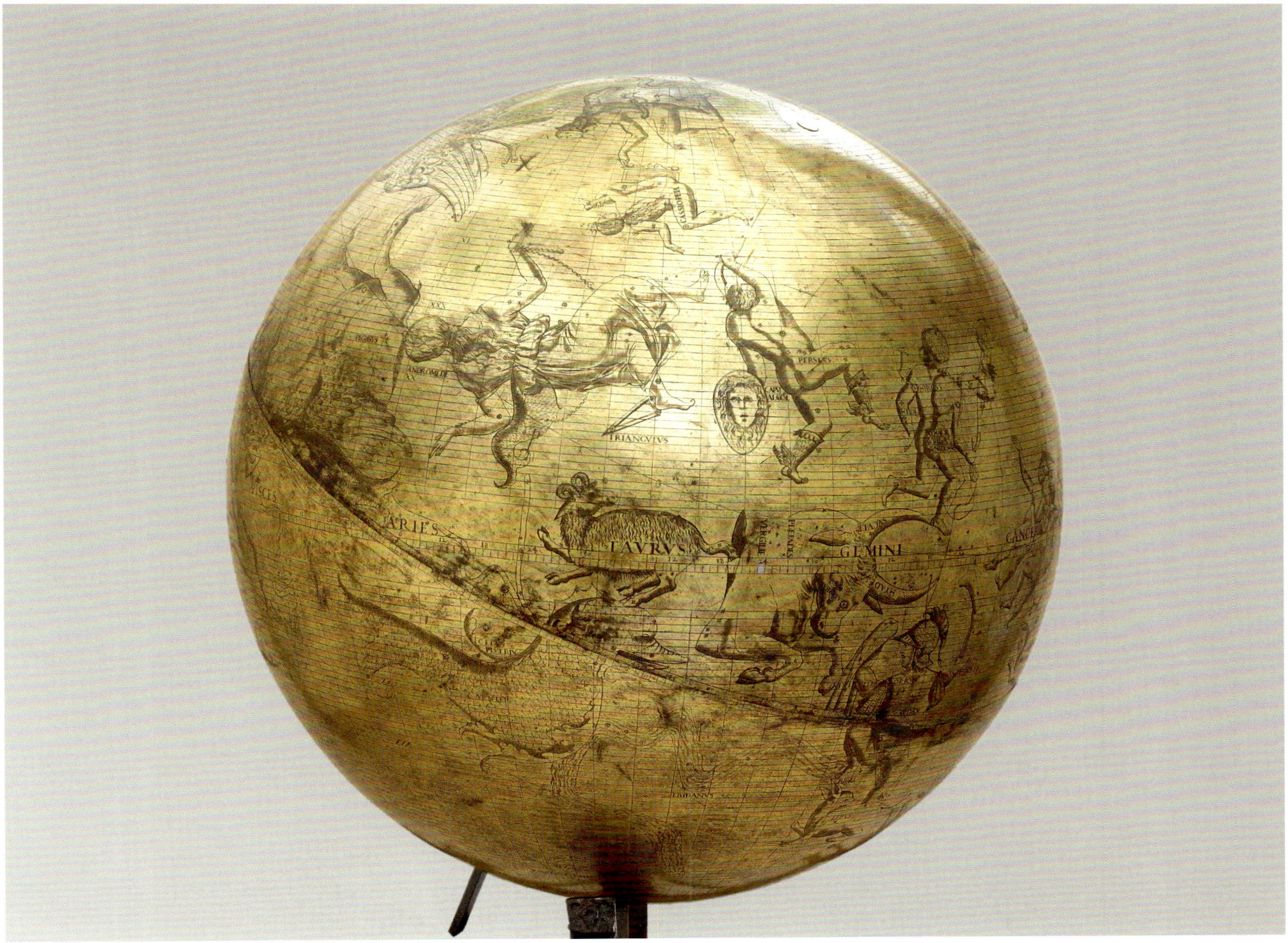
ARIES
TAVRVS
GEMINI
TRIANGVLVS

63

Gerard Mercator
Celestial Globe

Leuven

1551

Papier mâché, coloured copper engravings on paper, brass and wood, diameter 42 cm

Lüneburg, Museum für Fürstentum Lüneburg, inv. R.1785

BIBLIOGRAPHY

Muris & Saarmann 1961: 104, fig. 31; De Smet 1968; De Smet 1969: no. 3, 15-16; Warner 1979: 174-175; De Smet 1974a: 191; De Smet 1974b: 227; Van der Krogt 1984: 197-198; Pühlhorn & Laub 1993: 526-527, no. 1.24; Karrow 1993: 384-385, no. 56/10; Van der Krogt 1993: 57, 66; Van Gijsen 1994: 227; Dekker & Van der Krocht 1994: 242-267; Elkhadem, Heerbrant, Calcoen & Wellens-De Donder 1994: 144; De Nave, Imhof & Otte 1994: no. 15b, 101-103, 113; Bouza 1995: no. 42, 125-126; Van der Krogt 1995: 93, fig. 6; Dekker 1999: 94-95, 413-415; Dekker 2012: 296-299; Sumira 2014: no. 7, 59-61.

The terrestrial and celestial globes that Gerard Mercator built in Leuven in 1541 and 1551, respectively, are the most important pair of globes by far to emerge from the sixteenth century.[1] Like his astrolabe (cat. 71), they display a high level of ingenuity and dizzying refinement in the way they were designed and built. Consequently, they proved a major source of influence until well into the seventeenth century. Together with Gemma Frisius (cat. 59) and Gaspar van der Heyden, Mercator had already designed a highly innovative 'cosmographic globe' as early as 1536–1537 in Leuven.[2] His map of Palestine came in 1537, followed by a world map in 1538. In 1540, he published a treatise on Italic script as this enabled him to convey a great deal of legible information on the limited amount of space available on maps or globes. He used Italic script on his first terrestrial globes built in 1541, followed by this celestial globe of 1551.

The design of the celestial globe shows that Mercator was already incorporating the recent theories of Copernicus's *De revolutionibus* (1543), even though the classical model set out in Ptolemy's *Almagest* (*c*.150) remained at the root of his worldview. This tension between new research and a high degree of deference to classical knowledge was typical of Mercator's work.

A major innovation that had a huge impact was the pedestal of this globe, which was the result of the device's wide 42 cm diameter. Mercator's globes were much larger than was common at the time. They were able to deliver more information, which made them more accurate and much more legible. One implication was the fact that they were becoming too large for the customary tripod as used in Germany. Mercator designed an 'architectural' pedestal which consisted of four lathed wooden columns, clearly inspired by the Doric order, connected at the bottom by a wooden cross on which a large wooden disc rests. The four columns carry a wooden ring onto which the printed paper horizon has been glued. In the horizon ring are two holes for the meridian ring, in which the globe

1 On this topic see Van Raemdonck 1873–1875; Stevenson 1921; Dekker 1995; Bertsch 1997: in particular 149–150; Kugel, Van Cleempoel & Sabrier 2002: 19–41; Meskens 2012: 41.

2 This is a combination of a terrestrial and a celestial globe. The unique specimen is kept in Greenwich, National Maritime Museum, glb135 (Dekker 1999: 87).

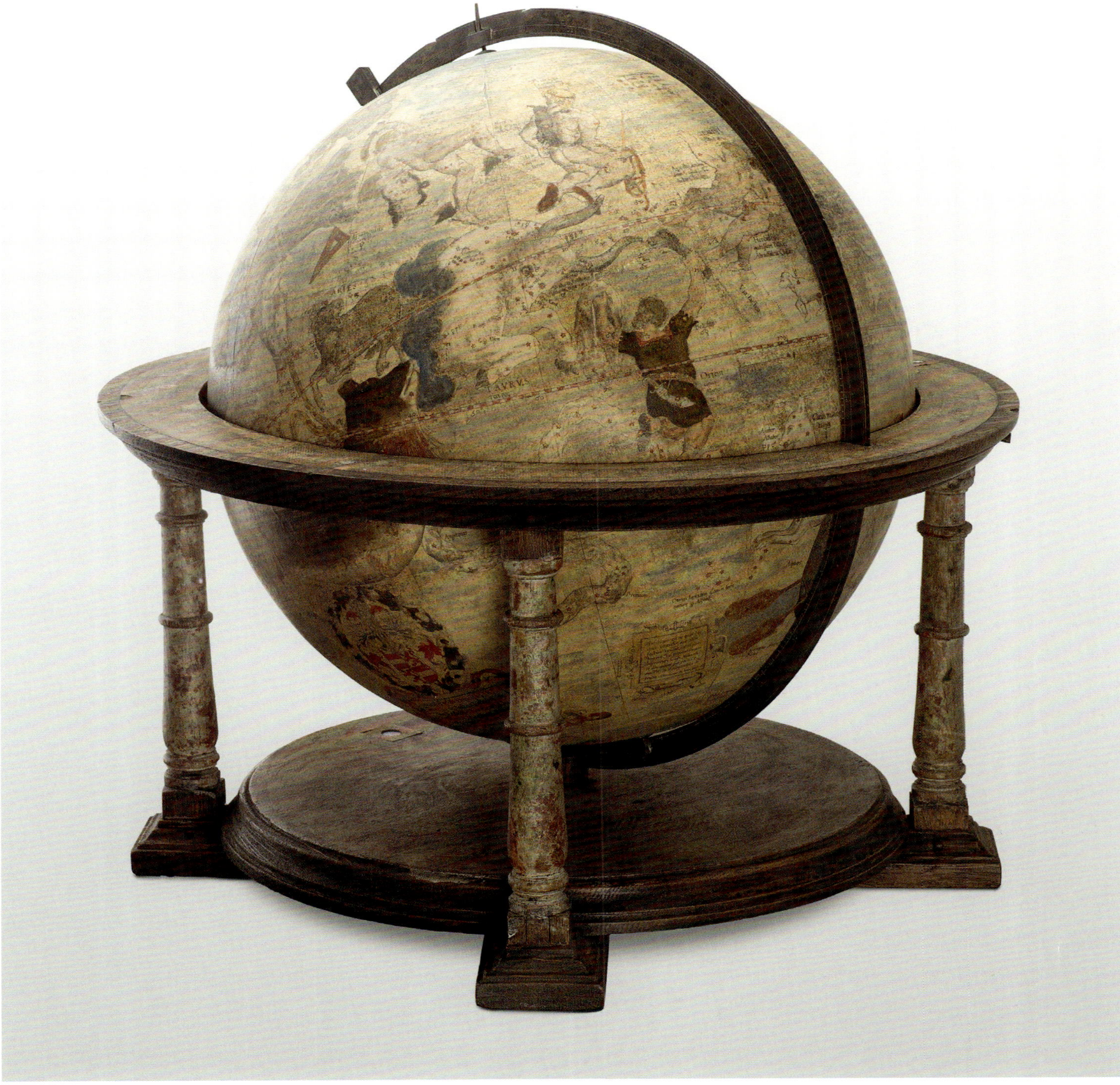

is ensconced around an axis. To brace the construction, the graduated meridian is made from brass; the globes themselves are in *papier mâché* and plaster onto which twelve printed paper bans or 'gores' have been glued. The twelve gores are prints of Mercator's copper engravings. The seams of the gores coincide with the hour circles, or every 30°. In this specimen, the pedestal on top is finely painted.

The globe shows the 48 classical star signs of Ptolemy, yet Mercator adopted a new style in drawing the figures, which also sets his creation apart from the famous star chart by Albrecht Dürer of 1515 (cat. 60A–B). The positions of the stars have been calculated in accordance with Copernicus's method, based on an epoch of 1550. In front of *Ursa Maior* sits a cartouche containing a text about the magnitude of the stars, on a scale from 1 to 6: from *nebulae* to *occultae*. The globe also shows the Milky Way (*Via Lactae*). Mercator also added two constellations that, even though they had been known since Antiquity, had never previously been shown: *Antinous* and *Coma Berenices*.

KVC

64

Jan Gossaert
Portrait of a Girl with an Armillary Sphere (Princess Dorothea of Denmark)

c.1530

Oil on oak, 38.2 x 29.1 cm.

London, National Gallery, inv. NG 2211

On the jewel pinned to her breast are the letters *.../IHESVS/M(ARIA?)... N.E..../...* On the armillary sphere are the letters *.I.I.A.A.N.N.R.R.R.P.E.E.E.S.S.S.(H?).G.T.Y* (fig. 64.1). This may be an anagram of the artist's name *IENNI[N] G[O]SSART PAI[NT]RE*,[1] the letters *R E S (H?) Y* remain to be explained.

PROVENANCE

By 1882, Sir Edmund Anthony Harley Lechmere (1826-1894), The Rhyd (Worcestershire); his son Sir Edmund Arthur Lechmere (1865-1937); sale London (Christie's), 27 April 1901, lot 75; Léon Gauchez (1825-1907), sale Paris (Hôtel Drouot), 16 December 1907, lot 29; Agnew; purchased in 1908.

BIBLIOGRAPHY

Pauwels, Hoetink & Herzog 1965: no. 19; Campbell 2014: vol. I, 344-351 (with further bibliography).

The child represented, who may be about nine or ten years old, must be of the highest rank: her clothes are sewn with hundreds of pearls. She is identified, by comparison with other portraits, as [2]Dorothea, daughter of the exiled King Christian II of Denmark, whose three children resided in the Low Countries. Both Margaret of Austria and Charles V gave Dorothea enormous quantities of jewellery, including huge numbers of pearls. In her left hand Dorothea holds, upside down, a small armillary sphere, made of brass or possibly gilded. Touching with her right hand the outer ring of the instrument, she indicates a point approximately 55° north of the equator. The latitude of Copenhagen is 56° north; the child may be directing attention to her father's lost kingdoms. The sphere, and the world, may be upside down because of the political disturbances that drove him and his family from Scandinavia.

Dorothea, born in November 1520, was the elder daughter of Christian II and Isabella of Austria and the sister of Christina, duchess of Milan, whose portrait by Holbein is in the National Gallery, London. Brought to the Low Countries in 1523 when their father was deposed, Dorothea, Christina and their brother were educated at the courts of Margaret of Austria and Mary of Hungary. Dorothea married in 1535 the Count Palatine Frederick II, who succeeded as Elector Palatine in 1544 and died in 1556. She died, childless, in 1580.

Though Gossaert's portrait has lost some of its opulence of colour, it remains a most arresting image, full of contrived contradictions. The blue parts of the sleeves and snood should be purple: the red lake component has faded. The painted frame appears to be in the same plane as the real frame but is overlapped

1 On 12 September 1532, Gossaert signed a quittance 'Jennin Gossart painter': see Steppe 1965: 36 and the facsimile on the back cover of the exhibition catalogue (Pauwels, Hoetink & Herzog 1965).

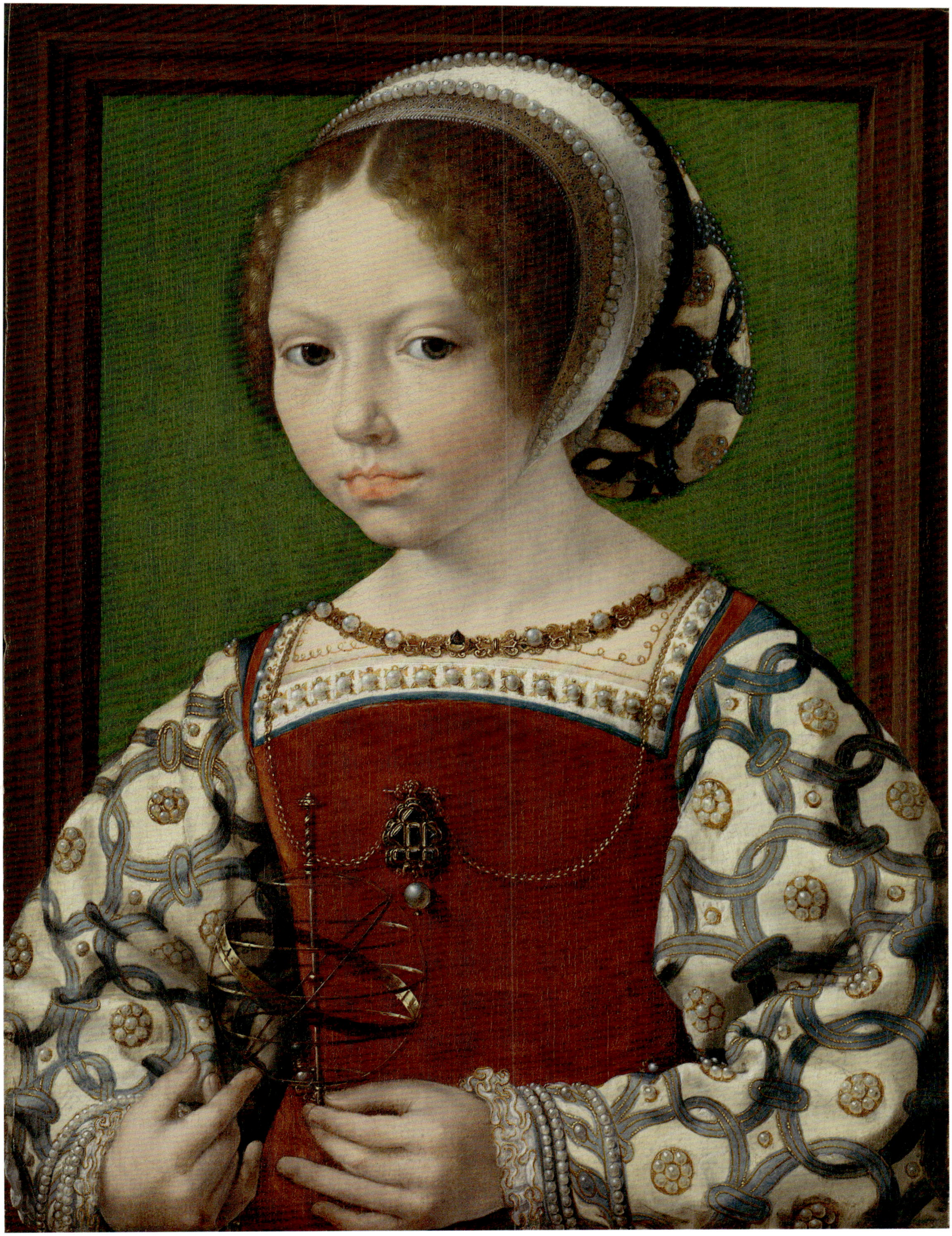

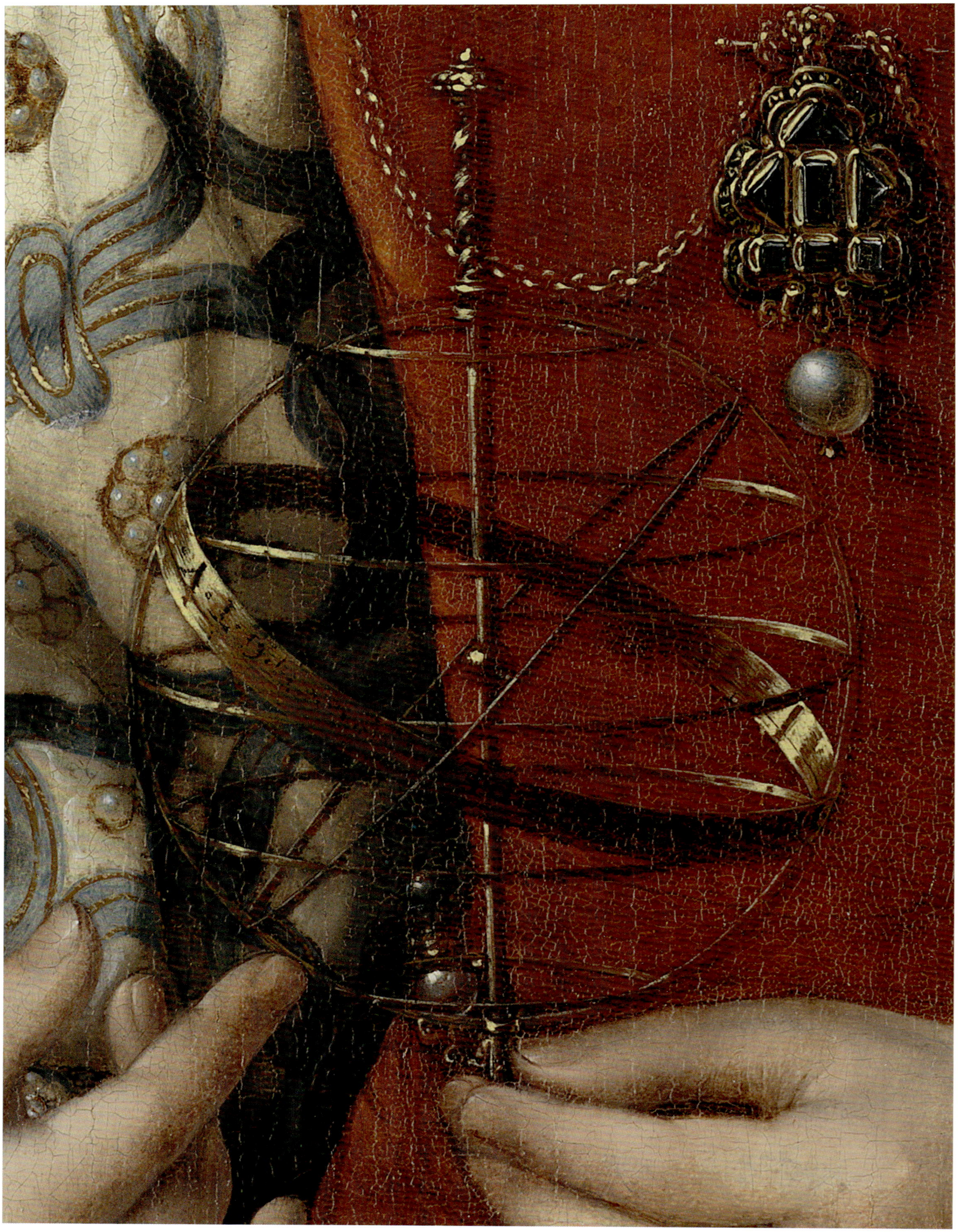

[cat. 64] detail

by the child, who is projected towards the spectator, in front of the frame. The portrait may have been commissioned by Margaret of Austria or another of the Habsburgs, for whom Gossaert often worked.

LC

65

Gualterus Arsenius
Ptolemaic Armillary Sphere

Leuven

1562

Signed on the edge of the base: *NEPOS GEMMAE FRISY FECIT ANNO 1562 GAL*

Brass, height 43.5 cm (pedestal included), diameter 26.5 cm

Chicago, Adler Planetarium, inv. M-1

BIBLIOGRAPHY

Fox 1932: 331; Fox 1933: 33; Zinner 1956: 237; Turner 1987: 44; Van Cleempoel 2002: 163-164, no. 37; Van Cleempoel 2009: 117.

The armillary sphere of 1562 from Chicago (cat. 65) is the earliest known armillary sphere by Gualterus Arsenius (1530–1580), a nephew of Gemma Frisius's and a pupil of Gerard Mercator's. It predates the Munich sphere by eleven years (cat. 66).[1] Like all Arsenius's instruments, they were both built in Leuven. Only the design of the pedestal is different: the sphere is supported by four caryatids, but with the Chicago exemplar, an additional circular openwork plate has been interposed, raising the whole construction in height. The pedestal of the Munich sphere is simpler and more tranquil in appearance. On their heads, the four figures are carrying a ring from which four intertwined bands ascend to support the horizon ring. Between the intertwined bands and the horizon are four lion's heads.

These exceptional spheres show the mechanism of the precession of the equinoxes (the intersection of the ecliptic and the equator). As early as the second century BCE, Hipparchus observed that each year the equinoxes shift a little to follow the Sun in its apparent path across the Zodiac. In the Middle Ages, this was referred to as the path of the eighth sphere, in reference to Ptolemy's geocentric model as described in his *Almagest* (*c.*150): a central Earth surrounded by ten spheres in concentric circles, consecutively (1) the Moon, (2) Mercury, (3) Venus, (4) the Sun, (5) Mars, (6) Jupiter and (7) Saturn. The eighth sphere holds the fixed stars. The ninth sphere is immobile and is called the 'crystalline' sphere. The tenth sphere, or primary sphere, is the *primum Mobile*.

Ptolemy too had described the model of the precession of the equinoxes. During the Middle Ages, astronomers expanded the notion into a theoretical concept that coalesced in the astronomical treatises (*Libros del saber de astrologia*) compiled under King Alfonso X, 'El Sabio', between 1276–1279. In this 'Alfonsine' theory, precessional movement of 1° occurs every 136 years, whereby the equinoxes perform a full cycle every 7,000 years. During the Renaissance, this model was re-examined and transformed into a series of mechanical models. Both of Arsenius's spheres specify this 7,000 years on the wheels between the eighth and the ninth sphere.

Only a handful of Renaissance spheres adopted this complex astronomical concept, including the four known specimens built by Arsenius.[2] However, Arsenius

1
On this topic, see inter alia Dekker 1999: 3–12; Kugel, Van Cleempoel & Sabrier 2002: 89–135.

2
Van Cleempoel 2002: 163–170. For an astronomical description of precession and the relation to instruments, see Dekker 1999: 96–99.

3
The original Mercator instrument remains to be uncovered. However, on 9 April 1997 Christie's in London auctioned (lot 37) an armillary sphere with precession, which – along with an engraved terrestrial and celestial sphere is part of a set of three instruments built in 1579 for Sultan Murad III. It seems safe to assume that this set was inspired by a model of Mercator's.

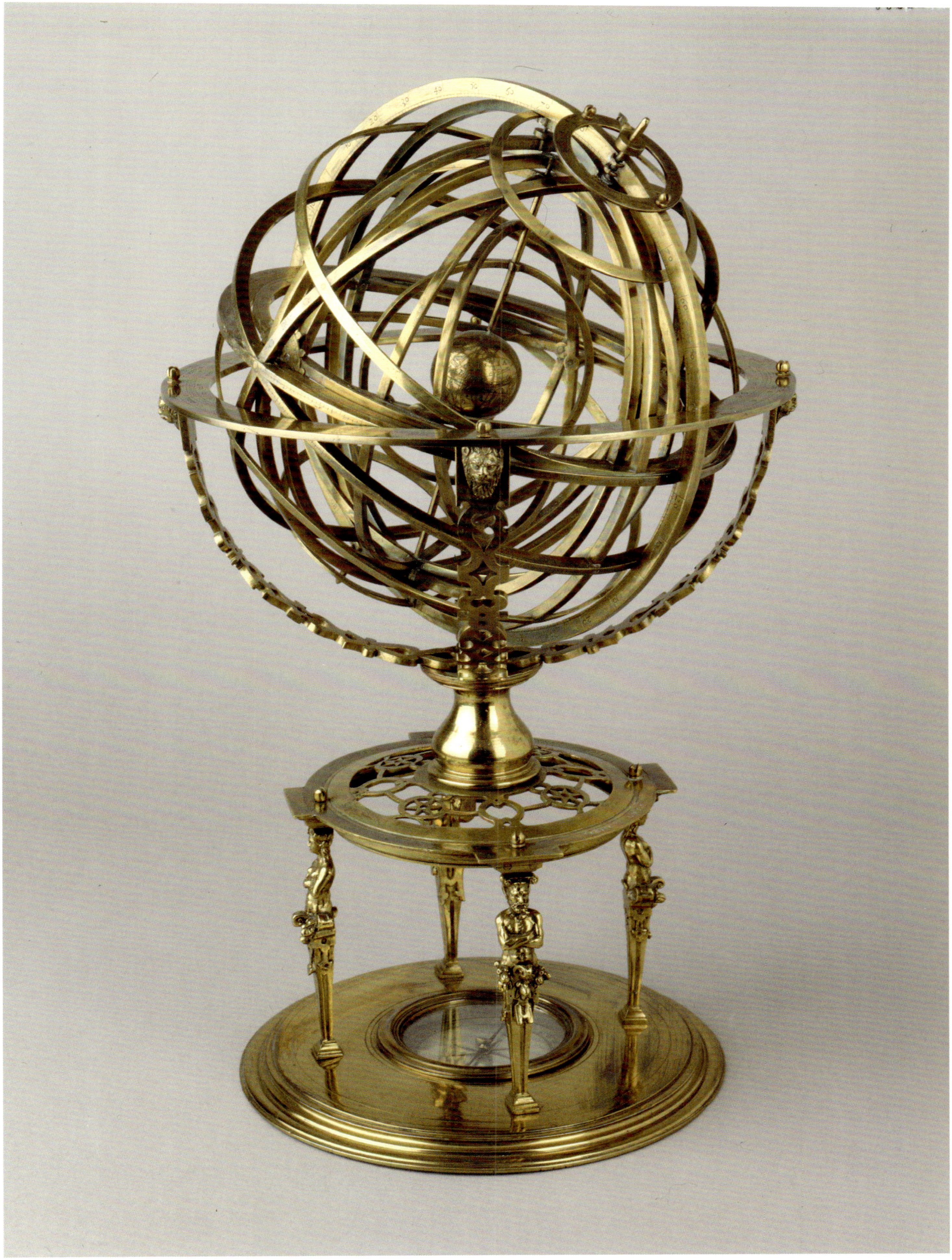

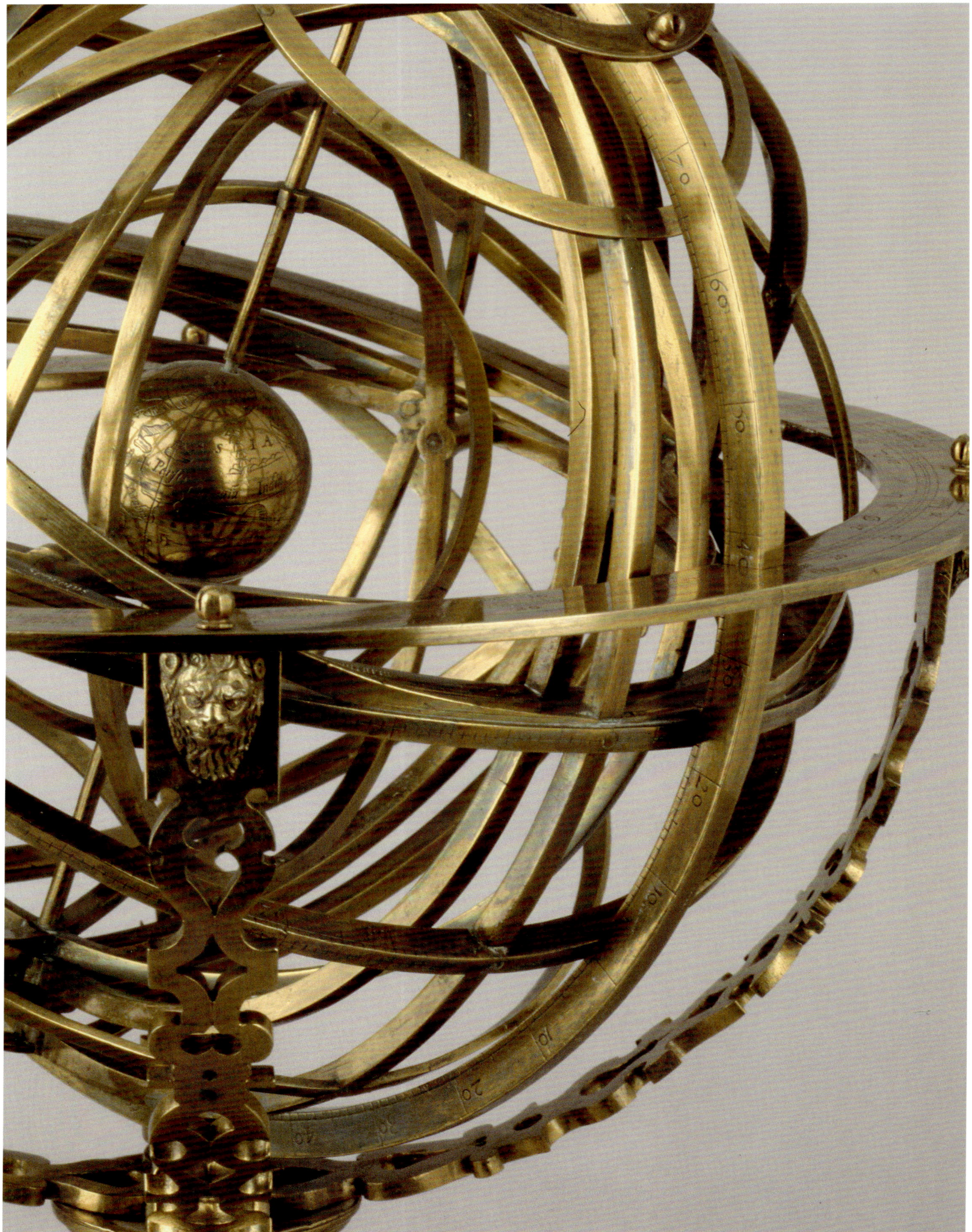

[cat. 65] detail

is likely to have operated in accordance with an as yet unknown model by Mercator, as he did for his other instruments.[3]

In the spheres from Chicago and Munich shown here, the primary sphere consists of eight rings, two of which are positioned at a right angle in relation to each other: the *colures*. Five parallel rings represent the polar circles, the tropical circles and the equator (graduated in 360°). Between the tropical circles, at an angle of 23.5°, sits a band that represents the Zodiac or the *ecliptic*; the apparent annual path of the Sun. This primary sphere also holds some fifteen stars, indicated by way of a star pointer and the star's name.

Inside this outer 'tenth' sphere is the 'ninth' sphere, made up of three rings positioned at a right angle in relation to each other and at an angle of 23.5° in between: the meridians and the Zodiac (*zodiacus nonae spherae*). They are connected with the outer sphere by two small axes at the N-S pole. Inside the 'ninth' sphere, sits the 'eighth' sphere, again made up of two rings at a right angle in relation to each other, with the Zodiac (*zodiacus octavae spherae*) positioned diagonally between them. The system of the eighth sphere is connected to the ninth by way of two wheels with 14 cogs. One wheel says *In 7000 annis circumuitur*, another wheel *In 7000 annis circumuoluitur*. The wheels are mounted on the equinoctial axes of the two spheres. Turning the wheels sets in motion the eighth sphere inside the ninth, with the latter showing the apparent path of the precession: the up- and downward motion of the eighth sphere.

Within this system, there is again a movable system of four rings that is attached to the axis of the ecliptic of the eighth sphere. One serves to actuate the Moon, the other drives the Sun. At the centre sits planet Earth, mounted onto the axis of the ecliptic to the effect that the N-S axis of the poles is in line with the N-S axis of the armillary sphere itself. The contours of the continents have been traced out on the terrestrial globe, showing America as separate from Asia.

KVC

66

Gualterus Arsenius
Ptolemaic Armillary Sphere

Leuven

1573

Signed on the base: *GUALTERUS ARSCHENIUS NEPOS GEMMAE FRISIJ 1573*

Brass, height 40 cm (pedestal included), diameter 26 cm

Munich, Bavarian National Museum, inv. 33/241

BIBLIOGRAPHY

Zinner 1956: 237; Von Bassermann-Jordan 1961: 14; Van Cleempoel 2002: 168-169, no. 40.

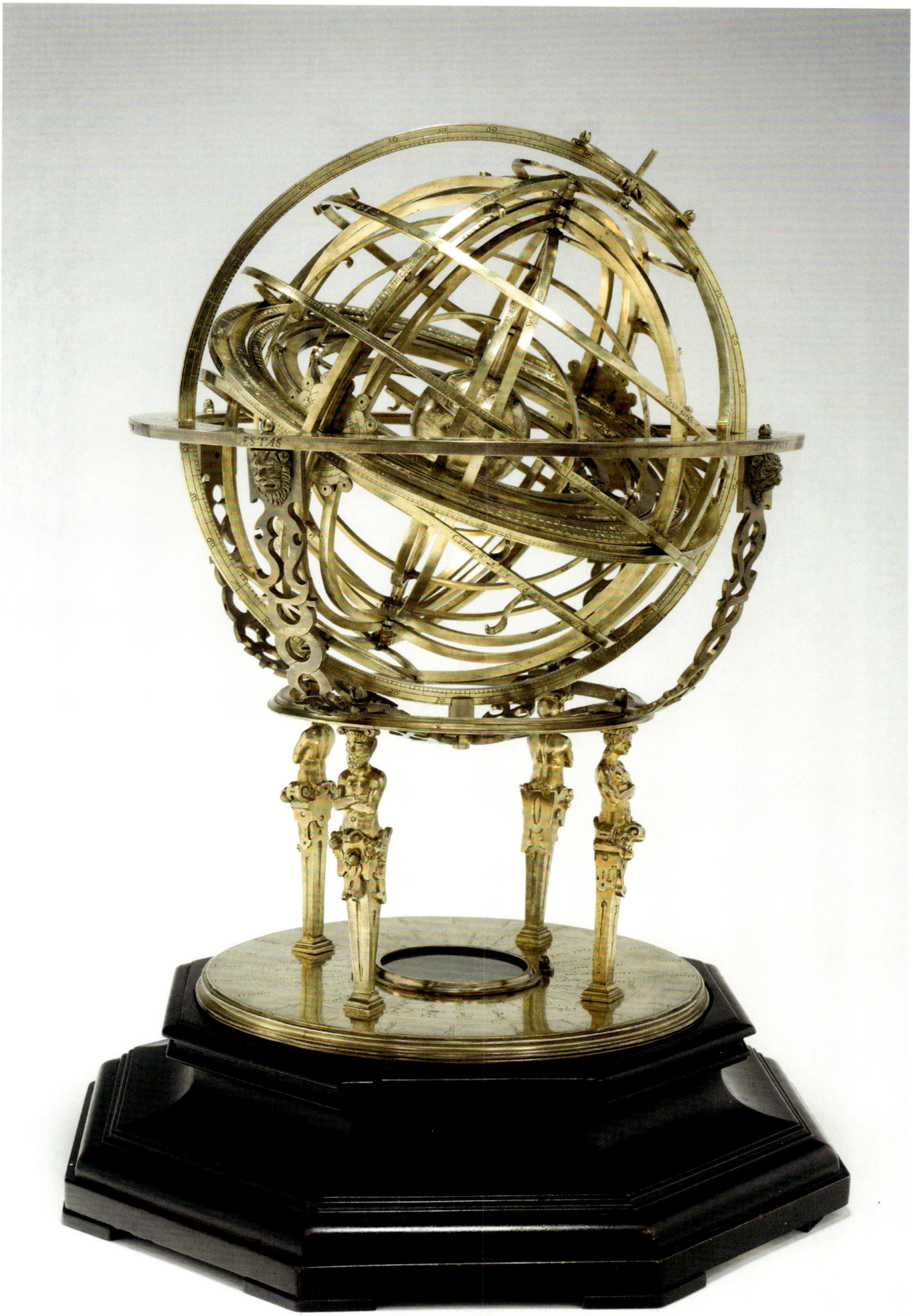

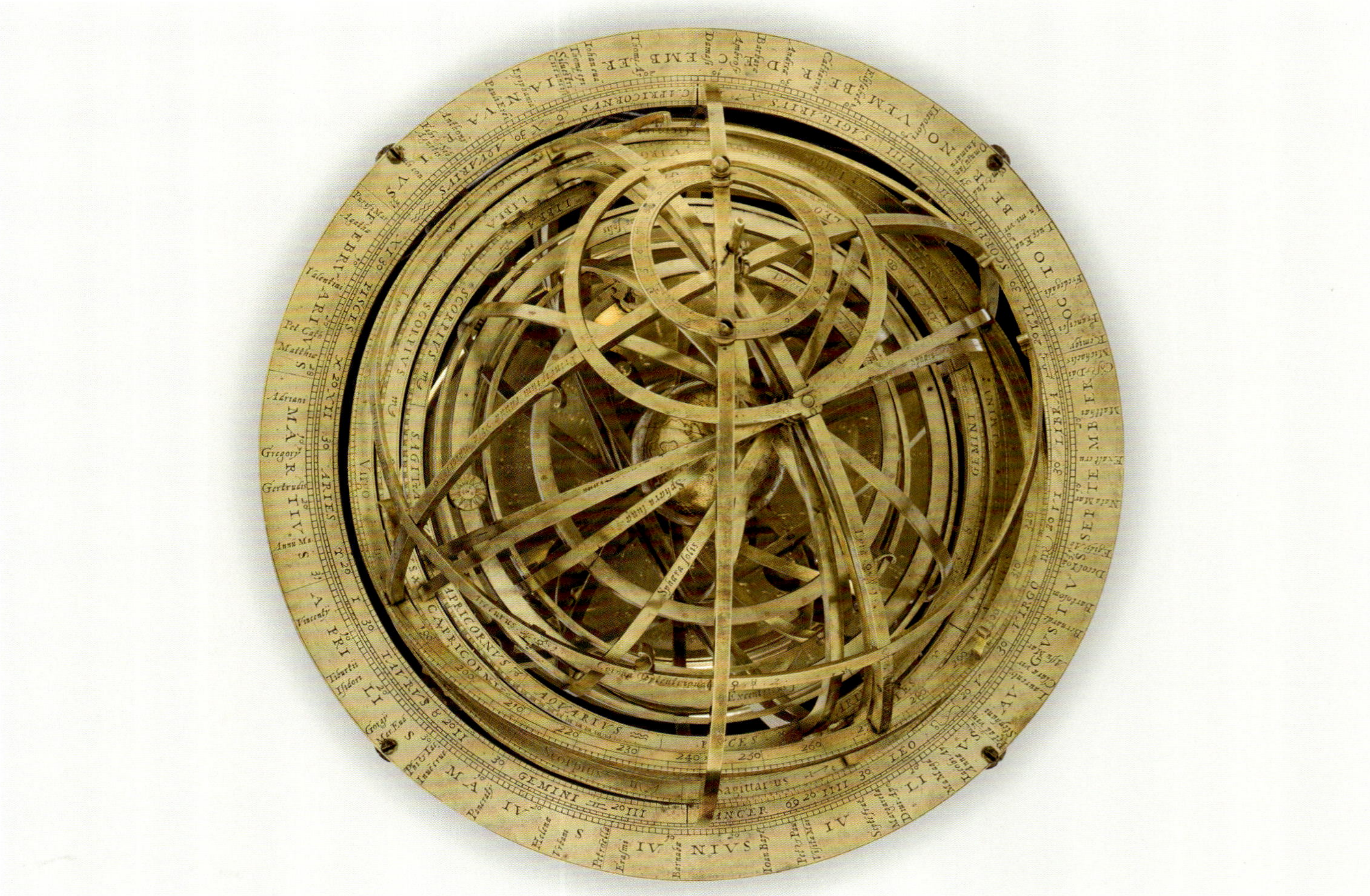

[cat. 66] Top view

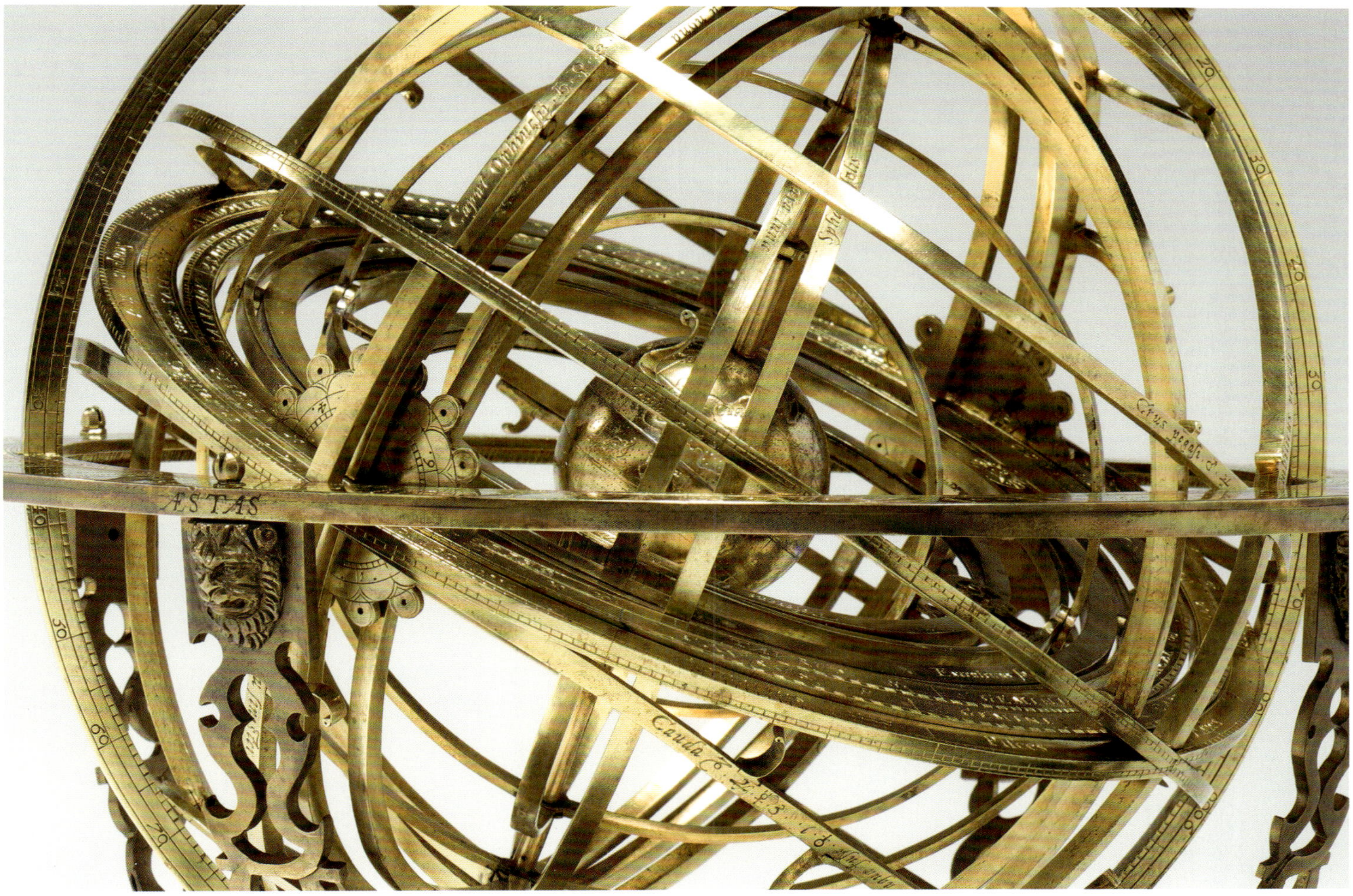

[cat. 66] detail

67

Adriaan Descrolières, attributed to *Ptolemaic Armillary Sphere*

Leuven
c.1575
Brass, height 31 cm, diameter 17.9 cm
Private collection

BIBLIOGRAPHY

Baptiste 1984: 33, no. 5; Van Cleempoel, Van Damme, L'Estrange Turner et al. 1997: 160-161, no. E.3; Van Cleempoel 2002: 196-197, no. 62 and plate p. 261.

Based on the engraving style, it is possible to attribute this armillary sphere to Adriaan Descrolières.[1] The pedestal is made up of three S-shaped flourishes that hold up a central baluster as the support of the four slender arms with profiled edges that carry the horizon ring. On the horizon, scales have been engraved for a calendar and the Zodiac. The onset of spring (1 Aries) coincides with 11 March, which would allow us to infer that the instrument was built before the 1582 calendar reform. From this time forward, spring started on 21 March.

In the horizon, two small openings have been made in which the meridian ring fits. The ring is graduated in 360° and numbered 4 x 0–90°. At the North Pole sits an hour ring, divided into 24 hours. The N-S poles are connected by way of axis on which the various spheres have been mounted. The outer, primary sphere consists of two rings that transect each other at a right angle, representing the *colures*. At a 90° angle against the *colures* are five parallel rings: two polar circles, both tropical circles and the equator in the middle. Between the tropical circles at a 23.5° angle is the Zodiac, a broad band with a scale (12 x 0–30°) and the twelve figurative representations of the signs of the Zodiac. Various stars have also been specified on the *colures* of the primary sphere, as a result of which this sphere is in fact also seen to represent the 'eighth' sphere, as described in the ancient model of Ptolemy.

Further inward are another two systems with rings, each made up of two rings positioned at a right angle in relation to each other: one for the Sun and one for the Moon. Centrally positioned on the axis sits planet Earth, on which the *colures*, the polar circles, the tropical circles and the equator have been indicated.

KVC

1 On this topic, see inter alia Dekker & Van der Krocht 1994; Dekker 1999: 3–12; Kugel, Van Cleempoel & Sabrier 2002: 89–135.

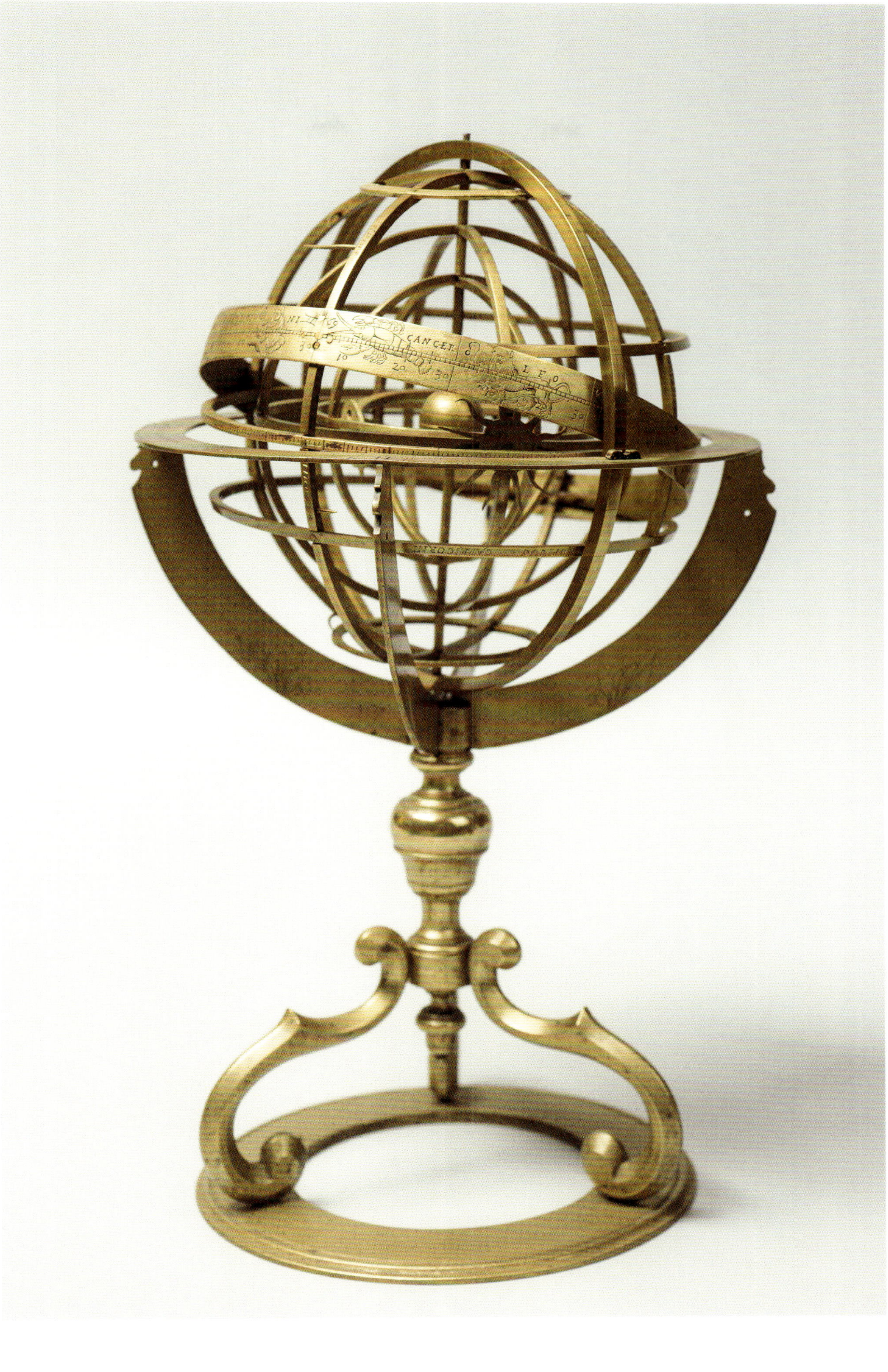
CANCER
LEO

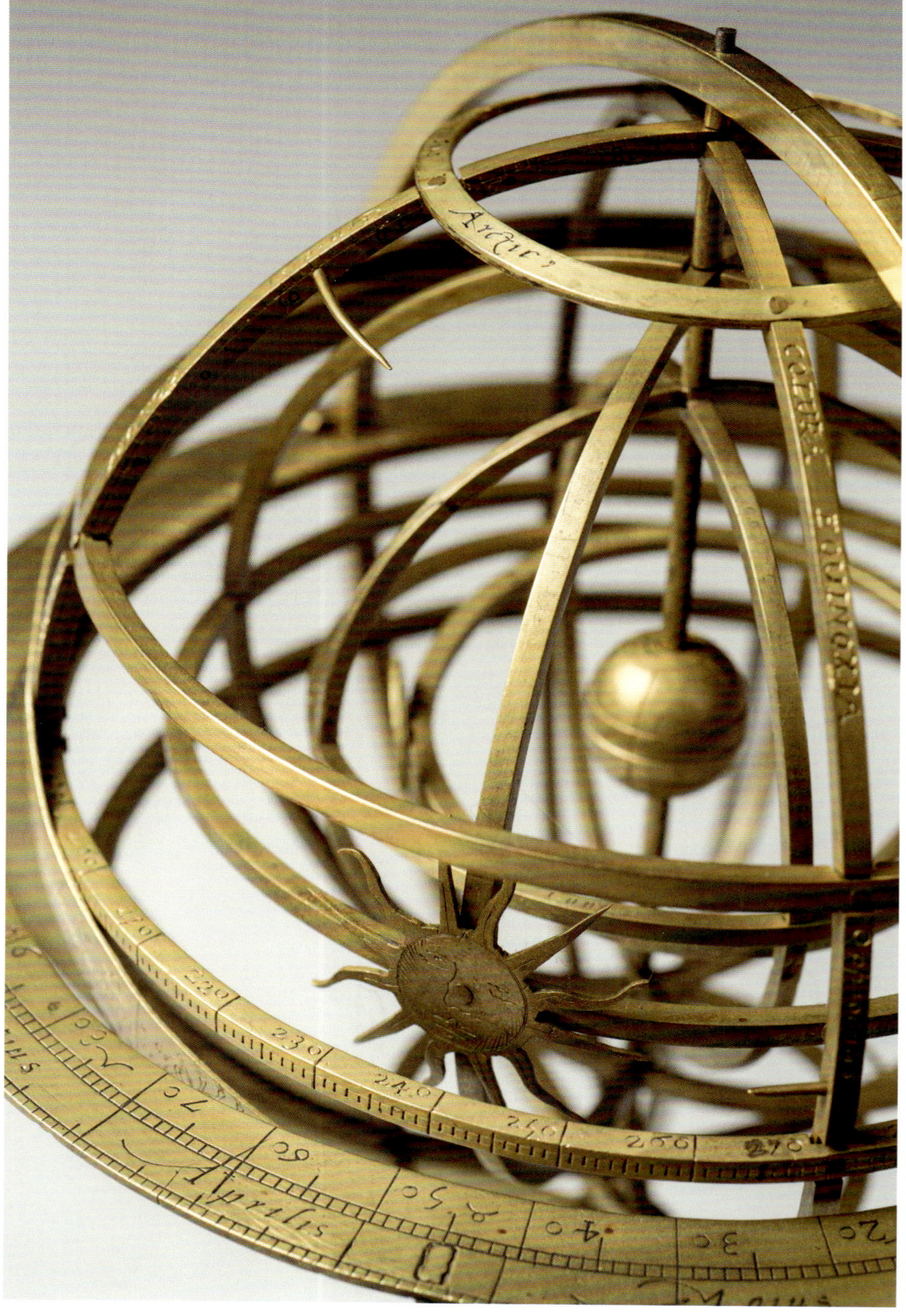

[cat. 67] detail

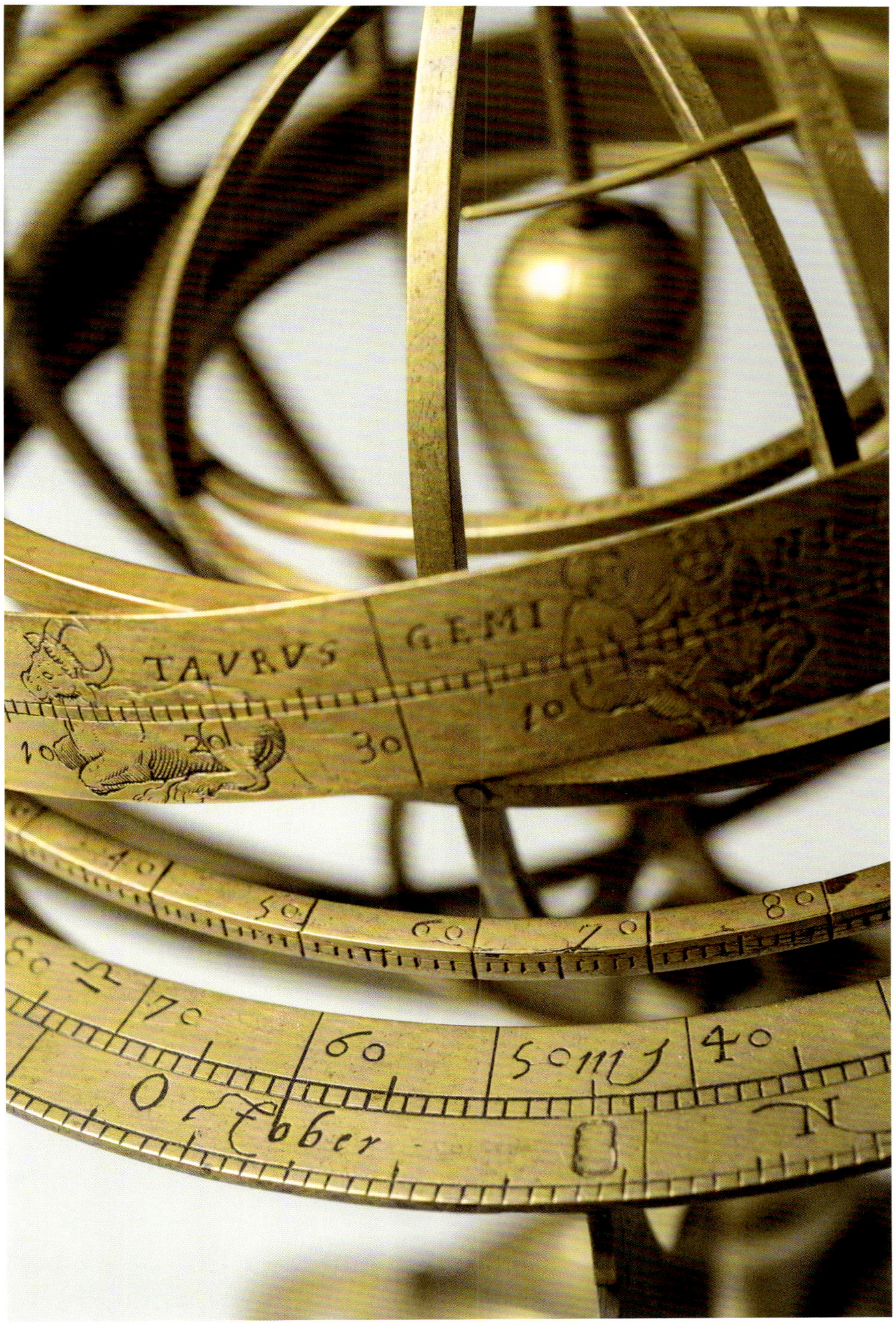

[cat. 67] detail

68

Gualterus Arsenius
Armillary Sphere

Leuven

1568

Signed on the pedestal: NEPOS GEMMAE FRISIJ LOVANIJ FECIT AN. 1568

Brass, height 27 cm, diameter 18.7 cm

Milan, Museo Poldi Pezzoli, inv. 3466

BIBLIOGRAPHY

Pirovano 1981: 22, no. 50, 62, fig. 66; Van Cleempoel, Van Damme, L'Estrange Turner et al. 1997: 159, no. E.1; Van Cleempoel 2002: 165-166, no. 38.

The instrument consists of two parts: (1) a fixed base with pedestal and horizon on which (2) the detachable sphere rests.[1] A circular base sits on a lathed wooden socle, with an inlaid compass in the middle that has the names of the winds engraved in a surround. Four intersecting, intertwined arms jointly carry a broad, flat horizontal ring representing the horizon. The horizon is engraved with the Zodiac and calendar details. A hole has been made at two points in which the vertical meridian ring is seen to rest. The assembly has been engineered in such a way that the sphere is able to tilt in the horizontal plane: the graduations on the meridian refer to the lines of latitude, so the sphere can be positioned at the width of the user and a picture is evoked of the position of the (imaginary) sphere.

The meridian ring holds the armillary sphere together. Between the polar circles, an axis is seen to run at a 23.5° angle in relation to the N-S axis, which in its middle has an engraved ball (globe) representing Earth, as described in Ptolemy's geocentric concept. The globe has engraved circles for the equator, the tropical circles, the poles and the meridians. Around the globe the Moon and Sun are seen to move in circular trajectories. Both of the latter are mounted on movable rings that in turn are connected to the central axis. The outer sphere is made up of two polar circles, two tropical circles and the equator in the middle, divided into in 360°. At a 23.5° angle in relation to the equator sits the *ecliptic*, or the apparent path traversed by the Sun each year. The broad band of the *ecliptic* is gracefully engraved with a central scale of 12 x 30°, specifying the name and symbol of each star sign. The *ecliptic*-band touches on the tropical circles on a *solstitium* or solstice: the longest day on 21 June and the longest night on 21 December. The equator is transected at a right angle by two rings: the *colures* that intersect at the Poles.

Between 1554 and 1579, Gualterus Arsenius was Leuven's most productive maker of scientific instruments. Over forty specimens have been preserved. His early instruments clearly evince of Mercator's influence, from whom he is believed to have learned the trade. The composition of the instruments, the construction and the engraving style are in keeping with those of Mercator's, who left for Duisburg in 1552. Arsenius stayed in Leuven and went on to work with Mercator and Gemma Frisius's network. As such, he built his first astrolabes for members of the Spanish court. He subsequently sold instruments through Christopher Plantin in Antwerp

1 On this topic, see inter alia Dekker 1999: 3–12; Kugel, Van Cleempoel & Sabrier 2002: 89–135.

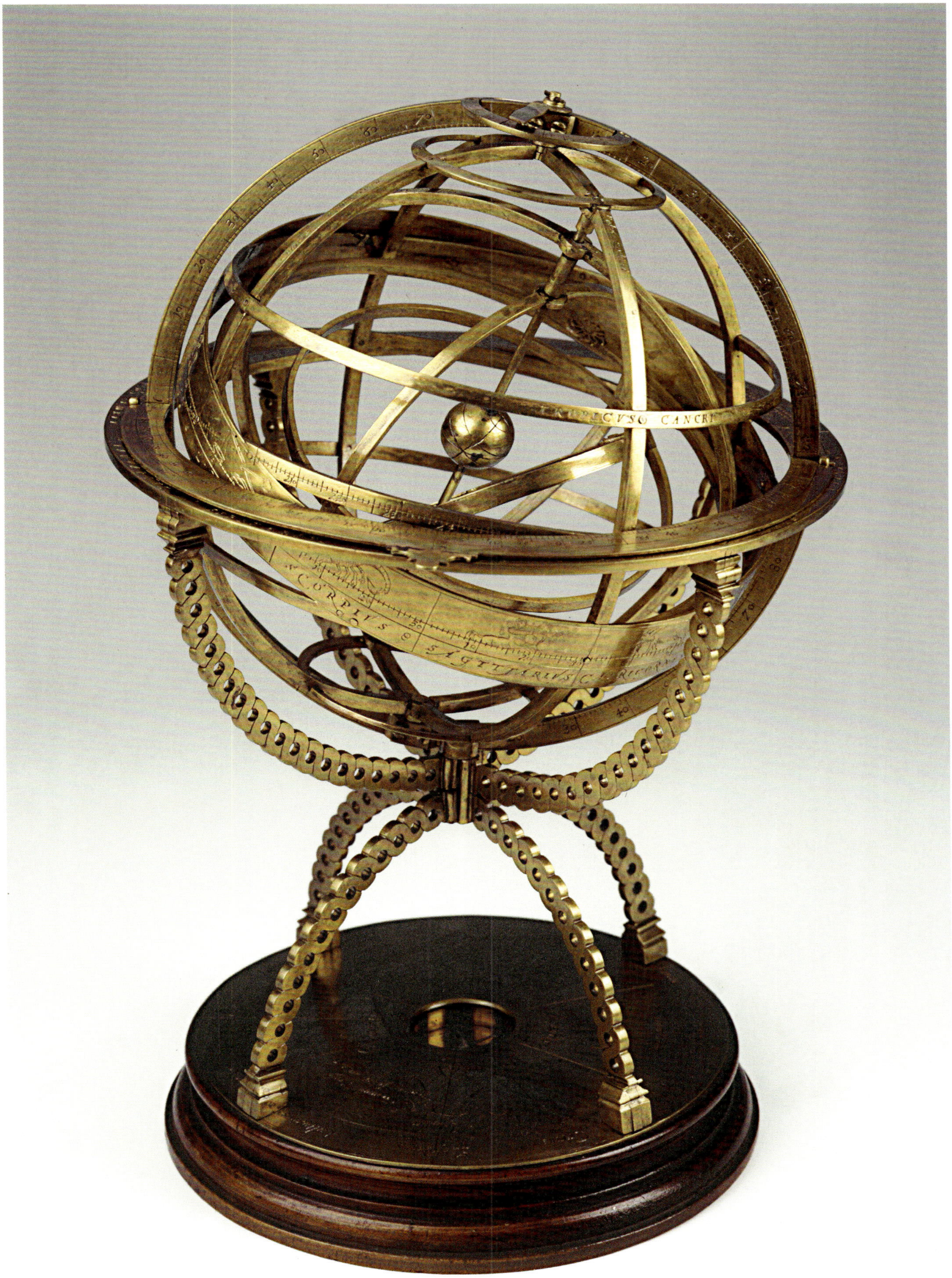
SAGITTARIVS

who also operated as an agent for Mercator. It is striking to see how Arsenius's signature invariably refers to Gemma Frisius, whose nephew (*nepos*) he may have been. For all that, there was likely to be a commercial reason, as the more famous Frisius had authored various treatises on scientific instruments that were known all across Europe by way of reprints. All of his instruments, including well over 25 astrolabes and five armillary spheres, stand out thanks to the highly refined construction and proportions, accurate scales and beautiful engraving style. Much like Mercator, Arsenius managed to arrive at a blend of the scientific and the artistic dimension in his instruments.

KVC

69

The Motion of the Universe

Flanders or Tournai
c.1490–1510
Tapestry, wool and silk, 415 x 800 cm
Toledo, Museo de tapices y textiles de Toledo (Catedral de Toledo: Colegio de Infantes)

PROVENANCE

According to a document of 1503, purchased by Diego López de Ayala at the *almoneda* (public sale) of the possessions of the Marquis of Priego.

BIBLIOGRAPHY

Florence 1980: 327, no. I. 1; Cortes Hernández 1982: 27-28, 30, 126-127; Revuelta Tubino 1987: 55-56, no. 87; Massing 1991a: 214-215, no. 111 (with previous literature); Cortes Hernández 2002: 80-107; Van Cleempoel 2006: 109; HALI 2009: 144.

This exceptional tapestry depicts the course of the cosmos as one imagined it around 1500 – that is, before the great astronomical discoveries of the sixteenth century. The central part is taken up by the celestial sphere in the form of an *astrolabium* (cat. 71, 72).[1] Its centre, and also the centre of the tapestry, is the Pole Star (*Polus arctic*[*us*]). Through this immobile star runs the axis of the cosmos, which is turned by an angel with a crank. The angel floats in a field of stars issuing from God the Father at the upper left. Thus it is his energy that powers the rotation of the universe. Near him we read: *Primu(m) movens* (Prime mover). Beneath the angel sits Atlas, who holds back the celestial sphere with his hands as a reminder of his cosmic capacity in antiquity. Also supporting the universe with her hands is a slender young woman to the right of God: she is the personification of 'Light movement' (*Agilitas mobilis*), a character that is evoked by the high-blown, open sleeves of her gown. Her counterpart on the other side of the sphere is *Astrologia*. This lady, who looks upwards as she raises her hand, also has a fluttering headdress and sleeve adornments as a sign of her 'enthusiasm'.

The celestial sphere is full of figuratively depicted constellations according to the schemes of Hyginus, Aratus and Manilius, who were known for their illuminated manuscripts in the late Middle Ages.

To the right of the sphere, several authorities gather around a majestic, rather haughty woman in an Eastern turban on a glittering throne: *Philosophia*. She is the true master of world knowledge. At her feet sit Geometry and Mathematics; at her right stands *Abrachis*, the Arabic bastardisation of *Hipparchus*. This was a famous astronomer from the second century before Christ. It was through Arabic scholars that part of the ancient knowledge of the world was transmitted to the medieval West. But scientists were not the only ones to acquire knowledge of the world – poets did so through intuition; therefore Virgil stands at the upper right. The text above reads: 'Hipparchus knew these (aspects) of astrology, of which the poet Virgil speaks, through philosophy and science. Others possess this knowledge

1 See also the essay by Koenraad Van Cleempoel, 'In Search of Utopian Perfection: Builders of Scientific Instruments – the "Leuven School"' in this publication (p. 331).

through geometry and mathematics, and their number is growing'. Scientific and poetic methods were not mutually exclusive: astrology belonged with astronomy (cf. cat. 77).

Originally, this tapestry was one of a series called *Astrología*, to which an *Allegory of Death* also belonged. The subject of the third tapestry is unknown. It was apparently an ambitious concept that situated life and death in a cosmological perspective.

PVDB

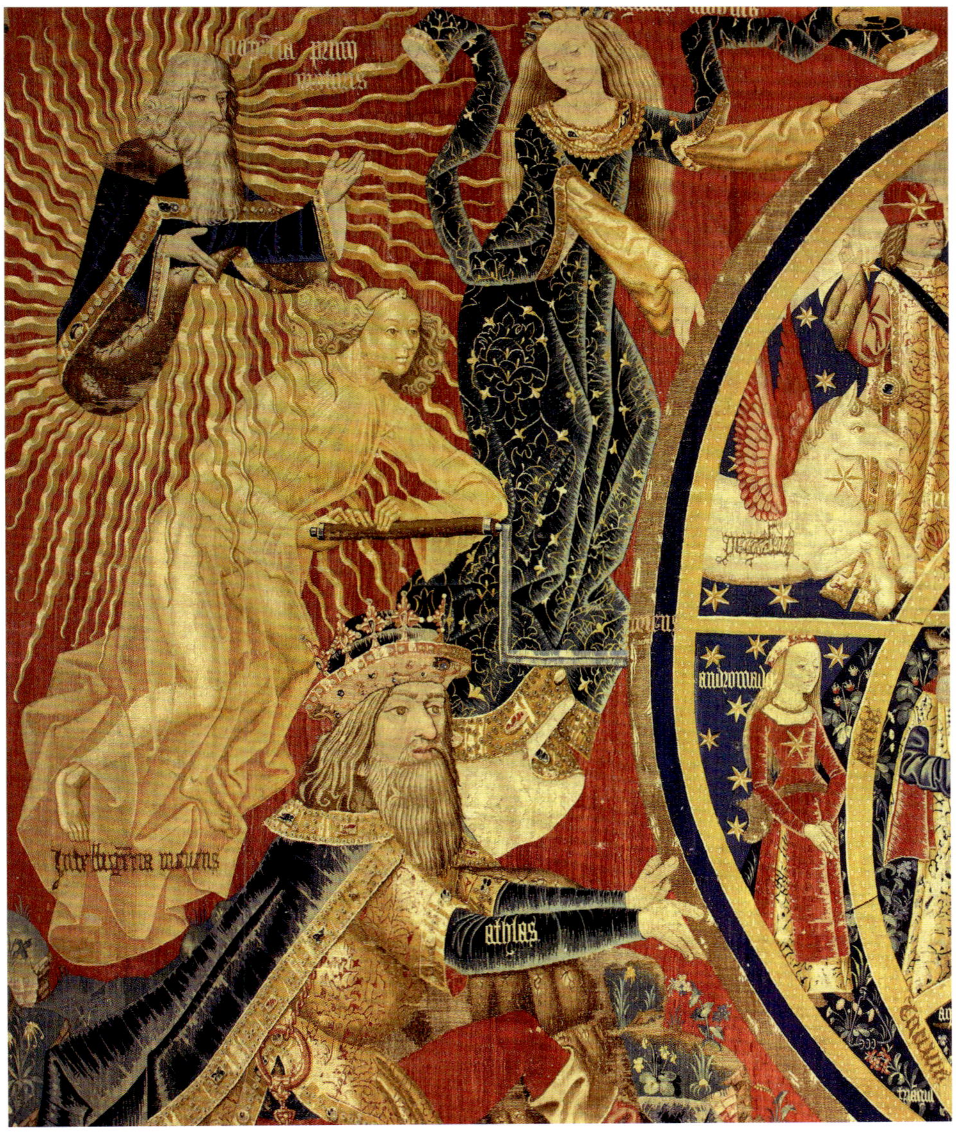

[cat. 69] detail

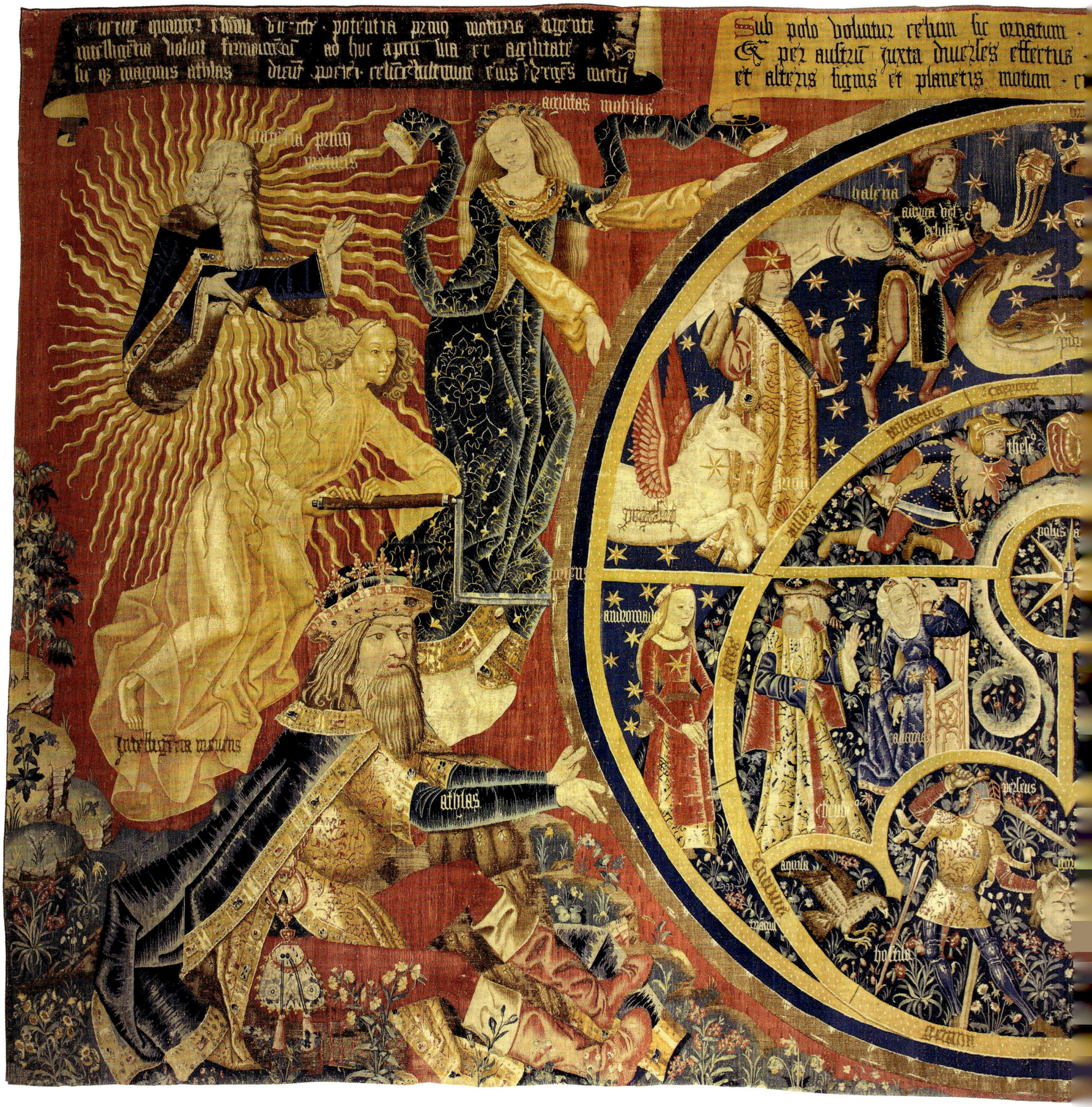
Sub polo volutur celum sic ornatum ·
Et per auctum juxta diversos effectus ·
et alternis signis et planetis motum ·
agilitas mobilis
intelligentia movens
athlas

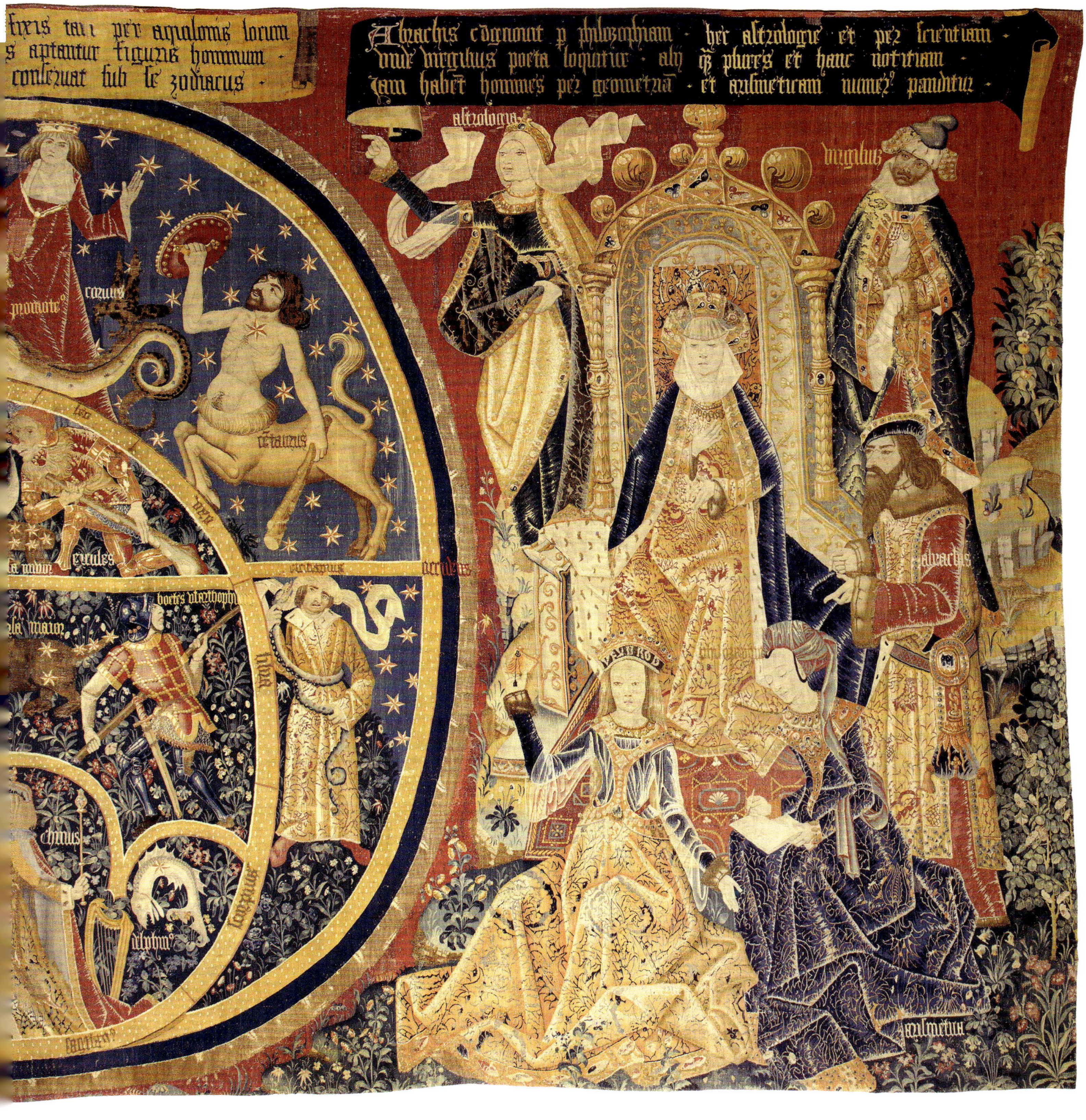

70

Master Bartolomé
Chaos

Spain
Late fifteenth century
Oil on panel, 154.2 x 111.1 cm
Tucson (AR), The University of Arizona Museum of Art, Gift of the Samuel H. Kress Foundation, inv. 1961.013.030

PROVENANCE

Made at the behest of Catedral de Ciudad Rodrigo (Spain) and kept there from 1488 to 1879; Jose Fallola, Madrid; probably sold by John Charles Robinson (1824-1913); Sir Francis Cock, 1st Bart. (1817-1901), Doughty House, Surrey, England, in 1882; by inheritance to Frederick Lucas Cook, 2nd Bart. (1844-1920), Doughty House; by inheritance to Sir Herbert Frederick Cook, 3rd Bart. (1907-1978), Doughty House; in 1952 possibly with Thomas Agnew & Sons, London; M. Knoedler & Co., London; sold to Samuel H. Kress (1863-1955) on 28 January 1954; given to the University of Arizona Museum of Art in 1961.

BIBLIOGRAPHY

London 1920: no. 16; London 1948: no. 6; Eisler 1977: 162-177, fig. 152-177; Williamsen 1994: 128, fig. 7; Silva Maroto 2004: 289; Silva Maroto 2006: 435-446; Roglán 2008: 52-67; Dotseth, Anderson & Roglán 2008: no. 25.

This panel belongs to an enormous, complex retable from Ciudad Rodrigo (Extremadura, Spain). It was painted by Fernando Gallego and several assistants (among them Meester Bartoloméo) and consisted of at least 26 panels. The retable depicted scenes from the history of salvation, from before Creation to the Last Judgment. With its extraordinary iconography, this painting attempts to visualise the unimaginable powers of the universe.

In the very centre of the panel one sees a disc in different shades of red with the inscription *hille* (=υλη, *hyle*, the Greek name for primordial matter). A polished marble or onyx plaque evokes the unimaginable but magnificent versatility of the creative forces; these forces were 'feminine', like that of the cosmic *matrix*.[1] Around the 'primordial space/matter' extend the circles of the nine angelic choirs, each with its own colour, and above them the *mandorla* of God, who holds a white globe in his hand as a symbol of the universe. A smaller *mandorla* with a bearded man hovers below; he is probably Chronos (Time), depicted with the gears of a clock as an image of the temporality of the cosmos. An aniconic dappled green space below shows what preceded the order of creation and beginning of time: CAHOS (Chaos).

Hyle is the classical Greek name for primordial matter, formless but bursting with energy, 'begging for form'. In the *Cosmographia* of the twelfth-century thinker Bernardus Silvestris, *Hyle* is 'the oldest manifestation of Nature, the inexhaustible matrix, the primordial ground of every form that exists … the foundation of substance. Her competence knows no bounds or limits and extends from the first core to the magnificent fields of growth, so wide as the totality of all that can be

1
Wetherbee 1973.

2
Ibid.

3
Already in the late antique period, some (such as Pseudo-Dionysius in the fifth century), were aware that with *hyle*, the artist could follow two paths. These paths follow two aesthetic principles: the 'mimetic' (*mimetikos*) and the 'form-creating' (*plattomenos*). The second was by far to be preferred from a religious and mystical perspective. Figuration, or mimetic form, is deceptive. 'God' (or whatever one wants to call it) is not a 'being', but a 'super-being'. Striving after an iconic 'likeness' is therefore a priori pointless. 'Irrational' and 'displaced' 'figures', however, bring our spirits closer to the mystery. They create a *figura infigurabilis*, an 'un-depictable guise', a *forma informis*, a 'formless form'. Therefore they refer to a fundamental 'otherness', that opens an unknown path of beauty – and ideally of the Sublime – for human beings. Cf. Didi-Huberman 1995.

4
Leclerc 1972.

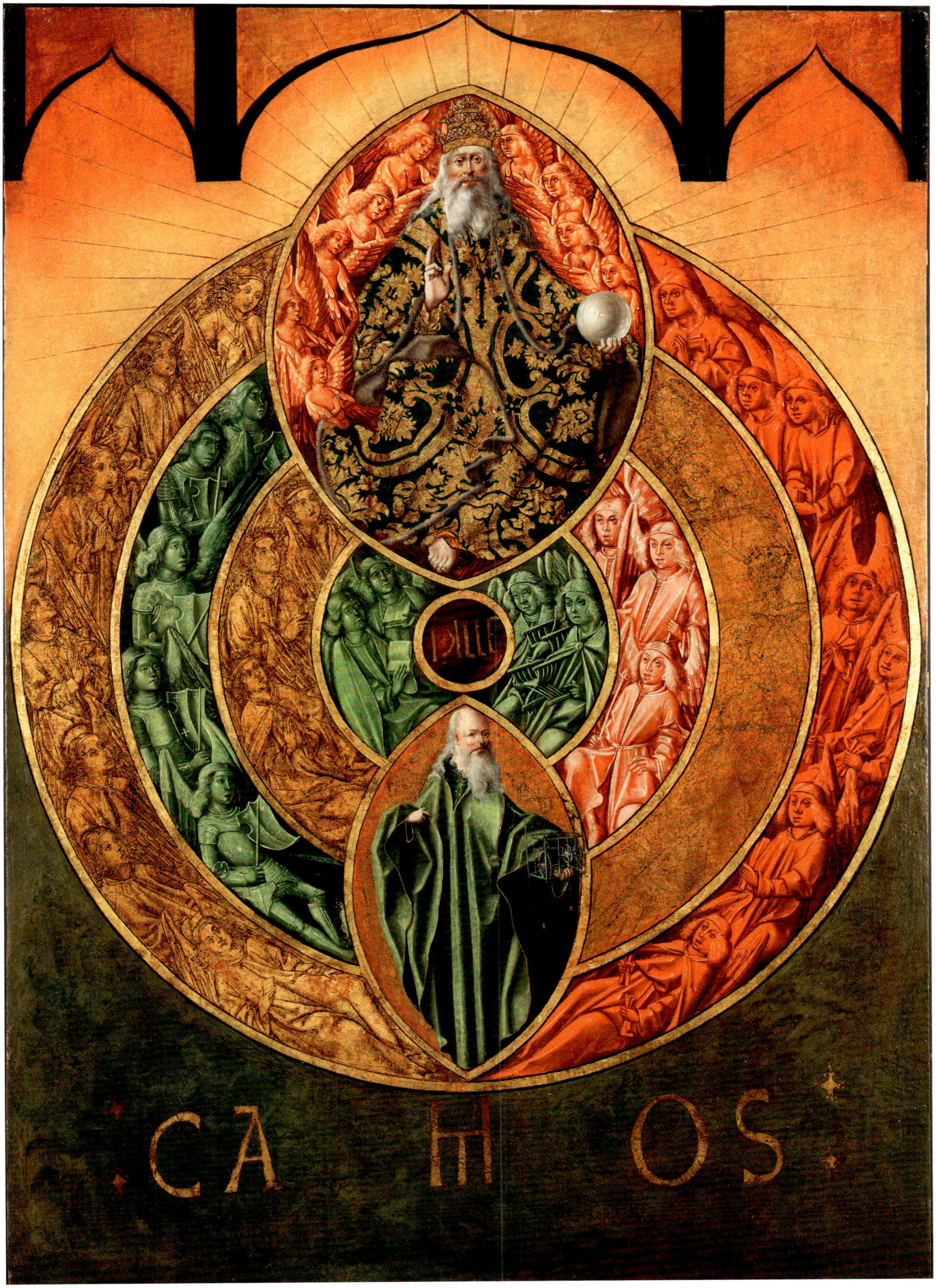

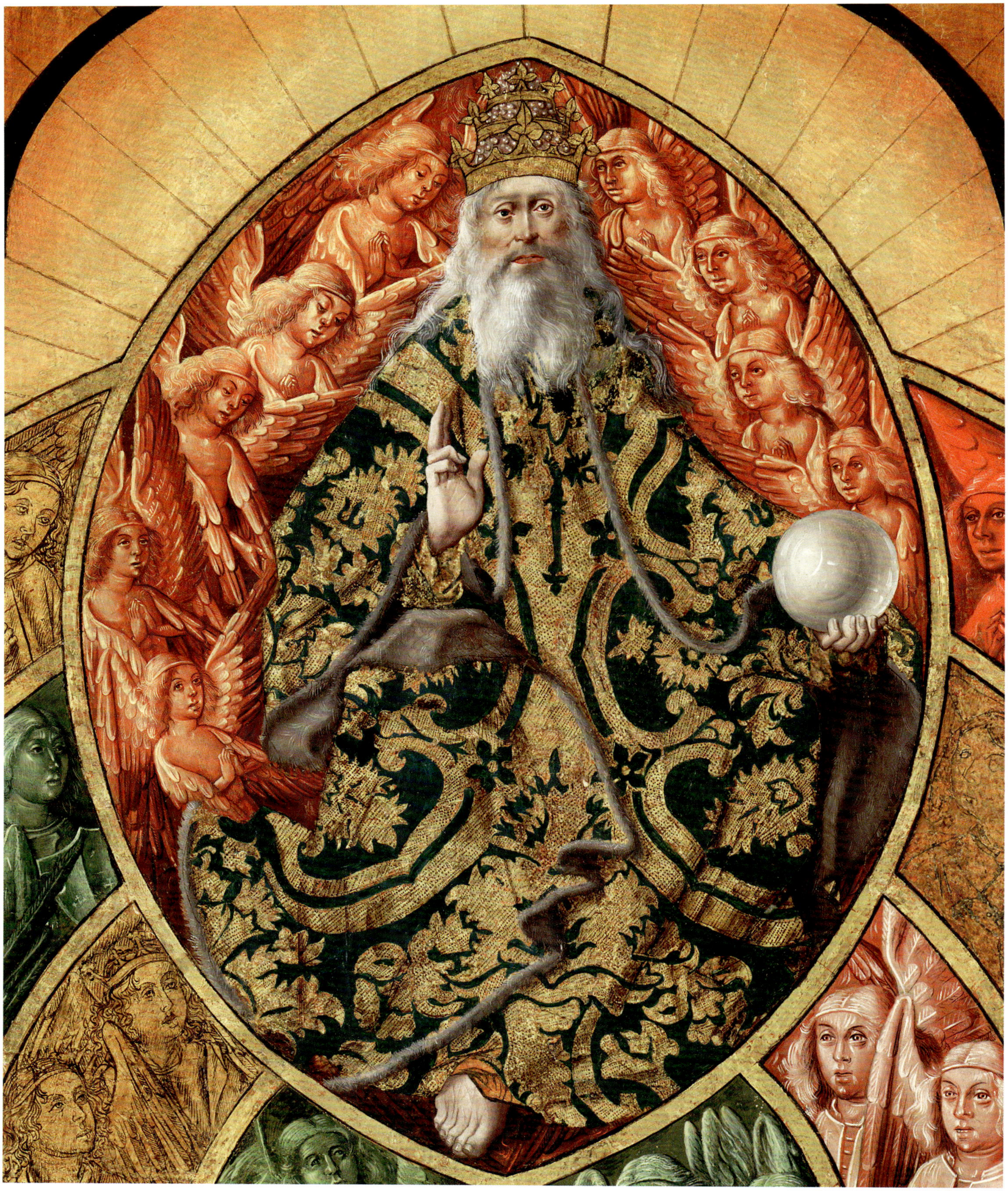

[cat. 70] detail

created. Given that she is permeated by the most diverse and inextinguishable urges … she often inclines towards confusion, because she is beleaguered in countless ways by all that exists.'[2] In medieval Latin, *hyle* was also called *silva* (wood/forest). *Silva* was the undifferentiated stage of matter, 'spun round in a single flux and hurled back in her primordial chaos'. Procreation from *silva* and its differentiation in the forms of the elements took place before Time, in a transitional zone between the divine and creation. 'This non-space is an interstice between the material and the spiritual. *Silva* is a wilful, formless form, a torn clumping, a motley apparition of being, contradictory in herself. In her turbulence she longs for a tempering force … begging to escape from her ancient confusion she seeks the creation of form … and it is in this cradle that the new-born universe crows …'[3]

The painting of Master Bartolomé tries to depict a literal version of utopia: HYLE and CHAOS are non-places (*ou-topoi*) out of which everything in time and space is created.[4]

PVDB

71

Gerard Mercator
Astrolabe

Leuven
c.1545
Monogram: GMR [= Gerardus Mercator Rupelmondanus]
Brass, diameter 27.8 cm
Brno, Moravská Galerie, inv. 24-385

BIBLIOGRAPHY

L'Estrange Turner & Dekker 1994: 181; L'Estrange Turner 1994a: 331, 339-341, 345, 351; L'Estrange Turner 1994b; L'Estrange Turner & Dekker 1995: 198 fig. 1, 201 fig. 4, 207 fig. 6; Van Cleempoel, Van Damme, L'Estrange Turner et al. 1997: 93, 119, 173-174 no. A2; Padmos & Vanpaemel 2000: 229 no. 122 and 273; Van Cleempoel 2002: 75-76 no. 1 and plate p. 245.

Gerard Mercator (1512–1594) built the most beautiful and accurate scientific instruments of his time. He worked in Leuven up until 1552 before moving to Duisburg, where he continued his work. He built this astrolabe in Leuven, presumably around 1545.

Mercator is the world's most prominent and influential designer and engraver of maps and globes (cat. 63). Nowadays, his instruments are less well known. Yet they are every bit as innovative, precise and elegant as his cartographic work. Charles V owned several of his instruments, built at the behest of Nicholas Perrenot de Granvelle.[1] Another link that connected Mercator to the Habsburg nobility was Gemma Frisius (cat. 59), personal physician to Charles V, professor of medicine at the old University of Leuven and designer of globes and scientific instruments (he signed as *medicus ac mathematicus*).[2] This is explained by the fact that a due understanding of the celestial bodies was considered an important skill for physicians, as people believed astronomy and medicine to be connected. Mercator was Gemma's most gifted pupil, whose exceptional intuition enabled him to combine mathematics and cosmographic expertise with an extremely refined level of mechanical skill. Together with Gasper van der Heyden and Gemma, he built the first globe to properly showcase his engraving skills in 1536 (fig. 71.1). This in turn engendered a Leuven school of instrument makers, who built Europa's most highly coveted scientific instruments.

The surround of this astrolabe bears the monogram GMR: Gerardus Mercator Rupelmondanus. Up until 1992, the monogram was still unidentified. In 1991, Mercator's first instrument was 'discovered' at the Museo Galileo in Florence and shortly afterwards a second specimen turned up in Augsburg.[3] The attribution was made based on Mercator's distinctive engraving style and a miscellany of astronomical information. For instance, Mercator was among the first to establish the positions of the stars based on the Copernican theories, thereby moving away from the theories first propounded by Ptolemy in Antiquity. Part of a set of three Mercator astrolabes, this specimen from Brno is unique in that it is the only astrolabe of Mercator's to have been built in Leuven; the other two are presumed to have been built in Duisburg after 1552. In spite of the fact that this type of instrument had been known for some 1,500 years before Mercator's time, he managed to add a few elements

1 See Walter Ghim, *Vita Mercatoris* (1595) in Osley 1969: 186.

2 Hallyn 2008: 23–53.

3 Turner 1994b.

4 *Literarum latinarum, quas Italicas, cursoriasque vocant, scribendarum ratio*, Leuven 1540.

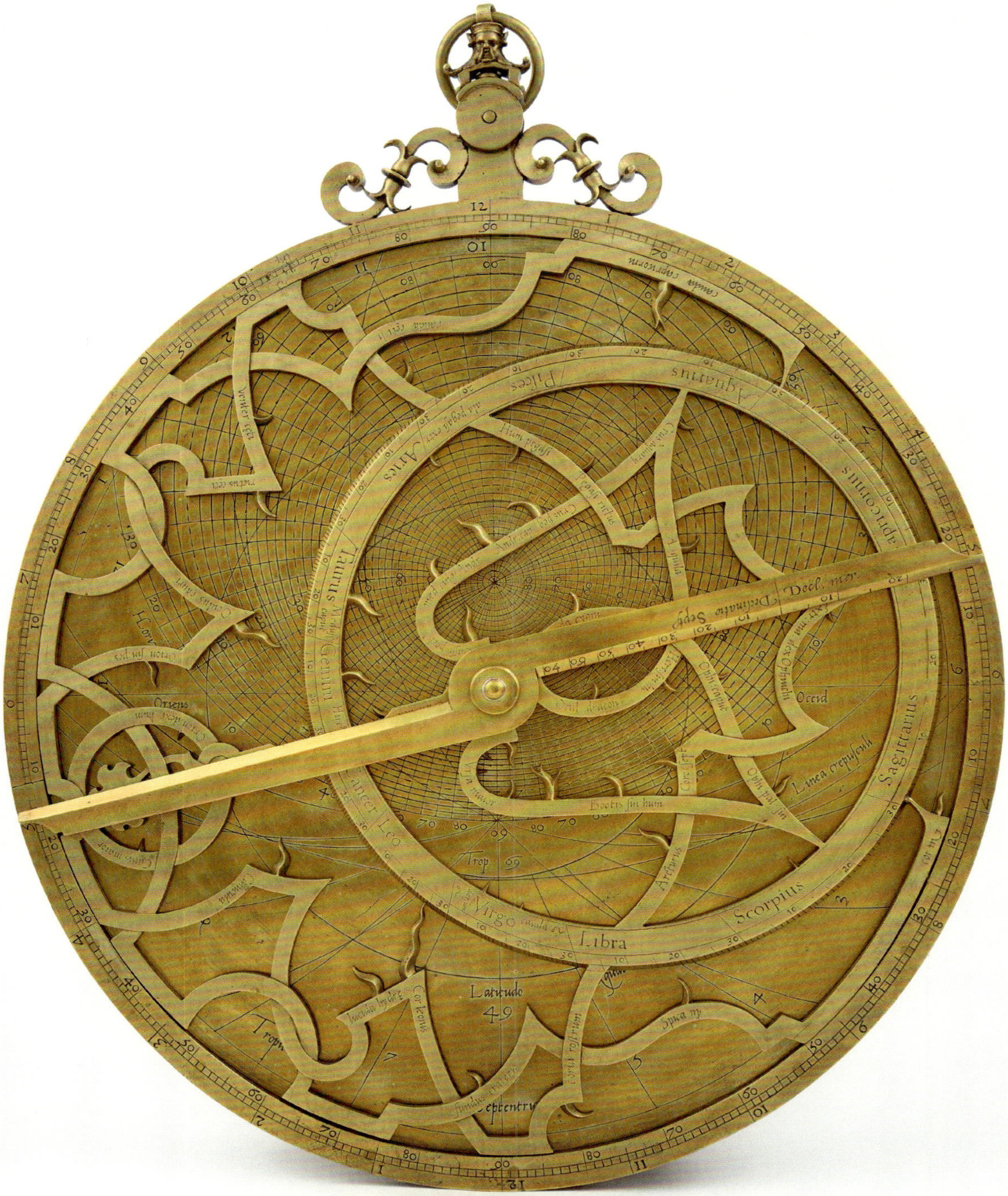

[cat. 71] Front with the stars and the Zodiac on the latticework *(rete)* above a stereographic projection for latitude 49

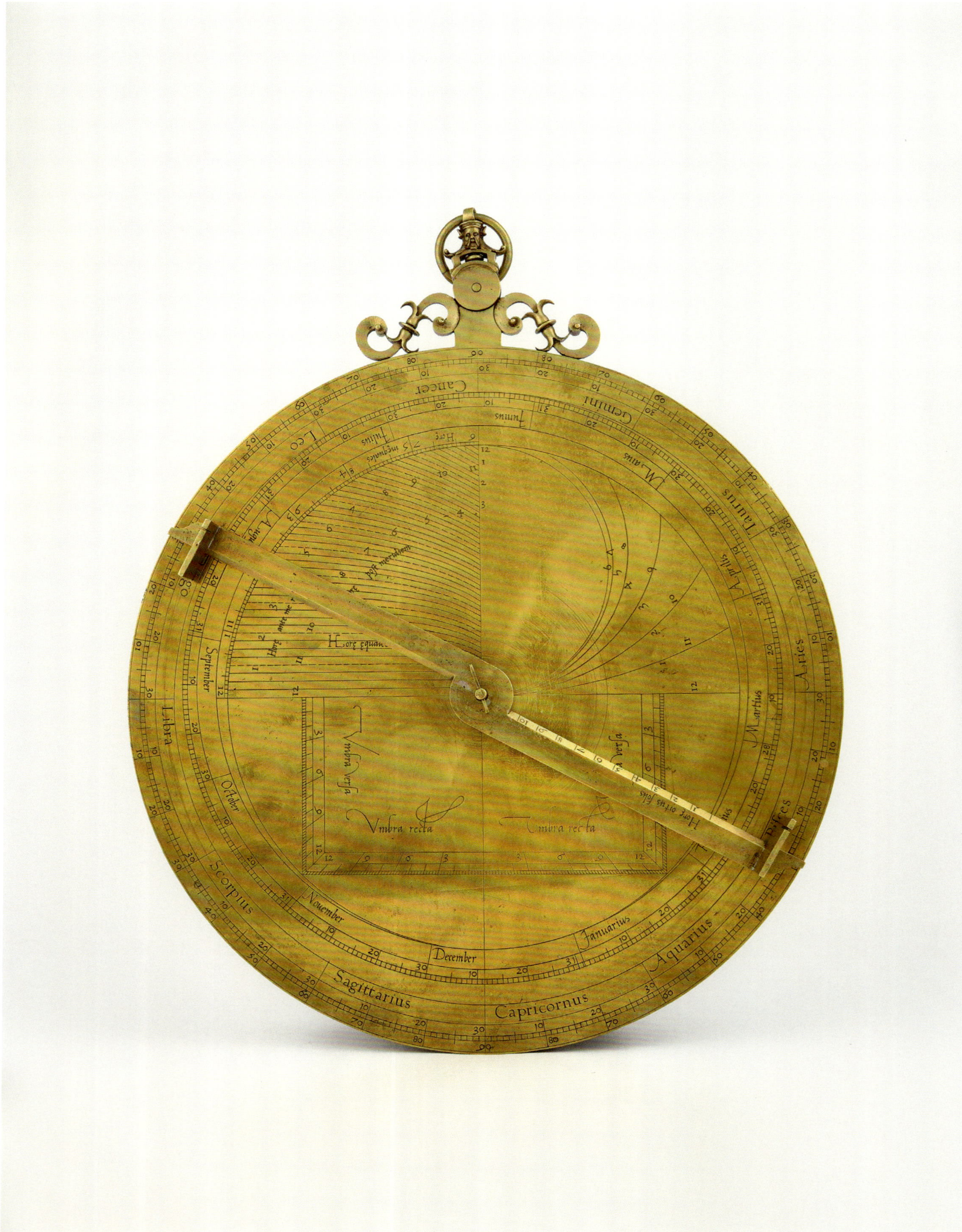

[cat. 71] Rear *(dorsum)* with scales for the Zodiac, calendar in the surround, above: the even and uneven hours and at the bottom: the 'shadow quadrant' for angle measurements.

that are highly typical of 'Leuven' astrolabes: (1) the diameter has been upsized to just under 30 cm, whereas previously this usually just one third of that; (2) the symmetrical latticework of the rete is not only highly aesthetic, it also provides room to accommodate 40 stars; (3) the engraving style is typified by the use of italics, as described by Mercator in his equally influential handbook on how to properly apply Italic script.[4]

This is a unique opportunity to see this early yet influential astrolabe along with other 'Leuven' instruments: the armillary spheres (cat. 65–68) and the astrolabe of Zeelst (cat. 72). All of which bears witness to the highly harmonious balance between art and science.

KVC

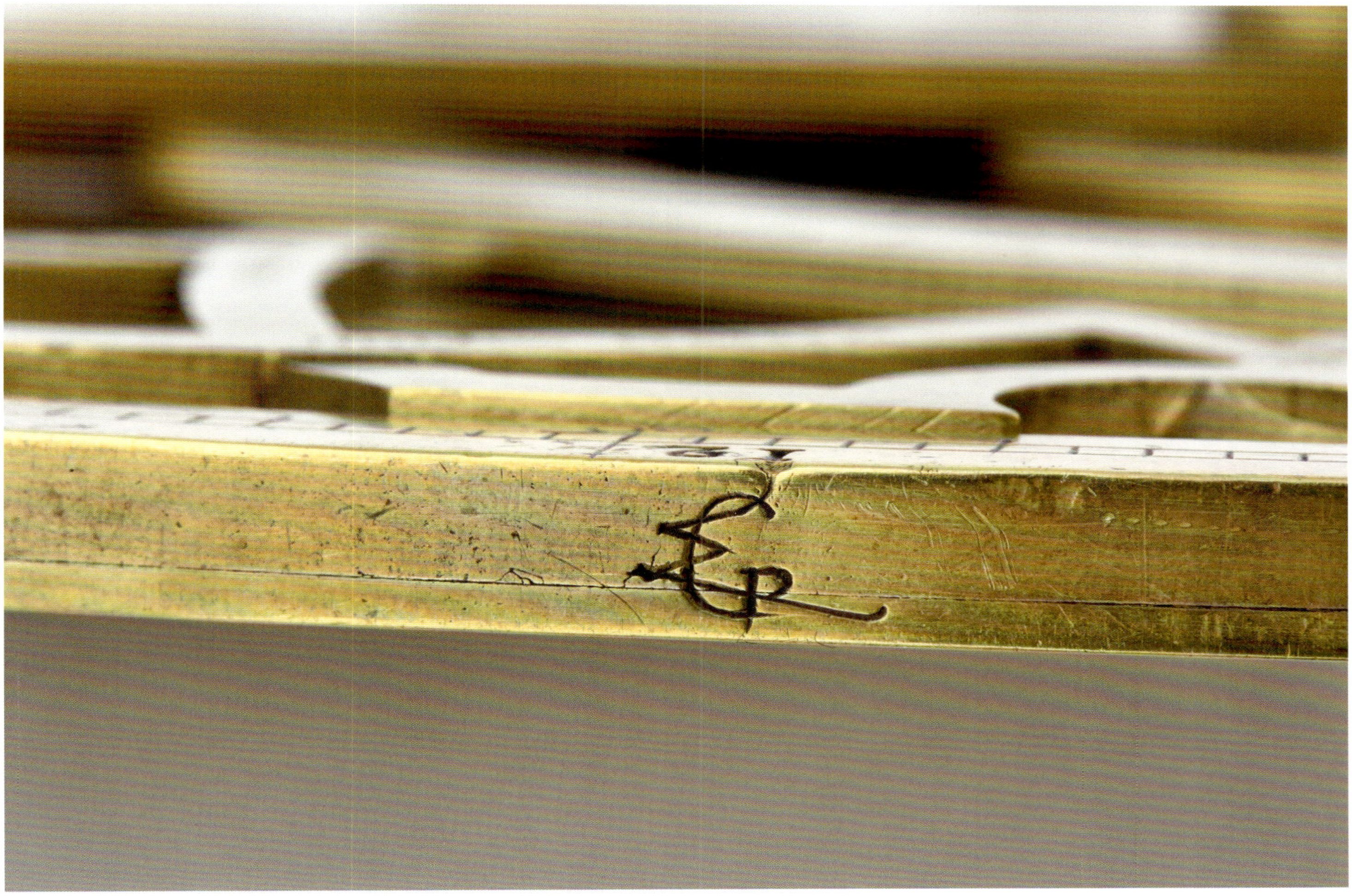

[cat. 71] Monogram: Gerardus Mercator Rupelmondanus

72

Adrianus Zeelst
Astrolabe

Leuven

1569

Signed: ADRIANUS ZEELST FACIEBAT 1569

Brass, diameter 34.2 cm

Madrid, Museo Arqueológico Nacional, inv. 52080

BIBLIOGRAPHY

García Franco 1945: 205-209, no. 8; Van Cleempoel, Van Damme, L'Estrange Turner et al. 1997: no. A.1; Van Cleempoel 2002: 209-213, no. 71.

Little is known about the life and times of Adrianus Zeelst (*c.*1530–1620). He was attached to Leuven University as a mathematician, where – among other things – he was consulted on the calendar reform when the switch was made from the Julian to the Gregorian calendar around 1580. His reputation was such that the calendar reform was postponed by one year because Rome was eager to hear his views. Around 1600, he was a member of the scientific household of Prince-Bishop Ernest of Bavaria in Liège. In 1602, whilst working in Liège, he published a treatise on astrolabes, co-authored with Gerard Stempel.[1]

The scientific instruments of Stempel's that have been preserved are of a very high standard and are in keeping with Mercator's body of work, in terms of both style and substance. Setting out from several instruments of Mercator's, which have since been lost and which we only know from descriptions, we are still able to get some idea of what they must have been like, because Zeelst built them as well, albeit one generation later.

This astrolabe is unique by dint of the fact that it is still on its original pedestal, consisting of a circular base with two arabesque arms that hold up the instrument at its centre.[2] In front the beholder's eye is caught straight away by the elegant latticework of the *rete*, with 48 star pointers for 36 star names (Ursa Major is made up of seven stars). The surround shows a combined 360° scale of 360°, for degrees (4 x 0–90°) along one end, and for the hours (2 x I–XII) along the other. The interior of the *mater* is completely engraved with a *quadratum Nauticum*, invented by Gemma Frisius and Mercator, which can also been seen in various Leuven-built astrolabes. The purpose of the nautical quadrant was to make it easy to plot one's course on board seafaring vessels. At the bottom, the construction is signed: *Adrianus Zeelst faciebat 1569*.[3]

This astrolabe displays the five characteristic elements of a 'Leuven' Renaissance astrolabe: (1) *rete* with symmetrical latticework; (2) a nautical quadrant on the interior; (3) one face of a plate with a full set of horizons and (4) a universal stereographic projection on the reverse; (5) a wide diameter of *c.*30 cm.

KVC

1 Gerard Stempel, or Stempelius (who worked from c.1580 to 1610), was a priest in Cologne and an electoral mathematician. Around 1580, he engraved copies of a number of prints by Dürer. In 1598, he drew a map of the valley of the river Meuse between Maastricht and Liège. He built celestial globes and conducted research into observation techniques for celestial bodies. In 1602, along with Adriaan Zeelst, he wrote a treatise on the astrolabe.

2 On this topic, see Gunther 1976; Van Cleempoel 2006; Sezgin & Neubauer 2010.

3 Inside the *mater* sit five plates, four of which are engraved with stereographic projections for lines of latitude: 39°/41°, 43°/45°, 47°/49° and 51°/52°. On one face, the fifth plate displays the Zodiac with a calendar in the surround; in the top half sits a conversion diagram for the even/uneven hours, whilst the bottom half holds a shadow quadrant to measure height and/or depth. The other face is completely engraved with stereographic projections of the horizon between 0–90°. The reverse features a universal stereographic projection (or *saphae*) with ten stars. The surround is divided into 360°.

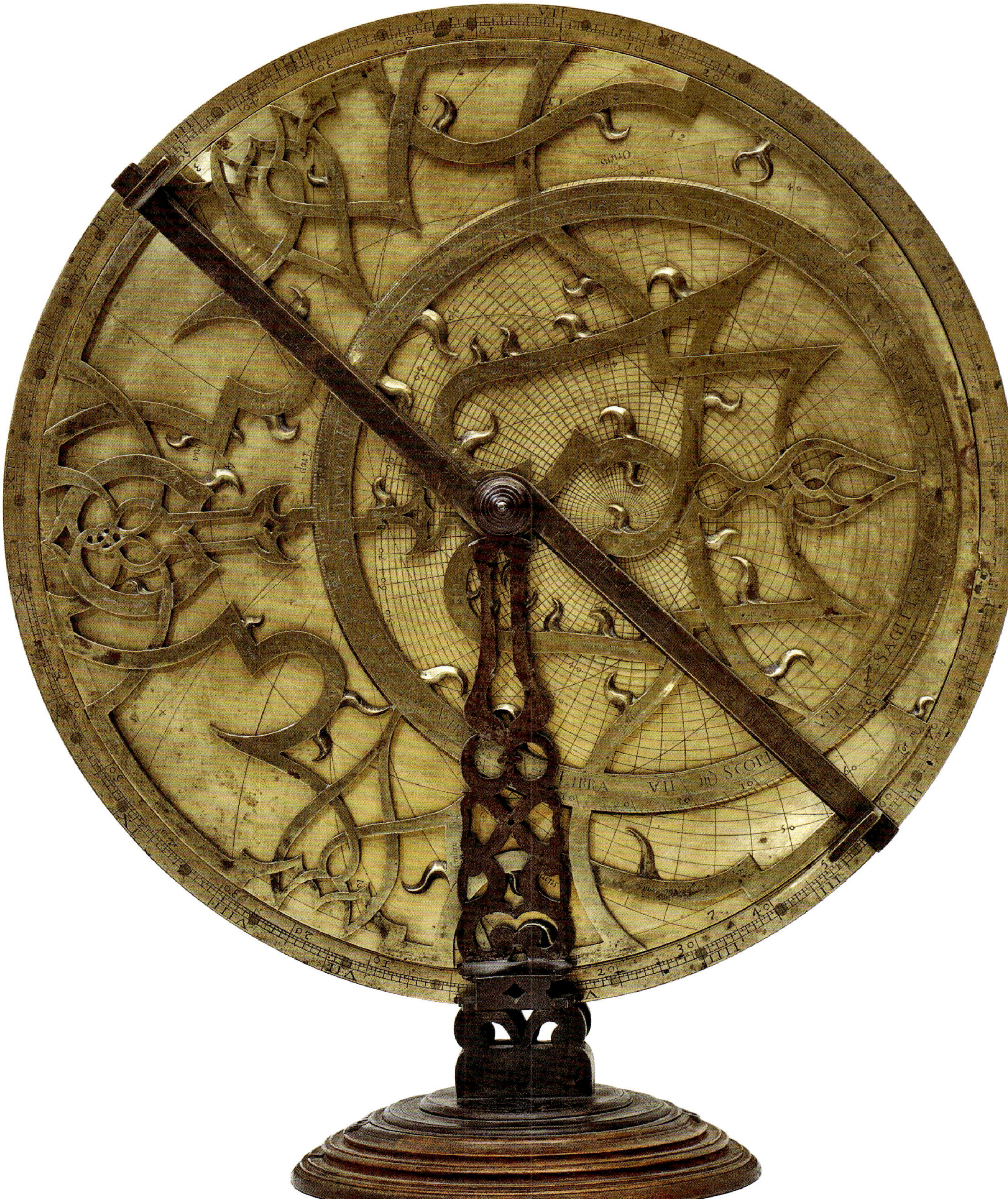

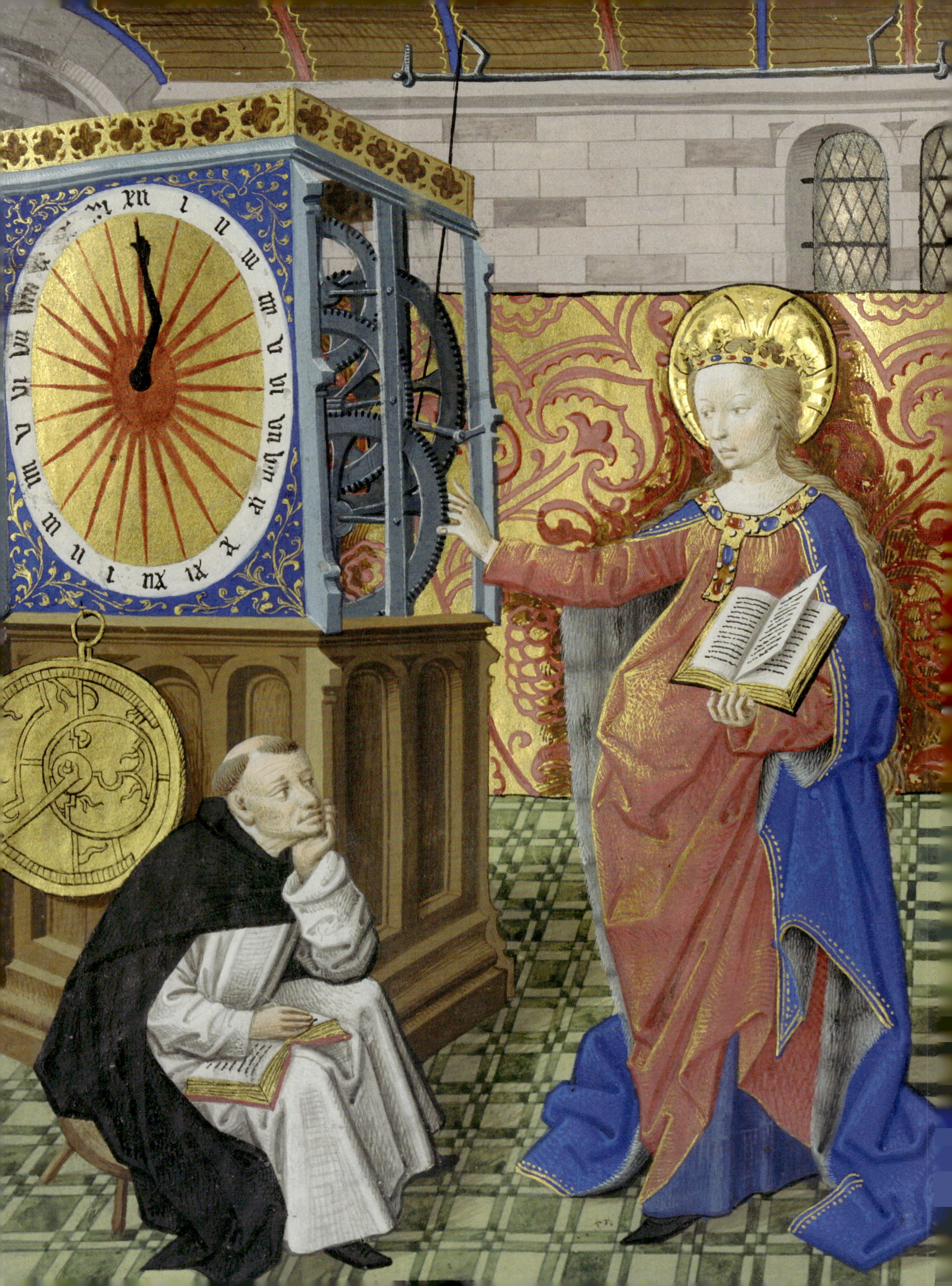

Eternity: Imaging Time

JULIE BECKERS

[cat. 76] detail

Eternity signifies an indeterminably long period of time. An everlasting continuum, which religion has continuously claimed as a promise; the righteous and devout can expect their reward in a blissful afterlife in the presence of an eternal and, hopefully, forgiving God, others will be less fortunate and are headed for certain damnation.

Time, a subjective experience incarnated by the flow of sand through the hourglass, was thus best spent anticipating what would come. For the early modern viewer, time was not so much a loss as a movement towards the fulfilment of Salvation history. Time had a beginning and an ending, a purpose and a plan, and was spent in a specific non-numerical space.

Early modern dwellers were guided by the liturgical year, the order of Saint's days, religious feast days, offices prescribed by the Church, and seen in popularised visual representations such as breviaries or books of hours (cat. 73). The circle, as a geometric shape, was a major way of representing and dictating cyclical time and was applied to the year, as well as to the larger span of human life. This same shape was therefore used as a functional object on which the months of the year, the seasons, the zodiac and other key annual events were visualised (cat. 74) employing picturesque images such as workers ploughing the land; agricultural or trade images not unlike the scenes one would imagine after reading More's account of his fictional island society, where simplicity was celebrated as the highest virtue in an idealised landscape.

By the fifteenth century, and alongside the rise of humanism, time - or rather its measurement - changed. By exploiting knowledge inherited by the Arabs through Muslim Spain, determining the exact time of day or night became more straightforward: all that was needed was to look at the sun, stars and moon. This development triggered the naissance of a number of complicated yet elegantly designed astronomical instruments, which calculated the position of stars and spherical frameworks in the sky, among other things (cat. 65-68, 70-71).

The renewed Ptolemaic interest in astronomy combined an attentiveness to its wider scientific grounding with a more local knowledge, leading to a certain zest for superstition: early modern life adapted to science, couples committed to marriage solely during a particular moon cycle, land could only be worked at given times, animals slaughtered at appropriate moments. Knowledge of the stars created a society led by Claudius Ptolemy's ancient expertise, while his muse, Astronomia, was celebrated by scribes and peasants alike (cat. 77).

Equally, the scientific method introduced a mechanism that led to the appearance of the mechanical clock (cat. 74-76). This introduction ensured a life more timely managed and reasserted the authority of the ever-present Church; by installing a clock on every bell tower of every square, the Church remained the centre of all activities and, by extension, the arbiter of time.

Metaphorically and visually, the developing concept of time and eternity is perhaps best captured by the Dominican friar Henry Suso, who as early as the fourteenth century devised a book he named 'The Little Book of Eternal Wisdom' (*Büchlein der ewige Weisheit*), which he later translated to Latin and aptly named 'The Clock of Wisdom' (*Horlogium Sapientiae*, cat. 76).

73

Simon Bening
The Months January, March, May, July, September and November

In: *Hennessy Book of Hours,* Use of Rome, Bruges, c.1530–1535,
fol. 1v–2, 3v–4, 5v–6, 7v–8, 9v–10, 11v–12

Tempera, gold colour and gold leaf on parchment, 117 x 87 mm

Brussels, Royal Library of Belgium, inv. Ms. II 158

PROVENANCE

Damião de Góis (?); Queen Catherine of Portugal (?); in 1852 in the collection of Pierre d'Hennessy, Ostend; acquired from his widow by Charles Ruelens, curator of the Belgian Royal Library, Brussels, in 1874.

BIBLIOGRAPHY

Destrée 1896; Leidinger 1912: 15-20; Gaspard 1943; Lyna 1989: 424-425, no. 259; Kren & Rathofer 1988; Testa 2000: 107-124; Kren & McKendrick 2003: 467-468; Watteeuw 2006: 75-86; Watteeuw 2009: 599-611.

The *Hennessy Book of Hours* is one of the most outstanding Bruges manuscripts from the first half of the sixteenth century. The miniatures are a cynosure in the body of work left by Simon Bening (1483/1484–1561) and, as such, the *ne plus ultra* of Flemish miniature art. Bening's works are eulogised as *eccellenti* by Vasari (1568) as well as Guicciardini (1588).[1]

In the early sixteenth century, illuminated books of hours were considered appealing showpieces for private devotion. The Bruges workshop of the Bening family produced this type of gem for the most prestigious of families in Europe. Bening's services were commissioned by the Habsburg court, among others, and various Flemish and Brabant families of note and distinction. Portuguese humanist and art collector Damião de Góis, or Damiaan van der Goes (1502–1574), also purchased illuminated manuscripts from Bening. Góis served as the go-to liaison for various Spanish and Portuguese noble families in the Low Countries. As a collector who owned works by Quinten Metsys and Hieronymus Bosch as part of his own collection, de Góis may have been the party who commissioned the *Hennessy Book of Hours*. It is through his ministrations that the book is believed to have ended up in the possession of Catharine of Portugal.[2]

The calendar miniatures depict the activities of city and countryside during the various months of the year. For centuries, the calendar and its related depictions were considered the benchmark for the seasons, months and days of the month. The perpetual calendar (*calendarium perpetuum*) was a collection of tables that listed the exact day of the week for any given date, past or future. As books of hours involved a hefty investment and were meant to last a lifetime, they invariably came with a calendar in the beginning, sometimes involving twelve depictions of the activities of the months. The months are linked to the signs of the Zodiac. Each calendar month comes with a full-page miniature (left) in which the depiction continues illusionistically into the depictions in the margin (right). The month of November,

1 Guicciardini 1588: 151; Vasari 1981: 181; Kren & McKendrick 2003: 468.

2 Kren & McKendrick 2003: 468.

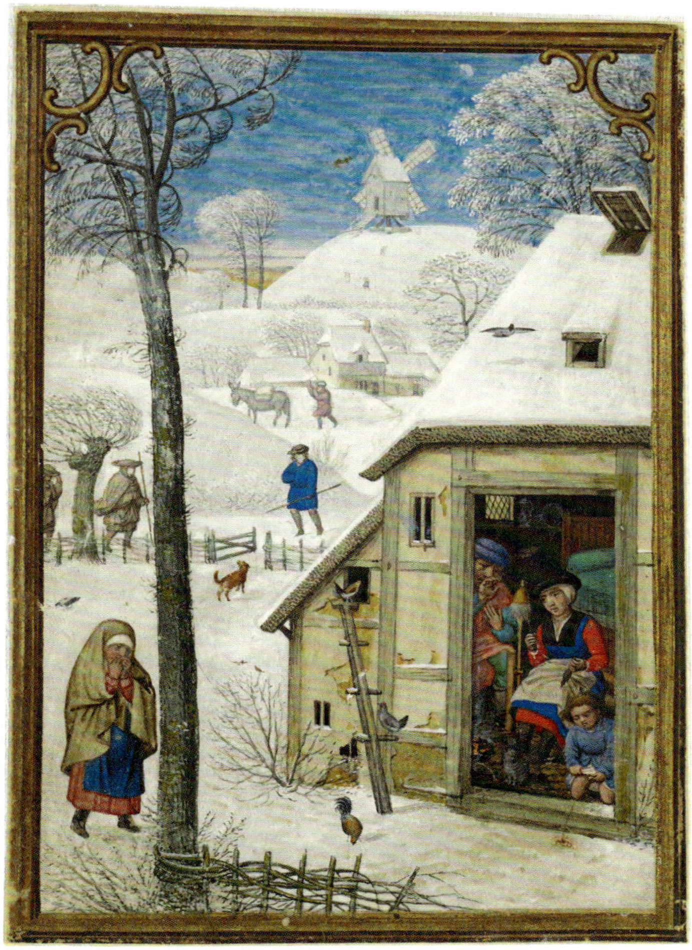

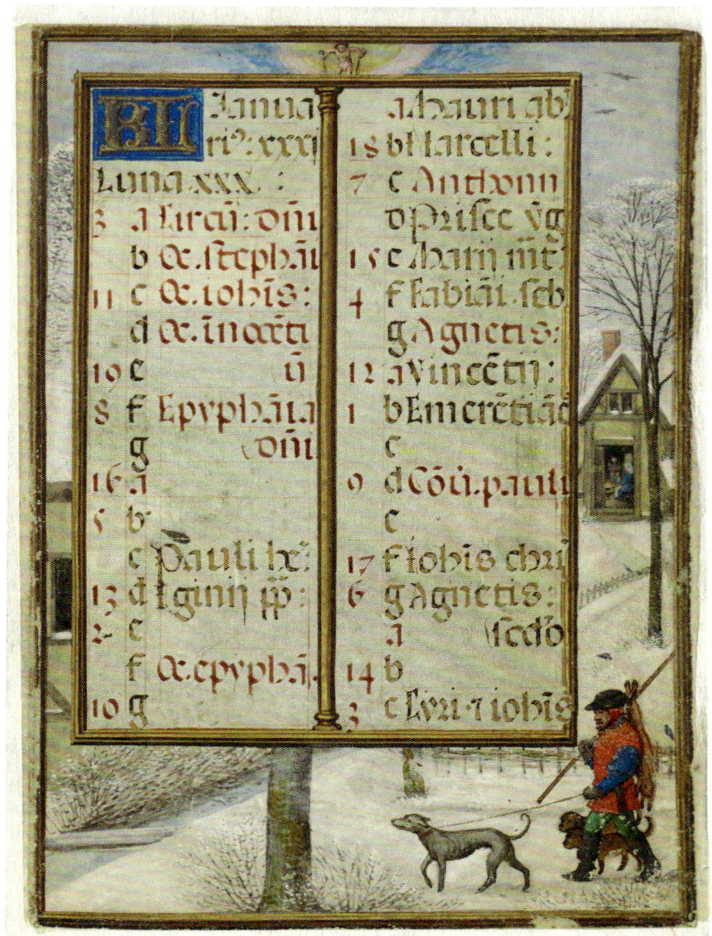

[cat. 73] fol. 1v-2 *January*

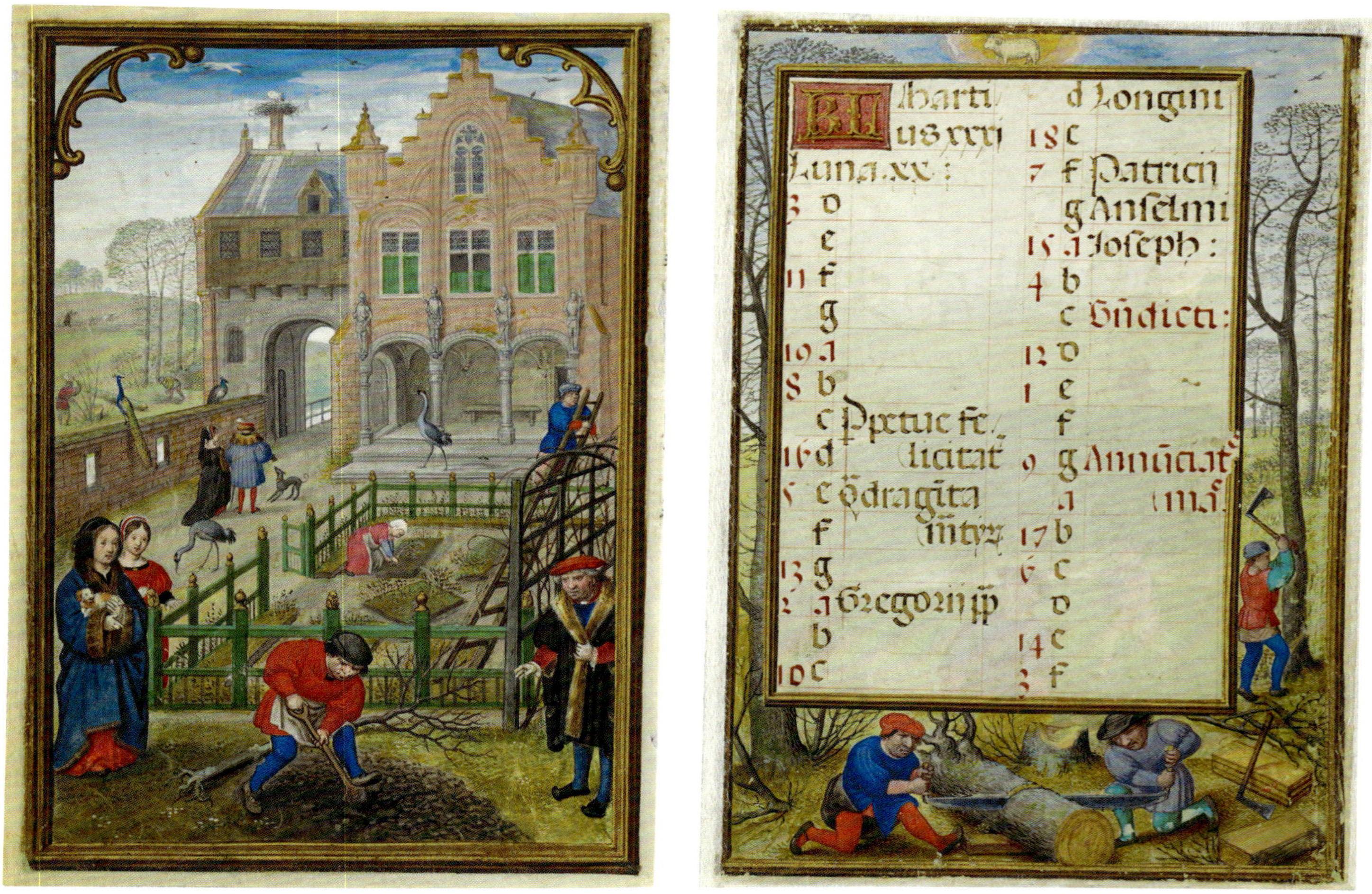

[cat. 73] fol. 3v-4 *March*

for instance, depicts archery, with a target range close to a walled city (the ramparts of Bruges?). A minute rendition of the Zodiac sign of Sagittarius can be seen at the top of the calendar list.

The images are the crowning achievement of the art of book illumination in the Netherlands. They are based on a centuries-old tradition: after all, there is no disputing the link between the miniatures in the *Très Riches Heures du Duc de Berry* (Chantilly, Musée Condé, Ms. 95) from the early fifteenth century (*c.*1410–1416), the calendar miniatures in the *Breviarium Grimani* (*c.*1517–1518) and the *Hennessy Book of Hours,* shown here. Yet by the same token Simon Bening was also an innovator. More than anyone, he reveals himself as a direct precursor of the art of Pieter Bruegel the Elder. An entire generation before Bruegel, he created deeply atmospheric landscapes populated by peasants and burghers. In the full-page miniature for the month of January, for instance – the snowy landscape with the windmill and the hunter – the colour of the freezing light of winter shimmers and sparkles to such an extent that, both through its theme and its palette, the sheet seems to herald Bruegel's winter landscapes of the second half of the sixteenth century.

LW

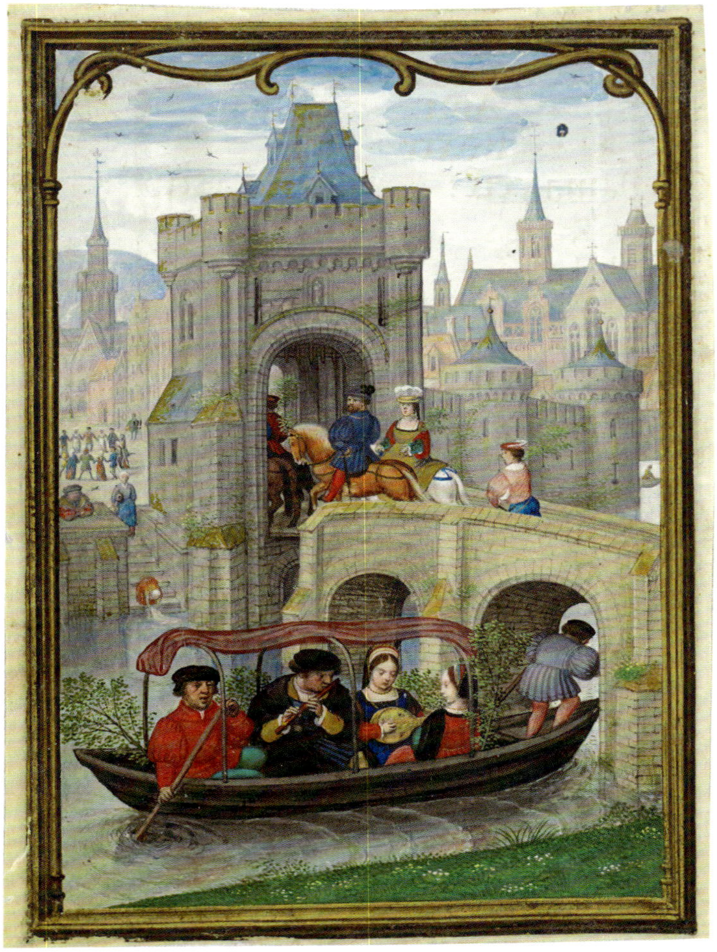

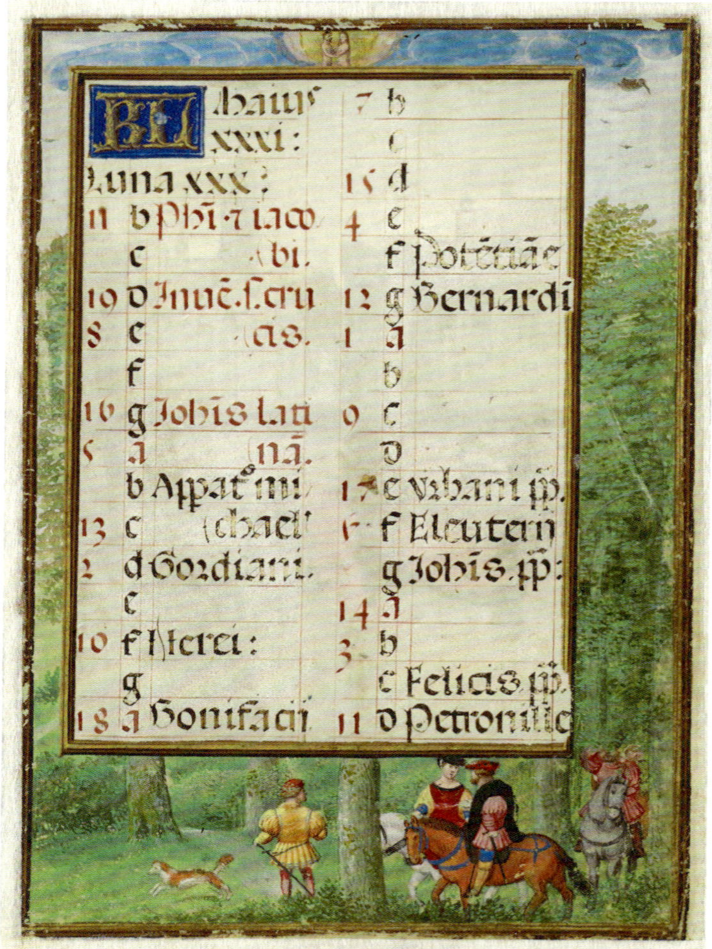

[cat. 73] fol. 5v–6 May

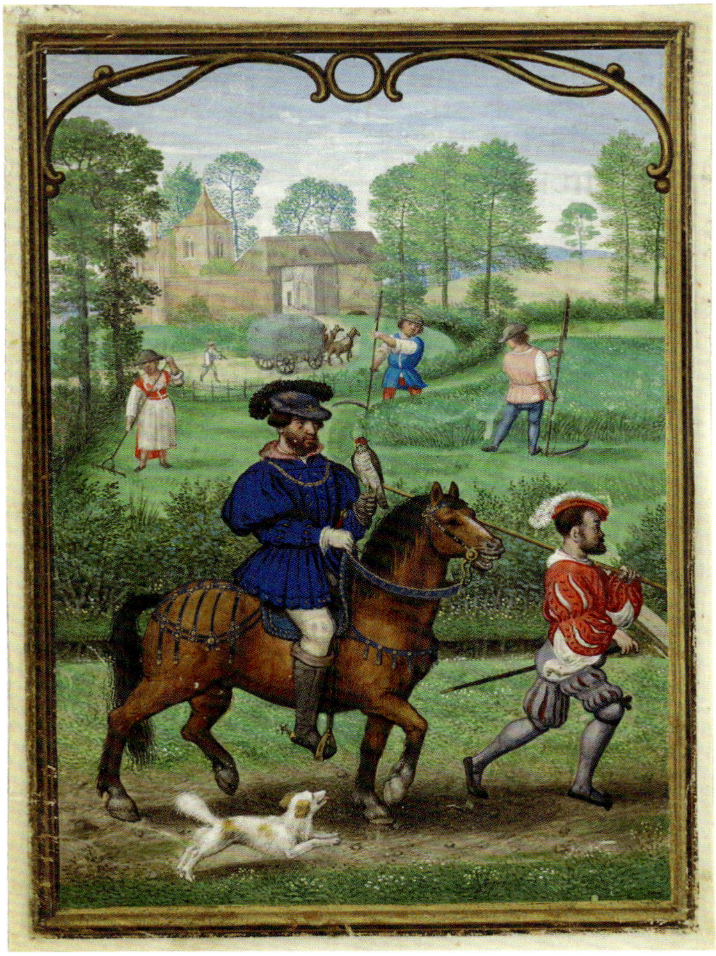

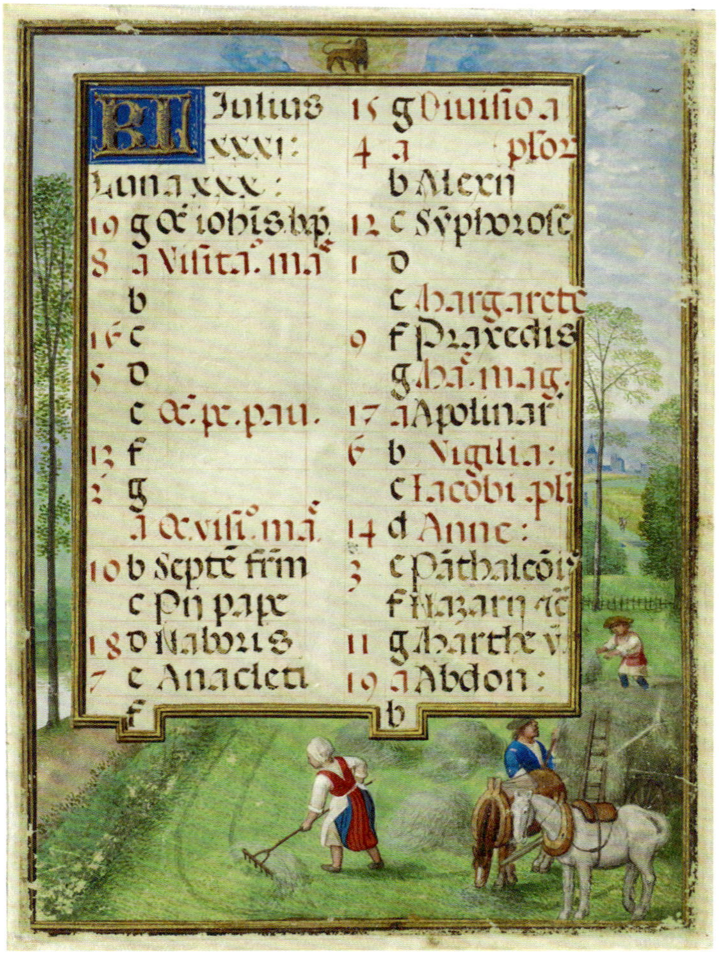

[cat. 73] fol. 7v-8 *July*

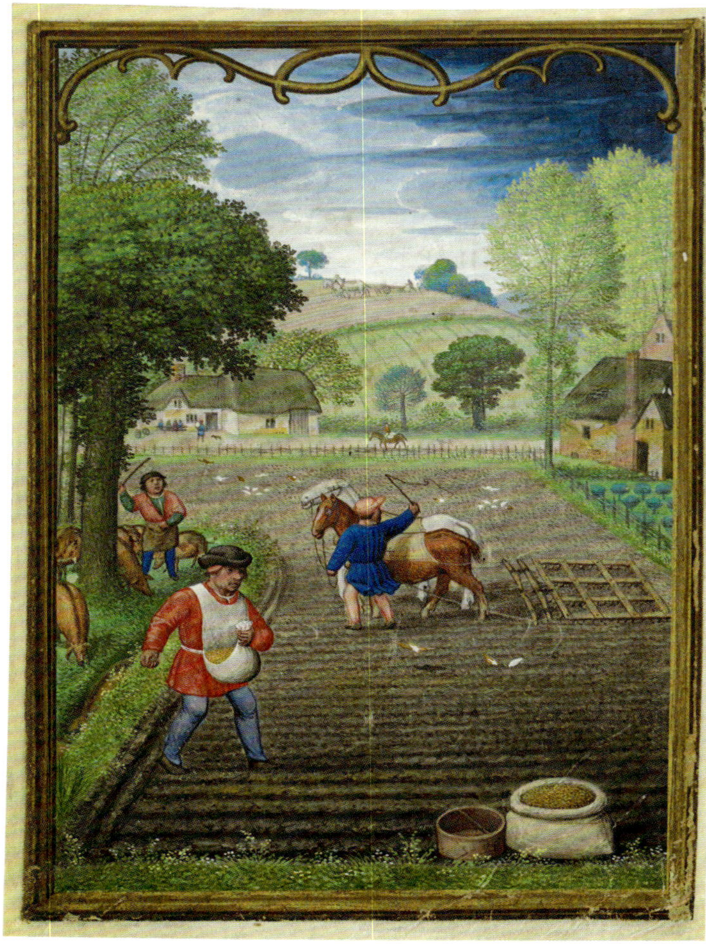

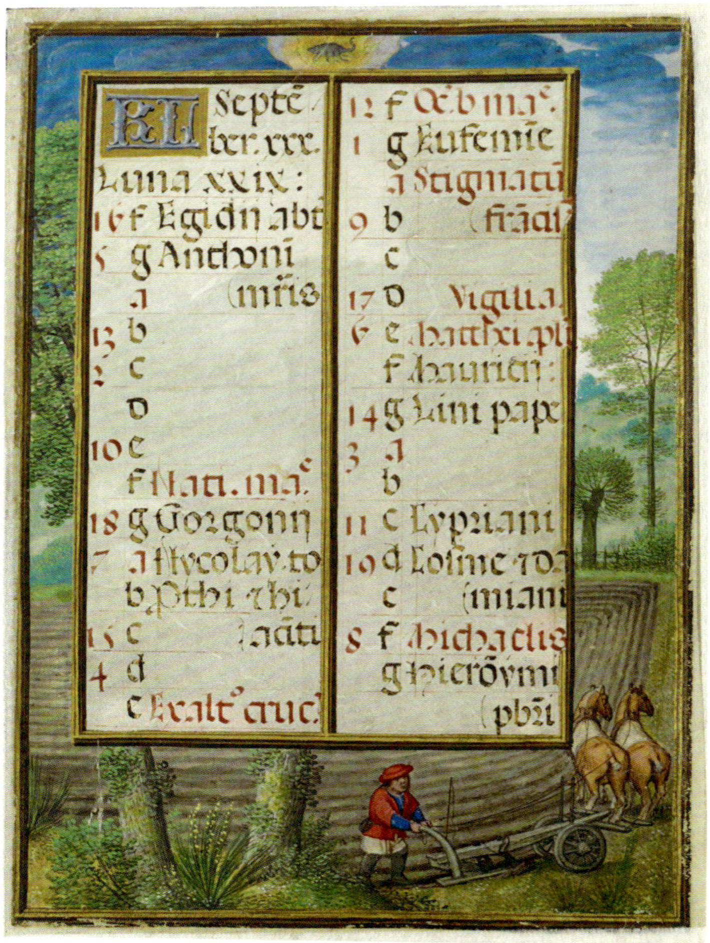

[cat. 73] fol. 9v-10 *September*

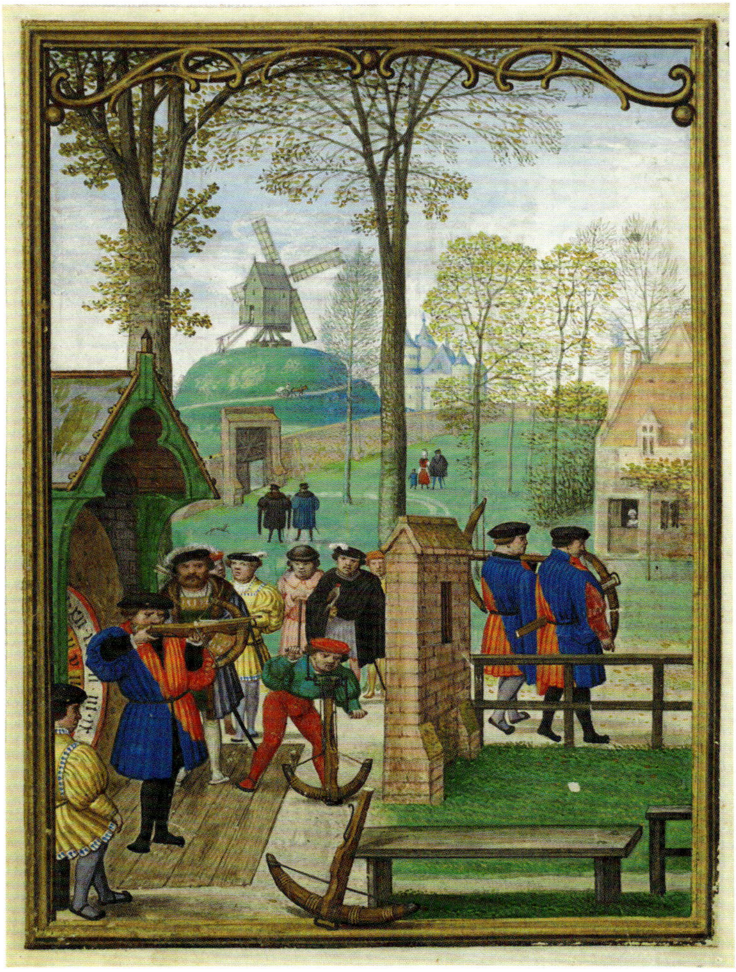

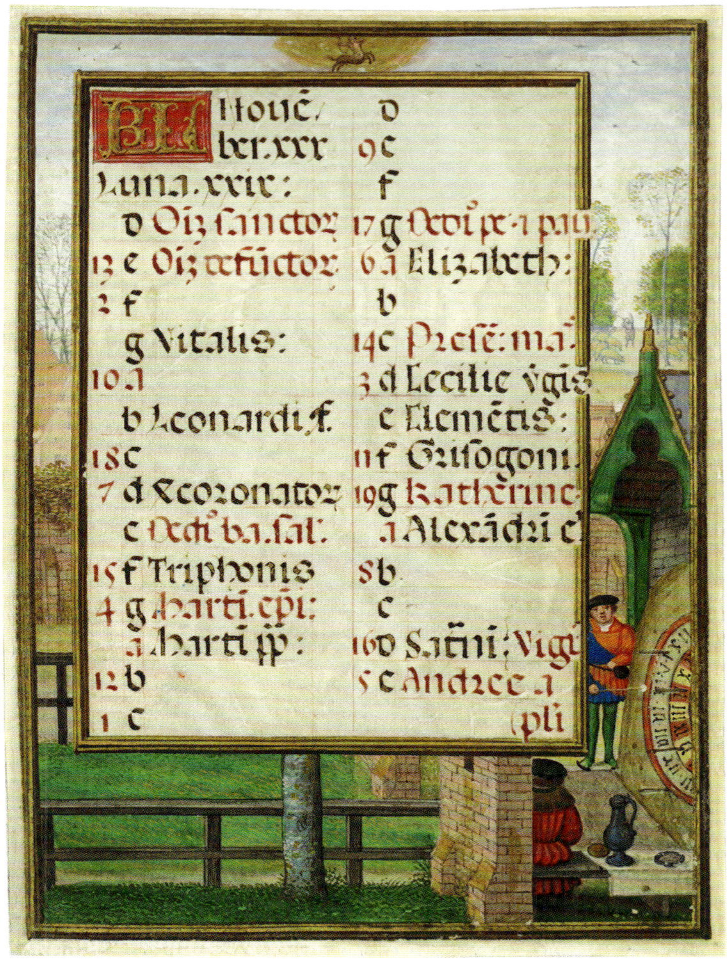

Nouē/ber.xxx
Luna.xxix:
d Oīs sanctor
13 e Oīs defūctor
2 f
g Vitalis:
10 a
b Leonardi.f.
18 c
7 d Coronator
e Dedi ba. sal.
15 f Triphonis
4 g Marti. epi:
a Marti pp:
12 b
1 c
d
9 e
f
17 g Dedi pe. & pau
6 a Elizabeth:
b
14 c Presē: ma.
3 d Cecilie vgis
e Clemētis:
11 f Grisogoni
19 g Katherine
a Alexādri e
8 b
c
16 d Satini: Vigi
5 e Andree a
pli

[cat. 73] fol. 11v-12 *November*

74

Time and Calendar Dial

Leuven (?)
c. 1500,
Tempera and oil on panel, 125.6 x 123.4 cm
Leuven, Museum M, inv. S/4/0

PROVENANCE

From the sixteenth up until the late eighteenth century: in Leuven, in unknown possession; in the nineteenth century in the possession of the Van Brussel family in Leuven; in 1890 sold to Edward van Even; in 1908 in possession of Victor Demunter; in 1938 bequeathed in last will and testament to the City Museum Vanderkelen-Mertens, Leuven.

BIBLIOGRAPHY

Leuven 1961: no. 17; Leuven 1962: 200-201, add/3; Van Uytven 1964: 60-61, no. 82; Roegiers 1969: no. 21; Crab et al. 1971: 220-222; Decavele 1976: 35-36, no. 5; Crab & Smeyers 1979: no. XVIII/8; Van der Stock 1991: no. 308; Smeyers 1993: 83-124; Vandendriessche 1993: 43-80; Smeyers 1997: 1-60; Madou 1997: 61-68; Vandendriessche & Smeyers 1998: 191-192, no. 41; Bruijnen & Janssen 2002: 105-107, no. 1.

In the late Middle Ages, it was in their own best interest for burghers, merchants and craftsmen to know exactly what time of day it was, and to be able to measure and manage time, if only for economic reasons. Many of them disposed of mechanical clocks that accurately showed the time. After all, time is money. This painting is a unique testimony of one such mechanical clock, even though the mechanism itself is missing, as are the construction which it was part of (possibly a tower-like building) and the dials (cf. cat. 76).

The panel is very special from an iconographic perspective. It depicts the objective measurement of time in that it shows the days and the time whilst it also illustrates the intangible influence the macrocosm wields over the human microcosm. In the middle are the twelve signs of the Zodiac. The surrounding circle shows the typical labour activities associated with each of the twelve months – starting bottom right with a festive banquet in January and ascending counter clockwise. The theme of the labour activities of the months was known in the Middle Ages and frequently featured in books of hours (cat. 73). Next, the daytime and night-time hours have been applied in Roman numerals in a narrow gold strip. The next circle contains 24 scenes, showing people engaged in a variety of occupations and activities. These are the so-called children of the planets, an element which was making headway around 1400: the lifestyle, behaviour and occupation of each individual is determined by the planet and the Zodiac sign under which he or she is born. This is followed by a calendar, divided into 365 tiny sections, marking out the holy days of the liturgical year in red and the foremost saints recorded in French. How many days each month has is specified in a narrow band. Finally, the four corners are packed with the personifications of the planets Jupiter (top left), Mars (top right), Venus (bottom left) and Mercury (bottom right).

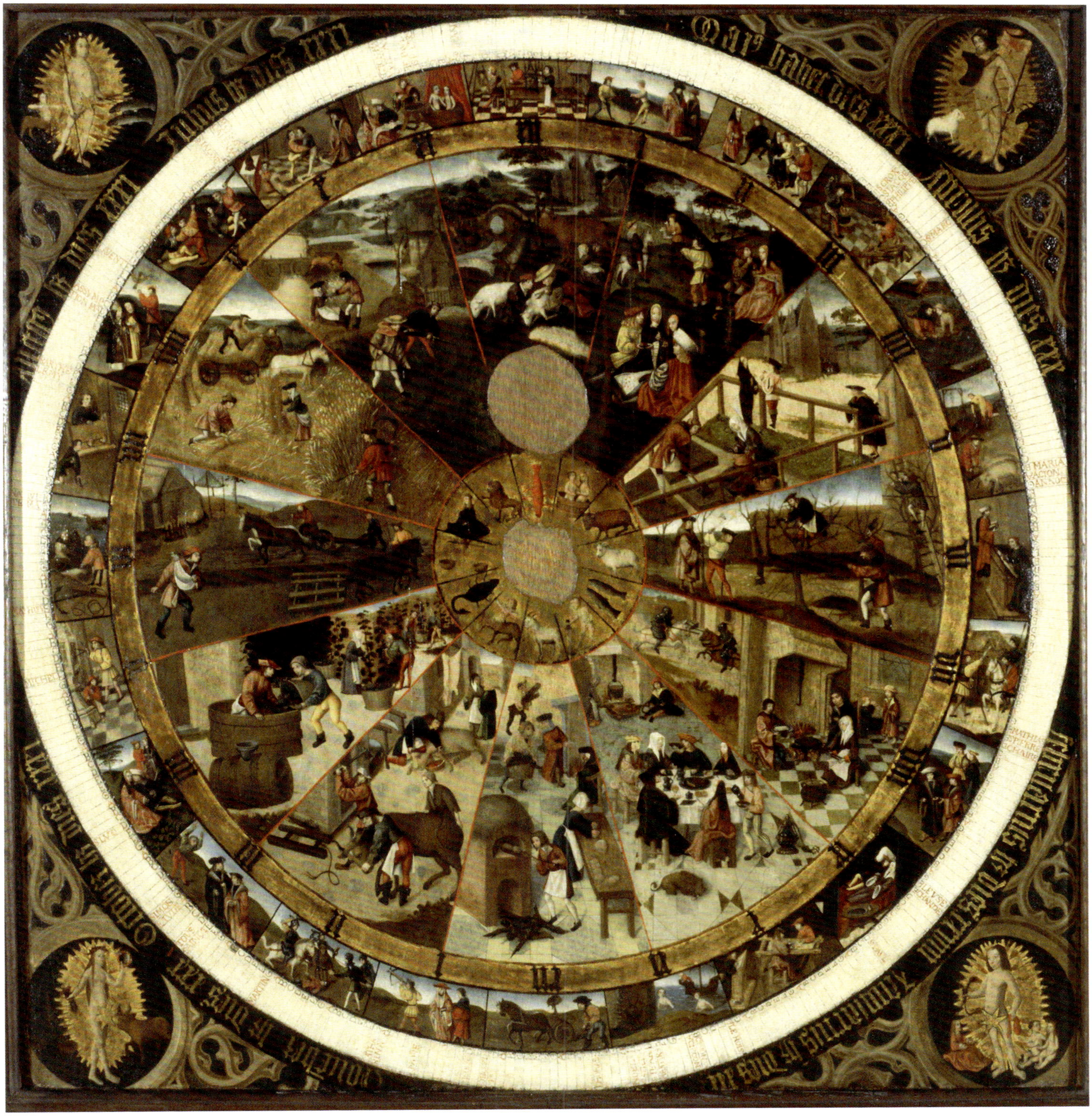

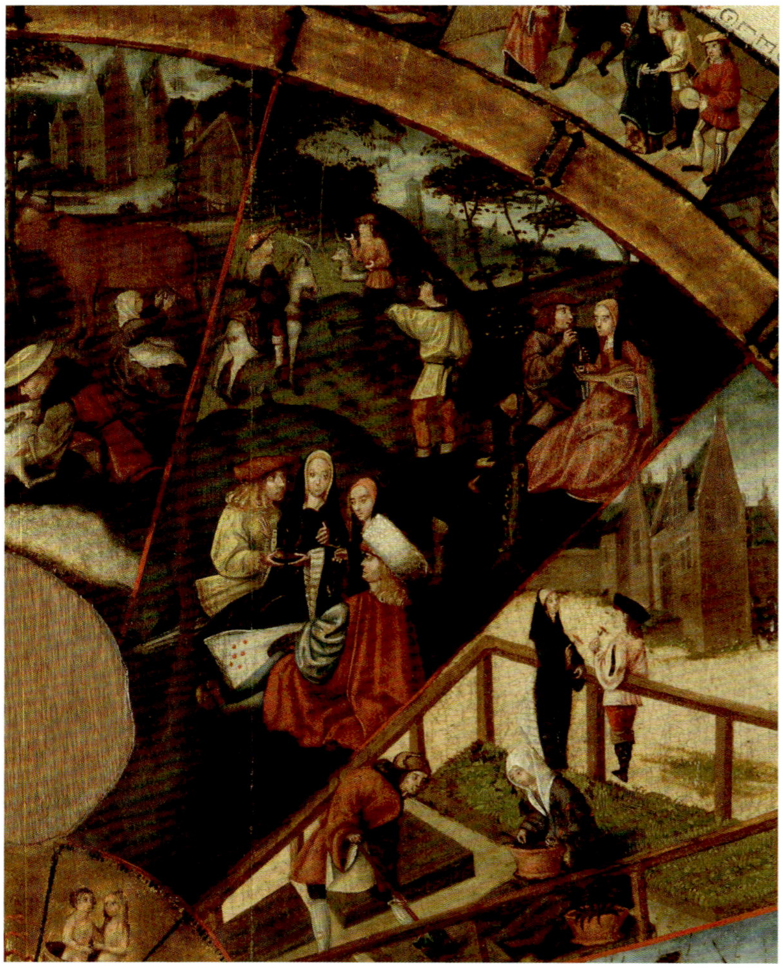

[cat. 74] detail

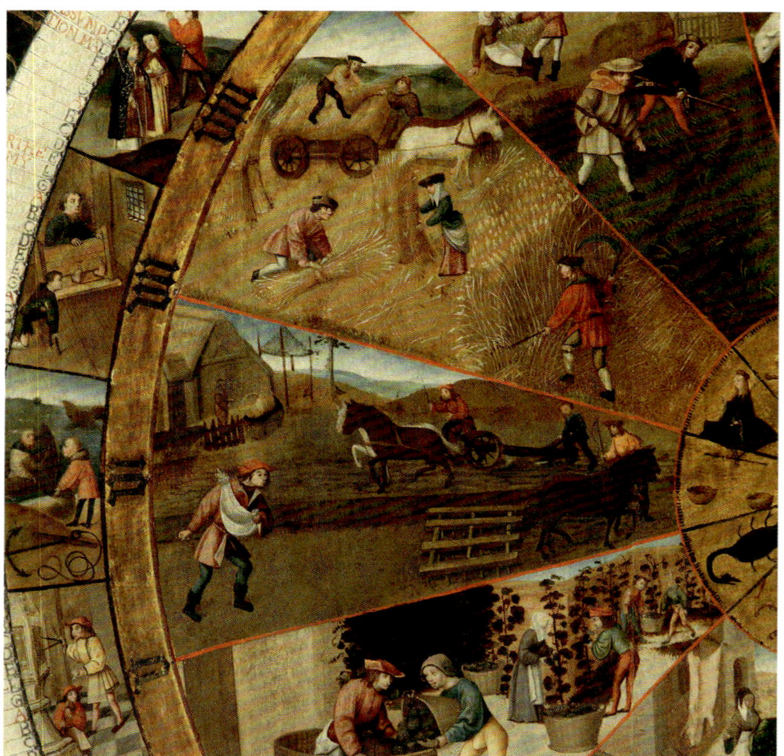

[cat. 74] detail

The (original) size, the French calendar and the lavish decoration of the piece suggest that the clock was built for a wealthy middle class customer. The detailed and painstakingly painted small scenes, both those of the children of the planets and the typical labour activities of the months, show sumptuously attired people and their activities: celebrating, making music, going out hunting, having picnics and visiting a brothel with its own bath house. The recreational pursuits are in marked contrast to the heavy labour performed by less fortunate mortals. As the land workers are seen pruning the vines (in March) or working the garden (in April), the land owner watches at his leisure.

According to Leuven town archivist Edward Van Even (1821–1905), the clock mechanism was built by Joos Metsys the Younger, who took over his father's blacksmith's workshop in 1483. Still according to Van Even, Metsys' brother, the famous painter Quinten Metsys (*c.*1465/1466–1530), is believed to have painted the scenes. However, there are no stylistic indications to underpin this suggestion (cf. cat. 13–16). From an iconographic angle, the typical labour activities of the months are certainly comparable to similar representations n the calendar miniatures in fifteenth and sixteenth century books of hours made in the Low Countries (cat. 73). For now, the painting remains the handiwork of an unknown – Leuven-based? – artist.

KS

75

Pierre de Fobis
Mechanical Sphere

Lyon
c. 1540–1550
Signed: PIERRE DE FOBIS[1]
Gilded bronze, silver and glass, height: 53.6 cm, diameter of the terrestrial globe 8 cm, diameter of the celestial sphere 15.2 cm, diameter of the outer sphere with the coordinates 16.5 cm
Private collection

PROVENANCE

Collection Nathaniel & Albert Rothschild, Vienna; Kunsthistorisches Museum, Vienna.

BIBLIOGRAPHY

Ferrari d'Occhieppo & Neumann 1959a: 195, fig. 2; Ferrari d'Occhieppo & Neumann 1959b: 144–146; Von Bertele 1961: 30–31, fig. 17; King 1978: 76–77, fig. 5.13; Brusa 1982: no. 65–67; Kugel, Van Cleempoel & Sabrier 2002: 145, no. S2.

This is probably one of the foremost mechanical spheres from the Renaissance period, showing the humanist ideal in a particularly refined and complex manner: the revival of Antique art and knowledge.[2] In *De Republica* (*c.* 60 BCE), Cicero chronicles how Archimedes invented two spheres (*spherae*): one a compact, massive globe surmounted by a series of stars found in the temple of Vesta, a second sphere that had been cut away, showing the trajectories of the Sun, the Moon and the planets: 'This kind of globe in which the Sun and the Moon are able to move, along with five other planets, could not have been made as a solid globe [like the one from the temple of Vesta] … When Gallus sets this sphere in motion, the Sun and the Moon move on bronze wheels of the machine like the days in the heavens, to the effect that the sphere also presents a solar eclipse when the Moon arrives at a certain point, like the shadow of the Earth did when the Sun shone from the opposite direction.'[3]

In keeping with the Renaissance spirit, Pierre de Fobis (who worked in Lyon between *c.* 1540–1575) was not looking merely to imitate this Antique text in his instrument (*imitatio*), but also to further refine it (*aemulatio*). On what looks like a rock-like base are three Corinthian pillars supporting an architrave with an articulated frieze as the foundation of the mechanical sphere. The signature is located at the bottom of the base: PIERRE DE FOBIS. The drive mechanism is visible in a glass cylinder between the three pillars. On the architrave sits a tholobate, from which four arms carry a horizon ring in which the primary sphere is ensconced. This primary sphere holds the mechanical sphere which consists of three components (starting from the centre and working outwards): planet Earth at the heart, encompassed by a gorgeous celestial sphere, which in turn is surrounded by a sphere that holds astrological coordinates. The latter sphere is unique and was used to compute astrological forecasts, a common practice in the sixteenth century.[4] The horizontal bands of the outer sphere are the polar circles, the tropical circles and the equator (divided into 2 x 12 hours). The purpose of the fine diagonal lines between the tropical circles was to determine the uneven hours.

1
This description is based on Kugel, Van Cleempoel & Sabrier 2002: 144, S.2. The comprehensive body of literature on this instrument is also included in this work.

2
On this topic, see Leopold 1977, among others.

3
Cicero, *De Republica* I:xiv, also in Cicero *Tusculanae Disputationes* I:63.

4
Nostradamus was not only a contemporary of Pierre de Fobis, he also lived and worked in Lyon.

The celestial sphere was the only sphere that rotated: one orbit every 23 hours and 56 minutes from East to West, driven by the mechanism between the three pillars. This is how the undulation (ascending and descending motion) of the stars and the Sun is represented. The sphere itself consists of two openwork hemispheres that meet on the *ecliptic*. Together, the 48 Ptolemaic constellations make up a delightful and delicate ensemble. Underneath, we can see the graduated equator (360°), the ecliptic (12 x 0–30°) and the twelve astrological houses. Pierre de Fobis drew inspiration for this mechanical sphere from the star chart by Johannes Honter (1541). The centrally positioned Earth likewise consists of two hemispheres that meet on the equator. The cartographic information specified was highly accurate for its time, and was inspired by the map produced by Oronce Finé (1536), in which North America still makes up a single landmass with Asia.

KVC

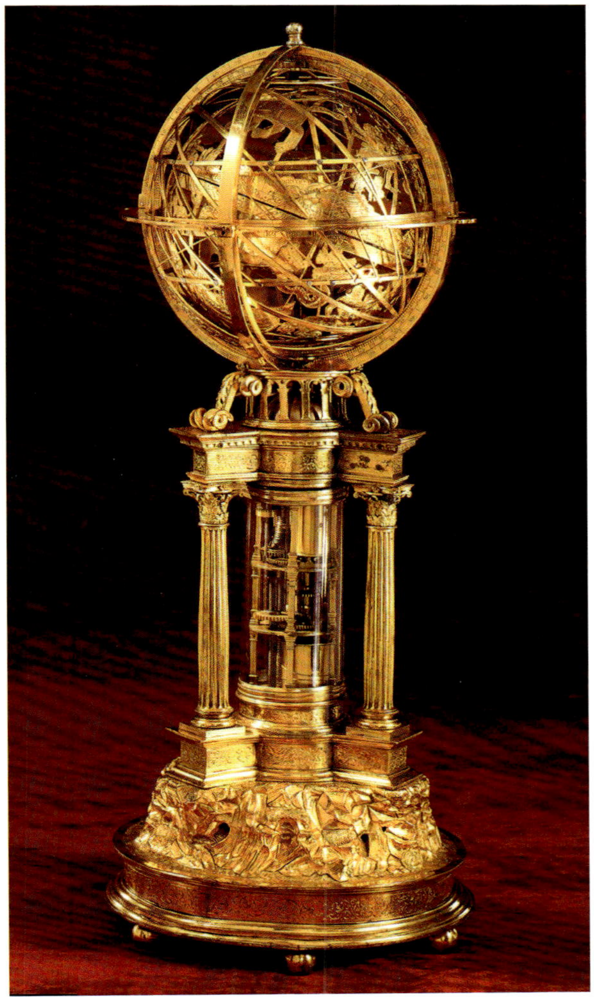

76

Master of Jean Rolin II, attributed to
Wisdom Teaching the Author Heinrich Seuse (Henricus Suso) the Precepts of Temperance

In: Henricus Suso, *Horloge de Sapience* (*Horologium Sapientiae*),
France, 1450–1460, fol. 13v
Manuscript on parchment, 370 x 225 mm
Brussels, Royal Library of Belgium, inv. Ms. IV 111

PROVENANCE

Acquired in 1961.

BIBLIOGRAPHY

Michel 1960: 294-297; Spencer 1963: 282-283; Brussels 1969: 83-86, no. 69 (with previous literature); Michel 1977: 125; White 1978: 196-197; Monks 1990: 53-58, 136-137; Dohrn-van Rossum 1997: 107; Rozenski Jr. 2010; Salzer 2013: 69.

The miniature, painted in the middle of the fifteenth century and attributed to Paris miniaturist Master of Jean Rolin II, opens the French translation of the treatise *Horologium Sapientiae*, composed by the German Dominican friar Henricus Suso (*c.*1334). This is a moral treatise in which Wisdom instructs the pupil (the author in this case) on *Temperantia* or Temperance. The text was translated into French in 1389 and became hugely popular in the fifteenth century. It is a virtuoso confluence of devotion, erudition and morality: 'Lady Wisdom' dictates the *Büchlein der ewigen Weisheit* to author Suso. In the opening miniature, they are depicted together, surrounded by a number of clocks and measuring instruments. Wisdom touches the wheel of time with her right hand.

The popular devotional text is divided into twenty-four chapters, just as the day is divided into twenty-four hours. It specifies the rhythm of meditation and prayer at the appropriate times of the day and prompts reflection on the passing of the hours. The miniature also shows various instruments for measuring time and establishing the position of the celestial bodies. This miniature is the earliest image to depict any of these instruments. Measuring time acquired a high moral status in both political and religious contexts and was associated with the virtue of Temperance.

The left part of the miniature shows a large, square, weight-driven timepiece or clock. The dial has just one hand, with two sets of numerals from I up to and including XII, thus dividing the day into twenty-four units. The mechanism is connected by a length of string to a bell, seen at the top between the two text columns. An enormous astrolabe is suspended from the clock case. On the right is a large, cogwheel-driven mechanism without hands that actuates the five small hammers of a set of chimes. On the table right are several quadrants: a horizontal quadrant, a portable equatorial quadrant and a quadrant with a plumb line (nautical quadrant). The small table clock, driven by springs, is one of the earliest depictions of a portable clock. What stands out is the gold-coloured, cylinder-shaped gnomon that

de leurs bonnes euvres de leur vie parfaitte et de leur sainte conversacion.

Ceulx ci queroient dieu en grant bonte et en parfaitte simplesse de cuer. Car toute leur entente et leur estude estoit les occupacions mondaines oublier, leurs pechiez gemir et plourer, les choses divines assavourer. Eulx dedens eulx contenir, et garder le corps et a servitude ramener. Et seulement a la divine sapience actendre et leur entente occuper.

Mais las dolent au jour d'ui le monde est envieilli en malice et l'amour divine est es cuers de plusieurs tellement refroidee quelle est pou s'en fault estainte. Car on en treuve pou qui estudient en devocion ne qui aient diligence de renouveller de la grace de dieu pour eulx eschauffer ne qui par desplaisance de leurs faultes aient souvent la lerme a l'ueil ne qui desirent la presence, la visitacion, la doulce alloqucion et plement de la grace divine. Ains se occupent et estudient en vanitez, en narracions [illegible] et hystoires vaines et temporelles et en delices corporelles esquelles ilz sont endormis par ung grief et pesant sommeil.

Et pourtant la divine sapience qui tousjours est songneuse du sauvement de toute humaine creature en desirant d'amender la vie de ses esleus et en voulant oster et anientir tous vices de leurs consciences veult et entent en ce livre principalment enlumer les estains et enflamer les les refroidiez, les pecheurs esmouvoir, les mal devots a devocion rappeller et promouvoir. Et les endormis par negligence a l'estude des euvres vertueuses esveiller

La .ij.e hystoire.

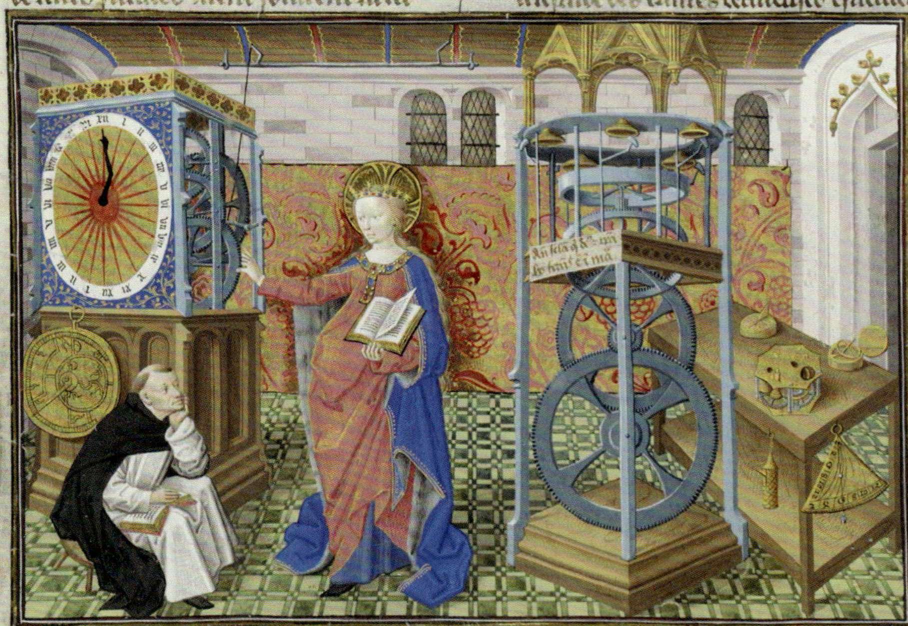

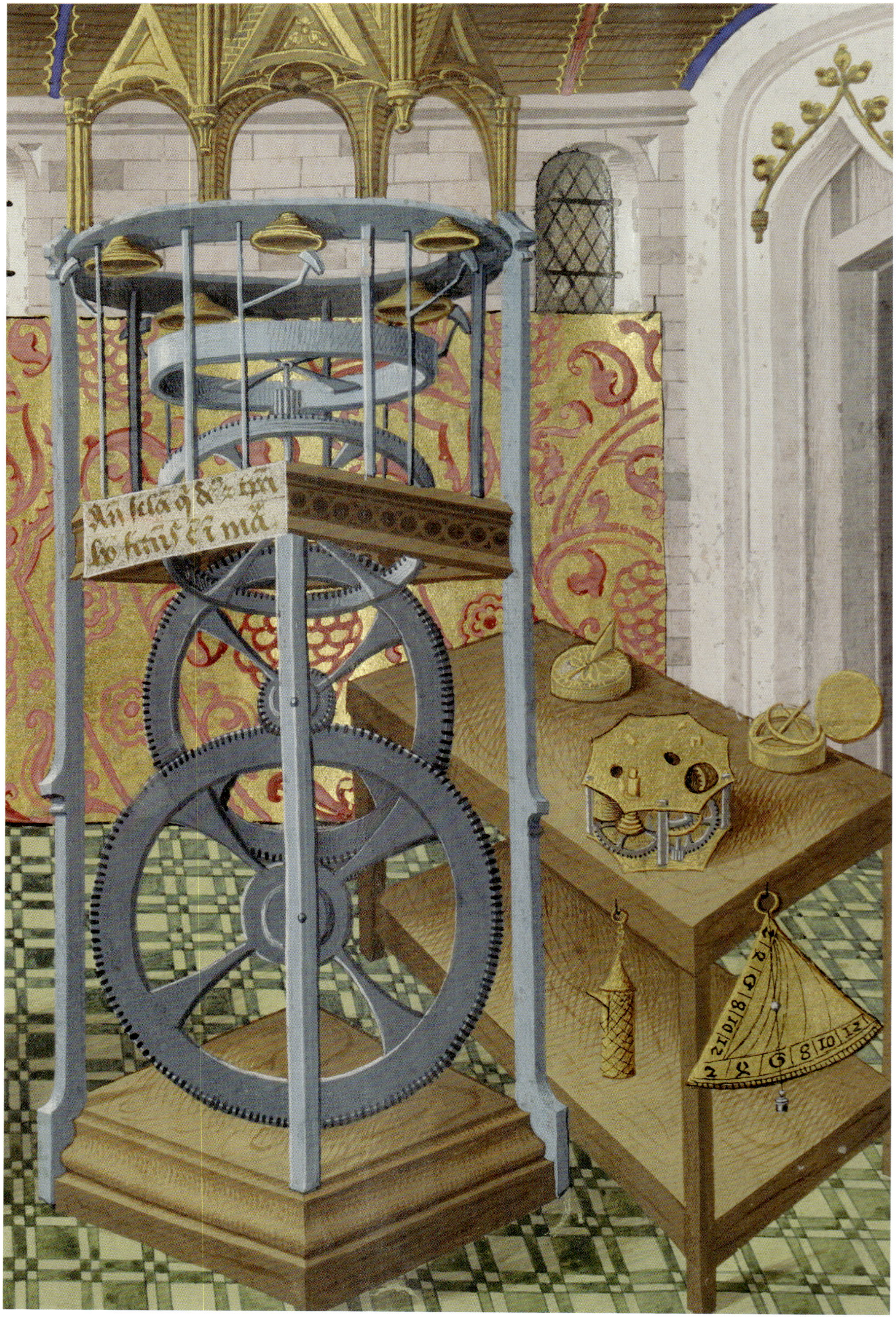

[cat. 76] detail

hangs from the table, an instrument that was used to determine the position of the sun and measure time.

The development of increasingly accurate measuring instruments had a major impact on cartography and maritime exploratory expeditions during the second half of the fifteenth century. Time and space were fast becoming manageable by man.

LW

77

Astronomy: The Study of Heavenly Bodies

Tournai (?)

c.1510

Tapestry, wool and silk, 240 x 330 cm

Göteborg, The Röhss Museum of Arts and Crafts, inv. RKM 79-1964

PROVENANCE

Gift of the Association of Friends of the Röhsska Museet

BIBLIOGRAPHY

Burger 1950: 859-873; Strömberg 1965: 14-28, 46-47; Souchal 1974: 157-158, no. 64; Massing & Adelson 1991: 225-226, no. 125; Wells 1991: 33-37; Brosens 2000: 259-261, no. 140; Charron 2011: 334, no. 4, 337-338.

The crowned and luxuriously dressed woman at the right, who points towards the sky with her right hand, is, according to the banderol in her left hand, Astronomy. From time immemorial, she was one of the sciences of the *quadrivium* ('Four ways'), together with mathematics, geometry and music. In the Middle Ages, the distinction between astrology and astronomy was vague, although Thomas More distinguishes explicitly between the two in his *Utopia*.[1]

Two astronomers observe the heavens, which contain the moon, stars, comets and flaming stars. The standing scientist, in 'antique' clothing, holds an *armillarium* in one hand and a roll of parchment in the other. His seated colleague, whose clothes are more contemporary, notes his findings in a book, while on the lectern near by stands an *astrolabium*. In late medieval iconography, astronomy was often represented by the ancient thinkers Pythagoras and Ptolemy; the letters on their gowns here, however, are illegible, and serve no more purpose than to represent exotic writing.

In the landscape at the left, three additional figures are depicted. They are shepherds, to judge by their staffs and flattened hats (one also sees a heard of sheep in a pen and the shelters used by the shepherds, such as the triangular hut and the wooden house on wheels). They, too, study the heavens. In the late Middle Ages, people were aware of the fact that even in the margins of society, among shepherds, for example, there was a profound living knowledge of the firmament. This knowledge was vulgarised in early editions such as the *Calendrier des Bergers* or the *Shepheardes calendar*.[2] These texts dealt with the cosmic-astrological coherence of human life, the seasons of the year, human types, good and evil, hell … shepherds are chosen people, because they were thought to be close to nature and because they were the first to whom the angels delivered the message of Christ's birth.

Thus, astronomy was nourished by two wellsprings: the popular knowledge of shepherds and the science of scholars, both from the past and the present. Two forms of utopia are here united: that of the 'carefree' pastoral life with its earthy but nevertheless profound existential knowledge, and that of scholarly scientific

1
'[They] were perfectly acquainted with the motions of the heavenly bodies; and have many instruments, well contrived and divided, by which they very accurately compute the course and positions of the sun, moon, and stars. But for the cheat of divining by the stars, by their oppositions or conjunctions, it has not so much as entered into their thoughts. They have a particular sagacity, founded upon much observation, in judging of the weather, by which they know when they may look for rain, wind, or other alterations in the air.' More 2008: 100.

2
See Le Rouge 1976; Braekman 1985; Johnson 1990.

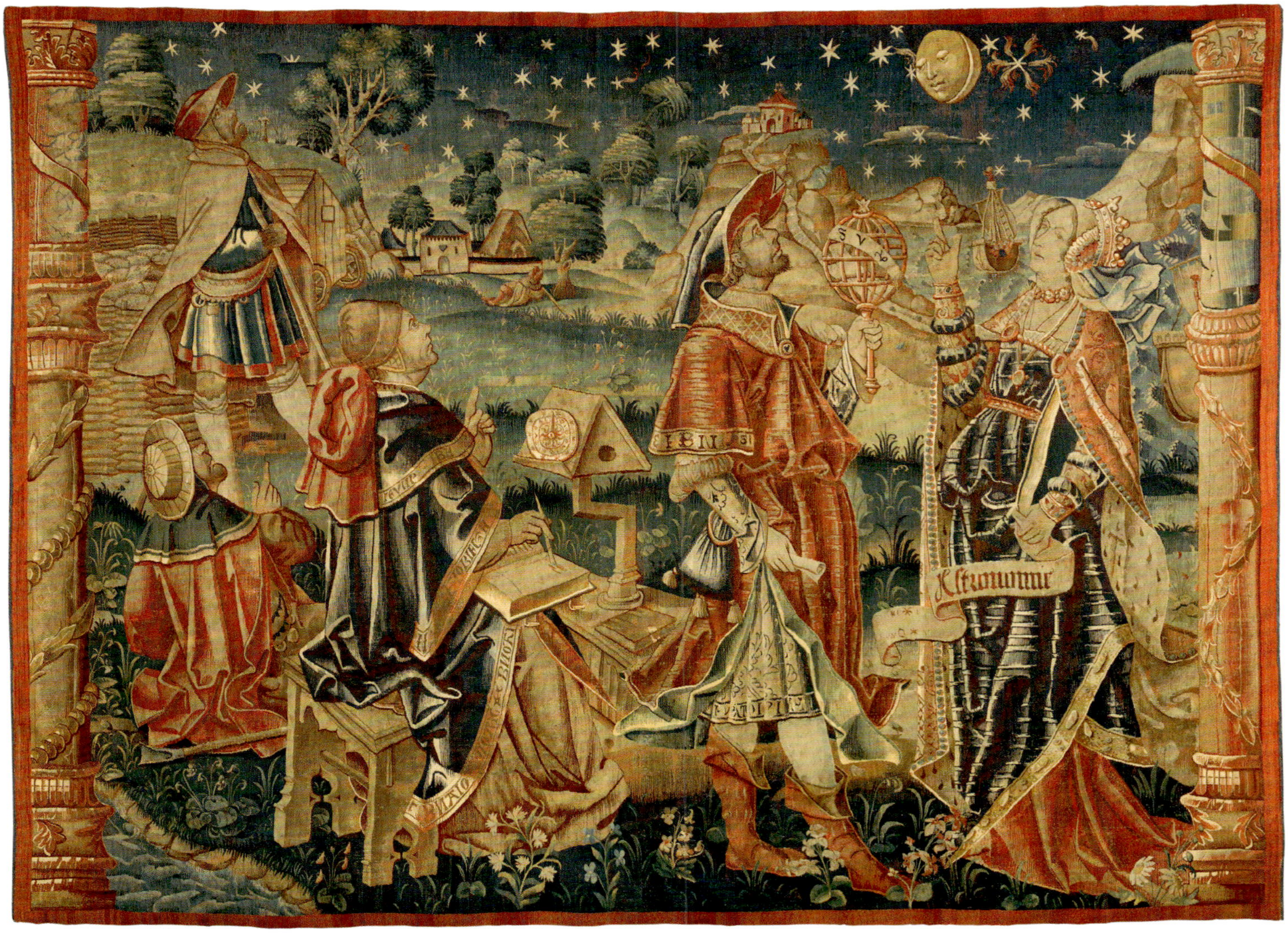
Astronomie

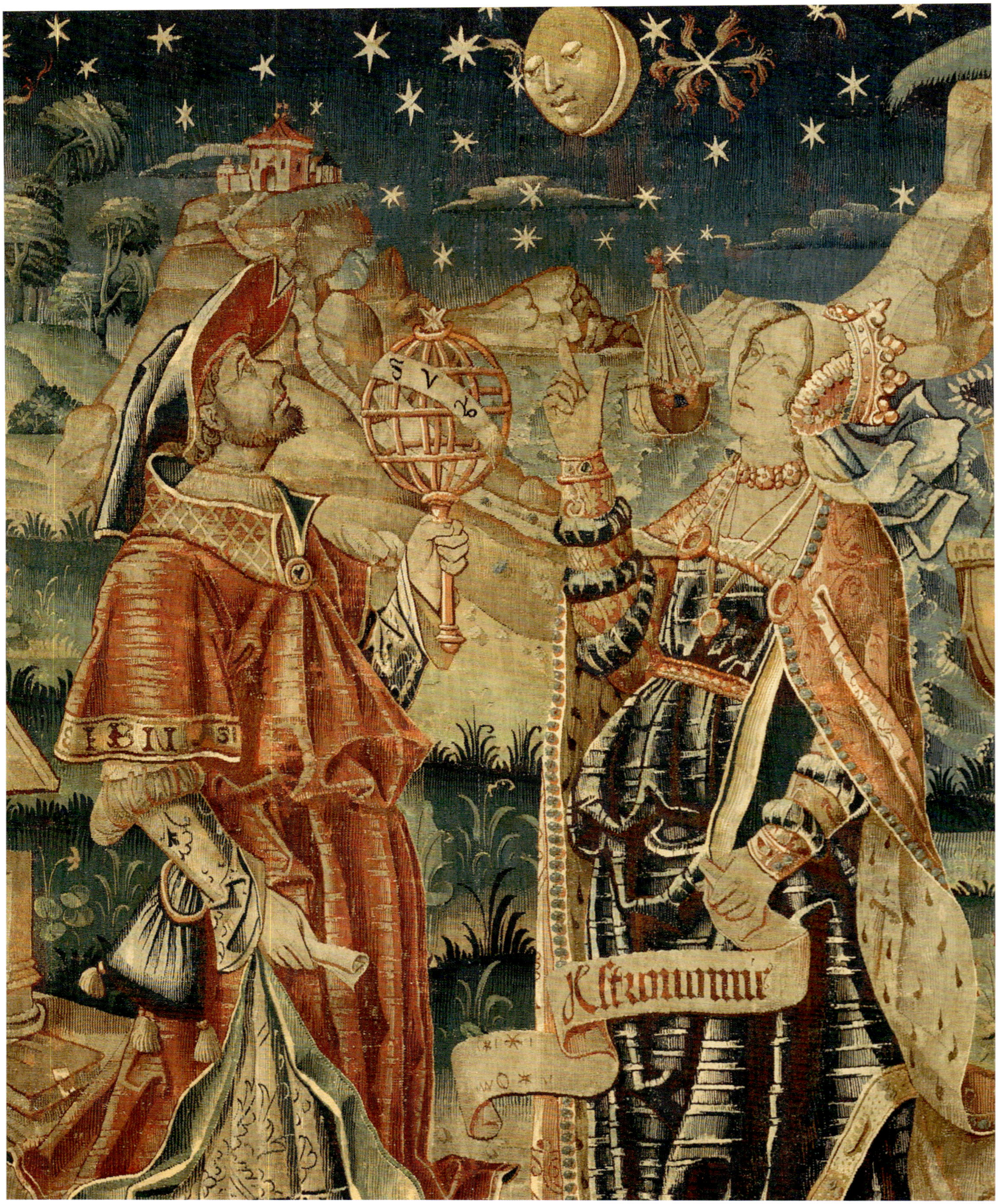

[cat. 77] detail

knowledge, developed apart from 'everyday life'. Around 1500, both strove for a holistic wisdom concerning humanity's place in a greater whole; soon enough the new astronomy of Copernicus and his kindred spirits would change this perspective completely.

PVDB

Bibliography

Ackroyd 1998
Peter Ackroyd, *The Life of Thomas More*, London 1998.

Ackroyd, Billinge, Campbell & Kirby 2003
Paul Ackroyd, Rachel Billinge, Lorne Campbell & Jo Kirby, 'The "Two Tax-Gatherers" by Marinus van Reymerswale: Original and Replica', *National Gallery Technical Bulletin* 24 (2003): 50-63.

Adam 1620
Melchior Adam, *Vitae Germanorum medicorum*, Heidelberg 1620.

Adam 2011
Renaud Adam, 'Le berceau d'une collection. Les incunables de Charles Van Hulthem (1764-1832)', *In Monte Artium* 4 (2011): 7-36.

Adam & Vanautgaerden 2009
Renaud Adam & Alexandre Vanautgaerden, *Thierry Martens et la figure de l'imprimeur humaniste (une nouvelle biographie). Avec une liste de ses éditions et les adresses aux lecteurs signées par l'imprimeur*, Nugæ humanisticæ 11.1, Turnhout 2009.

Adonias 1963
Isa Adonias, *La Cartografia da Região Amazônica: Catalogo descritivo (1500-1961)*, Rio de Janeiro 1963.

Aikema 2001
Bernard Aikema, 'Hieronymus Bosch and Italy?', *Hieronymus Bosch - New Insights into his Life and Work*, Jos Koldeweij, Bernard Vermet & Barbera van Kooij (eds.), Rotterdam 2001: 24-31.

Aikema 2014
Bernard Aikema, 'The Reception of Bosch in Italy: Macaronic Culture and the Taste for the Bizarre and Fanciful', *Jheronimus Bosch: his Patrons and his Public*, Jo Timmermans (ed.), 's-Hertogenbosch 2014: 30-43.

Ainsworth 2010
Maryan W. Ainsworth (ed.), *Man, Myth, and Sensual Pleasures: Jan Gossart's Renaissance: The Complete Works*, exh. cat. (New York, The Metropolitan Museum of Art/London, National Gallery), New Haven/London 2010.

Alegria, Daveau, Garcia & Relaño 2007
Maria Fernanda Alegria, Suzanne Daveau, João Carlos Garcia & Francesc Relaño, 'Portuguese Cartography in the Renaissance', *The History of Cartography - Volume 3: Cartography in the European Renaissance*, pt. 1, David Woodward (ed.), Chicago/London 2007: 975-1068.

Anthiaume 1916
Albert Anthiaume, *Cartes marines, constructions navales, voyages de découverte chez les Normands, 1500-1650*, Paris 1916.

Asiedu 2001
Felix B. A. Asiedu, 'The Song of Songs and the Ascent of the Soul: Ambrose, Augustine, and the Language of Mysticism', *Vigiliae Christianae* 55, 3 (2001): 299-317.

Alexander 1999
Jonathan J. G. Alexander, 'The Last Things: Representing the Unrepresentable, The Medieval Tradition', *The Apocalypse and the Shape of Things to Come*, Frances Carey (ed.), exh. cat. (London, British Museum), Toronto 1999: 43-98.

Alexander, Marrow & Sandler 2005
Jonathan J. G. Alexander, James H. Marrow & Lucy Freeman Sandler (eds.), *The Splendor of the Word. Medieval and Renaissance Illuminated Manuscripts at the New York Public Library*, exh. cat. (New York, New York Public Library), New York/London 2005.

Allart 2000
Dominique Allart, 'Henri Bles, un paysagiste à redécouvrir', *Autour de Henri Bles*, Jacques Toussaint (ed.), exh. cat. (Namur, Musée des Arts anciens du Namurois) 2000: 21-30.

Allen 1906
Percy Stafford Allen (ed.), *Opus epistolarum Des. Erasmi Roterodami*, vol. 1, Oxford 1906.

Allen 1963
Peter R. Allen, ' "Utopia" and European humanism. The Function of the Prefatory Letters and Verses', *Studies in the Renaissance* 10 (1963): 91-107.

Allen, Allen & Garrod 1906-1958
Percy Stafford Allen, Helen Mary Allen & Heathcote William Garrod (eds.), *Opus epistolarum Des. Erasmi Roterodami*, 12 vols., Oxford 1906-1958.

Almog 2014
Shulamit Almog, 'Dystopian Narratives and Legal Imagination: Tales of Noir Cities and Dark Laws', *Law and the Utopian Imagination*, Austin Sarat, Lawrence Douglas & Martha Merrill Umphrey (eds.), Stanford 2014: 155-178.

Alvin 1862
Louis-Joseph Alvin, *Rapport triennal sur la situation de la Bibliothèque royale, pendant les années 1858-1859, 1859-1860 et 1860-1861*, Brussels 1862.

Ambrosio & Brusa 1992
Franco Ambrosio & Francesca Brusa (eds.), *Art et Culture vers 1492*, exh. cat. (Seville, Exposition Universelle Seville), Seville 1992.

Andersson 1968
Armstrong Andersson, 'Copies of Pollaiuolo's Battling Nudes', *Art Quarterly* 31 (1968): 155-167.

Andersson & Talbot 1983
Christiane Andersson & Charles W. Talbot (eds.), *From a mighty fortress: prints, drawings, and books in the age of Luther (1483-1546)*, exh. cat. (Detroit, Detroit Institute of Arts/Ottawa, National Gallery of Canada/Coburg, Kunstsammlungen der Veste Coburg), Detroit 1983.

Angres 2003
Marcel Angres, *Triumphus Veneris. Ein allegorisches Epos von Heinrich Bebel. Edition, Übersetzung und Kommentar*, Münster 2003.

Antoine 2002
Élisabeth Antoine (ed.), *Sur la terre comme au ciel: jardins d'Occident à la fin du Moyen Âge*, exh. cat. (Paris, Musée national du Moyen Âge - Thermes de Cluny), Paris 2002.

Antwerp 1948
Kerkelijke Kunst, exh. cat. (Antwerp, Stedelijke Feestzaal Meir), Antwerp 1948.

Anzelewsky, Mende & Eeckhout 1977
Fedja Anzelewsky, Matthias Mende & Paul Eeckhout, *Albert Dürer aux Pays-Bas: son voyage (1520 - 1521), son influence*, exh. cat. (Brussels, Palais des beaux-arts), Brussels 1977.

Anzelewsky & Mielke 1984
Fedja Anzelewsky & Hans Mielke, *Albrecht Dürer, Kritischer Katalog der Zeichnungen*, mus. cat. (Berlin, Staatliche Museen preussischer Kulturbesitz - Kupferstichkabinett), Berlin 1984.

Appuhn 1976
Horst Appuhn, 'Zur Ikonographie der Riesenholzschnitte in Deutschland und den Niederlanden', *Riesenholzschnitte und Papiertapeten der Renaissance*, Horst Appuhn & Christian von Heusinger (eds.), Unterschneidheim 1976: 35-86.

Arnould 1991
Alain Arnould, *The Art Historical Context fort the Library of Raphael de Mercatellis*, PhD dissertation, Ghent University, Ghent 1991.

Arnould & Massing 1993
Alain Arnould & Jean Michel Massing, *Splendours of Flanders. Late Medieval Art in Cambridge Collections*, exh. cat. (Cambridge, Fitzwilliam Museum), Cambridge 1993.

Avril & Reynaud 1993
François Avril & Nicole Reynaud (eds.), *Les manuscrits à peintures en France 1440-1520*, exh. cat. (Paris, Bibliothèque nationale de France), Paris 1993.

Bach 1949
Otto Karl Bach, 'Bosch's Vision of Tondalys', *The Denver Art Museum Quarterly* (1949): 6-10.

Bachfischer 1998
Margit Bachfischer, *Musikanten, Gaukler und Vaganten. Spielmannskunst im Mittelalter*, Augsburg 1998.

Backhouse 1987
Janet Backhouse, 'Founders of the Royal Library: Edward IV and Henry VII as Collectors of Illuminated Manuscripts', *England in the Fifteenth Century* (proceedings of the 1986 Harlaxton Symposium), David Williams (ed.), Woodbridge 1987: 23-41.

Backhouse 1993
Janet Backhouse, 'A Salute to the Tudor Rose', *Miscellanea Martin Wittek: Album de codicologie et de paléographie offert à Martin Wittek*, Anny Raman & Eugène Manning (eds.), Leuven 1993: 1-14.

Backhouse 1997
Janet Backhouse, *The Illuminated Page: Ten Centuries of Manuscript Painting in the British Library*, Toronto 1997.

Backhouse 2000
Janet Backhouse, 'Charles of Orléans Illuminated', *Charles d'Orléans in England (1415-1440)*, Mary-Jo Arn (ed.), Cambridge 2000: 157-164.

Baert 2013
Barbara Baert, '"An Odour. A Taste. A Touch. Impossible to Describe": Noli me tangere and the Senses', *Religion and the Senses in Early Modern Europe*, Wietse de Boer (ed.), Leiden 2013: 111-152.

Baert 2014
Barbara Baert, 'Echoes of Liminal Spaces: Revisiting the Late Mediaeval "Enclosed Gardens" of the Low Countries (A Hermeneutical Contribution to Chthonic Artistic Expression)', *Jaarboek Koninklijk Museum voor Schone Kunsten Antwerpen 2012* (2014): 9-45.

Baert 2016a
Barbara Baert, 'Die spätmittelalterlichen eingefassten Gärten in den Niederlanden', *Zeitschrift für Medien- und Kulturforschung* 7, 1 (2016): 27-44.

Baert 2016b
Barbara Baert (with the collaboration of Hannah Iterbeke), 'Revisiting the Enclosed Gardens of the Low Countries (15th century onwards). Gender, Textile, and the Intimate Space as Horticulture', *Textile. Journal of Cloth and Culture* (2016): 1-31.

Baert 2016c
Barbara Baert, *Late Mediaeval 'Enclosed Gardens' of the Low Countries: Contributions to Gender Artistic Expression*, Studies in Iconology, 2, Leuven 2016.

Baert 2016d
Barbara Baert, *Locus Amoenus and the Sleeping Nymph*, Leuven/Paris/Bristol 2016.

Bagrow 1945
Leo Bagrow, 'The Origin of Ptolemy's Geographia', *Geografiska Annaler* 27 (1945): 318-387.

Baker-Smith 2011
Dominic Baker-Smith, 'Reading Utopia', *The Cambridge Companion to Thomas More*, George M. Logan (ed.), Cambridge 2011: 141-167.

Bakker 2004
Boudewijn Bakker, *Landschap en wereldbeeld. Van Van Eyck tot Rembrandt*, Bussum 2014.

Bakker 2012
Boudewijn Bakker, 'A Sequel to Patinir: Herri met de Bles', *Landscape and Religion from Van Eyck to Rembrandt*, Farnham 2012: 106-108.

Bakker 2015
Boudewijn Bakker, 'The Amsterdam 'Paradise' by Herri met de Bles and the Fountain of Life', *The Rijksmuseum Bulletin* 63 (2015): 114-132.

Baldass 1938
Ludwig von Baldass, 'Nachahmer des Hieronymus Bosch. Christus in der Vorhölle', *Katalog der Gemäldegalerie - Führer durch das Kunsthistorische Museum* 8, mus. cat. (Vienna, Die Gemäldegalerie des Kunsthistorischen Museums), Vienna 1938: 21.

Baligand 1999
Françoise Baligand, *Le Musée de la Chartreuse, Douai*, mus. cat. (Douai, Musée de la Chartreuse), Douai 1999.

Baligand 2001
Françoise Baligand, 'Reymerswaele Marinus van - Saint Jérôme en méditation', *La peinture nordique de 1400 à 1550 dans les musées du Nord-Pas-de-Calais*, online exh. cat. (musenor.com, Association des conservateurs des musées du Nord-Pas-de-Calais) 2001.

Bange 1928
Ernst Friedrich Bange, *Die Kleinplastik der deutschen Renaissance in Holz und Stein*, Munich 1928.

Baptiste 1984
Georges Baptiste (ed.), *De tijdmeting in Belgische verzamelingen*, exh. cat. (Brussels, Generale Bankmaatschappij), Brussels 1984.

Barber & Harper 2010
Peter Barber & Tom Harper, *Magnificent Maps. Power, Propaganda and Art*, exh. cat. (London, British Library), London 2010.

Bartelink 1994
Gerard Bartelink, 'De demonologie van het vroege christendom', *Duivelsbeelden. Een cultuurhistorische speurtocht door de Lage Landen*, Gerard Rooijakkers, Lène Dresen-Coenders & Margriet Geerdes (eds.), Baarn 1994: 21-70.

Barto 1916
Philip Barto, *Tannhäuser and the Mountain of Venus. A Study in the Legend of the Germanic Paradise*, New York 1916.

Bartrum 1995
Giulia Bartrum, *German Renaissance Prints 1490-1550*, exh. cat. (London, British Museum), London 1995.

Bartrum 2002
Giulia Bartrum (ed.), *Albrecht Dürer and his Legacy: The Graphic Work of a Renaissance Artist*, London 2002.

Bartsch 1803-1821
Adam Bartsch, *Le peintre-graveur*, 21 vols., Vienna 1803-1821.

Bartsch 1808
Adam Bartsch, *Le peintre-graveur - vol. 7: Dürer*, Vienna 1808.

Bass 2015
Marisa Bass, 'Cat. 14', *Beyond Bosch: The Afterlife of a Renaissance Master in Print*, Marisa Bass & Elizabeth Wyckoff (eds.), exh. cat. (Saint Louis, Saint Louis Art Museum), Saint Louis 2015: 136-139.

Bassani & Fagg 1988
Ezio Bassani & William Fagg, *Africa and the Renaissance. Art in Ivory*, Susan Vogel (ed.), exh. cat. (New York, Center for African Art), New York 1988.

Bassani 1997
Ezio Bassani, *Réceptacles anciens*, exh. cat. (Paris, Musée Dapper), Paris 1997.

Bassani 2000
Ezio Bassani, *African Art and Artefacts in European Collections 1400-1800*, Malcom McLeod (ed.), London 2000.

Bassani 2003
Ezio Bassani (ed.), *Africa. Capolavori da un Continento, Mille Anni d'Arte*, exh. cat. (Turin, GAM Galleria d'Arte Moderna), Milan 2003.

Bassani 2008
Ezio Bassani (ed.), *Ivoires d'Afrique dans les anciennes collections francaises*, Paris 2008.

Bätschmann 1981
Oskar Bätschmann, ' "Lot und seine Töchter" im Louvre. Metaphorik, Antithetik und Ambiguität in einem niederländischen Gemälde des frühen 16. Jahrhunderts', *Städel-Jahrbuch* N.F. 8 (1981): 59-185.

Bätschmann 2013
Oskar Bätschmann, 'Migrationen: Holbeins Bildnis Thomas Morus', *Synergies in Visual Culture - Bildkulturen im Dialog*, Manuela de Giorgi, Annette Hoffmann & Nicola Suthor (eds.), Munich 2013: 359-370.

Bauer & Haupt 1976
Rotraud Bauer & Herbert Haupt, 'Das Kunstkammerinventar Kaiser Rudolfs II., 1607-1611', *Jahrbuch der Kunsthistorischen Sammlungen in Wien* 72 (1976): 1-191.

Baverel, Goutagny & Méasson 2011
Danièle Baverel, Pascale Goutagny & Josette Méasson, *Les cartographes & les nouveaux mondes: une représentation normande des grandes découvertes: une lecture historique, symbolique et mathématique des cartes normandes des XVIe et XVIIe siècles*, Bonsecours 2011.

Bax 1949
Dirk Bax, *Ontcijfering van Jeroen Bosch*, The Hague 1949.

Bax 1983
Dirk Bax, *Hieronymus Bosch and Lucas Cranach. Two Last Judgement Triptychs. Description and exposition*, Amsterdam 1983.

Bayot 1928
Alphonse Bayot, *Martin le Franc. L'estrif de fortune et de vertu: étude du manuscrit 9510 de la Bibliothèque Royale de Belgique, provenant de l'ancienne 'librairie' des Croÿ de Chimay*, Paris 1928.

Bejczy 1994
István Pieter Bejczy, *Pape Jansland en Utopia. De verbeelding van middeleeuwen en renaissance. Een wetenschappelijke proeve op het geboed van de Letteren*, PhD dissertation, Katholieke Universiteit Nijmegen, Nijmegen 1994.

Bellavitis 2010
Maddalena Bellavitis, *Telle depente forestiere: quadri nordici nel Veneto: le fonti e la tecnica*, Padova 2010.

Bergmans 1904
Paul Bergmans, 'David de Gand', *Bulletin de la société d'histoire et d'archéologie de Gand* 12 (1904): 151.

Bergmans 1963
Simone Bergmans, 'Gilles I Mostaert. La Descente du Christ aux Limbes', *Le Siècle de Bruegel. La Peinture en Belgique au XVIe Siècle*, exh. cat. (Brussels, Musées Royaux des Beaux-Arts de Belgique), Brussels 1963: 136-137.

Berk 1982
Philip R. Berk, ' "Et Adam dilexit Abel": Jan Mostaert's First Family', *Oud Holland* 96, 4 (1982): 201-212.

Bernheimer 1952
Richard Bernheimer, *Wild Men in the Middle Ages: A Study in Art, Sentiment and Demonology*, Cambridge 1952.

Bertsch 1997
Gerhard Bertsch, 'Mercator als Mathematicus - Landvermessung und Kugelgeometrie im 16. Jahrhundert', *Gerhard Mercator und seine Welt*, Rienk Vermij (ed.), Duisburg 1997: 132-160.

Bessemans, Honoré, Smeyers et al. 1998
Lutgarde Bessemans, Inès Honoré, Maurits Smeyers et al. (eds.), *Leven te Leuven in de late Middeleeuwen*, exh. cat. (Leuven, Stedelijk Museum Vander Kelen-Mertens), Leuven 1998.

Beumer 1954
Jurjen Beumer, 'Die marianischen Deutung des Hohen Liedes in der Frühscholastik', *Zeitschrift für katholische Theologie* 76 (1954): 411-439.

Bezzel 1983
Irmgard Bezzel, *Verzeichnis der im deutschen Sprachbereich verschienenen Drucke des XVI. Jahrhunderts. VD 16*, Stuttgart 1983.

Biot 1995
Brigitte Biot, 'Barthélémy Aneau lecteur de l'*Utopie*', *Moreana* 121 (1995): 11-28.

Bishop 2005
Malcolm Bishop, 'Ambrosius Holbein's *memento mori* map for Sir Thomas More's *Utopia*. The meanings of a Masterpiece of Early Sixteenth-Century Graphic Art', *British dental journal* 199 (2005): 107-112.

Black 2012
Maggie Black, *The Medieval Cookbook*, Los Angeles 2012.

Blackmun 1992
Barbara Blackmun, *The Elephant. The Animal and its Ivory in African Culture*, Doran H. Ross (ed.), Los Angeles 1992.

Blanc 2004
Monique Blanc, *Voyages en enfer: De l'art paléochrétien à nos jours*, Paris 2004.

Blöcker 1993
Susanne Blöcker, *Studien zur Ikonographie der sieben Todsünden in der niederländischen und deutschen Malerei und Graphik von 1450-1560*, Münster 1993.

Blockmans 2000
Wim Blockmans, *Keizer Karel V. De utopie van het keizerschap*, Leuven/Amsterdam 2000.

Bloom 2002
James Bloom, *The Rise of the Painted Panel in Early Modern Netherlandish Art or How Antwerp Stole the Idea of Popular Culture*, PhD dissertation, Duke University, Durham 2002.

Blöte 1898
J. F. D. Blöte, 'Das Aufkommen des Clevischen Schwanritters', *Zeitschrift für deutsches Altertum* 42 (1898): 1-53.

Bock 1921
Elfried Bock, *Staatliche Museen zu Berlin. Die Deutschen Meister. Beschreibendes Verzeichnis sämtlicher Zeichnungen*, 2 vols., Berlin 1921.

Bode 1888
Wilhelm Bode, *Die Grossherzogliche Gemälde-Galerie in Oldenburg*, Vienna 1888.

Boeke 1974
Enno Boeke, *Rondom het paradijs verhaal*, Wassenaar 1974.

Boeren 1979
Petrus Cornelis Boeren, *Catalogus van de handschriften van het Rijksmuseum Meermanno-Westreenianum*, mus. cat. (The Hague, Rijksmuseum Meermanno-Westreenianum), The Hague 1979.

Bonnell 1917
John K. Bonnell, 'The Serpent with a Human Head in Art and in Mystery Play', *American Journal of Archaeology* 21, 3 (1917): 255-291.

Bonney 1991
Richard Bonney, *The European Dynastic States. 1494-1660*, Oxford 1991.

Boon, Bruyn, Lemmens et al. 1967
K. G. Boon, J. Bruyn, G. Lemmens et al. (eds.), *Jheronimus Bosch*, exh. cat. ('s-Hertogenbosch, Noordbrabants Museum), 's-Hertogenbosch 1967.

Boorsch & Orenstein 1997
Suzanne Boorsch & Nadine Orenstein, 'The Print in the North: The Age of Albrecht Dürer and Lucas van Leyden', *The Metropolitan Museum of Art Bulletin* 54, 4 (1997): 13-60.

Borchert 2010
Till-Holger Borchert (ed.), *Van Eyck tot Dürer: de Vlaamse Primitieven en Centraal-Europa 1430-1530*, exh. cat. (Bruges, Groeningemuseum), Tielt 2010.

Borchert 2016
Till-Holger Borchert, 'Hieronymus Bosch and Workshop, *c.*1505/15', *El Bosco. La exposición del V Centenario*, Pilar Silva Maroto (ed.), exh. cat. (Madrid, Museo del Prado) 2016.

Borchert & Jonckheere 2015
Till-Holger Borchert & Koenraad Jonckheere (eds.), *Renaissanceportretten uit de Lage Landen*, exh. cat. (Brussels, Bozar), Brussels 2015.

Borenius 1939
Tancred Borenius, *A Catalogue of the Pictures at Corsham Court*, London 1939.

Borenius 1943
Tancred Borenius, 'Two Little-Known Pictures by Holbein in England', *The Burlington Magazine for Connoisseurs* 83 (1943): 285-288.

Bott 1992
Gerhard Bott (ed.), *Focus Behaim Globus*, exh. cat. (Nuremberg, Germanischen Nationalmuseums), Nuremberg 1992.

Böttiger 1800
Karl August Böttiger, 'Wilhelm Tischbein', *Der Neue Teutsche Merkur vom Jahre 1800*, vol. 3, 9, C.M. Wieland (ed.), Weimar 1800: 61-76.

Bourdillon 1906
Francis William Bourdillon, *The Early Editions of the Roman de la Rose*, London 1906.

Bousmanne & Delcourt 2011
Bernard Bousmanne & Thierry Delcourt (eds.), *Miniatures flamandes 1404-1482*, exh. cat. (Brussels, Bibliothèque royale de Belgique/Paris, Bibliothèque nationale de France), Paris 2011.

Bousmanne & Van Hoorebeeck 2000
Bernard Bousmanne & Céline van Hoorebeeck (eds.), *La Librairie des ducs de Bourgogne: manuscrits conservés à la Bibliothèque royale de Belgique*, vol. 1, Turnhout 2000.

Bouza 1995
Fernando Bouza (ed.), *De Mercator a Blaeu: España y la edad de oro de la cartografía en las Diecisiette Provincias de los Países Bajos*, exh. cat. (Madrid, Fundación Carlos de Amberes), Madrid 1995.

Bowen & Imhof 2008
Karen L. Bowen & Dirk Imhof, *Christopher Plantin and Engraved Book Illustrations in Sixteenth-Century Europe*, Cambridge 2008.

Braekman 1985
Willy Louis Braekman (ed.), *Der scaepherders kalengier: een Vlaams volksboek, naar het unieke exemplaar van de Antwerpse druk door Willem Vorsterman van 1513*, Bruges 1985.

Braesel 2004
Michaela Braesel, 'The Influence of Medieval Illuminated Manuscripts on the Pre-Raphaelites and the Early Poetry of William Morris', *Journal of William Morris Studies* 15 (2004): 41-54.

Braet 1989
Herman Braet, 'Der Roman der Rose, Raum im Blick', *Träume im Mittelalter: Ikonographische Studien*, Agostino Paravicini Bagliani & Giorgio Stabile (eds.), Stuttgart 1989: 183-208.

Braet 1994
Herman Braet, 'Aux Sources du Roman de la Rose', *Medieval Codicology, Iconography, Literature, and Translation: Studies for Keith Val Sinclair*, Peter Rolfe Monks & Douglas David Roy Owen (eds.), Leiden 1994: 110-116.

Braet 1996
Herman Braet, 'L'illustration de l'illustration: l'exemple et l'image dans le Roman de la Rose', *Ensi firent li ancessor: Mélanges de philologie médiévale offerts à Marc René Jung*, Luciano Rossi (ed.), Alessandria 1996: 491-504.

Branham 1985
R. B. Branham, 'Utopian Laughter: Lucian and Thomas More', *Thomas More and the Classics - Moreana* 86 (1985): 23-43.

Brayer & Korteweg 2003
Édith Brayer & Anne S. Korteweg (eds.), *Catalogue of French-language Medieval Manuscripts in the Koninklijke Bibliotheek and Meermanno-Westreenianum Museum, The Hague*, cat. on microfiche, Amsterdam 2003.

Bredius & Schmidt-Degener 1906
Abraham Bredius & Frederik Schmidt-Degener (eds.), *Die Großherzogliche Gemälde-galerie im Augusteum zu Oldenburg: 41 Reproductionen in Photogravure*, Oldenburg 1906.

Brewer 1862-1932
J. S. Brewer (ed.), *Letters and papers, foreign and domestic of the reign of Henry VIII. preserved in the Public Record Office, the British Museum and elsewhere in England*, addenda I, vol.1, London 1862-1932.

Brinkmann 1997
Bodo Brinkmann, *Die flämische Buchmalerei am ende des Burgunderreichs: Der Meister des Dresdener Gebetbuchs und die Miniaturisten seiner Zeit*, Turnhout 1997.

Brokken 1990
Ann Brokken (ed.), *Tussen heks en heilige: het vrouwbeeld op de drempel van de moderne tijd, 15de/16de eeuw*, exh. cat. (Sint-Niklaas, Stedelijk Museum), Brussels 1990.

Broos 1988
Ben P. J. Broos (ed.), *Mauritshuis, 's-Gravenhage: gids van het Koninklijk Kabinet van Schilderijen /Mauritshuis, The Hague: guide to the Royal Cabinet of Paintings*, The Hague 1988.

Brosens 2000
Koenraad Brosens, 'Astronomia', *De geleerde wereld van Keizer Karel*, Tineke Padmos & Geert Vanpaemel (eds.), exh. cat. (Leuven, Stedelijk Museum Vander Kelen-Mertens), Leuven 2000: 259-261.

Brown 1952
L. A. Brown (ed.), *The World encompassed: an Exhibition of the History of Maps*, exh. cat. (Baltimore, Baltimore Museum of Art & Walters Art Gallery), Baltimore 1952.

Brown 2000
Wesley A. Brown, *The World Image expressed in the Rudimentum Novitiorum*, Washington 2000.

Brown 2008
Michelle P. Brown, *The Lion Companion to Christian Art*, Oxford 2008.

Bruges 1960
De eeuw der Vlaamse Primitieven, exh. cat. (Bruges, Groeningemuseum), Bruges 1960.

Bruijnen & Janssen 2002
Yvette Bruijnen & Paul Huys Janssen (eds.), *De vier jaargetijden in de kunst van de Nederlanden 1500-1750*, exh. cat. ('s-Hertogenbosch, Noordbrabants Museum/Leuven, Stedelijk Museum Van der Kelen-Mertens), Zwolle 2002.

Brusa 1982
Guiseppe Brusa, *L'arte dell'orologeria in Europa: sette secoli di orologi meccanici*, Bramante 1982.

Brussels/Mechelen 1954
Tentoonstelling Oude Mechelse Kunstnijverheden, exh. cat. (Brussels/Mechelen, Koninklijke musea voor kunst en geschiedenis), Brussels 1954.

Brussels 1969
Quinze années d'acquisitions de la pose de la première pierre à l'inauguration officielle de la Bibliothèque, exh. cat. (Brussels, Bibliothèque royale Albert Ier), Brussels 1969.

Brussels 1973
Le cinquième centenaire de l'imprimerie dans les anciens Pays-Bas, exh. cat. (Brussels, Bibliothèque royale Albert Ier), Brussels 1973.

Bruyn 1987
Josua Bruyn, 'Oude en nieuwe elementen in de 16de-eeuwse voorstellingswereld', *Bulletin van het Rijksmuseum* 35, 3 (1987): 138-163.

Bruyn 1988
Josua Bruyn, 'Old and new Elements in 16th-Century Imagery', *Oud Holland* 102 (1988): 90-113.

Bull 2009-2010
Duncan Bull, 'Jan Gossaert and Jacopo Ripanda on the Capitoline', *Simiolus: Netherlands Quarterly for the History of Art* 34, 2 (2009-2010): 89-94.

Burger 1950
Dycnis Burger, 'Old Tapestries Representing the Seven Liberal Arts', *Archives Internationales d'Histoire des Sciences* 3 (1950): 859-873.

Burckhardt 1923
Rudolf F. Burckhardt, *Gewirkte Bilateppiche des XV. und XVI. Jahrhunderts im Historischen Museum zu Basel*, Leipzig 1923.

Burk 2004
Jens Ludwig Burk, 'Conrat Meit', *The Encyclopedia of Sculpture*, vol. 2, Antonia Boström (ed.), New York/Dearborn 2004: 1036-1039.

Burk 2016
Jens Ludwig Burk, 'Meit (Meijt; Meyt; Mit; Myt), Conrad(t)', *Allgemeines Künstlerlexikon. Die Bildenden Künstler aller Zeiten und Völker - Bd. 88 (Matijin – Meixner)*, Andreas Beyer, Bénédicte Savoy & Wolf Tegethoff (eds.), Berlin/Boston 2016: 536-538.

Burroughs 1927
Bryson Burroughs, 'The Descent of Christ into Hell by Hieronymus Bosch', *Metropolitan Museum of Art Bulletin* 22 (1927): 272-274.

Büttner 2011
Nils Büttner, 'Weil die Erinnerung an diesen famosen Mann nicht sterben darf: das Grab- und Ehrenmal des Malers Quentin Massys († 1530)', *Künstlergrabmäler. Genese – Typologie – Intention – Metamorphosen*, Birgit Ulrike Münch, Markwart Herzog & Andreas Tacke (eds.), Petersberg 2011: 60-79.

Büttner 2014
Nils Büttner, 'Fälschung, Plagiat und Kopie nach Hieronymus Bosch', *Fälschung - Plagiat - Kopie: Künstlerische Praktiken in der Vormoderne*, Birgit Ulrike Münch, Andreas Tacke, Markwart Herzog & Sylvia Heudecker (eds.), Petersberg 2014: 27-40.

Butts & Hendrix 2000
Barbara Butts & Lee Hendrix, *Painting on Light. Drawings and Stained Glass in the Age of Dürer and Holbein*, exh. cat. (Los Angeles, J. Paul Getty Museum /St. Louis, The Saint Louis Art Museum), Los Angeles 2000.

Buyens 2009
Vincent Buyens, 'Over edele tegenvoeters en welvarende Hollanders: ideaal en droom in de zeventiende-eeuwse literatuur', *Paradijzen van papier. Utopie in de Nederlandse literatuur*, Anne Decelle, An Faems & Tom Sintobin (eds.), Leuven 2009: 43-68.

Bynum 1992
Carolyn Walker Bynum, *Fragmentation and Redemption: Essays on Gender and the Human Body in Medieval Religion*, New York 1992.

Bynum 2011
Carolyn Walker Bynum, *Christian Materiality. An Essay on Religion in Late Medieval Europe*, New York 2011.

Byron 2011
John Byron, *Cain and Abel in Text and Tradition. Jewish and Christian Interpretations of the First Sibling Rivalry*, Themes in Biblical Narrative, 14, Leiden 2011.

Cambridge 1955
Illuminated & Calligraphic Manuscripts, exh. cat. (Cambridge, Fogg Art Museum & Houghton Library), Cambridge 1955.

Camille 1998
Michael Camille, *Medieval art of love: objects and subjects of desire*, New York 1998.

Campbell, Phillips, Schulte Herbrüggen & Trapp 1978
Lorne Campbell, Margaret Mann Phillips, Hubertus Schulte Herbrüggen & Joseph Burney Trapp, 'Quentin Matsys, Desiderius Erasmus, Pieter Gillis and Thomas More', *Burlington Magazine* 120 (1978): 716-725.

Campbell 1985
Lorne Campbell, *The Early Flemish Pictures in the Collection of Her Majesty The Queen*, Cambridge 1985.

Campbell 1987
Tony Campbell, *The Earliest Printed Maps 1472-1500*, London 1987.

Campbell 1990
Lorne Campbell, *Renaissance Portraits: European Portrait-Painting in the 14th, 15th and 16th Centuries*, New Haven/ London 1990.

Campbell 2002
Thomas P. Campbell (ed.), *Tapestry in the Renaissance: Art and Magnificence*, exh. cat. (New York, Metropolitan Museum of Art), New Haven/London 2002.

Campbell et al. 2008
Lorne Campbell et al., *Renaissance Faces: from Van Eyck to Titian*, exh. cat. (London, National Gallery) 2008.

Campbell 2014
Lorne Campbell, *National Gallery Catalogues, The Sixteenth Century Netherlandish Paintings with French Paintings before 1600*, 2 vols., London 2014.

Cantimpratensis 1973
Thomas Cantimpratensis, *Liber de naturis rerum creatarum*, Berlin/New York 1973.

Caraci & Destombes 1938
Giuseppe Caraci & Marcel Destombes, 'Lopo Homem and the Miller Atlas of 1519', *The Geographical Journal* 91, 3 (1938): 263-266.

Cardon 1996
Bert Cardon, *Manuscripts of the Speculum Humanae Salvationis in the Southern Netherlands (c.1410 - c.1470). A Contribution to the Study of the 15th-Century Book Illumination and of the Function and Meaning of Historical Symbolism*, Corpus of Illuminated Manuscripts, 6, Leuven 1996.

Cardon, Van der Stock & Vanwijnsberghe 2002
Bert Cardon, Jan Van der Stock & Dominique Vanwijnsberghe (eds.), *'Als Ich Can.' Liber Amicorum in Memory of Professor Dr. Maurits Smeyers*, 2 vols., Corpus of Illuminated Manuscripts, 11, Leuven 2002.

Carroll 2011
Margaret D. Carroll, 'Bosch's Iron Age', *Masculinities, Childhood, Violence: Attending to Early Modern Women - and Men: Proceedings of the 2006 Symposium*, Amy E. Leonard & Karen L. Nelson (eds.), Newark 2011: 239-266.

Cavagna 2008
Mattia Cavagna (ed.), *La Vision de Tondale: Les versions françaises de Jean de Vignay, David Aubert, Regnaud le Queux*, Paris 2008.

Cavalli-Björkman 1993
Görel Cavalli-Björkman, 'Hieronymus in der Studierstube und das Vanitasstilleben', *Leselust: Niederländische Malerei von Rembrandt bis Vermeer*, Ann J. Adams & Sabine Schulze (eds.), Frankfurt am Main/ Stuttgart 1993: 47-53.

Cavallo 1993
Adolfo Salvatore Cavallo, *Medieval Tapestries in The Metropolitan Museum of Art*, New York 1993.

Cave 1991
Alfred A. Cave, 'Thomas More and the New World', *Albion. A Quarterly Journal concerned with British Studies* 23 (1991): 209-229.

Cave 2008a
Terence Cave (ed.), *Thomas More's Utopia in Early Modern Europe. Paratexts and contexts*, Manchester 2008.

Cave 2008b
Terence Cave (ed.), 'The English translation: thinking about the commonwealth', *Thomas More's Utopia in Early Modern Europe. Paratexts and contexts*, Terence Cave (ed.), Manchester 2008: 87-103.

Cecchi, Hersant & Rabbi-Bernard 2013
Alessandro Cecchi, Yves Hersant & Chiara Rabbi-Bernard (eds.), *La Renaissance et le Rêve - Bosch, Véronèse, Greco*, exh. cat., (Florence, Palazzo Pitti/Paris, Musée du Luxembourg), Paris 2013.

Chamberlain 1913
Arthur B. Chamberlain, *Hans Holbein the Younger*, vol. 1, London 1913.

Chambers 1935
R. W. Chambers, *Thomas More*, London 1935.

Chapiro, Meslin-Perrier & Turner 1989
Adolphe Chapiro, Chantal Meslin-Perrier & Anthony Turner, *Catalogue de l'horlogerie et des instruments de précision: du début du XVIe au milieu du XVIIe siècle*, Paris 1989.

Charron 2011
Pascale Charron, 'Les Arts libéraux dans la tapisserie à la fin du Moyen Âge: entre iconographie savante et pratiques d'atelier', *L'allégorie dans l'art du Moyen Age. Formes et fonctions. Héritages, créations, mutations* (conference proceedings, RILMA, Institut Universitaire de France (INHA), Paris, 27-29 May 2010), Christian Heck (ed.), Turnhout 2011: 331-344.

Charron & Gil 1999
Pascale Charron & Marc Gil, 'Les enluminures des manuscrits de David Aubert', *Les manuscrits de David Aubert*, Danielle Quéruel (ed.), Paris 1999: 81-100.

Chatelain 1967
M. J. Chatelain (ed.), *Musée de Douai, Peintures*, Douai 1967.

Checa Cremades 1999
Fernando Checa Cremades, *Carlos V - La imagen del poder en el Renacimiento*, Madrid 1999.

Checa Cremades 2000
Fernando Checa Cremades (ed.), *Carolus: quinto centenario del nacimiento de Carlos V*, exh. cat. (Toledo, Museo de Santa Cruz), Madrid 2000.

Chipps Smith 1983
Jeffrey Chipps Smith, *Nuremberg, A Renaissance City, 1500-1618*, exh. cat. (Austin, The University of Texas - the Archer M. Huntington Art Gallery), Austin 1983.

Chipps Smith 1994
Jeffrey Chipps Smith, *German Sculpture of the Later Renaissance 1520-1580. Art in an Age of Uncertainty*, Princeton/New Jersey 1994.

Chipps Smith 2011
Jeffrey Chipps Smith, 'Panofsky's Dürer', *Dürer, l'Italia e l'Europa* (conference proceedings, Rome, 23-25 April 2007), Sybille Ebert-Schifferer & Kristina Hermann Fiore (eds.), Milano 2011: 186-193.

Chirol 1980-1981
Elisabeth Chirol, 'L'Isle de Brésil,' *La Renaissance à Rouen*, exh. cat. (Rouen, Musée des Beaux-Arts), Rouen 1980-1981.

Chirol 1984
Elisabeth Chirol, *La Renaissance et le Nouveau Monde*, exh. cat. (Québec, Musée du Québec), Québec 1984.

Cicéron 1980
Cicéron, *La République - Tome I: Livre I*, Esther Bréguet (transl.), Paris 1980.

Claerhout 1976
Adriaan Claerhout (ed.), *Stad Antwerpen. Oudheidkundige musea. Etnografisch museum. Tentoonstelling aanwinsten /A. Claerhout*, exh. cat. (Antwerp, Hessenhuis), Antwerp 1976.

Claerhout 1989
Adriaan Claerhout (ed.), 'Een Afro-Portugees ivoorsnijwerk', *Openbaar Kunstbezit in Vlaanderen* 27, 1 (1989): 4.

Claeys 2011
Gregory Claeys, *Searching for Utopia. The History of an Idea*, London 2011.

Clemen 1899
Paul Clemen, *Die Kunstdenkmaeler der Rheinprovinz - vol. 2, pt. 3: Kreis Bergheim*, Düsseldorf 1899.

Clobus & Hébert 2009
Jean-Claude Clobus & Brigitte Hébert, 'Le monde, image par image: la "Chronique de Nuremberg"', *Pouvoirs de l'image aux 15e, 16e et 17e siècles: pour un nouvel éclairage sur la pratique des Lettres à la Renaissance*, Marie Couton et al. (eds.), Clermont-Ferrand 2009: 35-56.

Cloulas-Brousseau 1963
Annie Cloulas-Brousseau, 'Le Jubé de la Cathédrale de Limoges,' *Bulletin de la société archéologique et historique du Limousin*, Limoges 1963: 101-188.

Cockshaw & Van den Bergen-Pantens 1996
Pierre Cockshaw & Christiane van den Bergen-Pantens, *L'ordre de la Toison d'or de Philippe le Bon à Philippe le Beau (1430-1505). Idéal ou reflet d'une société?*, Turnhout/Brussels 1996.

Cockx-Indestege, Glorieux & Op de Beeck 1968-94
Elly Cockx-Indestege, Geneviève Glorieux & Bart Op de Beeck, *Belgica Typographica 1541-1600. Catalogus librorum impressorum ab anno MDXLI ad annum MDC in regionibus quæ nunc Regni Belgarum partes sunt*, 4 vols., Nieuwkoop 1968-94.

Cole & Gealt 1991
Bruce Cole & Adelheid M. Gealt, *Art of the Western World. From Ancient Greece to Post-modernism*, New York 1991.

Colin 1987
Susi Colin, 'The Wild Man and the Indian in Early 16th Century Book Illustration', *Indians and Europe: an Interdisciplinary Collection of Essays*, Christian Feest (ed.), Aachen 1987: 5-36.

Colin 1988
Susi Colin, *Das Bild des Indianers im 16. Jahrhundert*, Idstein 1988.

Collobi Ragghianti 1990
Licia Collobi Ragghianti, *Dipinti Fiamminghi in Italia 1420-1570. Catalogo*, Bologna 1990.

Cologne 1862
Illustrierte Katalog der reichen Gemälde-Galerie des Herrn J.P. Weyer, Stadtbaumeister des Leopold-Ordens, auction cat., Cologne 1862.

Cooper 2008
Tarnya Cooper, *A Guide to Tudor and Jacobean Portraits*, London 2008.

Cordemans de Bruyne 1895
H. Cordemans de Bruyne, 'Histoire de l'Art Typografique à Malines', *Handelingen van de Mechelse Kring voor Oudheidkunde, Letteren en Kunst* 6 (1895): 20-111.

Cortes Hernández 1982
Susana Cortes Hernández, *Dos series de tapices flamencos en el Museo de Santa Cruz de Toledo*, Madrid 1982.

Cortes Hernández 2002
Susana Cortes Hernández, *Tapices flamencos en Toledo. Catedral y Museo de Santa Cruz*, Madrid 2002.

Cortesão & Teixeira da Mota 1987
Armando Cortesão & Avelino Teixeira da Mota, *Portugaliae monumenta cartographica, Descoberta do mundo*, Lisbon 1987.

Corwin 1976
Nancy Corwin, *The Fire Landscape: Its Sources and Its Development from Bosch Through Jan Brueghel I, With Special Emphasis on the Mid-Sixteenth Century Bosch "Revival"*, PhD dissertation, University of Washington, Washington 1976.

Costa 1964
Dominique Costa (ed.), *Bois sculptés: l'art du bois du XIIIe au XIXe siècle*, exh. cat. (Nantes, Musée Dobrée), Nantes 1964.

Crab et al. 1971
Jan Crab et al., *Aspecten van de Laatgotiek in Brabant*, exh. cat. (Leuven, Stedelijk Museum Vander Kelen-Mertens), Leuven 1971.

Crab & Smeyers 1979
Jan Crab & Maurits Smeyers (eds.), *Het laatgotisch beeldsnijcentrum Leuven*, exh. cat. (Leuven, Stedelijk Museum Vander Kelen-Mertens), Leuven 1979.

Cro 1994
Stelio Cro, 'Classical Antiquity, America, and the Myth of the Noble Savage,' *The Classical Tradition and the Americas - Volume I: European Images of the Americas and the Classical Tradition*, Wolfgang Haase & Reinhold Meyer (eds.), Berlin 1994: 379-418.

Crowther 2010
Kathleen M. Crowther, *Adam and Eve in the Protestant Reformation*, Cambridge 2010.

Curnow 2007
Katy Curnow, *Benin. Kings and Rituals. Court Arts from Nigeria*, Kathleen Bickford Berzock (ed.), exh. cat. (Vienna, Museum für Völkerkunde - Kunsthistorisches Museum/ Paris, Musée du quai Branly/Berlin, Ethnologisches Museum - Staatliche Museen/Chicago, The Art Institute of Chicago), New Haven 2007.

Cuttler 1969
Charles D. Cuttler, 'Two Aspects of Bosch's Hell Imagery', *Scriptorium* 23, 2 (1969): 313-319.

CWE
Collected Works of Erasmus, 86 vols., Toronto/Buffalo/London 1974- .

Dackerman 2002
Susan Dackerman, *Painted Prints: The Revelation of Color in Northern Renaissance and Baroque Engravings, Etchings and Woodcuts*, exh. cat. (Baltimore, Museum of Art/Saint Louis, Saint Louis Art Museum), Pennsylvania 2002.

Dackerman 2011
Susan Dackerman (ed.), *Prints and the Pursuit of Knowledge in Early Modern Europe*, exh. cat. (Cambridge, Arthur M. Sackler Museum/Evanston, the Mary and Leigh Block Museum of Art), New Haven/London 2011.

Dallas 1947
30 Masterpieces: An Exhibition of Paintings from the Collection of the Metropolitan Museum of Art, exh. cat. (Dallas, Dallas Museum of Fine Arts), Dallas 1947.

Danesi Squarzina 1995
Silvia Danesi Squarzina, *"Fiamenghi che vanno e vengono non li si puol dar regola". Paesi Bassi e Italia fra Cinquecento e Seicento: pittura, storia e cultura degli emblem*, Sant'Oreste 1995.

Davenport & Cabanillas Cárdenas 2008
Randi Lise Davenport & Carlos F. Cabanillas Cárdenas, 'The Spanish Translations: Humanism and Politics', *Thomas More's Utopia in Early Modern Europe. Paratexts and contexts*, Terence Cave (ed.), Manchester 2008: 110-127, 233-271.

Davies 1993
Alice I. Davies, *16th- and 17th-Century Dutch and Flemish Paintings in the Springfield Museum of Fine Arts*, Springfield 1993.

Davies 1998
Cliff S. L. Davies, 'Tournai and the English Crown, 1513-1519', *The Historical Journal* 41 (1998): 1-26.

Davies 2012
Surekha Davies, 'Depictions of Brazilians on French Maps, 1542-1555,' *The Historical Journal* 55, 2 (2012): 317-348.

Davies 2016
Surekha Davies, *Renaissance Ethnography and the Invention of the Human - New Worlds, Maps and Monsters*, Cambridge 2016.

Daynes-Diallo 2009
Sophie Daynes-Diallo, 'Globe céleste', *Album Musée National de la Renaissance Château d'Ecouen*, mus. cat. (Musée national de la Renaissance, Ecouen), Paris 2009: 88.

Debae 1995
Marguerite Debae, *La bibliothèque de Marguerite d'Autriche. Essai de reconstitution d'après l'inventaire de 1523-1524*, Leuven 1995.

De Borghgrave d'Altena 1959
J. De Borghgrave d'Altena, 'Statuettes Malinoise', *Bulletin des Musées Royaux d'Art et d'Histoire* 31, 4 (1959): 2-98.

De Bosque 1975
Andrée de Bosque, *Quinten Metsys*, Brussels 1975.

De Bruijne & Van der Coelen 2012
Rozanne de Bruijne & Peter van der Coelen (eds.), *Maarten van Heemskerck: Het oude Rome herleeft = Ancient Rome relives*, exh. cat. (Rotterdam, Museum Boijmans Van Beuningen), Rotterdam 2012.

Decavele 1976
Johan Decavele (ed.), *Eenheid en scheiding in de Nederlanden, 1555-1585*, exh. cat. (Ghent, Centrum voor Kunst en Cultuur), Ghent 1976.

De Farcy 1913
Louis de Farcy, 'Tapisserie tournaisienne de 1502 à 1504', *Les arts anciens de Flandre* 6 (1913): 105-111.

De Kesel 2010
Lieve de Kesel, 'Heritage and Innovation in Flemish Book Illumination at the Turn of the Sixteenth Century', *Books in Transition at the Time of Philip the Fair*, Hanno Wijsman (ed.), Turnhout 2010: 93-130.

De Kesel 2011
Lieve de Kesel, 'None is More Spendid than the *Roman de la Rose* of Count Engelbert II of Nassau. Some Considerations of the Silencing of Male Bouche in Harley 4425 of the British Library in London', *Nouvelles de la Rose: Actualité et perspectives du Roman de la Rose*, Dulce-María Gonsálvex-Doreste & María del Pilar Mendoza-Ramos (eds.), La Laguna 2011: 61-77.

De Kesel 2013
Lieve de Kesel, *The Master of the Prayer Books of around 1500 in the Spotlight. An Investigation of the Artist(s), the Oeuvre, their Colleagues and their Patrons*, PhD dissertation, Ghent University, Ghent 2013.

Dekker 1995
Elly Dekker, 'Conspicuous Features on Sixteenth Century Celestial Globes', *Der Globusfreund* 43-44 (1995): 77-106.

Dekker 1999
Elly Dekker (ed.), *Globes at Greenwich: A Catalogue of the Globes and Armillary Spheres in the National Maritime Museum, Greenwich*, mus. cat. (Greenwich, National Maritime Museum), Oxford/Greenwich 1999.

Dekker & Van der Krocht 1994
Elly Dekker & Peter van der Krocht, 'De Globes', *Gerardus Mercator Rupelmundanus*, Marcel Watelet (ed.), Antwerp 1994: 242-267.

Dekker 2012
Elly Dekker, *Illustrating the phaenomena: celestial cartography in Antiquity and the Middle Ages*, Oxford 2012.

De la Garanderie 1989
Marie-Madeleine de la Garanderie, 'Guillaume Budé, lecteur de l'*Utopie*', *Miscellanea Moreana. Essays for Germain Marc'hadour - Moreana* 100, 26 (1989): 327-338.

Delaissé 1959
Léon Marie Joseph Delaissé (ed.), *Le Siècle d'Or de la miniature Flamande: Le mécénat de Philippe le Bon*, exh. cat. (Brussels, Palais des Beaux-Arts), Brussels 1959.

De la Quérière 1821
E. de la Quérière, *Description historique des maisons de Rouen, les plus remarquables par leur décoration extérieure et par leur ancienneté*, Paris 1821.

De Léry 1994
Jean de Léry, *Histoire d'un voyage faict en la terre du Brésil (1578)*, Frank Lestringant (ed.), Paris 1994.

Delmarcel 1976
Guy Delmarcel (ed.), *Tapisseries bruxelloises de la pré-Renaissance*, exh. cat. (Brussels, Musées Royaux d'Art et d'Histoire), Brussels 1976.

Delmarcel 1999
Guy Delmarcel, *Het Vlaamse wandtapijt van de 15de tot de 18de eeuw*, Tielt 1999.

De Lorris & De Meun 1995
Guillaume de Lorris & Jean de Meun, *The Romance of the Rose*, Charles Dahlberg (transl.), New York 1995.

Demmler 1918
Theodor Demmler, *Kunstwerke aus dem besetzten Nordfrankreich ausgestellt im Museum zu Valenciennes*, Munich 1918.

De Moraes Belluzo 1994
'A Ilha do Brasil: O Embarque, Corte e Transporte do Pau-Brasil, 1530 c.', *O Brasil dos Viajantes*, Ana Maria de Moraes Belluzo (ed.), exh. cat. (Sao Paulo, Museu de Arte de São Paulo/Lisbon, Centro Cultural de Belém), Sao Paulo 1994.

Demus 1981
Klaus Demus, *Bruegel, Pieter d. Ältere. Flämische Malerei von Jan Van Eyck bis Pieter Bruegel d. Ä. Katalog der Gemäldegalerie der Kunsthistorischen Museum Wien*, mus. cat. (Vienna, Gemäldegalerie der Kunsthistorischen Museum), Vienna 1981.

De Nave, Imhoff & Otte 1994
Francine de Nave, Dirk Imhoff & Els Otte (eds.), *Gerard Mercator en de geografie in de Zuidelijke Nederlanden (16e eeuw)*, exh. cat. (Antwerp, Museum Plantin-Moretus), Antwerp 1994.

De Nijn 2000
Heidi de Nijn, *Het Mechelse meubel 1500 - 2000: van houtsnijwerk tot design*, exh. cat. (Mechelen, Stedelijke Musea Mechelen), Mechelen 2000.

De Nijn & Eeman 1998
Heidi de Nijn & Michèle Eeman, *800 jaar Onze-Lieve-Vrouwegasthuis. Uit het erfgoed van de Mechelse gasthuiszusters en het OCMW*, exh. cat. (Mechelen, Stedelijk Museum Hof van Busleyden), Mechelen 1998.

Denucé 1908
Jean Denucé, *Les origines de la cartographie portugaise et les cartes des Reinel*, Ghent 1908.

De Palmenaer 1999
Els de Palmenaer, *Carolus, Charles Quint 1500-1558*, Hugo Soly & Johan Van de Wiele (eds.), exh. cat. (Ghent, Kunsthal De Sint-Pietersabdij), Ghent 1999.

De Palmenaer & Parmentier 2015
Els de Palmenaer & Jan Parmentier (eds.), *De wereld gespiegeld: wereldkaarten van de middeleeuwen tot nu*, exh. cat. (Antwerp, Museum aan de Stroom), Kontich 2015.

De Pontfarcy 2010
Yolande de Pontfarcy, *L'au-delà au Moyen Age*, Bern 2010.

Deproost, Van Parijs, Nyns & Vielle 2015
Chemins d'Utopie. Thomas More à Louvain, 1506-2016, Paul-Augustin Deproost, Philippe van Parijs, Charles-Henri Nyns & Christophe Vielle (eds.), Louvain-la-Neuve 2015.

Derez 1999
Mark Derez, 'Prologue: an attempt at topography', *Leuven in books. Books in Leuven. The oldest University of the Low Countries and its Library*, Christian Coppens (ed.), exh. cat. (Philadelphia, Van Pelt Library), Leuven 1999: 7-12.

Derez & Verbrugge 2005
Mark Derez & Anne Verbrugge (eds.), *De Celestijnenpriorij te Heverlee. Van klooster tot bibliotheek*, exh. cat. (Heverlee, Campusbibliotheek Arenberg KU Leuven), Leuven 2005.

Derolez 1979
Albert Derolez, *The Library of Raphael de Mercatellis, Abbot of St. Bravo's, Ghent, 1437-1508*, Ghent 1979.

Derolez, De Kok & De Schryver 1984
Albert Derolez, H. de Kok & Antoine de Schryver (eds.), *Sint-Arnoldus en de Sint-Pietersabdij te Oudenburg 1084–1984*, exh. cat. (Oudenburg, Sint-Pietersabdij), Oudenburg 1984.

De Roy 2011
Judy de Roy, *Borstbeeld Keizer Karel, toegeschreven aan Conrat Meit, materiaaltechnisch vooroderzoek* (Internal Research Report for the Royal Institute for Cultural Heritage), Brussels 2011.

De Roy 2013
Judy de Roy, *Technical examination of a group of similar portrait busts of Charles V: Preliminary results* (paper presented at the 2013 Conference 'Copia e invencion. Modelos, réplicas, series y citas en la escultura Europea', Museo Nacional de Escultura, (Valladolid), Valladolid 2013.

De Roy 2016
Judy de Roy, 'Onderzoek van een groep gelijkaardige portretbustes van Keizer Karel', *Bulletin van het Koninklijk Instituut voor het Kunstpatrimonium* 34 (2016): 97-107.

De Roy & Van Audenaeren 2015
Judy de Roy & Kristel van Audenaeren, 'De portretbuste van Karel V: Een materiaaltechnische studie', *Museum Bulletin - Brugge Musea* 35 (2015): 10-13.

Derrett 1962
J. Duncan M. Derrett, 'Thomas More and Joseph the Indian', *Journal of the Royal Asiatic Society* (1962): 18-34.

Derrett 1966
J. Duncan M. Derrett, 'The Utopian alphabet', *Moreana* 12 (1966): 61-66.

De Selliers 2006
Diane de Selliers, *Le Moyen Âge flamboyant: Poésie et peinture*, Paris 2006.

De Smet 1964
Antoine de Smet, 'Gaspard van der Heyden', *Nationaal Biografisch Woordenboek*, pt. 1, Brussels 1964.

De Smet 1968
Antoine de Smet, *Les sphères Terrestre et Céleste de Gérard Mercator 1541 et 1551. Reproductions anastatiques des fuseaux originaux gravés par G. Mercator et conservés à la Bibliothèque royale à Bruxelles. Préface de M. Antoine De Smet*, Brussels 1968.

De Smet 1969
Antoine de Smet, *Tentoonstelling van globen in het raam van het 3de internationaal symposium van de Coronelli-Weltbund der Globusfreunde en van de 3de internationale conferentie over de geschiedenis van de cartografie*, exh. cat. (Brussels, Bibliothèque royale Albert Ier), Brussels 1969.

De Smet 1974a
Antoine de Smet, 'Das Interesse für Globen in den Niederlanden in der ersten Hälfte des 16.Jahrhunderts', *Album Antoine de Smet*, Brussels 1974: 183-191.

De Smet 1974b
Antoine de Smet, 'Mercator à Louvain, 1530-1552', *Album Antoine de Smet*, Brussels 1974: 193-250.

De Solla Price 1975
Derek de Solla Price, 'Science and technology in the fifteenth century', *The Secular Spirit: Life and Art at the End of the Middle Ages*, Timothy B. Husband & Jane Hayward (eds.), New York 1975: 174-203.

Desonay 1930
Antoine de la Sale, *Le Paradis de la Reine Sibilla*, F. Desonay (ed.), Paris 1930.

Destombes 1937
Marcel Destombes, 'Lopo Homem's Atlas of 1519', *The Geographical Journal* 90, 5 (1937): 460-464.

Destombes 1939
Marcel Destombes, 'La Mappemonde Lopo Homem et l'Atlas Miller' *The Geographical Journal* 94, 6 (1939): 485-492.

Destombes 1968
Marcel Destombes, 'Un globe céleste inédit de l'Année 1502', *Actes du XI Congrès International d'Histoire des Sciences* 3 (1968): 73-81.

Destombes & Gernez 1987
Marcel Destombes & Desiré Gernez, 'Un atlas nautique du XVIème siècle à la Bibliothèque Royale de La Haye (Pays-Bas)', *Marcel Destombes (1905-1983). Contributions selectionnées à l'histoire de la cartographie et des instruments scientifiques*, vol. 3, Günther Schilder, Peter van der Krogt & Steven de Clercq (eds.), Utrecht 1987: 141-151.

Destrée 1896
Joseph Destrée, *Les Heures de Notre-Dame dites de Hennessy*, Brussels 1896.

De Tolnay 1937
Charles de Tolnay, *Hieronymus Bosch*, Basel 1937.

De Tolnay 1965
Charles de Tolnay, *Hieronymus Bosch*, Baden-Baden 1965.

Devereux 1975
E. J. Devereux, 'Thomas More and his printers', *A Festschrift for Edgar Ronald Seary. Essays in English language and literature presented by colleagues and former students*, A. A. MacDonald, P. A. O'Flaherty & G. M. Story (eds.), St. John's 1975: 40-57.

Devliegher 1960
Luc Devliegher, 'Het beeld van Keizer Karel in het Brugse Gruuthusemuseum', *Handelingen van het Genootschap voor Geschiedenis* 97 (1960): 240-241.

De Vries 1934
A. B. de Vries, *Het Noord-Nederlandsch portret in de tweede helft van de 16de eeuw*, Amsterdam 1934.

De Vrij 2006
Marc Rudolf de Vrij, *Jan Mandyn*, Zwanenburg 2006.

De Vrij 2012
Marc Rudolf de Vrij, *Jheronimus Bosch. An Exercise in Common Sense*, Hilversum 2012.

Deuter 2001
Jörg Deuter, *Johann Heinrich Wilhelm Tischbein als Sammler*, Oldenburg 2001.

Dickason 1984
Olive Patricia Dickason, 'The Brazilian Connection: A Look at the Origin of French Techniques for Trading with Amerindians,' *Revue française d'histoire d'outre-mer* 71, 264-265 (1984): 129-146.

Didi-Huberman 1995
Georges Didi-Huberman, *Dissemblance and figuration: Fra Angelico*, Chicago 1995.

Dienst 2002
Barbara Dienst, *Der Kosmos des Peter Flötner: eine Bildwelt der Renaissance in Deutschland*, Munich/Berlin 2002.

Dinzelbacher 2002
Peter Dinzelbacher, *Himmel, Hölle, Heilige: Visionen und Kunst im Mittelalter*, Darmstadt 2002.

Dodgson 1907
Campbell Dodgson, 'Hans Lützelburger and the Master N.H.', *The Burlington Magazine for Connoisseurs* 47, 10 (1907): 319-322.

Dodgson 1911
Campbell Dodgson, *Catalogue of Early German and Flemish Woodcuts Preserved in the British Museum*, vol. 2, London 1911.

Dodgson 1926
Campbell Dodgson, 'Some of Dürer's Studies for Adam and Eve', *The Burlington Magazine for Connoisseurs* 48, 279 (1926): 308-311.

Dodgson 1933
Campbell Dodgson, 'Rare Woodcuts in the Ashmolean Museum. II', *The Burlington Magazine for Connoisseurs* 63, 366 (1933): 114-117 & 120-121.

Dogaer 1987
Georges Dogaer, *Flemish Miniature Painting in the 15th and 16th Centuries*, Amsterdam 1987.

Dogaer & Debae 1967
Georges Dogaer & Marguerite Debae, *La librairie de Philippe le Bon*, exh. cat. (Brussels, Bibliothèque royale Albert Ier), Brussels 1967.

Dohrn-van Rossum 1997
Gerhard Dohrn-van Rossum (ed.), *L' Histoire de l'heure: L'horlogerie et l'organisation moderne du temps*, Paris 1997.

Donath 1926
Adolph Donath (ed.), 'Der neuentdeckte Hieronymus Bosch', *Kunstwanderer* 8 (1926): 378-379.

Dörnhöffer 1904
Friedrich Dörnhöffer, 'Röttinger, H. – Hans Weiditz, der Petrarcameister', *Kunstgeschichtliche Anzeigen* 1904: 51-66.

Dotseth, Anderson & Roglán 2008
Amanda Dotseth, Barbara Anderson & Mark Roglán (eds.), *Fernando Gallego and his Workshop: The Altarpiece from Ciudad Rodrigo, Paintings from the Collection of the University of Arizona Museum of Art*, exh. cat. (Dallas, Meadows Museum), Dallas 2008.

Douai 1953
Les Trésors Artistiques du Musée et de la Bibliothèque de Douai, exh. cat. (Douai, Bibliothèque de Douai), Douai 1953.

Doutrepont 1909
Georges Doutrepont, *La littérature française a la cour des Ducs de Bourgogne*, Paris 1909.

Dubois 2013
Anne Dubois, 'Les Histoires que l'on peut raisonnablement faire sur les livres… Un cycle iconographique écrit pour listoire de madame sainte Katherine', *Art de l'enluminure* 45 (2013): 36-37.

Dubois 2016
Anne Dubois, 'Some technical notes on the "Fleur des Histoires" (Brussels, KBR, ms 9231-9232) by the Mansel Master, the Thérouanne Master and Simon Marmion', *Inside Illuminations. Art Technical Research & the Illuminated Manuscript. Proceedings of the study day held on 5 June 2014*, Anne Dubois, Lieve Watteeuw & Marina Van Bos (eds.), Leuven 2016.

Dubosc 1921
Georges Dubosc, *Rouen et son port a travers les âges*, exh. cat. (Rouen, Palais des Consuls), Rouen 1921.

Duchâteau & Van Damme 1990
Armand Duchâteau & Wilfried van Damme, *Benin: Hofkunst uit Afrika*, Brussels 1990.

Duprat 1973
Gabrielle Duprat, 'Les globes terrestres et celestes en France', *Der Globusfreund* 21 (1973): 198-225.

Du Sommerard 1883
E. du Sommerard, *Musée des Thermes et de l'Hôtel de Cluny, Catalogue et description des objets d'art de l'Antiquité, du Moyen Age et de la Renaissance*, Paris 1883.

Durrieu 1921
Paul Durrieu, *La Miniature flamande au temps de la cour de Bourgogne (1415-1530)*, Brussels/Paris 1921.

Duverger 1934
Jozef Duverger, *Conrat Meijt (ca. 1480-1551)*, Ghent 1934.

Duverger 1984
Erik Duverger (ed.), *Antwerpse kunstinventarissen uit de zeventiende eeuw*, vol. 1: *1600-1617*, Fontes historiae artis Neerlandicae. Bronnen voor de kunstgeschiedenis van de Nederlanden, 1, Brussels 1984.

Eaton 2001
Ruth Eaton, *Ideal Cities. Utopianism and the (Un) Built Envoronnement*, Antwerp 2001.

Edgerton Jr. 1987
Samuel Y. Edgerton Jr., 'From Mental Matrix to Mappamundi to Christian Empire: The Heritage of Ptolemaic Cartography in the Renaissance', *Art and Cartography: Six Historical Essays*, David Woodward (ed.), Chicago/London 1987: 10-50.

Edson 2007
Evelyn Edson, *The World Map, 1300-1492: The Persistence of Tradition and Transformation*, Baltimore 2007.

Eeckhout 1985
Paul Eeckhout, 'Conrad Meyt (toegeschreven aan). Borstbeeld van Keizer Karel', *Luister van Spanje en de Belgische steden 1500-1700*, 2 vols., Jean-Marie Duvosquel, Ignace Vandevivere et al. (eds.), exh. cat. (Brussels, Palais des Beaux-Arts), Brussels 1985: 400.

Egmond 2005
Florike Egmond, *Het Visboek. De wereld volgens Adriaen Coenen 1514-1587*, Zutphen 2005.

Egmond & Mason 2000
Florike Egmond & Peter Mason, 'These are the People who Eat Raw Fish: Contours of the Ethnographic Imagination in the Sixteenth Century', *Viator. Medieval and Renaissance Studies* 31 (2000): 311-360.

Eichberger 1992
Dagmar Eichberger, 'Image follows Text? The Visions of Tondal and its Relationship to Depictions of Hell and Purgatory in Fifteenth Century Illuminated Manuscripts', *Margaret of York, Simon Marmion, and The Visions of Tondal* (papers delivered at the 1990 Symposium organized by the Department of Manuscripts of the J. Paul Getty Museum in collaboration with the Huntington Library and Art Collections), Thomas Kren (ed.), Malibu 1992: 129-140.

Eichberger 2002
Dagmar Eichberger, *Leben mit Kunst, Wirken durch Kunst: Sammelwesen und Hofkunst unter Margarete von Österreich, Regentin der Niederlande*, Turnhout 2002.

Eichberger 2005
Dagmar Eichberger (ed.), *Women of Distinction: Margaret of York, Margaret of Austria*, exh. cat. (Mechelen, Lamot), Leuven 2005.

Eikelmann & Burk 2006
Renate Eikelmann & Jens Ludwig Burk (eds.), *Conrat Meit. Bildhauer der Renaissance*, exh. cat. (Munich, Bayerisches Nationalmuseum), Munich 2006.

Eisler 1977
Colin Eisler, *Paintings from the Samuel H. Kress Foundation, European Schools excluding Italian*, Oxford 1977.

Elkhadem, Heerbrant, Calcoen & Wellens-De Donder 1994
Hossam Elkhadem, Jean-Paul Heerbrant, Roger Calcoen & Liliane Wellens-De Donder (eds.), *Gerard Mercator: cartograaf 1512-1594*, exh. cat. (Brussels, Bibliothèque royale Albert Ier), Brussels 1994.

Elkhadem 2005
Hossam Elkhadem, 'Claude Ptolémée, Geographia et Cosmographia', *Cent trésors de la Bibliothèque royale de Belgique*, Pierre Delsaerdt, Jean-Marie Duvosquel et al. (eds.), exh. cat. (Brussels, Bibliothèque royale Albert Ier), Brussels 2005: 86-87.

Elsig 2004
Frédéric Elsig, 'Le contexte artistique des Songes drolatiques', *Les Songes drolatiques de Pantagruel*, Michel Jeanneret (ed.), Genève 2004: 175-196.

Elton 1972
G. R. Elton, 'Thomas More, councillor', *St. Thomas More: Action and Contemplation*, R. S. Sylvester (ed.), New Haven 1972: 87-122.

Erasmus 1962
Desiderius Erasmus, *The Epistles of Erasmus* , vol. 3, Francis M. Nichols (ed.), New York, 1962.

Erasmus 2006a
Desiderius Erasmus, *De correspondentie van Desiderius Erasmus, 4: Brieven 446-593*, M. J. Steens (transl.), Rotterdam 2006.

Erasmus 2006b
Desiderius Erasmus, *De correspondentie van Desiderius Erasmus, 7: Brieven 993–1121*, István Bejczy & John Piolon (transl.), Rotterdam 2006.

Erasmus 2008
Desiderius Erasmus, *De correspondentie van Desiderius Erasmus, 5: Brieven 446–593*, M. J. Steens (transl.), Rotterdam 2008.

Ernst 1995
Ilse Ernst, 'Peter Apian und die Schwarze Kunst in Ingolstadt', *Peter Apian: Astronomic, Kosmographie und Mathematik am Beginn der Neuzeit*, Karl Röttel (ed.), Buxheim 1995: 33-38.

Estrada de Gerlero 2015
Elena Isabel Estrada de Gerlero, 'The Amantecayotl, Transfigured Light', *Images Take Flight: Feather Art in Mexico and Europe*, Alessandra Russo, Gerhard Wolf & Diana Fane (eds.), Munich 2015: 299-309.

Ewing 2006
Dan Ewing, 'Magi and Merchants: The Force behind the Antwerp Mannerists' Adoration Pictures', *Jaarboek Koninklijk Museum voor Schone Kunsten Antwerpen 2004-2005* (2006): 274-299.

Falke 1937
Otto von Falke, 'Ein unbekanntes Werk des Meisters Conrat Meit von Worms', *Pantheon* 19 (1937): 111-115.

Falkenburg 1988
Reindert Falkenburg, *Joachim Patinir: Landscape as an Image of the Pilgrimage of Life*, Amsterdam/Philadelphia 1988.

Falkenburg 1994
Reindert Falkenburg, *The Fruit of Devotion: Mysticism and the Imagery of Love in Flemish Paintings of the Virgin and the Child, 1450-1550*, Amsterdam 1994.

Falkenburg 2007
Reindert Falkenburg, 'The Devil is in the Detail: Ways of seeing Joachim Patinir's "World Landscapes"', *Patinir. Essays and Critical Catalogue*, exh. cat. (Madrid, Museo del Prado), Madrid 2007: 61-79.

Falkenburg 2011
Reindert Falkenburg, *The Land of Unlikeness: Hieronymus Bosch, The Garden of Earthly Delights*, Zwolle 2011.

Fauvel, Flood & Wilson 2013
John Fauvel, Raymond Flood & Robin Wilson (eds.), *Oxford Figures. Eight Centuries of the Mathematical Sciences*, Oxford 2013.

Feest 1987
Christian F. Feest, 'Koloniale Federkunst aus Mexico', *Gold und Macht, Spanien und der Neuen Welt*, Christian F. Feest & Peter Kann (eds.), exh. cat. (Vienna, Künstlerhaus/Cologne, Josef-Haubrich-Kunsthalle/Budapest, Nationalgalerie), Vienna 1987: 173-177.

Feest 1990
Christian F. Feest, 'Vienna's Mexican treasures. Aztec, Mixtec, and Tarascan works from 16th century Austrian collections', *Archiv für Völkerkunde* 44 (1990): 1-64.

Feest 2013
Christian F. Feest, 'From Calicut to America: Albrecht Dürer and the "wondrous artificial things" from the "new golden land"', *Albrecht Dürer: his art in context*, Jochen Sander (ed.), exh. cat. (Frankfurt am Main, Städel Museum) Munich 2013: 366-371.

Feest 2014
Christian F. Feest, 'The people of Calicut: objects, texts, and images in the Age of Proto-Ethnography', *Boletim do Museu Paraense Emílio Goeldi. Ciências Humanas* 9, 2 (2014): 287-303.

Ferguson & McConica 1975
Wallace K. Ferguson & James K. McConica (eds.), R.A.B. Mynors & D.F.S. Thomson (transl.), *The Correspondence of Erasmus - Letters 142 to 297 (1501 to 1514)*, vol. 2, Toronto/Buffalo 1975.

Fernández-Armesto 2006
Felipe Fernández-Armesto, *Amerigo. The Man Who Gave his Name to America*, London 2006.

Ferrari d'Occhieppo & Neumann 1959a
Konradin Ferrari d'Occhieppo & Erwin Neumann, 'Armillarsphären mit Uhrwerksantrieb', *Das Österreichische Uhren- und Schmuck-Journal* 10 (1959): 194-197.

Ferrari d'Occhieppo & Neumann 1959b
Konradin Ferrari d'Occhieppo & Erwin Neumann, 'Astronomische und kunstgeschichtliche Bedeutung zweier Armillarsphären des Kunsthistorischen Museums in Wien', *Mitteilungen des Instituts für österreichische Geschichtforschungen* 67 (1959): 135-148.

Filedt Kok 2008
Jan Piet Filedt Kok, '"Het Paradijs" van Herri Bles', *De verbeelde wereld. Liber amicorum voor Boudewijn Bakker*, Jaap Evert Abrahamse et al. (eds.), Bussum 2008: 104-111.

Fiorani 2005
Francesca Fiorani, *The Marvel of Maps - Art, Cartography and Politics in Renaissance Italy*, New Haven/London 2005.

Firmenich-Richartz, Clemen & Hartmann 1904
Eduard Firmenich-Richartz, Paul Clemen & Paul Hartmann (eds.), *Kunsthistorische Ausstellung Düsseldorf 1904: Katalog*, exh. cat. (Düsseldorf, Kunstpalast), Düsseldorf 1904.

FitzRoy 2015
Charles FitzRoy, *The Rape of Europa. The Intriguing History of Titian's Masterpiece*, London 2015.

Flechsig 1928-1931
Eduard Flechsig, *Albrecht Dürer – Sein Leben und seine künstlerische Entwicklung*, 2 vols., Berlin 1928-1931.

Fleischhauer 1976
Werner Fleischhauer, *Die Geschichte der Kunstkammer der Herzöge von Württemberg in Stuttgart*, Stuttgart 1976.

Fleming 1969
John Vincent Fleming, *The Roman de la Rose: A Study in Allegory and Iconography*, Princeton 1969.

Florence 1980
Firenze e la Toscana dei Medici nell'Europa del Cinquecento. La corte, il mare, i mercanti. La rinascita della scienza. Editoria e società. Astrologia, magia e alchimia, exh. cat. (Florence, Biblioteca Medicea Laurenziana), Florence 1980.

Flores 1981
Nona Cecilia Flores, *"Virgineum vultum habens": The woman-headed serpent in art and literature from 1300 to 1700*, PhD dissertation, University of Illinois, Urbana-Champaign 1981.

Foister 2004
Susan Foister, *Holbein and England*, New Haven/London 2004.

Fortunati & Trousson 2000
Vita Fortunati & Raymond Trousson (eds.), *Dictionary of literary utopias*, Paris 2000.

Foucart 1965
Jacques Foucart, *Le XVIe siècle européen. Peintures et dessins dans les collections publiques Françaises*, exh. cat. (Paris, Petit Palais), Paris 1965.

Fox 1932
Philip Fox, 'The Adler Planetarium and Astronomical Museum of Chicago', *Popular Astronomy* 40, 6 (1932): 321-350.

Fox 1933
Philip Fox, *Adler Planetarium and Astronomical Museum of Chicago*, Chicago 1933.

Fraeters 2013
Veerle Fraeters, 'The Appearance of Queen Reason: Construction and Pragmatics of the Imagery in Vision 9 of Hadewijch', *Speaking to the Eye. Sight and Insight through Text and Image (1150-1650)*, Medieval Identities: Socio-CulturalSpaces, 2, T. de Hemptinne, V. Fraeters & M. E. Gongora (eds.), Turnhout 2013: 65-94.

Fricke 2011
Beate Fricke, 'Zeugen und Bezeugen. Vom Anfangen, Blicken und Enden. Zu Lot und seine Töchter von Joachim Patinir', *Zeugnis und Zeugenschaft. Perspektiven aus der Vormoderne*, Heike Schlie & Wolfram Drews (eds.), Munich 2011: 271-300.

Friedländer 1908
Max J. Friedländer, 'Review of Julius Janitsch - Das Bildnis Sebastian Brants von Albrecht Dürer', *Repertorium für Kunstwissenschaft* 31 (1908): 79-80.

Friedländer 1916
Max J. Friedländer, *Von Eyck bis Bruegel, Studien zur Geschichte der niederländischen Malerei*, Berlin 1916.

Friedländer 1917
Max J. Friedländer, 'Der Meister von Frankfurt', *Jahrbuch der Königlich Preussischen Kunstsammlungen* 38 (1917): 135-150.

Friedländer 1924–1937
Max J. Friedländer, *Die altniederländische Malerei*, 14 vols., Berlin/Leiden 1924–1937.

Friedländer 1926
Max J. Friedländer, *Die altniederländische Malerei - vol. 4: Hugo van der Goes*, Berlin/Leiden 1926.

Friedländer 1927
Max J. Friedländer, *Die altniederländische Malerei - vol. 5: Geertgen van Haarlem und Hieronymus Bosch*, Berlin/Leiden 1927.

Friedländer 1936
Max J. Friedländer, *Die altniederländische Malerei - vol. 13: Anthonis Mor und seine Zeitgenossen*, Berlin/Leiden 1936.

Friedländer 1967–1976
Max J. Friedländer, *Early Netherlandish Painting*, 14 vols., Leiden/Brussels 1967–1976.

Friedländer 1969a
Max J. Friedländer, *Early Netherlandish Painting - vol. 4: Hugo van der Goes*, Leiden/Brussels 1969.

Friedländer 1969b
Max J. Friedländer, *Early Netherlandish Painting - vol. 5: Geertgen tot Sint Jans and Jerome Bosch*, New York 1969.

Friedländer 1975
Max J. Friedländer, *Early Netherlandish Painting - vol. 12: Jan van Scorel and Pieter Coeck van Aelst*, Leiden/Brussels 1975.

Friedländer 1976
Max J. Friedländer, *Early Netherlandish Painting - vol. 13: Antonis Mor and his contemporaries*, Leiden/Brussels 1976.

Friedman 1981
John Block Friedman, *The Monstrous Races in Medieval Art and Thought*, Cambridge 1981.

Friedman 1986
John Block Friedman, 'The Marvels-of-the-East Tradition in Anglo-Saxon Art', *Sources of Anglo-Saxon Culture*, Paul E. Szarmach (ed.), Kalamazoo 1986: 320-341.

Füssel 1994
Stephan Füssel, 'Die Welt-chronik - Eine Nürnberger Gemeinschaftsleistung', *500 Jahre Schedelsche Weltchronik. Akten des interdisziplinären Symposions vom 23./24. April 1993 in Nürnberg*, Stephan Füssel (ed.), Nuremberg 1994: 7-30.

Füssel 2001
Stephan Füssel, *Le livre des Chroniques. L'édition de Nuremberg de 1493 coloriée et commentée. D'après l'exemplaire de la Bibliothèque de la duchesse Anna Amalia, Weimar* (comments to the facsimile edition, Cologne 2001.

Ganz 1936
Paul Ganz, 'Zwei Werke Hans Holbeins d. J. aus der Frühzeit des ersten englischen Aufenthalts', *Festschrift zur Eröffnung des Kunstmuseums*, Basel 1936: 141-158.

Ganz 1950
Paul Ganz, *The Paintings of Hans Holbein*, London 1950.

García Franco 1945
Salvador García Franco, *Catálogo crítico de los astrolabios en España*, Madrid 1945.

Gaspard 1943
Camille Gaspard, *De kalender van het Hennessy-getijdenboek*, Brussels 1943.

Gautier Dalché 2013
Patrick Gautier Dalché (ed.), *La Terre. Connaissance, représentations, mesure au Moyen Âge*, Turnhout 2013.

Gee 1933
John Archer Gee, 'Cuthbert Tunstall's copy of the first edition of Utopia', *Yale University Library Gazette* 7 (1933): 87-88.

Gee 1941
John Archer Gee, 'The second edition of the Utopia, Paris 1517', *Yale University Library Gazette* 15 (1941): 77-83.

Geelen & Steyaert 2009
Ingrid Geelen & Delphine Steyaert, 'Considération sur la vente et la diffusion des statuettes malinoises de la première moitié du XVIe siècle', *Il collezionismo locale adesioni e rifiuti*, Ranieri Varese & Federica Veratelli (eds.), Ferrara 2009: 19-49.

Geelen & Steyaert 2011
Ingrid Geelen & Griet Steyaert, *Imitation and Illusion: Applied Brocade in the Art of the Low Countries in the Fifteenth and Sixteenth Centuries*, Brussels 2011.

Geisberg & Strauss 1974
Max Geisberg & Walter L. Strauss, *The German Single-leaf Woodcut: 1500-1550*, vol. 3, New York 1974.

Gentil Quina 1998
Maria Antónia Gentil Quina, *À maneira de Portugal e da India. Uma Série de Tapeçaria Quinhentista*, Lisbon 1998.

Geritz 1992
Albert Geritz, 'Recent studies in More (1977-1990)', *English Literary Renaissance* 22 (1992): 112-140.

Geritz 1998
Albert Geritz, *Thomas More. An Annotated Bibliography of Criticism, 1935-1997*, Westport (Connecticut)/London 1998.

Geritz 2005
Albert Geritz, 'Recent studies in More (1990-2003)', *English Literary Renaissance* 35 (2005): 123-155.

Gibb 1977
Paul Allen Gibb, *Wonders of the East: A Critical Edition*, PhD dissertation, Duke University, North Carolina 1977.

Gibson 1961
R. W. Gibson, *St. Thomas More. A Preliminary Bibliography of his Works and of Moreana to the year 1750. With a Bibliography of Utopiana compiled by R. W. Gibson and J. Max Patrick*, New Haven/London 1961.

Gibson 1989
Walter S. Gibson, *Mirror of the Earth: The World Landscape in Sixteenth-Century Flemish Painting*, Princeton 1989.

Gibson 1992
Walter S. Gibson, 'Bosch's Dreams: A Response to the Art of Bosch in the Sixteenth Century', *The Art Bulletin* 74, 2 (1992): 205-218.

Gielis 2009
Gert Gielis, *Verdoelde schaepkens, bytende wolven. Inquisitie in de Lage Landen*, Leuven 2009.

Gingerich 1971
Owen Gingerich, 'Apianus' Astronomicum Caesareum and Its Leipzig Facsimile', *Journal for the History of Astronomy* 2 (1971): 168-177.

Gingerich 1994
Owen Gingerich, 'Early Astronomical Books with Moving Parts', *A History of Book Illustration: 29 Points of View*, Bill Katz (ed.), London 1994: 288-296.

Gingerich 1995
Owen Gingerich, 'A Survey of Apian's Astronomicum Caesareum', *Peter Apian: Astronomic, Kosmographie und Mathematik am Beginn der Neuzeit*, Karl Röttel (ed.), Buxheim 1995: 113-122.

Giorgi 2005
Rosa Giorgi, *Angels and Demons in Art*, Los Angeles 2005.

Gjerpe 2008
Kristin Gjerpe, 'The Italian *Utopia* of Lando, Doni and Sansovino: Paradox and Politics', *Thomas More's Utopia in Early Modern Europe. Paratexts and contexts*, Terence Cave (ed.), Manchester 2008: 47-66.

Göbel 1933
Heinrich Göbel, *Wandteppiche: Die germanischen und slawischen Länder: Deutschland einschließlich Schweiz und Elsass (Mittelalter), Süddeutschland (16. bis 18. Jahrhundert)*, vol. 3, pt. 1, Berlin 1933.

Goddard 1984
Stephen H. Goddard, *The Master of Frankfurt and his Shop*, Brussels 1984.

Goddard 1985
Stephen H. Goddard, 'Brocade Patterns in the Shop of the Master of Frankfurt: An Accessory to Stylistic Analysis', *The Art Bulletin* 67, 3 (1985): 401-417.

Godenne 1957
Willy Godenne, 'Préliminaires à l'inventaire général des statuettes d'origine malinoise, présumées des XVe et XVIe siècles', *Handelingen van de Koninklijke Kring voor Oudheidkunde, Letteren en Kunst van Mechelen* 66 (1957): 47-127.

Godenne 1972
Willy Godenne, 'Préliminaires à l'inventaire général des statuettes d'origine malinoise, présumées des XVe et XVIe siècles', *Handelingen van de Koninklijke Kring voor Oudheidkunde, Letteren en Kunst van Mechelen* 76, 2 (1972): 3-80.

Godenne 1976
Willy Godenne, 'Préliminaires à l'inventaire général des statuettes d'origine malinoise, présumées des XVe et XVIe siècles', *Handelingen van de Koninklijke Kring voor Oudheidkunde, Letteren en Kunst van Mechelen* 80, 1 (1976): 71-105.

Goldhill 2002
Simon Goldhill, *Who needs Greek? Contests in the Cultural History of Hellenism*, Cambridge 2002.

Gombert 2005
Florence Gombert, 'Un réseau de collaborations d'artistes', *Le Maître au Feuillage brodé - Secrets d'ateliers*, Florence Gombert & Didier Martens (eds.), exh. cat. (Lille, Palais des Beaux-Arts), Lille 2005: 29-38.

Gosman, MacDonald & Vanderjagt 2003
Martin Gosman, Alasdair MacDonald & Arjo Vanderjagt (eds.), *Princes and Princely Culture, 1450-1650*, Leiden 2003.

Göttler 2010
Christine Göttler, *Last Things: Art and The Religious Imagination in the Age of Reform*, Turnhout 2010.

Grabmayer 2009
Johannes Grabmayer, *Das Königreich der Narren: Fasching im Mittelalter*, Friesach 2009.

Grandsart 1987
Hervé Grandsart, 'L'enfer selon Marmion', *Connaissance des Arts* 429 (1987): 69-71.

Greenblatt 2001
Stephen Greenblatt, *Hamlet in Purgatory*, Princeton/Oxford 2001.

Greene 2001
Joseph A. Greene, 'Sodom and Gomorrah', *The Oxford Guide to People & Places of the Bible*, Bruce M. Metzger & Michael D. Coogan (eds.), Oxford 2001: 294.

Greengrass 2014
Mark Greengrass, *Christendom Destroyed. Europe 1517-1648*, London 2014.

Grimm 1977
Reinhold R. Grimm, *Paradisus coelestis, paradisus terrestris. Zur Auslegungsgeschichte der Paradieses im Abendland bis um 1200*, Medium Aevum, 33, Munich 1977.

Grosshans 1980
Rainald Grosshans, *Maerten van Heemskerck. Die Gemälde*, Berlin 1980.

Guernelli 2014
Daniele Guernelli, 'Le prince d'enluminure', *Alumina Pagine Miniate* 12, 44 (2014): 24-33.

Guicciardini 1588
Lodovico Guicciardini, *Descrittione di tutti I Paesi Bassi*, (ed. Christoffel Plantin) Antwerp 1588.

Guillouet 1956
Jacques Guillouet (ed.), *Chefs-d'oeuvre du Musée de Douai*, exh. cat. (Paris, Galerie Heim), Paris 1956.

Guillouet 1967
Jacques Guillouet, 'Catalogue pour le musée de Douai. Ses peintures des écoles du Nord', *Connaissance des Arts* 184 (1967).

Gullick 1990
Michael Gullick, *Calligraphy*, London 1990.

Gunther 1976
Robert T. Gunther, *The Astrolabes of the World*, vol. 1, London 1976.

Guy 1980
John A. Guy, *The Public Career of Sir Thomas More*, Brighton 1980.

Guy 1988
John A. Guy, *Tudor England*, Oxford 1988.

Haak 1967
Bob Haak, 'De vergankelijkheidssymboliek in 16de eeuwse portretten en 17de eeuwse stillevens in Holland', *Antiek* 1, 7 (1967): 23-30.

HALI 2009
HALI, 'The motions of the universe', *HALI. The International Journal of Oriental Carpets and Textiles* 160 (2009): 144.

Hallowell 1960
Robert E. Hallowell, 'Jean Le Blond's defense of the French language (1549)', *Romanic Review* 51 (1960): 86-92.

Hallyn 2008
F. Hallyn, *Gemma Frisius, arpenteur de la terre et du ciel*, Paris 2008.

Hamann 1971
Günther Hamann, 'Albrecht Dürers Erd- und Himmelskarten', *Albrecht Dürers Umwelt. Festschrift zum 500. Geburtstag Albrecht Dürers am 21. Mai 1971*, Gerhard Hirschmann & Fritz Schnellbögl (eds.), Nüremberg 1971: 152-177.

Hamburger 1990
Jeffrey F. Hamburger, *The Rothschild Canticles: Art and Mysticism in Flanders and the Rhineland circa 1300*, New Haven 1990.

Hamburger 1997
Jeffrey F. Hamburger, *Nuns as Artists: The Visual Culture of a Medieval Convent*, Berkeley/London 1997.

Hamburger 2004
Jeffrey F. Hamburger, 'Body vs. Book: The Trope of Visibility in Images of Christian-Jewish Polemic', *Ästhetik des Unsichtbaren. Bildtheorie und Bildgebrauch in der Vormoderne, David Ganz & Thomas Lentes* (eds.), Münster 2004: 113-145.

Hamburger & Suckale 2005
Jeffrey F. Hamburger & Robert Suckale, 'Zwischen Diesseits und Jenseits. Die Kunst der geistlichen Frauen im Mittelalter', *Krone und Schleier. Kunst aus mittelalterlichen Frauenklöstern*, exh. cat. (Essen, Ruhrlandmuseum/Bonn, Kunst- und Ausstellungshalle der Bundesrepublik Deutschland), Munich 2005: 21-39.

Hand 1990
John Oliver Hand, '"Saint Jerome in His Study" by Joos van Cleve', *Record of the Art Museum, Princeton University* 49, 2 (1990): 2-10.

Hand, Metzger & Spronk 2006
John Oliver Hand, Catherine A. Metzger & Ron Spronk, *Prayers and Portraits: Unfolding the Netherlandish Diptych*, exh. cat. (Washington, National Gallery of Art/Antwerp, Koninklijk Museum voor Schone Kunsten), Washington/New Haven/London 2006.

Hand & Wolff 1986
John Oliver Hand & Martha Wolff, *Early Netherlandish Painting*, Cambridge 1986.

Hardingham 1884
G. G. Hardingham (ed. & transl.), *The Republic of Cicero reprinted from the edition of Cardinal Mai, Rome 1864*, London 1884.

Harlizius-Klück 2004
Ellen Harlizius-Klück, *Weberei als episteme und die Genese der deduktiven Mathematik*, Berlin 2004.

Harrison 1987
Jefferson C. Harrison, *The Paintings of Maerten van Heemskerck. A Catalogue Raisonné*, vol. 1, PhD dissertation, University of Virginia, Virginia 1987.

Hartlaub 1912
Gustav Friedrich Hartlaub, *Die Großherzogliche Gemälde-Galerie im Augusteum zu Oldenburg. 84 Tafeln in unveränderlicher Photographie*, bnd. 2, Oldenburg 1912.

Hassall 1970
William O. Hassall, *The Holkham Library: Illuminations and Illustrations in the Manuscript Library of the Earl of Leicester*, Oxford 1970.

Hausberger & Biedermann 1973
Isolde Hausberger & Rolf Biedermann (eds.), *Hans Burgkmair: Das Graphische Werk*, exh. cat. (Augsburg, Graphische Sammlung Staatsgalerie), Stuttgart 1973.

Heard, Whitaker et al. 2011
Kate Heard, Lucy Whitaker et al., *The Northern Renaissance: Dürer to Holbein*, exh. cat. (Edinburgh, The Queen's Gallery/London, The Queen's Gallery) London 2011.

Heawood 1930
Edward Heawood, 'An Undescribed Map of Lopo Homem, 1519', *The Geographical Journal* 76, 2 (1930): 159-160.

Heawood 1931
Edward Heawood, 'Lopo Homem's Map of 1519', *The Geographical Journal* 77, 3 (1931): 250-255.

Heck 2003
Christian Heck (ed.), *L'art flamand et hollandais: Le siècle des Primitifs 1380-1520*, Paris 2003.

Heck 2005
Christian Heck, *Répertoire des peintures flamandes des quinzième et seizième siecles. Collections du Nord-Pas-de-Calais. La peinture de Flandre et de France du Nord au XVe et au début du XVIe siècle*, Brussels 2005.

Heireman 1973
Kamiel Heireman (ed.), *Dirk Martens 1473-1973*, exh. cat. (Aalst, Stedelijk Museum-Oud Hospitaal), Aalst 1973.

Held 1931
Julius S. Held, *Dürers Wirkung auf die Niederlandische Kunst seiner Zeit*, The Hague 1931.

Helman 2013
Susannah Helman, 'Heavenly Woodcuts', *Mapping Our World: Terra Incognita to Australia*, exh. cat., (Canberra, National Library of Australia) Canberra 2013: 40-41.

Hemming 1978
John Hemming, *Red Gold: The Conquest of the Brazilian Indians*, Cambridge 1978.

Hendrikmann & Filedt Kok 2009a
Lars Hendrikmann and Jan Piet Filedt Kok, 'Herri met de Bles, *Paradise*, Antwerp, *c.*1541 – *c.*1550', *Early Netherlandish Paintings in the Rijksmuseum*, J.P. Filedt Kok (ed.), online mus. cat. (Amsterdam, Rijksmuseum) 2009: hdl.handle.net/10934/RM0001.COLLECT.6063 (consulted 29 February 2016).

Hendrikmann & Filedt Kok 2009b
Lars Hendrikmann & Jan Piet Filedt Kok, 'Marinus van Reymerswale, *St Jerome in his Study*, *c.*1535 - *c.*1545', *Early Netherlandish Paintings in the Rijksmuseum*, J.P. Filedt Kok (ed.), online mus. cat. (Amsterdam, Rijksmuseum) 2009: hdl.handle.net/10934/RM0001.COLLECT.5274 (consulted 15 June 2016).

Heremans 2015
Stephanie Heremans, *Atlas de Dauphin: Een kritische status quaestionis. Brazilianen in de beeldcultuur van de Noord-Franse koopvaardijlieden en het Franse koningshuis*, Bachelor paper, KU Leuven, Leuven 2015.

Herreman, Holsbeke & Van Alphen 1991
Frank Herreman, Mireille Holsbeke & Jan van Alphen, *Etnografisch Museum Antwerpen*, Brussels 1991.

Herrmann-Fichtenau 2007
Elisabeth Herrmann-Fichtenau (ed.), *Das Kunsthistorische Museum in Wien*, mus. cat. (Vienna, Kunsthistorisches Museum), Munich 2007.

Herz 1898
W. Herz, *Perzival*, Stuttgart 1898.

Hess 1971
Günter Hess, *Deutsch-lateinische Narrenzunft: Studien zum Verhältnis von Volkssprache und Latinität in der satirischen Literatur des 16. Jahrhunderts*, Munich 1971.

Hessels 1887
H. J. Hessels, *Abrahami Ortelii epistulae*, Cambridge 1887.

Hester 1998
Nathalie Hester, 'Stolen texts? Gabriel Chappuys' "L'estat, description et gouvernement des royaumes et républiques du monde"', *Sondaggi sulla riscrittura del Cinquecento*, P. Cerchi (ed.), Ravenna 1998: 133-148.

Hexter 1952
J. H. Hexter, *More's "Utopia": The Biography of an Idea*, Princeton 1952.

Hiatt 2008
Alfred Hiatt, *Terra Incognita. Mapping the Antipodes before 1600*, Chicago/London 2008.

Hindman 1977
Sandra Hindman, 'The Case of Simon Marmion: Attributions & Documents', *Zeitschrift für Kunstgeschichte* 40, 3/4 (1977): 185-204.

Hindman, D'Ancona, Palladino & Saffiotti 1997
Sandra Hindman, Mirella Levi D'Ancona, Pia Palladino & Maria Francesca Saffiotti, *The Robert Lehman Collection - vol. 4: Illuminations*, New York/Princeton 1997.

His 1870
Eduard His, 'Die Basler Archive über Hans Holbein, den Jüngern, seine Familie und einige zu ihm in Beziehung stehende Zeitgenossen', *Jahrbücher für Kunstwissenschaft*, vol. 3, Albert von Zahn (ed.), Leipzig (1870): 113-173.

Hoffman 1969
Edith Warren Hoffman, 'Simon Marmion Re-considered', *Scriptorium* 23, 2 (1969): 243-271.

Hoffman 1973
Edith Warren Hoffman, 'Simon Marmion or "The Master of the Altarpiece of Saint-Bertin": A Problem in Attribution', *Scriptorium* 27, 2 (1973): 263-290.

Hoflack 1998
Marijke Hoflack et al., *800 jaar Onze-Lieve-Vrouwegasthuis: uit het erfgoed van de Mechelse gasthuiszusters en het OCMW*, exh. cat., (Mechelen, Stedelijk Museum Hof van Busleyden), Mechelen 1998.

Hofmann, Richard & Vagnon 2012
Catherine Hofmann, Hélène Richard & Emmanuelle Vagnon (eds.), *L'âge d'or des cartes marines. Quand l'Europe découvrait le monde*, exh. cat. (Paris, Bibliothèque nationale de France), Seuil 2012.

Hollstein Dutch
F.W.H. Hollstein, *Dutch and Flemish Etchings, Engravings and Woodcuts ca. 1450-1700*, Amsterdam [etc.], 1949-… .

Hollstein German
F.W.H. Hollstein, *German Engravings and Woodcuts ca. 1400-1700*, Amsterdam [etc.], 1954-…

Holtze 1927
Otto Holtze, 'Die Neuordnung der alten Galerie im Oldenburger Landesmuseum', *Oldenburger Jahrbuch des Vereins für Altertumskunde und Landesgeschichte* 31 (1927): 291-301.

Honigmann 1990
Ernst A. J. Honigmann, 'The Play of Sir Thomas More and Some Contemporary Events', *Shakespeare and the Elizabethans, Shakespeare Survey* 42 (1990): 77-84.

Honour 1976
Hugh Honour (ed.), *L'Amérique vue par l'Europe*, exh. cat. (Paris, Grand Palais), Paris 1976.

Hoogewerff 1941-1942
Godefridus J. Hoogewerff, *De Noord-Nederlandsche schilderkunst*, vol. 4, The Hague 1941-1942.

Hoogewerff 1947
Godefridus J. Hoogewerff, *De Noord-Nederlandsche schilderkunst*, vol. 5, The Hague 1947.

Hoogvliet 2007
Margriet Hoogvliet, *Pictura et scriptura. Textes, images et herméneutique des Mappae Mundi (XIIIe – XVIe siècle) (Terrarum Orbis. History of Representations of Space in Text and Image*, Patrick Gautier Dalché (ed.), Turnhout 2007.

Horst 2011
Thomas Horst (ed.), *De wereld in kaart: Gerard Mercator (1512-1594) en de eerste wereld atlas*, Brussels 2011.

Hosington 1984
Brenda Hosington, 'Early French Translations of Thomas More's Utopia 1550-1730', *Humanistica Lovaniensia* 33 (1984): 116-134.

Hoskins 1976
W. G. Hoskins, *The Age of Plunder: The England of Henry VIII*, London 1976.

Hottois 1982
Isabelle Hottois, *L'iconographie musicale dans les manuscrits de la Bibliothèque royale Albert Ier*, exh. cat. (Brussels, Bibliothèque royale Albert Ier), Brussels 1982.

Hoyer 1981
Siegfried Hoyer, 'Utopia deutsch. Zu den Gleichheitsvorstellungen im Basler Humanistenkreis', *Jahrbuch für Geschichte des Feudalismus* 5 (1981): 237-254.

Husband 1980
Timothy B. Husband, *The Wild Man: Medieval Myth and Symbolism*, exh. cat. (New York, The Metropolitan Museum of Art), New York 1980.

Hunt 1991
Lydia Hunt, 'The first Spanish Translation of Utopia (1637)', *Moreana* 105 (1991): 21-41.

Huvenne 1991
Paul Huvenne, 'Beeld van de stad vanuit cartografisch perspectief', *Stad in Vlaanderen: cultuur en maatschappij 1477-1787*, Jan Van der Stock (ed.), exh. cat. (Brussels, Galerij van het Gemeentekrediet/Schallaburg, Renaissanceschloß) Brussels 1991: 45-58.

Huvenne 2003
Paul Huvenne, *Het museumboek: hoogtepunten uit de verzameling*, Ghent 2003.

Ilsink & Koldeweij 2016
Matthijs Ilsinck & Jos Koldeweij, *Jheronimus Bosch: Visioenen van een genie*, exh. cat. ('s Hertogenbosch, Noordbrabants Museum) 2016.

Inneh 2007
Daniel E. Inneh, 'The Guilds working for the Palace', *Benin. Kings and Rituals. Court Arts from Nigeria*, Barbara Plankensteiner (ed.), Vienna/Paris/Berlin/Chicago 2007: 102-117.

Israel 2001
Jonathan I. Israel, *Radical Enlightenment. Philosophy and the Making of Modernity 1650-1750*, Oxford 2001.

Isaacson 1998
Richard Isaacson, 'The Admiral and the Wild Man', *HALI. The International Journal of Oriental Carpets and Textiles* 98 (1998): 78-83.

Iterbeke 2015
Hannah Iterbeke, *Devotie tot de zijden bloem: tussen techniek en betekenis - de zijden bloemen uit de Besloten Hofjes van de Lage Landen in de zestiende eeuw*, Master dissertation, KU Leuven, Leuven 2015.

J. Paul Getty Museum 1988
'Acquisitions/1987', *The J. Paul Getty Museum Journal* 16 (1988): 135-201.

Jackson 2006
Peter Jackson, 'Christians, Barbarians and Monsters: The European Discovery of the World Beyond Islam', *The Medieval World*, Peter Linehan & Janet Laughland Nelson (eds.), London 2006: 93-110.

Jacobsson 2002
Augustine, *De Musica Liber VI: A Critical Edition with a Translation and an Introduction. Acta Universitatis Stockholmiensis. Studio Latina Stockholmiensia*, M. Jacobsson (trans. & ed.), Stockholm 2002.

Jaffe, Viola & Rovigatti 1991
R. A., 'The Island of Brazil', *Imagining the New World: Columbian Iconography*, Irma B. Jaffe, Gianni Eugenio Viola & Franca Rovigatti (eds.), exh. cat. (New York, Historical Society Museum), Rome/New York 1991.

James 1926
Montague R. James, *The Apocryphal New Testament. Being the Apocryphal Gospels, Acts, Epistles, and Apocalypses, with other Narratives and Fragments*, Oxford 1926.

James 1929
Montague R. James (ed.), *Marvels of the East*, Oxford 1929.

Janitsch 1906a
Julius Janitsch, 'Ein Bildnis Sebastian Brants von Albrecht Dürer im Königlichen Kupferstichkabinett zu Berlin', *Jahrbuch der Königlich Preussischen Kunstsammlungen* 27, 2 (1906): 75-78.

Janitsch 1906b
Julius Janitsch, 'Das Bildnis Sebastian Brants von Albrecht Dürer', *Studien zur deutschen Kunstgeschichte* 7 (1906): 5-18.

Jansen & Herck 1948
A. Jansen & Ch. Herck (eds.), *Kerkelijke Kunstschatten*, exh. cat. (Antwerp, Stadsfeestzaal), Antwerp 1948.

Jardine 1993
Lisa Jardine, *Erasmus, Man of Letters: The Construction of Charisma in Print*, Princeton 1993.

Jeon 2005
Hanho Jeon, *Meditatio Mortis: zur Ikonographie des heiligen Hieronymus mit dem Totenschädel unter besonderer Berücksichtigung des Lissaboner Gemäldes von Albrecht Dürer*, PhD dissertation, Westfälische Wilhelms-Universität Münster, Münster 2005.

Johns 1989
Karl Johns, 'St. Jerome in Meditation by Anthonie Blocklandt', *Oud Holland - Quarterly for Dutch Art History* 103, 1 (1989): 46-50.

Johnson 1990
Lynn Staley Johnson (ed.), *The Shepheardes Calender, An Introduction*, Pennsylvania 1990.

Jones 1998
Peter Murray Jones, *Medieval Medicine in Illuminated Manuscripts*, London 1998.

Jubinal 1840
Achille Jubinal, *Recherches sur l'usage et l'origine des tapisseries à personnages, dites historiées*, Paris 1840.

Junquera de Vega & Herrero Carretero 1986
Paulina Junquera de Vega & Concha Herrero Carretero, *Catálogo de tapices del Patrimonio Nacional. I. Siglo XVI*, Madrid 1986.

Justo Guedes & Lombardi 1990
Max Justo Guedes & Gerald Lombardi (eds.), *Portugal/Brazil: The Age of Atlantic Discoveries*, exh. cat. (New York, Brazilian Cultural Foundation/New York, Public Library), New York/Milan/Lisbon 1990.

Kaegi 1924
Werner Kaegi, 'Hutten und Erasmus: Ihre Freundschaft und ihr Streit', *Historische Vierteljahrsschrift* 22 (1924): 200-278 & 461-514.

Kalff 1884
G. Kalff, *Het lied in de Middeleeuwen*, Leiden 1884.

Kalkoff 1905
Paul Kalkoff, 'Zur Lebensgeschichte Albrecht Dürers. 3. Albrecht Dürer, Sebastian Brant und Konrad Peutinger in Antwerpen im Sommer 1520', *Repertorium für Kunstwissenschaft* 28 (1905): 474-485.

Kammerer 1938
Albert Kammerer, 'The Lopo Homem Map Once More', *The Geographical Journal* 91, 5 (1938): 450-453.

Kanas 2014
Nick Kanas, *Solar System Maps. From Antiquity to the Space Age*, New York 2014.

Karr Schmidt 2011
Suzanne Karr Schmidt, 'Peter Apian and Michael Ostendorfer, "Astronomicum Caesareum"' *Prints and the Pursuit of Knowledge in Early Modern Europe*, Susan Dackerman (ed.), exh. cat. (Cambridge, Arthur M. Sackler Museum/Evanston, the Mary and Leigh Block Museum of Art), New Haven/London 2011: 104-107.

Karrow 1993
Robert W. Karrow, *Mapmakers of the Sixteenth Century and Their Maps. Bio-Bibliographies of the Cartographers of Abraham Ortelius, 1570*, Chicago 1993.

Kaspersen 2004
Sören Kaspersen (ed.), *Images of Cult and Devotion. Function and Reception of Christian Images in Medieval and Post-Medieval Europe*, Copenhagen 2004.

Katz, Auersperg & Zock 2000
Daniel Katz, Johannes Auersperg & Katherine Zock (eds.), *European Sculpture*, exh. cat. (New York, Simon Dickinson Inc./London, Simon Dickinson Inc.), New York 2000.

Keiser 1967
Herbert Wolfgang Keiser, *Gemäldegalerie Oldenburg*, Munich 1967.

Kelly 1953-1954
Gerald Kelly (ed.), *Flemish Art, 1300-1700*, exh. cat., (London, Royal Academy of Art), London 1953-1954.

Kelly 1972
Henry Ansgar Kelly, 'The Metamorphoses of the Eden Serpent during the Middle Ages and Renaissance', *Viator* 2 (1972): 301-328.

Kenner 1893
Friedrich Kenner, 'Die Porträtsammlung des Erzherzogs Ferdinand von Tirol', *Jahrbuch der Kunsthistorischen Sammlungen des Allerhöchsten Kaiserhauses* 14 (1893): 37-186.

Kier & Zehnder
Hiltrud Kier & Frank Günter Zehnder (eds.), *Lust und Verlust II. Corpus-Band zu Kölner Gemäldesammlungen (1800-1860)*, Cologne 1998.

King 1978
Henry C. King, *Geared to the Stars. The Evolution of Planetariums, Orreries, and Astronomical Clocks*, Toronto 1978.

Kisch 1970
Guido Kisch, *Claudius Cantiuncula. Ein Basler Jurist und Humanist des 16. Jahrhunderts*, Basel 1970.

Klawiter 1977
Randolph J. Klawiter, *The Polemics of Erasmus of Rotterdam and Ulrich von Hutten*, Indiana 1977.

Klein 2001
Peter Klein, 'Dendrochronological Analysis of Works by Hieronymus Bosch and his Followers', *Hieronymus Bosch. New Insights into his Life and Work*, Jos Koldeweij, Bernard Vermet & Barbera van Kooij (eds.), Rotterdam 2001: 120-131.

Klein 2007
Holger A. Klein, *Sacred Gifts and Worldly Treasures: Medieval Masterworks from the Cleveland Museum of Art*, exh. cat. (Los Angeles, J. Paul Getty Museum), Cleveland 2007.

Kloek 2000
Wouter Kloek, 'Herri met de Bles, "Het Paradijs"', *Netherlandish Art in the Rijksmuseum 1400-1600*, Henk van Os et al. (eds.), Zwolle/Amsterdam 2000.

Kloek & Ubl 2015
Wouter Kloek & Matthias Ubl, '"Het Paradijs," Herri met de Bles', *1100-1600*, Frits Scholten (ed.), Amsterdam 2015: 200-203.

Kluge 1908
Friedrich Kluge, 'Der Venusberg', *Bunte Blätter. Kulturgeschichtliche Vorträge und Aufsätze*, Freiburg 1908.

Knecht 1994
Robert J. Knecht, *Renaissance Warrior and Patron. The Reign of Francis I*, Cambridge 1994.

Koch 1965
Robert A. Koch, 'The Salamander in Van der Goes' "Garden of Eden"', *Journal of Warburg and Courtauld Institutes* 28 (1965): 323-326.

Koch 1968
Robert A. Koch, *Joachim Patinir*, Princeton 1968.

Köhler 1947
Walther Köhler (ed.), *Eramus von Rotterdam, Briefe*, Wiesbaden 1947.

Koldeweij 2000
Jos Koldeweij, 'Lijfelijke en geestelijke pelgrimage: materiële "souvenirs" van spirituele pelgrimage', *Geen povere schoonheid. Laat-middeleeuwse kunst in verband met de Moderne Devotie*, Kees Veelenturf (ed.), Nijmegen 2000: 222-252.

Koldeweij 2006
Jos Koldeweij (ed.), *Geloof en geluk: Sieraad en devotie in middeleeuws Vlaanderen*, exh. cat. (Bruges, Bruggemuseum – Gruuthuse), Tielt 2006.

Korteweg 1998
Anne S. Korteweg (ed.), *Boeken van Oranje-Nassau: De bibliotheek van de graven van Nassau en prinsen van Oranje in de vijftiende en zestiende eeuw*, exh. cat. (The Hague, Museum van het Boek /Museum Meermanno-Westreeenianum), The Hague 1998.

Korteweg 2002
Anne S. Korteweg, *Praal, ernst en emotie. De wereld van het Franse middeleeuwse handschrift*, exh. cat. (The Hague, Museum Meermanno-Westreenianum), Zwolle 2002.

Kraus 1978
Hans P. Kraus, *In Retrospect: A Catalogue of 100 Outstanding Manuscripts sold in the last four decades by H. P Kraus*, New York 1978.

Kren 1983
Thomas Kren (ed.), *Renaissance Painting in Manuscripts. Treasures from the British Library*, exh. cat. (Malibu, J. Paul Getty Museum/New York, Pierpont Morgan Library/London, British Library), New York 1983.

Kren 1992
Thomas Kren (ed.), *Margaret of York, Simon Marmion, and The Visions of Tondal* (papers delivered at the 1990 Symposium organized by the Department of Manuscripts of the J. Paul Getty Museum in collaboration with the Huntington Library and Art Collections), Malibu 1992.

Kren 2010
Thomas Kren (ed.), *Illuminated Manuscripts from Belgium and the Netherlands in the J. Paul Getty Museum*, exh. cat. (Los Angeles, J. Paul Getty Museum), Los Angeles 2010.

Kren 2013
Thomas Kren, 'La Vie de sainte Catherine illustré par Simon Marmion', *Art de l'enluminure* 45 (2013): 4-65.

Kren & Ainsworth 2003
Thomas Kren & Maryan W. Ainsworth, 'Illuminators and Painters: Artistic Exchanges and Interrelationships', *Illuminating the Renaissance: the Triumph of Flemish Manuscript Painting in Europe*, Thomas Kren & Scot McKendrick (eds.), exh. cat. (Los Angeles, J. Paul Getty Museum/London, Royal Academy of Arts), Malibu 2003: 35-58.

Kren & McKendrick 2003
Thomas Kren & Scot McKendrick (eds.), *Illuminating the Renaissance: The Triumph of Flemish Manuscript Painting in Europe*, exh. cat. (Los Angeles, J. Paul Getty Museum/London, Royal Academy of Arts), Los Angeles 2003.

Kren & Rathofer 1988
Thomas Kren & Johannes Rathofer, *Simon Bening. Flämisch Calendar. Clm. 23638 Bayerische Staatsbibliothek. München*, 2 pts., Luzern 1988.

Kren & Wieck 1990
Thomas Kren & Roger S. Wieck, *The Visions of Tondal from the Library of Margaret of York*, Malibu 1990.

Kristeller 1980
Paul Oskar Kristeller, 'More as a Renaissance Humanist', *Moreana* 65-66 (1980): 5-22.

Kroll & Schade 1974
Renate Kroll & Werner Schade, *Hans Burgkmair 1473-1531: Holzschnitte, Zeichnungen, Holzstöcke*, exh. cat. (Berlin, Staatliche Museen zu Berlin - Kupferstichkabinett und Sammlung der Zeichnungen), Berlin 1974.

Kronenberg 1967
Marie E. Kronenberg, 'Some notes on the first edition of the *Utopia*, Louvain, Dirck Martens', *Moreana* 15-16 (1967): 134-136.

Kruip 2006a
Marjolein Kruip, *In beeld gebracht: de sculptuur in de laatmiddeleeuwse Besloten Hofjes*, PhD dissertation, Radboud Universiteit, Nijmegen 2006.

Kruip 2006b
Marjolein Kruip, 'Het Besloten Hofje: een spirituele pelgrimage', *Geloof en geluk: Sieraad en Devotie in Middeleeuws Vlaanderen*, exh. cat. (Bruges, Bruggemuseum – Gruuthuse), Jos Koldeweij (ed.), Tielt 2006: 231-234.

Kruse 2000
Petra Kruse (ed.), *Kaiser Karl V. (1500 - 1558). Macht und Ohnmacht Europas*, exh. cat. (Bonn, Kunst- und Ausstellungshalle der Bundesrepublik Deutschland/Vienna, Kunsthistorischen Museum), Bonn 2000.

Kruyfhooft 1981
Cécile Kruyfhooft, 'A Recent Discovery: Utopia by Abraham Ortelius', *Map Collector* 16 (1981): 10-14.

Kugel, Van Cleempoel & Sabrier 2002
Alexis Kugel, Koenraad Van Cleempoel & Jean-Claude Sabrier (eds.), *Sphères: l'art des mécaniques célestes*, Paris 2002.

Kunsthistorisches Museum 1977
Die Kunstkammer - Kunsthistorisches Museum, Sammlungen Schloss Ambras, Innsbruck 1977.

Kunsthistorisches Museum Wien 1988
Kunsthistorisches Museum Wien. Führer durch die Sammlungen, mus. cat. (Vienna, Kunsthistorisches Museum), Vienna 1988.

Kurth 1926
Betty Kurth, *Die deutschen Bildteppiche des Mittelalters*, Vienna 1926.

Kurz 1967
Otto Kurz, 'Four Tapestries after Hieronymus Bosch', *Journal of the Warburg and Courtauld Institutes*, 30 (1967): 150-162.

Lachièze-Rey & Luminet 1998
Marc Lachièze-Rey & Jean-Pierre Luminet (eds.), *Figures du ciel: de l'harmonie des sphères à la conquête spatiale*, exh. cat. (Paris, Bibliothèque nationale de France), Paris 1998.

Lakowski 1995
Romuald Ian Lakowski, 'A Bibliography of Thomas More's Utopia', *Early Modern Literary Studies* 1.2 (1995) 6.1-10: last access 31/08/2015, http://www3.telus.net/lakowski/Utopbibo.html (consulted 31 October 2015).

Lammertse 1994
Friso Lammertse, *Van Eyck to Bruegel. Dutch and Flemish painting in the collection of the Museum Boymans-van Beuningen, 1400-1550*, Rotterdam 1994.

Lammertse & Van der Coelen 2015
Friso Lammertse & Peter van der Coelen (eds.), *Van Bosch tot Bruegel - De ontdekking van het dagelijks leven*, exh. cat. (Rotterdam, Museum Boijmans Van Beuningen), Rotterdam 2015.

Landau & Parshall 1994
David Landau & Peter Parshall, *The Renaissance Print: 1470-1550*, New Haven/London 1994.

Landesmuseum für Kunst und Kulturgeschichte Oldenburg 2003
Landesmuseum für Kunst und Kulturgeschichte Oldenburg (ed.), *Das Augusteum. Museen im Nordwesten*, vol. 8 (2nd ed.), Oldenburg 2003.

Landesmuseum für Kunst und Kulturgeschichte Oldenburg 2012
Landesmuseum für Kunst und Kulturgeschichte Oldenburg (ed.), *Das Augusteum. Museen im Nordwesten*, vol. 8, Oldenburg 2012

Laneyrie-Dagen 2008
Nadeije Laneyrie-Dagen, *L'invention de la nature: Les quatre éléments à la Renaissance ou le peintre premier savant*, Paris 2008.

Larsen 1960
Erik Larsen, *Les Primitifs Flamands au Musée Metropolitan de New York*, Utrecht 1960.

Larsen 1998
Erik Larsen, *Hieronymus Bosch*, New York 1998.

Laube 2011
Stefan Laube, *Von der Reliquie zum Ding. Heiliger Ort - Wunderkammer - Museum*, Berlin 2011.

Leclerc 1972
Ivor Leclerc, 'Chapter 9: The Concept of Matter: medieval *materia*', *The Nature of Physical Existence*, London 1972: 122-129.

Lecoq-Ramond 2003
Sylvie Lecoq-Ramond, *Histoire du Musée d'Unterlinden et de ses collection - De la révolution à la première guerre mondiale*, Colmar 2003.

Lecouteux 1982
Claude Lecouteux, 'Aspects mythiques de la montagne au Moyen-Âge', *Le monde alpin et rhodanien* (1982): 43-54.

Lecouteux 1987
Claude Lecouteux, *Geschichte der Gespenster und Wiedergänger im Mittelalter*, Cologne/Vienna 1987.

Leemans 2012
Annemie Leemans, *Albrecht Dürer and 'The Fall of Man' (1504): A Copper Engraving in Sign of Artistic Authority*, Master dissertation, KU Leuven, Leuven 2012.

Le Goff 1977
Jacques le Goff, 'L'Occident médiéval et l'océan Indien: un horizon onirique', *Pour un autre Moyen Âge. Temps, travail et culture en Occident*, Paris 1977: 280-298.

Legaré 1991
Anne-Marie Legaré, 'Splendeurs de la miniature en Hainaut', *Le livre des Échecs amoureux*, Anne-Marie Legaré, Françoise Guichard Tesson & Bruno Roy (eds.), Paris 1991: 80-95.

Legaré 2005
Anne-Marie Legaré, ' "La Librairye de madame": Two Princesses and their Libraries', *Women of Distinction: Margaret of York, Margaret of Austria*, Dagmar Eichberger (ed.), exh. cat. (Mechelen, Lamot), Leuven 2005: 207-219.

Lehmann 2010
Martin Lehmann, *Die Cosmographiae Introductio Matthias Ringmanns und die Weltkarte Waldseemüllers aus dem Jahr 1507. Ein Meilenstein frühneuzeitlicher Kartographie*, Munich 2010.

Le Guay 1998
Laetitia le Guay, *Les princes de Bourgogne lecteurs de Froissart. Les rapports entre le texte et l'image dans les manuscrits enluminés du livre IV des "Chroniques"*, Paris 1998.

Leidinger 1912
Georg Leidinger (ed.), *Miniaturen aus Handschriften der Kgl. Hof- und Staatsbibliothek in München. Flämischer Kalender*, pt. 2, Munich 1912.

Leitch 2009
Stephanie Leitch, 'Burgkmair's Peoples of Africa and India (1508) and the Origins of Ethnography in Print', *The Art Bulletin* 91, 2 (2009): 134-159.

Le Loup 1981
Willy le Loup (ed.), *Vlaamse kunst op perkament: Handschriften en miniaturen te Brugge van 12de tot de 16de eeuw*, exh. cat. (Bruges, Museum Gruuthuse), Bruges 1981.

Lemaire 1973
Claudine Lemaire, *Le cinquième centenaire de l'imprimerie dans les anciens Pays-Bas*, exh. cat. (Brussels, Bibliothèque royale Albert Ier), Brussels 1973.

Lemaire 2002
Claudine Lemaire, 'Justifications remarquables dans des manuscrits à miniatures du XVe siècle conservés à la Bibliothèque Royale à Bruxelles', *'Als Ich Can.' Liber Amicorum in Memory of Professor Dr. Maurits Smeyers*, 2 vols., Bert Cardon, Jan Van der Stock & Dominique Vanwijnsberghe (eds.), Corpus of Illuminated Manuscripts, 11, Leuven 2002: 795-813.

Lemaire & Henry 1991
Claudine Lemaire & Michèle Henry, *Isabelle de Portugal. Duchesse de Bourgogne, 1397-1471*, exh. cat. (Brussels, Bibliothèque royale Albert Ier), Brussels 1991.

Leopold 1977
John H. Leopold, 'Mechanical Globes circa 1500-1650', *Bulletin of the Scientific Instruments Society* 53 (1977): 5-8.

Le Rouge 1976
Nicolas le Rouge, *Le grant Kalendrier et compost des Bergiers avecq leur Astrologie. Et plusieurs aultres choses. Imprimé nouvellement à Troyes par Nicolas le Rouge. - (A la fin) "Cy finist le Calendrier et compost des bergiers" [Edition de 1529]*, Bertrand Guégan (ed.), Paris 1976.

Leroy 1937
Stéphane Leroy, *Catalogue des peintures, sculptures, dessins et gravures exposés dans les galeries du musée de Douai*, mus. cat. (Douai, Musée de la Chartreuse), Douai 1937.

Leslie 1998
Marina Leslie, *Renaissance Utopias and the Problem of History*, Ithaca 1998.

L'Estrange Turner 1994a
Gérard L'Estrange Turner, 'The three Astrolabes of Gerard Mercator', *Annals of Science* 51 (1994): 329-353.

L'Estrange Turner 1994b
Gérard L'Estrange Turner, '"To find the mind's construction in the face": The Newly-Discovered Astrolabes of Mercator', *Bulletin of the Scientific Instrument Society* 43 (1994): 16-21.

L'Estrange Turner & Dekker 1994
Gérard L'Estrange Turner & Elly Dekker, 'De Astrolabia', *Gerardus Mercator Rupelmundanus*, Marcel Watelet (ed.), Antwerp 1994: 178-191.

L'Estrange Turner & Dekker 1995
Gérard L'Estrange Turner & Elly Dekker, 'The first Instruments of Gerard Mercator to be identified: three Astrolabes', *Gerhard Mercator: Europa und die Welt*, Ruth Löffler & Gernot Tromnau (eds.), exh. cat. (Duisburg, Kultur- und Stadthistorischen Museum), Duisburg 1995: 197-210.

Lestringant 2005
Frank Lestringant, *Jean de Léry ou l'invention du sauvage*, Paris 2005.

Leuven 1961
500 jaar kunst te Leuven, exh. cat. (Leuven, Stedelijk Museum Vander Kelen-Mertens), Leuven 1961.

Leuven 1962
Ars sacra antiqua, exh. cat. (Leuven, Stedelijk Museum Vander Kelen-Mertens), Leuven 1962.

Levenson 1991
Jay A. Levenson (ed.), *Circa 1492: Art in the Age of Exploration*, exh. cat. (Washington, National Gallery of Art), New Haven/London 1991.

Levenson 2007
Jay A. Levenson, (ed.), *Encompassing the Globe: Portugal and the World in the 16th and 17th Centuries*, exh. cat. (Washington, Arthur M. Sackler Gallery & National Museum of African Art), Washington 2007.

Lewis 1998
Lesley Lewis, *The Thomas More Family Group Portraits After Holbein*, Leominster 1998.

Lewisohn 1937
Sam A. Lewisohn, *Painters and Personality. A Collector's View of Modern Art*, New York 1937.

Limentani Virdis 1992
Caterina Limentani Virdis, *Le Delizie dell'inferno: dipinti di Jheronimus Bosch e altri fiamminghi restaurati*, Venice 1992.

Limentani Virdis 2002
Caterina Limentani Virdis, 'Considérations à propos de la fortune de Henri Bles en Italie', *Actes du colloque 'Autour de Henri Bles' 9-10 octobre 2000*, Jacques Toussaint (ed.), Namur 2002: 75-84.

Linehan & Nelson 2006
Peter Linehan & Janet Laughland Nelson, *The Medieval World*, London 2006.

Lippmann 1899-1900
Friedrich Lippmann, *Engravings and Woodcuts by Old Masters*, vol. 2, London 1899-1900.

Lisbon 1983
XVII Exposição Europeia de Arte, Ciência e Cultura: Os descobrimentos portugueses e a Europa do Renascimento /XVII European Exhibition of Art, Science, and Culture: The Portuguese discoveries and Renaissance Europe, exh. cat. (Lisbon, Museu Nacional de Arte Antiga), Lisbon 1983.

Logan 1983
George M. Logan, *The Meaning of More's "Utopia"*, Princeton 1983.

Lohmeyer 1919
Ernst Lohmeyer, *Vom göttlichen Wohlgeruch*, Heidelberg 1919.

London 1865
Catalogue of the special Exhibition of Portrait Miniatures on Loan at the South Kensington Museum, exh. cat. (London, South Kensington Museum), London 1865.

London 1890
Exhibition of the Royal House of Tudor, exh. cat. (London, New Gallery), London 1890.

London 1920
Exhibition of Spanish Paintings, exh. cat. (London, The Royal Academy of Arts), London 1920.

London 1948
Spanish and Dutch Painting from the Cook Collection, exh. cat. (London, The Guildhall Art Gallery), London 1948.

López Estrada 1965
Francisco López Estrada, 'La primera versión española de la *Utopía* de Moro, por Jerónimo Antonio de Medinilla (Córdoba, 1637)', *Collected Studies in Honour of Américo Castro's Eightieth Year*, M.P. Hornik (ed.), Oxford 1965: 291-309.

López Estrada 1992a
Francisco López Estrada, 'Una temprana traducción española de la Utopía de Tomás Moro', *Hispanic studies in honour of Geoffrey Ribbans*, Ann L. Mackenzie & Dorothy S. Severin (eds.), Liverpool 1992: 43-45.

López Estrada 1992b
Francisco López Estrada, 'Une traduction espagnole précoce de l'*Utopie* de Thomas More', *Moreana* 111-112 (1992): 15-18.

Lowe 2005
Kate J. P. Lowe, 'The stereotyping of Black Africans in Renaissance Europe', *Black Africans in Renaissance Europe*, Thomas F. Earle & Kate J. P. Lowe (eds.), Cambridge 2005: 17-47.

Lowenthal 1976
Constance Lowenthal, *Conrat Meit*, PhD dissertation, New York University, New York 1976.

Luaces 1998
Joaquín Yarza Luaces, *El Jardín de las Delicias de El Bosco*, Madrid 1998.

Luyckx 2013
Jeroen Luyckx, *Civitas Lovaniensis. Een onderzoek naar het ontstaan van het gedrukte stadsportret in de Duitse gebieden en de Nederlanden in de 15de en 16de eeuw & een casestudie en contextualisering van het oudst bewaarde gezicht op de stad Leuven*, 2 vols., Master dissertation, KU Leuven, Leuven 2013.

Lyna 1989
Frédéric Lyna, *Les principaux manuscrits à peintures de la Bibliothèque Royale de Belgique - Tome III, Deuxième partie*, Christiane Pantens (ed.), Brussels 1989.

Lyna & Pantens 1984-1989
Frédéric Lyna & Christiane Pantens, *Les principaux manuscrits à peintures de la Bibliothèque Royale de Belgique*, 3 vols., Brussels 1984-1989.

Lyna & Van Eeghem 1930
J. Lyna & W. van Eeghem, *Jan van Styevoorts Refereinenbundel anno MDXXIV*, vol. 2, Antwerp 1930.

Macagy & Sourian 1964
Jermayne Macagy & Etienne Sourian (eds.), *Out of this World: An Exhibition of Fantastic Landscapes from the Renaissance to the Present*, exh. cat. (Houston, University of St. Thomas), Houston 1964.

Mackor 1995
Adri Mackor, 'Marinus van Reymerswaele: Painter, Lawyer and Iconoclast?', *Oud Holland - Quarterly for Dutch Art History* 109 (1995): 191-200.

MacLeod 1996
Catherine MacLeod, *Tudor Portraits in the National Portrait Gallery Collection*, London 1996.

MacLeod 1999
Catherine MacLeod, *Tudor and Jacobean Portraits in the National Portrait Gallery Collection at Montacute House*, London 1999.

Madou 1997
Mireille Madou, ' "Tabbaerts met bree lappen." De datum van de uur- en kalenderwijzerplaat', *Het rad van de tijd. De uur- en kalenderwijzerplaat in het Stedelijk Museum te Leuven (ca. 1500) - Arca Lovaniensis artes atque historiae reserans documenta* 23-24 (1997): 61-68.

Madrid 2000
El jardín de las delicias de El Bosco: copias, estudio técnico y restauración, exh. cat. (Madrid, Museo del Prado), Madrid 2000.

Mandabach 2010
Marisa Mandabach, 'Holy Shit: Bosch's Bluebird and the Junction of the Scatological and the Eschatological in Late Medieval Art', *Marginalia* 11 (2010): 28-49.

Marc'hadour 1982
Germain Marc'hadour, 'M for Map: Maps of Utopia', *Moreana* 19 (1982): 103-107.

Margolin 1989
J.-C. Margolin, 'Sur l'insularité d'*Utopia*: Entre l'érudition et la rêverie', *Miscellanea Moreana. Essays for Germain Marc'hadour - Moreana* 100, 26 (1989): 303-321.

Marlier 1963
Georges Marlier, 'Saint Jérôme', *Le siècle de Bruegel. La peinture en Belgique au XVIe siècle*, exh. cat. (Brussels, Musées royaux des beaux arts de Belgique), Brussels 1963: 163-164.

Marlier & Reis-Santos 1954
Georges Marlier & Luis Reis-Santos, *Erasme et la peinture Flamande de son temps*, Damme 1954.

Marques 1994
Alfredo Pinheiro Marques, 'L'Atlas Miller: un problème résolu. Histoire de l'art dans la cartographie portugaise', *Revue de la Bibliothèque nationale de France* 4 (1994): 53-57.

Marques 2005
Alfredo Pinheiro Marques, *Os Descobrimentos e o 'Atlas Miller'*, PhD dissertation, Universidade de Coimbra, Coimbra 2005.

Marques 2009
Alfredo Pinheiro Marques, 'La scienza dei principi l'Atlante Miller', *Alumina: pagine miniate* 7, 24 (2009): 68-73.

Martens 1998
Maximiliaan Martens, *Brugge en de Renaissance: Van Memling tot Pourbus*, exh. cat. (Bruges, Memlingmuseum), Ghent 1998.

Marti, Borchert & Keck 2009
Susan Marti, Till-Holger Borchert & Gabrielle Keck (eds.), *Charles the Bold (1433-1477): Splendour of Burgundy*, exh. cat. (Bern, Historisches Museum/ Bruges, Groeningemuseum/Vienna, Kunsthistorisches Museum), Sage 2009.

Martin & Vezin 1990
Henri-Jean Martin & Jean Vezin (eds.), *Mise en page et mise en texte du livre manuscrit*, Paris 1990.

Massing 1991a
Jean Michel Massing, 'The Movement of the Universe', *Circa 1492: Art in the Age of Exploration*, Jay A. Levenson (ed.), exh. cat. (Washington, National Gallery of Art), New Haven/London 1991: 214-215.

Massing 1991b
Jean Michel Massing, 'Marvels of the East', *Circa 1492: Art in the Age of Exploration*, Jay A. Levenson (ed.), exh. cat. (Washington, National Gallery of Art), New Haven/London 1991: 125.

Massing 1991c
Jean Michel Massing, 'Map of the World and People from Faraway Lands', *Circa 1492: Art in the Age of Exploration*, Jay A. Levenson (ed.), exh. cat. (Washington, National Gallery of Art), New Haven/London 1991: 124.

Massing 2003
Jean Michel Massing, 'La mappemonde de Pierre Desceliers de 1550', *Henri II et les arts* (conference proceedings, Ecole du Louvre - Musée national de la Renaissance, Ecouen, 25-27 September 1997), Hervé Oursel & Julia Fritsch (eds.), Paris 2003: 231-248

Massing 2005
Jean Michel Massing, 'The Image of Africa and the Iconography of Lipplated Africans in Pierre Desceliers's World Map of 1550', *Black Africans in Renaissance Europe*, Thomas Foster Earle & Kate J. P. Lowe (eds.), Cambridge 2005: 48-69.

Massing & Adelson 1991
Jean Michel Massing & Candace Adelson, 'Astronomy', *Circa 1492: Art in the Age of Exploration*, Jay A. Levenson (ed.), exh. cat. (Washington, National Gallery of Art), New Haven/ London 1991: 225-226.

Matilda & Barclay 1909
Helen Matilda & William Barclay, *Catalogue of the Pictures in the Collection of the Earl of Radnor*, London 1909.

Mauduech 1976
Gérard Mauduech (ed.), *Des conquistadores aux lumières Normandie Antilles Amériques du centre et du sud, XVIe-XVIIIe siècles*, exh. cat. (Rouen, ADSM) 1976.

McConica 1977
James K. McConica (ed.), R.A.B. Mynors & D.F.S. Thomson (transl.), *The Correspondence of Erasmus - Letters 446 to 593 (1516-1517)*, vol. 4, Toronto 1977.

McDonald 2003
Mark P. McDonald, 'Burkgmair's Woodcut Frieze of Natives of Africa and India', *Print Quarterly* 20, 3 (2003): 227-244.

McGrath 1968
Robert L. McGrath, 'Satan and Bosch: The Visio Tundali and the Monastic Vices', *Gazette des Beaux Arts* 110, 1188 (1968): 45-50.

Mechelen 1961
Mechelen. Vier eeuwen aartsbisschoppelijke stad 1561-1961, exh. cat. (Mechelen, Kultureel Centrum), Mechelen 1961.

Mechelen 1987
De Habsburgers & Mechelen, exh. cat. (Mechelen, Stedelijk Museum Hof van Busleyden), Brussels 1987.

Meder 1932
Joseph Meder, *Dürer-Katalog. Ein Handbuch über Albrecht Dürers Stiche, Radierungen, Holzschnitte, deren Zustände, Ausgaben und Wasserzeichen*, Vienna 1932.

Meganck 2014
Tine Luk Meganck, *Pieter Bruegel the Elder. Fall of the Rebel Angels: Art, Knowledge and Politics on the Eve of the Dutch Revolt*, Brussels 2014.

Mellinkoff 1981
Ruth Mellinkoff, *The Mark of Cain*, Oakland 1981.

Meridith 2014
Martin Meridith, *The Fortunes of Africa: A 5000 - Year History of Wealth, Greed, and Endeavor*, New York 2014.

Merlo 1864
Johann Jakob Merlo, *Anton Woensam von Worms, Maler und Xylograph zu Köln. Sein Leben und seine Werke*, Leipzig 1864.

Merlo 1895
Johann Jakob Merlo, 'Kölnische Künstler in alter und neuer Zeit', *Publikationen der Gesellschaft fuer Rheinische Geschichtskunde*, vol. 9, Eduard Firmenich-Richartz (ed.), Düsseldorf 1895: 1088-1089.

Meskens 2012
Ad Meskens, 'Instrumenten', *Mercator: reizen in het onbekende*, exh. cat. (Antwerp, Museum Plantin-Moretus), Schoten 2012: 37-49.

Meurer 2008
Peter H. Meurer, 'Europa Regina. 16th Century Maps of Europe in the Form of a Queen', *Belgeo. Belgisch Tijdschrift voor Geografie* 3-4 (2008): 355-370.

Meurer 1998
Peter H. Meurer, 'Ortelius as the Father of Historical Cartography', *Abraham Ortelius and the First Atlas. Essays Commemorating the Quadricentennial of His Death, 1598-1998*, Marcel van den Broecke, Peter van der Krogt & Peter H. Meurer (eds.), Utrecht 1998: 133-159.

Meyer 1984
Paul H. Meyer, 'Nicolas Gueudeville and his Work (1652-172?)', *The French Review* 58, 2 (1984): 289-290.

Mezger 1979
Werner Mezger, 'Der Ambraser Narrenteller von 1528: Ein Beitrag zur Ikonographie der spätmittelalterlichen Narrenidee', *Volkskunde* 75 (1979): 161-180.

Mezger 1991
Werner Mezger, *Narrenidee und Fastnachtsbrauch: Studien zum Fortleben des Mittelalters in der europäischen Festkultur*, Konstanz 1991.

Mezger 2009
Werner Mezger, 'Närrische Gesellschaften und ihre Quasi-Reiche in der urbanen Fastnacht des 15. und 16. Jahrhunderts - Versuch einer Annäherung', *Das Königreich der Narren: Fasching im Mittelalter*, Johannes Grabmayer (ed.), Friesach 2009: 145-212.

Michalove 2004
Sharon D. Michalove, 'Women as Book Collectors and Disseminators of Culture in Late Medieval England and Burgundy', *Reputation and Representation in Fifteenth Century Europe*, Douglas L. Biggs, Sharon D. Michalove & Albert C. Reeves (eds.), Leiden 2004: 57-79.

Michel 1960
H. Michel, 'L'horloge de sapience et l'histoire de l'horlogerie', *Physis* 2 (1960): 291-298.

Michel 1977
Henri Michel (ed.), *Images des sciences. Les anciens instruments scientifiques vus par les artistes de leur temps*, Rhode-St-Genèse 1977.

Michel & Sternath 2012
Eva Michel & Maria Luise Sternath (eds.), *Emperor Maximilian I and the Age of Dürer*, exh. cat. (Vienna, Albertina), Munich 2012.

Michelant 1872
Henri Michelant, 'Inventaire des joyaux, ornements d'église, vaisselles, tapisseries, livres, tableaux etc. de Charles-Quint, dressé à Bruxelles au mois de mai 1536', *Bulletin de la Commission Royale* 3, 13 (1872): 199-368.

Milkers 1985
Anne Milkers, *Omtrent Marinus Van Reymerswaele*, Master dissertation, Ghent University, Ghent 1985.

Miner, Carlson & Filby 1965
Dorothy E. Miner, Victor I. Carlson & P. William Filby (eds.), *Two Thousand Years of Calligraphy*, exh. cat. (Baltimore, Baltimore Museum of Art/ Peabody Institute Library/Walters Art Gallery), New York 1965.

Minerva 1992
Nadia Minerva, 'D'une définition à l'autre. Sur quelques préfaciers français d'*Utopia* de Thomas More', *Per una definizione dell' utopia. Metodologi e discipline a confronto*, Nadia Minerva (ed.), Ravenna 1992: 51-61.

Miraeus 1609
Aubertus Miraeus, *Elogia Belgica sive Illustrium Belgii scriptorum qui, nostra patrumque memoria, vel Ecclesiam Dei propugnarunt, vel disciplinas illustrarunt, vitae breviter commemoratae*, Antwerp 1609.

Mitjans 2007
Frank Mitjans, 'Non sum Oedipus, sed Morus: A Paper on the Portrait of Sir Thomas More and his Family', *Moreana* 168-170 (2007): 12-67.

Möller 1963
Lise Lotte Möller, *Die wilden Leute des Mittelalters*, exh. cat. (Hamburg, Museum für Kunst und Gewerbe), Hamburg 1963.

Monks 1990
Peter Rolfe Monks, *The Brussels Horloge de Sapience: Iconography and Text of Brussels, Bibliothèque Royale, MS IV 111*, Leiden 1990.

Moormann & Uitterhoeve 1997
Eric M. Moormann & Wilfried Uitterhoeve, *Van Achilles tot Zeus. Thema's uit de klassieke mythologie in literatuur, muziek, beeldende kunst en theater*, Nijmegen 1997.

Mora-Lebrun 1983
Antoine de la Sale, *Le paradis de la reine Sibylle*, Francine Mora-Lebrun (ed. & transl.), Paris 1983.

More 1965
Edward Surtz & J. H. Hexter (eds.), *The complete works of St. Thomas More. Volume 4* [*Utopia*], New Haven/London 1965.

More 1966
Thomas More, *Utopia [1516]*, Scolar Press facsimile, Menston 1966.

More 1970
Thomas More, *La description de l'Isle d'Utopie. Traduction française de 1550, introduction par Michel Jeanneret*, Wakefield/New York/Paris 1970.

More 2008
Thomas More, *Utopia*, Paul Silverentand (transl.), Amsterdam 2008.

Moreau 1977
Brigitte Moreau, *Inventaire chronologique des éditions parisiennes du XVIe siècle, d'après les manuscrits de Philippe Renouard. II: 1511-1520*, Paris 1977.

Morison 1957
Stanley Morison, 'The Portraiture of Thomas More by Hans Holbein and After', *Lecture to the Thomas More Society 18 October 1957*, Cambridge 1957.

Morison 1963
Stanley Morison, *The Likeness of Thomas More, an Iconographical Survey of Three Centuries*, London 1963.

Morison 1971
Samuel Eliot Morison, *The European Discovery of America: the Northern Voyages A.D. 500-1600*, New York 1971.

Morrison 2007
Elizabeth Morrison, *Beasts: Factual & Fantastic*, Los Angeles/London 2007.

Moseley-Christian 2012
Michelle Moseley-Christian, 'From Page to Print: the Transformation of the 'Wild Woman' in Eearly Modern Northern Engravings', *Word & Image: A Journal of Verbal/Visual Enquiry* 27, 4 (2012): 429-442.

Mout 2015
Nicolette Mout, 'Desiderius Erasmus (1466-1536). Opvoeding, filosofie van Christus en politiek', *Een nieuwe wereld. Denkers uit de Nederlanden over politiek en maatschappij (1500-1700)*, Erik de Bom (ed.), Antwerp 2015: 55-79.

Muir 2012
Carolyn Diskant Muir, *Saintly Brides and Bridegrooms: The Mystic Marriage in Northern Renaissance Art*, London 2012.

Mulder 2004
Herman Mulder, 'Middelnederlandse rijmspreuken: iets over vloertegels, doodkisten en Latijnse voorbeelden', *Queeste. Tijdschrift over middeleeuwse letterkunde in de Nederlanden* 1 (2004): 44-55.

Müller 2007
Jürgen Müller, 'Italienverehrung als Italienverachtung. Hans Sebald Behams "Jungbrunnen" von 1536 und die italiensche Kunst der Renaissance', *Bild/Geschichte: Festschrift für Horst Bredekamp*, Philine Helas, Maren Polte, Claudia Rückert & Bettina Uppenkamp (eds.), Berlin 2007: 309-318.

Müller 2016
Linda M. Müller, 'Follower of Hieronymus Bosch. Christ's Descent into Hell', *Early Netherlandish Paintings in the Metropolitan Museum of Art*, Maryan W. Ainsworth (ed.), online mus. cat. (New York, The Metropolitan Museum of Art) 2016: http://www.metmuseum.org/art/collection/search/435725?sortBy=Relevance&ft=bosch&pg=1&rpp=20&pos=1 (consulted April 2016).

Müller, Kemperdick et al. 2006
Christian Müller, Stephan Kemperdick et al., *Hans Holbein the Younger, The Basel Years 1515-1532*, exh. cat. (Basel, Kunstmuseum), Munich/ Berlin/London/New York 2006.

Müller-Wulckow 1961
Walter Müller-Wulckow, *Vier Bildnisse des Petrus Ägidius von Gossart, Massys, Dürer, Holbein* (Conference Proceedings, Deutscher Museumsbund, Oldenburg, 26-30 September 1961), Oldenburg 1961.

Mund 1977
Hélène Mund, *Jan Mandyn – Pieter Huys. Etude de l'oeuvre signée et datée*, PhD dissertation, Université Catholique de Louvain, Louvain 1977.

Mund-Dopchie 1994
Monique Mund-Dopchie, 'De Antieke Fabelwereld', *Gerardus Mercator Rupelmundanus*, Marcel Watelet (ed.), Antwerp 1994: 23-36.

Muris & Saarmann 1961
Oswald Muris & Gert Saarmann, *Der Globus im Wandel der Zeiten - Eine Geschichte der Globen*, Berlin/ Stuttgart 1961.

Musper 1943
Helmut Theodor Musper, 'Die Holzschnitte des Nicolaus Hogenberg aus München', *Jahrbuch der preussischen Kunstsammlungen* 63, vol. 4, Berlin (1943).

Musper 1952
Helmut Theodor Musper, *Albrecht Dürer: Der gegenwärtige Stand der Forschung*, Stuttgart 1952.

Nagler 1858
Georg Kaspar Nagler, *Die Monogrammisten*, vol. 1, Munich 1858.

Nascimento & Miranda 2000
Aires Augusto Nascimento & Maria Adelaide Miranda, *The Image of Time: European Manuscript Books*, exh. cat. (Lisbon, Calouste Gulbenkian Foundation), Lisbon 2000.

Nebenzahl 1995
Kenneth Nebenzahl, *Atlas zum Heiligen Land: Karten der Terra Sancta durch zwei Jahrtausende*, Stuttgart 1995.

Neeffs 1891
Emmanuel Neeffs, *Tableaux, sculptures et objets d'art conservés dans les édifices religieux et civils de Malines*, Mechelen 1891.

Nelson 2004
Eric Nelson, 'Greek Nonsense in More's *Utopia*', *The Greek Tradition in Republican Thought*, Eric Nelson (ed.), Cambridge 2004: 19-48.

Nichols 1873
John Gough Nichols, 'Remarks on some Pictures of Quintin Matsys and Holbein, in the Collection of the Earl of Radnor, at Longford Castle', *Archaeologia* 44 (1873): 435-458.

Niessen 2012
Judith Niessen, *Cornelis van Haarlem (1562-1638)*, exh. cat. (Haarlem, Frans Hals Museum), Rotterdam 2012.

Nijhoff & Kronenberg 1923-1971
Wouter Nijhoff & M. E. Kronenberg, *Nederlandsche bibliographie van 1500 tot 1540*, 3 parts in 8 vols. The Hague 1923-1971.

Ninane 1963
Lucie Ninane, 'Pieter Huys - Le Jugement dernier', *Le siècle de Bruegel. La peinture en Belgique au XVIe siècle*, exh. cat. (Brussels, Musées royaux des beaux arts de Belgique), Brussels 1963: 123.

Nink 1993
Rudolf Nink, 'Titelei und Ausstattung von Thomas More's Utopia, *Literatur und Typographie. Wort-Bild-Synthesen in der englischen Prosa des 16. bis 20. Jahrhunderts*, Buchwissenschaftliche Beiträge aus dem Deutschen Bucharchiv München, Bd. 45, Wiesbaden 1993: 28-43.

Nijs 2002
Greet Nijs, 'Typology of the Border Decoration in the Manuscripts of the Ghent-Bruges School', *'Als Ich Can.' Liber Amicorum in Memory of Professor Dr. Maurits Smeyers*, 2 vols., Bert Cardon, Jan Van der Stock & Dominique Vanwijnsberghe (eds.), Corpus of Illuminated Manuscripts, 11, Leuven 2002: 1007-1036.

Noever 2000
Peter Noever (ed.), *Kunst und Industrie. Die Anfänge des Museums für angewandte Kunst in Wien*, exh. cat. (Vienna, Österreichischen Museums für Angewandte Kunst), Ostfildern-Ruit 2000.

Nordenskiold 1973
Adolf Erik Nordenskiold, *Facsimile-atlas to the Early History of Cartography with Reproductions of the most important Maps printed in the XV and XVI Centuries*, New York 1973.

Norrington 1983
Ruth Norrington, *In the Shadow of a Saint: Lady Alice More*, Waddeston 1983.

Nuremberg 1971
Albrecht Dürer, 1471-1971, exh. cat. (Nuremberg, Germanisches Nationalmuseum), Munich 1971.

O'Malley 2013
John W. O'Malley, *Trent. What happened at the Council*, Cambridge/ London 2013.

Oncken 1913
Wilhelm Oncken, 'Die Großherzogliche Gemälde-Galerie im Augusteum', *Heimatkunde des Herzogtums Oldenburg*, vol. 2, Wilhelm Schwecke & Wilhelm von Busch (eds.), Bremen 1913: 534-542.

O'Neill & Schultz 1986
John P. O'Neill & Ellen Schultz (eds.), *Gothic and Renaissance Art in Nuremberg, 1300–1550*, exh. cat. (New York, Metropolitan Museum of Art/Nuremberg, Germanisches Nationalmuseum), Munich 1986.

Orchard 1997
Andy Orchard, *Pride and Prodigies: Studies in the Monsters of the Beowulf-Manuscript*, Cambridge 1997.

Orloff-Govaerts 1998
Nicolas & Odette Orloff-Govaerts (eds.), *Jean Mansel. La Fleur des histoires, premier livre*, 3 vols., Saint-Pierre-de-Chignac 1998.

Osley 1969
Arthur Sidney Osley, *Mercator, a Monograph on the Lettering of Maps, etc. in the 16th Century Netherlands, with a Facsimile and Translation of his Treatise on the Italic Hand and a Translation of Ghim's "Vita Mercatoris"*, London 1969.

Osterwold & Pollig 1987
Tilman Osterwold & Hermann Pollig (eds.), *Exotische Welten. Europäische Phantasien*, exh. cat., (Stuttgart, Institut für Auslandsbeziehungen und Württembergischer Kunstverein), Stuttgart 1987.

Pabel 2002
Hilmar M. Pabel, 'Reading Jerome in the Renaissance: Erasmus' Reception of the "Adversus Jovinianum" ', *Renaissance Quarterly* 55, 2 (2002): 470-497.

Pabel 2013
Hilmar M. Pabel, 'Praise and Blame: Peter Canisius's Ambivalent Assessment of Erasmus', *The Reception of Erasmus in the Early Modern Period*, Karl A. E. Enenkel (ed.), Leiden 2013: 129-162.

Pächt 1963
Otto Pächt, 'Zur Entstehung des Hieronymus im Gehäus', *Pantheon* 21, 3 (1963): 131-142.

Padmos & Vanpaemel 2000
Tineke Padmos & Geert Vanpaemel (eds.), *De geleerde wereld van Keizer Karel*, exh. cat. (Leuven, Stedelijk Museum Vander Kelen-Mertens), Leuven 2000.

Pagden 2001
Anthony Pagden, *Peoples and Empires. Europeans and the Rest of the World, from Antiquity to the Present*, London 2001.

Palmer 1982
Nigel F. Palmer, *Visio Tnugdali the German and Dutch translations and their Circulation in the Later Middle Ages*, Munich 1982.

Panofsky 1943
Erwin Panofsky, *Albrecht Dürer*, 2 vols., Princeton 1943.

Panofsky 1955
Erwin Panofsky, *The Life and Art of Albrecht Dürer*, Princeton 1955.

Papy 2000
Jan Papy, 'Politiek als humanistische wetenschap. De manifesten voor een betere wereld van Erasmus en Vives', *De geleerde wereld van Keizer Karel*, Tineke Padmos & Geert Vanpaemel (eds.), exh. cat. (Leuven, Stedelijk Museum Vander Kelen-Mertens), Leuven 2000: 118-128.

Papy 2015
Jan Papy, 'Juan Luis Vives (1492/3 – 1540). Christelijke *humanitas* voor een sociaal Europa van eenheid en vrede', *Een nieuwe wereld. Denkers uit de Nederlanden over politiek en maatschappij (1500-1700)*, Erik de Bom (ed.), Antwerp 2015: 81-113.

Paris 1904
Gaston Paris, *Légendes du Moyen-Âge*, Paris 1904.

Paris 1936
Instruments et outils d'autrefois, exh. cat. (Paris, Musée des arts décoratifs/ Palais du Louvre/Pavillon de Marsan), Paris 1936.

Paris 1955
France et Brésil, exh. cat. (Paris, Hôtel de Rohan), Paris 1955.

Paris 1973
Coligny - Protestants et catholiques en France au XVIe siècle, exh. cat. (Paris, Bibliothèque Nationale), Paris 1973.

Parker 1945
K. T. Parker, *The Drawings of Hans Holbein in the Collection of His Majesty the King at Windsor Castle*, London 1945.

Parthey 1863
Gustav Parthey, *Deutscher Bildersaal: Verzeichniss der in Deutschland vorhandenen Oelbilder verstorbener Maler aller Schulen*, vol. 1, Berlin 1863.

Parshall 1989
Peter Parshall, 'Reading the Renaissance Woodcut: The Odd Case of the Master N.H. & Hans Lützelburger', *The Register of the Spencer Museum of Art* 6, 6 (1989): 30-43.

Passavant 1833
Johann David Passavant, *Kunstreise durch England und Belgien*, Frankfurt 1833.

Passavant 1836
Johann David Passavant, *Tour of a German artist in England*, vol. 1, Cambridge 1836.

Passavant 1860-1864
Johann David Passavant, *Le Peintre-Graveur*, 6 parts in vols., Leipzig 1860-1864.

Passavant 1862
Johann David Passavant, *Le Peintre-Graveur*, vol. 3, Leipzig 1862.

Pauli 1901
Gustav Pauli, *Hans Sebald Beham. Ein kritisches Verzeichniss seiner Kupferstiche, Radierungen und Holzschnitte*, Strasbourg 1901.

Pauwels 1963
Henri Pauwels, 'Fête des archers', *Le siècle de Bruegel. La peinture en Belgique au XVIe siècle*, exh. cat. (Brussels, Musées royaux des beaux arts de Belgique), Brussels 1963: 180.

Pauwels, Hoetink & Herzog 1965
Henri Pauwels, H.T. Hoetink & Sadja Herzog (eds.), *Jean Gossaert dit Mabuse*, exh. cat. (Rotterdam, Museum Boymans Van Beuningen/Bruges, Groeningemuseum), Amsterdam 1965.

Pearson 2017
Andrea Pearson, 'Sensory Piety as Social Intervention in a Mechelen Besloten Hofje', *Journal of Historians of Netherlandish Art* (2017): forthcoming.

Peggram 1940
Reed Edwin Peggram, 'The first French and English Translations of Sir Thomas More's Utopia', *The modern language review* 35 (1940): 330-340.

Pelletier 1998
Monique Pelletier (ed.), *Couleurs de la terre: des mappemondes médiévales aux images satellitales*, exh. cat. (Paris, Bibliothèque nationale de France), Paris 1998.

Petit 1877
Jules Petit, 'Vue de Louvain', *Documents iconographiques et typographiques de la Bibliothèque Royale de Belgique*, Brussels/Amsterdam 1877: 47-65.

Pettegree 2005
Andrew Pettegree, *Reformation and the Culture of Persuasion*, Cambridge 2005.

Pettegree, Walsby & Wilkinson 2007
Andrew Pettegree, Malcolm Walsby & Alexander Wilkinson, *French Vernacular Books. Books published in the French Language before 1601: Livres vernaculaires français. Livres imprimés en français avant 1601*, Leiden/Boston 2007.

Pettegree & Walsby 2011
Andrew Pettegree & Malcolm Walsby, *Netherlandish Books. Books published in the Low Countries and Dutch Books printed abroad before 1601*, Leiden/Boston 2011.

Pettegree & Walsby 2012
Andrew Pettegree & Malcolm Walsby, *Books published in France before 1601 in Latin and Languages other than French*, Leiden/Boston 2012.

Philippa, Debrabandere, Quak et al. 2007
Marlies Philippa, Frans Debrabandere, Arend Quak et al., *Etymologisch woordenboek van het Nederlands, Deel 2 (Ke -R)*, Amsterdam 2007.

Pirovano 1981
Carlo Pirovano (ed.), *Museo Poldi Pezzoli: Orologi Oreficerie*, mus. cat. (Milan, Musei E Gallerie Di Milano), Milan 1981.

Pisters 1991
Ine Pisters (ed.), *America, bride of the sun. 500 years Latin America and the Low Countries*, exh. cat. (Antwerp, Koninklijk Museum voor Schone Kunsten), Ghent 1991.

Pons 1930
Émile Pons, 'Les langues imaginaires dans le voyage utopique – Un précurseur: Thomas Morus', *Revue de la littérature comparée* 10 (1930): 589-607.

Porteman 1995
Karel Porteman, 'Het embleem als *genus iocosum*, Theorie en praktijk bij Cats en Roemer Visscher', *De Zeventiende Eeuw* 11 (1995): 184-197.

Porter 2003
Pamela Porter, *Courtly Love in Medieval Manuscripts*, London 2003.

Poupeye 1912
Camille Poupeye, 'Les Jardins clos et leurs rapports avec la Sculpture malinoise', *Bulletin du Cercle Archéologique de Malines* 22 (1912): 51-114.

Pouvreau 2015
Florent Pouvreau, *Du poil et de la bête. Iconographie du corps sauvage en Occident à la fin du Moyen Age (XIIIe-XVIe siècle)*, Paris 2015.

Prevenier 1998
Walter Prevenier (ed.), *Prinsen en poorters. Beelden van de laat-middeleeuwse samenleving in de Bourgondische Nederlanden, 1384-1530*, Antwerp 1998.

Prevenier & Blockmans 1983
Walter Prevenier & Wim Blockmans, *De Bourgondische Nederlanden*, Antwerp 1983.

Prevenier & Blockmans 1986
Walter Prevenier & Wim Blockmans, *The Burgundian Netherlands*, Cambridge 1986.

Prévost 1978
André Prévost, *L'Utopie de Thomas More. Présentation, texte original, apparat critique, exégèse, traduction et notes. Préface de Maurice Schumann*, Paris 1978.

Ptolemaei 1990
Claudii Ptolemaei, *Cosmographia Tabulae - Edition des planches du manuscrit latin VF.32 (XVe siècle) de la Bibliothèque Nationale de Naples* (introduction by Lelio Pagani), Torriana/Paris 1990.

Pühlhorn & Laub 1993
Wolfgang Pühlhorn & Peter Laub (eds.), *Focus Behaim Globus*, 2 vols., exh. cat. (Nuremberg, Germanischen Nationalmuseums), Nuremberg 1993.

Quilliet 1986
Bernard Quilliet, *Louis XII, père du peuple*, Paris 1986.

Raaflaub & Talbert 2010
Kurt Raaflaub & Richard Talbert, *Geography and Ethnography: Perceptions of the World in Pre-Modern Societies*, New York 2010.

R.A.S. & Destombes 1964
R.A.S. & Marcel Destombes. 'A Portuguese SeaÐatlas of the 16th Century', *Imago Mundi: The International Journal for the History of Cartography* 18,1 (1964): 89.

Rabel 2007
Claudia Rabel, 'Memores estote uxoris Loth. Fortune d'une métamorphose biblique dans l'iconographie médiévale', *Quand la peinture était dans les livres. Mélanges en l'honneur de François Avril*, Mara Hofmann & Caroline Zöhl (eds.), *Ars nova: Studies in Late Medieval and Renaissance Northern Painting and Illumination* 15, Turnhout 2007: 259-272.

Racault 1991
Jean-Michel Racault, *L'utopie narrative en France et en Angleterre 1675-1761*, Oxford 1991.

Raitiere 1973
Martin N. Raitiere, 'More's *Utopia* and *The City of God*' *Studies in the Renaissance* 20 (1973): 144-168.

Rapp 1976
Anna Rapp, *Der Jungbrunnen in Literatur und bildender Kunst des Mittelalters*, Zürich 1976.

Rapp Buri & Stucky-Schürer 1990
Anna Rapp Buri & Monica Stucky-Schürer, *Zahm und Wild: Basler und Strassburger Bildteppiche des 15. Jahrhunderts*, Mainz am Rhein 1990.

Rasmussen 1973
Jörg Rasmussen, 'Eine Gruppe kleinplastischer Bildwerke aus dem Stilkreis Conrat Meits', *Städel Jahrbuch* 4 (1973): 121-144.

Recklinghausen 1967
Meisterwerke aus Douai: von Bellegambe bis Bonnard, exh. cat. (Recklinghausen, Städtische Kunsthalle), Recklinghausen 1967.

Reed 1926
A. W. Reed, 'John Clement and his Books', *The Library* 4, 6 (1926): 329-339.

Rehnig 2006
Jeanne E. Rehnig, *Todesmutig. Das siebte Werk der Barmherzigkeit*, Düsseldorf 2006.

Reidemeister 2006
Johann Reidemeister, *Superbia und Narziß. Personifikation und Allegorie in Miniaturen mittelalterlicher Handschriften*, Turnhout 2006.

Reiffenberg & Borgnet 1846-1859
Reiffenberg & Borgnet, *Le Chevalier au Cygne et Godefroid de Bouillon*, Brussels 1846-1859.

Reinitzer 1982
Heimo Reinitzer, *Der verschlossene Garten. Der Garten Marias im Mittelalter*, Wolfenbütteler Hefte 12, Wolfenbüttel 1982.

Reske 2000
Christoph Reske, *Die Produktion der Schedelschen Weltchronik in Nürnberg*, Wiesbaden 2000.

Revuelta Tubino 1987
Matilde Revuelta Tubino, *Museo de Santa Cruz de Toledo*, vol. 1, Ciudad Real 1987.

Rice 1985
Eugene F. Rice, *Saint Jerome in the Renaissance*, Baltimore/London 1985.

Roegiers 1969
Jan Roegiers (ed.), *Erasmus en Leuven*, exh. cat. (Leuven, Stedelijk Museum Vander Kelen-Mertens), Leuven 1969.

Roegiers 1976
Jan Roegiers (ed.), *550 jaar Universiteit Leuven*, exh. cat. (Leuven, Stedelijk Museum Vander Kelen-Mertens), Leuven 1976.

Roegiers 2008
Jan Roegiers (ed.), *Meten en weten. Gemma Frisius 1508-2008*, exh. cat. (Leuven, Universiteitsbibliotheek KU Leuven), Leuven 2008.

Rogers 1947
Elizabeth Frances Rogers (ed.), *The Correspondence of Sir Thomas More*, Princeton 1947.

Rogers 1961
Elizabeth Frances Rogers (ed.), *St Thomas More: Selected Letters*, New Haven/London 1961.

Roggen 2008
Vibeke Roggen, 'A Protean text: *Utopia* in Latin, 1516-1631', *Thomas More's Utopia in Early Modern Europe. Paratexts and contexts*, Terence Cave (ed.), Manchester 2008: 14-31.

Roglán 2008
Mark Roglán, 'Los secretos del retablo', *Ars. Revista de arte/Coleccionismo* 1 (2008): 52-67.

Roller 1965
Hans-Ulrich Roller, 'Der Nürnberger Schembartlauf. Studien zum Fest- und Maskenwesen des späten Mittelalters', *Volksleben*, pt. 11, Tübingen (1965).

Rosenberg 1982
Aubrey Rosenberg, *Nicolas Gueudeville and his work (1652-172?)*, The Hague/Boston/London 1982.

Röttel 1995
Karl Röttel (ed.), *Peter Apian: Astronomic, Kosmographie und Mathematik am Beginn der Neuzeit*, exh. cat. (Ingolstadt, Stadtmuseum Ingolstadt), Buxheim 1995.

Rotterdam 1936
Jeroen Bosch. Van Geertgen tot Scorel. Noord-Nederlandsche Primitieven, exh. cat. (Rotterdam, Museum Boymans), Rotterdam 1936.

Rotterdam 1969
Erasmus en zijn tijd, 2 vols., exh. cat. (Rotterdam, Museum Boijmans Van Beuningen), Rotterdam 1969.

Rouen 1932
La Normandie exploratrice et colonisatrice du XVe au XVIIIe siècle, exh. cat. (Rouen, Musée des Beaux-Arts), Rouen 1932.

Rouzet 1975
Anne Rouzet, *Dictionnaire des imprimeurs, libraires et éditeurs des XVe et XVe siècles dans les limites géographiques de la Belgique actuelle*, Nieuwkoop 1975.

Rowlands 1985
John Rowlands, *Holbein: The Paintings of Hans Holbein the Younger*, Oxford 1985.

Rozenski Jr. 2010
Steven Rozenski Jr., 'Henry Suso's Horologium Sapientiae in fifteenth-century France: Images of Reading and Writing in Brussels Royal Library MS IV 111', *Word & Image* 26, 4 (2010): 364-380.

Rücker 1973
Elisabeth Rücker, *Die Schedelsche Weltchronik: Das grösste Buchunternehmen der Dürer-Zeit*, Munich 1973.

Rudy 2011
Kathryn M. Rudy, *Virtual Pilgrimages in the Convent: Imagining Jerusalem in the Late Middle Ages*, Turnhout 2011.

Rudy 2015
Kathryn M. Rudy, *Postcards on Parchment: The social Lives of Medieval Books*, New Haven/London 2015.

Rupprich 1956-1969
Hans Rupprich (ed.), *Dürer. Schriftlicher Nachlass*, 3 vols., Berlin 1956-1969.

Russo 2014
Alessandra Russo, *The Untranslatable Image: A Mestizo History of the Arts in New Spain, 1500-1600*, Austin 2014.

Russo, Wolf & Fane 2015
Alessandra Russo, Gerhard Wolf and Diana Fane (eds.), *Images Take Flight: Feather Art in Mexico and Europe*, Munich 2015.

Sachs 1884
Hans Sachs, *Ausgewählte poetische Werke*, Karl Pannier (ed.), Leipzig 1884.

Sachs 1893
Hans Sachs, *Sämtliche Fabeln und Schwänke. In chronologischer Ordnung nach den Originalen herausgegeben von Edmund Goetze*, vol. 1, Halle an der Saale 1893.

Salberg 2008
Trond Kruke Salberg, 'The German Translations: Humanist Politics and Literary Journalism', *Thomas More's Utopia in Early Modern Europe. Paratexts and Contexts*, Terence Cave (ed.), Manchester 2008: 32-46, 149-169.

Salvini & Grohn 1971
Roberto Salvini & Hans Werner Grohn, *L'opera pittorica completa di Holbein il Giovane*, Milan 1971.

Salzer 2013
Ronald Salzer, 'Mobility Ahead of its Time: A Fifteenth-Century Austrian Pocket Sundial as a Trailblazing Instrument for Time Measurement on Travels', *Archaeologies of Mobility and Movement*, Mary C. Beaudry & Travis G. Parno (eds.), New York 2013.

Sander 2002
Jochen Sander (ed.), *Niederländische Gemälde im Städel 1400-1550. Katalog der Gemälde im Städelschen Kunstinstitut Frankfurt am Main II. Herausgegeben von Klaus Gallwitz und Jochen Sander. Unter Mitarbeit von Stephan Knobloch bei der Gemäldetechnologischen Dokumentation und mit einem Beitrag von Peter Klein zu den Ergebnissen der dendrochronologischen Untersuchungen*, mus. cat. (Frankfurt am Main, Kataloge der Gemälde im Städelschen Kunstinstitut Frankfurt am Main), Mainz am Rhein 2002.

Sander 2005
Jochen Sander, *Hans Holbein D.J., Tafelmaler in Basel, 1515-1532*, Munich 2005.

Sander, Knobloch & Klein 1993
Jochen Sander, Stephan Knobloch & Peter Klein (eds.), *Niederländische Gemälde im Städel 1400-1550*, mus. cat. (Frankfurt am Main, Kataloge der Gemälde im Städelschen Kunstinstitut Frankfurt am Main), Mainz am Rhein 1993.

Sasikumar 2011
Samyuktha Sasikumar, *Society and Culture under the Rulers of Kochi since 1500*, PhD dissertation, Kannur University, Kerala 2011.

Scafi 2006
Alessandro Scafi, *Mapping Paradise: A History of Heaven on Earth*, London 2006.

Scafi 2013
Alessandro Scafi, 'Paradise and Utopia', *Maps of Paradise*, Chicago 2013: 160-171.

Schaer & Sargent 2000
Roland Schaer & Lyman Tower Sargent (eds.), *Utopie: la quete de la societe ideale en Occident*, exh. cat. (Paris, Bibliotheque nationale de France/ New York, Public Library), Paris 2000.

Schaer, Claeys & Sargent 2000
Roland Schaer, Gregory Claeys & Lyman Tower Sargent (eds.), *Utopia. The Search for the Ideal Society in the Western World*, tent. cat. (New York, Public Library), Oxford 2000.

Schäfer 1912
Karl Schäfer, *Führer durch die Großherzogliche Gemaelde-Galerie im Augusteum zu Oldenburg*, Oldenburg 1912.

Scheibler & Bode 1881
Ludwig Scheibler & Wilhelm Bode, 'Verzeichniss der Gemälde des Jan van Scorel', *Jahrbuch der Königlich Preussischen Kunstsammlungen* 2 (1881): 211-214.

Scherer 2001
Annette Scherer, 'Vom Andachtsbild zum Sammlerbild: Varianten des "Sündenfalls" von Hugo van der Goes', *Opere e Giorni: Studi su mille anni di arte europea dedicati a Max Seidel*, Klaus Bertold & Giorgio Bonsanti (eds.), Venice 2001: 363-370.

Scherer 2007
Annette Scherer, 'Adam et Eve dans l'atelier du Maître au Feuillage brodé et la question de l'anonymat', *Le Maître au Feuillage brodé: démarches d'artistes et méthodes d'attribution d'oeuvres à un peintre anonyme des anciens Pays-Bas du XVème siècle = The Master of the Embroidered Foliage: artist's processes and attribution methods to an anonymus flemish painter of the XVth century*, Florence Gombert & Didier Martens (eds.), Lille 2007: 31-44.

Schinkel & Schmidt-Rutsch 2014
Eckhard Schinkel & Olaf Schmidt-Rutsch, Über Unterwelten Zeichen und Zauber des anderen Raums, exh. cat. (Essen, LWL-Industriemuseum), Essen 2014.

Schmeidler 1995
Felix Schmeidler, 'Die Scheiben in Peter Apians Astronomicum Caesareum', *Peter Apian: Astronomic, Kosmographie und Mathematik am Beginn der Neuzeit*, Karl Röttel (ed.), Buxheim 1995: 107-112.

Schmitt 2004
Jean-Claude Schmitt (ed.), *Femmes, art et religion au Moyen Âge*, Colmar 2004.

Schneebalg-Perelman 1971
Sophie Schneebalg-Perelman, 'Richesses du garde-meuble parisien de François Ier. Inventaires inédits de 1542 et 1551', *Gazette des Beaux-Arts* 6, 78 (1971): 253-304.

Schneider 2004
Ute Schneider, *Die Macht der Karten: eine Geschichte der Kartographie vom Mittelalter bis heute*, Darmstadt 2004.

Schoch 1986
Rainer Schoch, 'A Century of Nuremberg Printmaking', *Gothic and Renaissance Art in Nuremberg, 1300-1550*, John P. O'Neill & Ellen Schultz (eds.), exh. cat. (New York, Metropolitan Museum of Art/Nuremberg, Germanisches Nationalmuseum), Munich 1986: 93-100.

Schoch, Mende & Scherbaum 2002
Rainer Schoch, Matthias Mende & Anna Scherbaum (eds.), *Albrecht Dürer. Das druckgraphische Werk - Holzschnitte*, vol. 2, Munich 2002.

Schoch, Mende & Scherbaum 2004
Rainer Schoch, Matthias Mende & Anna Scherbaum (eds.), *Albrecht Dürer. Das druckgraphische Werk - Buchillustrationen*, vol. 3, Munich 2004.

Schoeck 1977
Richard J. Schoeck, 'More, Plutarch, and King Agis: Spartan History and the Meaning of Utopia', *Essential Articles for the Study of Thomas More*, R. S. Sylvester & G. P. Marc'hadour (eds.), Hamden 1977: 366-375.

Schoeck 1990-1993
Richard J. Schoeck, *Erasmus of Europe. 1: The Making of a Humanist 1467-1500; 2: The Prince of Humanists 1501-1536*, 2 vols., Edinburgh 1990-1993.

Schoeck 2002
Richard J. Schoeck, 'Thomas More and the Book : *Utopia* as Book', *Thomas More… and More. Freundesgabe für/ Liber Amicorum for Hubertus Schulte Herbrüggen*, Christoph M. Peters & Friedrich K. Unterweg (eds.), Frankfurt am Main 2002: 89-106.

Schoen 2001
Christian Schoen, *Albrecht Dürer. Adam und Eva. Die Gemälde, ihre Geschichte und Rezeption bei Lucas Cranach d. Ä. und Hans Baldung Grien*, Berlin 2001.

Scholten 1996
Simone Scholten, *Besloten hofjes des 16. Jahrhunderts in Mechelen*, PhD dissertation, Rheinische Friedrich-Wilhelms-Universität, Bonn 1996.

Schröder & Sternath 2003
Klaus Albrecht Schröder & Maria Luise Sternath (eds.), *Albrecht Dürer*, exh. cat. (Vienna, Albertina), Ostfildern-Ruit 2003.

Schulte Herbrüggen 1997a
Hubertus Schulte Herbrüggen, 'Morus ad Craneveldium Litterae Balduinianae novae /More to Cranevelt: New Baudouin Letters', *Supplementa Humanistica Lovaniensia* 11, Leuven 1997.

Schulte Herbrüggen 1997b
Hubertus Schulte Herbrüggen, '"Utopiae insulae figura": The Title Woodcut in Thomas Morus' *Utopia*, 1516', *Ut granum sinapis. Essays on Neo-Latin Literature in Honour of Jozef IJsewijn*, Gilbert Tournoy & Dirk Sacré (eds.), Leuven 1997: 215-230.

Schwarz 1953
Heinrich Schwarz, 'Jan Gossaert's Adam and Eve Drawings', *Gazette des Beaux-Arts* 6, 42 (1953): 146-168.

Schwarz 2012
Martin Schwarz, 'Visions of the Afterlife', *Apollo* 175, 598 (2012): 34-39.

Sciacca 2010
Christine Sciacca, *Building the Medieval World*, Los Angeles/London 2010.

Scott 1980
Margaret Scott, *The History of Dress: Late-Gothic Europe, 1400-1500*, London 1980.

Scott 2007
Margaret Scott, *Medieval Dress & Fashion*, London 2007.

Scott 2011
Margaret Scott, *Fashion in the Middle Ages*, Los Angeles 2011.

Scrivano 1972
Riccardo Scrivano, 'Ortensio Lando traduttore di Thomas More', *Studi sulla cultura lombarda in memoria di Mario Apollonio* 1, Milano 1972: 99-108.

Seelig 2015
Gero Seelig, 'Massys, Quentin', *Allgemeines Künstlerlexikon, Die Bildenden Künstler aller Zeiten und Völker*, Bd. 87, Berlin/ Boston 2015: 501.

Seidel Menchi 1996
Silvana Seidel Menchi, 'Ortensio Lando cittadino di Utopia. Un esercizio di lettura', *La fortuna dell'Utopia di Thomas More nel dibattito politico europeo del'500*, Florence 1996: 95-119.

Seipel 2008
Wilfried Seipel (ed.), *Meisterwerke der Sammlungen Schloss Ambras*, Milan 2008.

Sell & Chapman 2015
Stacey Sell & Hugo Chapman (eds.), *Drawing in Silver and Gold. Leonardo to Jasper Johns*, exh. cat. (Washington, National Gallery of Art/London, British Museum), Princeton/ Oxford 2015.

Sellevold 2008a
Kirsti Sellevold, 'The French Versions of Utopia: Christian and Cosmopolitan Models', *Thomas More's Utopia in Early Modern Europe. Paratexts and contexts*, Terence Cave (ed.), Manchester 2008: 67-86.

Sellevold 2008b
Kirsti Sellevold, 'Some "hardis repreneurs" in Sixteenth-Century France. Du Bellay, Aneau, Chappuys', *Borrowed feathers. Plagiarism and the limits of imitation in Early Modern Europe*, Hall Bjørnstad (ed.), Oslo 2008: 53-65.

Sellink 1997
Manfred Sellink, *Philips Galle (1537-1612): Engraver and Print Publisher in Haarlem and Antwerp*, PhD dissertation, Amsterdam 1997.

Sellink & Leesberg 2001
Manfred Sellink & Marjolein Leesberg, *The New Hollstein Dutch & Flemish Etchings, Engravings and Woodcuts, 1450-1700 - PHILIPS GALLE, Part IV*, Rotterdam 2001.

Serck 1990
Luc Serck, *Henri Bles et la peinture de paysage sans les pays-bas meridionaux avant Bruegel*, PhD dissertation, Université Catholique de Louvain, Louvain 1990.

Serck 2000
Luc Serck, 'Catalogue de l'exposition', *Autour de Henri Bles*, Jacques Toussaint (ed.), exh. cat. (Namur, Musée des Arts anciens du Namurois) 2000: 147-275.

Sezgin & Neubauer 2010
Fuat Sezgin & Eckhard Neubauer, 'Part III: Astrolabes (plane and spherical), Quadrants, Miscellaneous Instruments, including Mechanical-Astronomical Calendars', *Science and Technology in Islam - vol. 2: Catalogue of the Instrument of the Institute for the History of Arabic and Islamic Sciences*, Frankfurt am Main 2010: 78-172.

Shawe-Taylor & Scott 2007
Desmond Shawe-Taylor & Jennifer Scott, *Bruegel to Rubens: Masters of Flemish Painting*, exh. cat. (Edinburgh, The Queen's Gallery/Brussels, Musées royaux des Beaux-Arts de Belgique/ Koninklijke Musea voor Schone Kunsten van België/London, The Queen's Gallery), London 2007.

Shirley 1984
Rodney W. Shirley, *The Mapping of the World: Early Printed World Maps 1472–1700*, London 1984.

Shoham-Steiner 2008
Ephraim Shoham-Steiner, *Involuntary Marginals. Marginal Individuals in Medieval Northern European Jewish Society*, Jerusalem 2008.

Siewert & Bartetzky 1994
Antje Siewert & Arnold Bartetzky, 'Bildnisse', *Sébastien Brant – 500e anniversaire de la Nef des Folz / Das Narren Schyff zum 500 jährigen Jubiläum des Buches von Sebastian Brant: 1494-1994*, exh. cat. (Universitätsbibliotheken Basel & Freiburg im Breisgau/Karlsruhe, Badische Landesbibliothek/ Basel, Bibliothèque Nationale et Universitaire de Strassbourg), Basel 1994: 41-46.

Silva Maroto 2000
Pilar Silva Maroto (ed.), *El jardín de las delicias de El Bosco: copias, estudio técnico y restauración*, exh. cat. (Madrid, Museo Nacional del Prado), Madrid 2000.

Silva Maroto 2004
Pilar Silva Maroto, *Fernando Gallego*, Salamanca 2004.

Silva Maroto 2006
Pilar Silva Maroto, 'Pinturas del antiguo retablo mayor de la catedral de Ciudad Rodrigo', *Kyrios*, Fundación Las Edades del Hombre (ed.), exh. cat. (Ciudad Rodrigo, Cathedral of Saint Mary), Salamanca 2006.

Silver 1983
Larry Silver, 'Forest Primeval: Albrecht Altdorfer and the Early German Wilderness Landscape', *Simiolus* 13 (1983): 4-43.

Silver 1984
Larry Silver, *The Paintings of Quinten Massys, with Catalogue Raisonné*, Totowa/Montclair 1984.

Silver 1986
Larry Silver, 'Figure nude, historie e poesie, Jan Gossaert and the Renaissance Nude in the Netherlands', *Nederlands Kunsthistorisch Jaarboek* 37 (1986): 1-40.

Silver 1999
Larry Silver, 'Second Bosch. Family Resemblance and the Marketing of Art', *Nederlands Kunsthistorisch Jaarboek* 50 (1999): 31-56.

Silver 2001
Larry Silver, 'God in the Details: Bosch and Judgment(s)', *The Art Bulletin* 83, 4 (2001): 626-650.

Silver 2006a
Larry Silver, *Hieronymus Bosch*, New York 2006.

Silver 2006b
Larry Silver, *Peasant Scenes and Landscapes. The Rise of Pictorial Genres in the Antwerp Art Market*, Philadelphia 2006.

Silver 2010
Larry Silver, 'India Ink: Europe's Earliest Views of the Subcontinent', *The Power of Things and the Flow of Cultural Transformations: Art and Culture between Europe and Asia*, Lieselotte E. Saurma-Jeltsch & Anja Eisenbeiß (eds.), Berlin/Munich 2010: 217-247.

Silver & Smith 1978
Larry Silver & Susan Smith, 'Carnal Knowledge, the late Engravings of Lucas van Leyden', *Nederlands Kunsthistorisch Jaarboek* 29 (1978): 239-298.

Silver & Wyckoff 2008
Larry Silver & Elizabeth Wyckoff (eds.), *Grand scale. Monumental Prints in the Age of Dürer and Titian*, exh. cat. (Wellesley, Davis Museum and Cultural Center/New Haven, Yale University Art Gallery/Philadelphia, Philadelphia Museum of Art), New Haven/London 2008.

Simon 2007
Elliott M. Simon, *The Myth of Sisyphus. Renaissance Theories of Human Perfectibility*, Madison 2007.

Simpson & Fawcett 1987
William Douglas Simpson & Richard Fawcett, *Edzell Castle*, Edinburgh 1987.

Sintobin 1998
Véronique Sintobin, 'Style of Hieronymus Bosch. Christ's Descent into Hell', *From Van Eyck to Bruegel. Early Netherlandish Painting in The Metropolitan Museum of Art*, Maryan W. Ainsworth & Keith Christiansen (eds.), exh. cat. (New York, The Metropolitan Museum of Art), New York 1998: 254-256.

Sivefors 2014
Per Sivefors, 'Utopian English. Transferring and adapting the Text of *Utopia* in Early Modern England', *Approaches to the Text. From Pre-Gospel to Post-Baroque*, Roy Eriksen & Peter Young (eds.), Pisa 2014: 155-169.

Skinner 1978
Quentin Skinner, *The Foundations of Modern Political Thought*, 2 vols., Cambridge 1978.

Smets 1974
Leon Smets, *Het "Liber de naturis rerum creatarum", een laat-middeleeuws handschrift uit de librije van Rafaël Mercatel: Bijdrage tot de studie van de Gentse boekverluchting*, PhD dissertation, KU Leuven, Leuven 1974.

Smeyers 1993
Katharina & Maurits Smeyers, 'De kalenderwijzerplaat in het Leuvens Stedelijk Museum en de kalendericonografie in laat-middeleeuwse getijdenboeken', *Arca Lovaniensis artes atque historiae reserans documenta* 22 (1993): 83-124.

Smeyers 1994
Maurits Smeyers, 'Het Marianum of Onze-Lieve-Vrouw-in-de-zon. Getuige van een laat-middeleeuwse devotie in de Nederlanden en in Duitsland', *Nederlands Kunsthistorisch Jaarboek* 45 (*Beelden in de late Middeleeuwen en Renaissance*), Zwolle 1994: 270-299.

Smeyers 1996
Maurits Smeyers, 'Flemish Miniatures for England', *The Low Countries: Art and Society in Flanders and the Netherlands, A Yearbook. 1995-96*, Rekkem 1996: 240-250.

Smeyers 1997
Maurits Smeyers, 'Macrocosmos & microcosmos. De uur- en kalenderwijzerplaat en het laatmiddeleeuwse wereldbeeld', *Het rad van de tijd. De uur- en kalenderwijzerplaat in het Stedelijk Museum te Leuven (ca. 1500) - Arca Lovaniensis artes atque historiae reserans documenta* 23-24 (1997): 1-60.

Smeyers 1998
Maurits Smeyers, *L'Art de la miniature flamande du VIIIe au XVIe siècle*, Tournai 1998.

Smeyers 1999
Maurits Smeyers, *Flemish Miniatures from the 8th to the mid-16th Century: The Medieval World on Parchment*, Turnhout 1999.

Smith 1971
Constance Smith, 'Additional Locations for Thomas More's *Utopia*', *Moreana* 31-32 (1971): 261-262.

Smith 1981
Constance Smith, 'Locations for the updated Gibson [nos. 1-44]', *An updating of R. W. Gibson's St. Thomas More: A Preliminary Bibliography*, Sixteenth century bibliography 20, St. Louis 1981: 20-29.

Smith 2001
Paul J. Smith, 'The Melancholic Elk. Animal Symbolism and Linguistic Ambiguity in Albrecht Dürer's 'The Fall of Man' (1504) and Aegidius Sadeleer's 'Theatrum Morum' (1609)', *Quitte ou double sens: articles sur l'ambiguïté offerts à Ronald Landheer*, Paul Bogaards, Johan Rooryck & Paul J. Smith (eds.), Amsterdam 2001: 322-333.

Smith 2005
David R. Smith, 'Portrait and Counter-Portrait in Holbein's "The Family of Sir Thomas More"', *The Art Bulletin* 87, 3 (2005): 484-506.

Smith 2011
Alison Smith (ed.), *Watercolour*, exh. cat. (London, Tate Gallery), London 2011.

Snyder 1984
George S. Snyder, *Oude hemelkaarten - Hoogtepunten uit tien eeuwen cartografie*, Amsterdan 1984.

Société historique et archéologique de Tournai 1897
Société historique et archéologique de Tournai, *Annales de la Société historique et archéologique de Tournai*, Nouvelle Série - Tome 2, Tournai 1897.

Soly & Van de Wiele 1999
Hugo Soly & Johan van de Wiele (eds.), *Carolus, Charles Quint 1500-1558*, exh. cat. (Ghent, Kunsthal De Sint-Pietersabdij), Ghent 1999.

Soil de Moriamé 1892
Eugène Soil de Moriamé, *Les tapisseries de Tournai. Tapissiers et hautelisseurs de cette ville. Recherches et documents*, Tournai/ Antwerp Lille 1892.

Souchal 1974
Geneviève Souchal, *Chefs-d'oeuvre de la tapisserie du XIVe au XVIe siècle*, exh. cat. (Paris, Grand Palais), Paris 1974.

Spaans & Cave 2008
Ronny Spaans & Terence Cave, 'The Dutch Translation: Austerity and Pragmatism', *Thomas More's Utopia in Early Modern Europe. Paratexts and Contexts*, Terence Cave (ed.), Manchester 2008: 104-109, 219-231.

Spanier 1894
Thomas Murners Narrenbeschwörung. Text und Bilder der ersten Ausgabe, Max Spanier (ed.), Halle 1894.

Spencer 1963
Eleanor Spencer, 'L'Horloge de Sapience, Bruxelles, Bibliothèque Royale, MS IV.111', *Scriptorium* 17 (1963): 277-299.

Springer 1509
Balthasar Springer, *Die Merfart vn[d] erfarung nüwer Schiffung vnd Wege zu viln onerkanten Jnseln vnd Künigreichen von dem großmechtigen Portugalische[n] Kunig Emanuel Erforscht funden bestritten vnnd Jngenomen*, Nuremberg 1509.

Squellati Brizio 2013
Paola Squellati Brizio, 'Cat. 60', *La Renaissance et le rêve: Bosch, Véronèse, Greco...*, Alessandro Cecchi, Yves Hersant, Chiara Rabbi Bernard et al. (eds.), exh. cat. (Paris, Musée du Luxembourg/Florence, Palazzo Pitti), Paris 2013: 133-135.

Stammler 1962
Wolfgang Stammler, 'Der allegorische Garten', *Wort und Bild. Studien zu den Wechselbeziehungen zwischen Schrifttum und Bildkunst in Mittelalter*, Wolfgang Stammler (ed.), Berlin 1962: 106-116.

Stapleton 1928
Thomas Stapleton, 'The Life and Illustrious Martydom of Sir Thomas More', *Tres Thomae*, vol. 3, Philip E. Hallett (transl.), London 1928.

Stechow 1967
Wolfgang Stechow, 'Seventeen Great Prints: A Gift of the Max Kade Foundation', *Allen Memorial Art Museum Bulletin* 25, 1 (1967): 3-10.

Steppe 1965
Jan Karel Steppe, 'La date de décès de Jean Gossaert', *Jean Gossaert dit Mabuse*, Henri Pauwels, H.T. Hoetink & Sadja Herzog (eds.), exh. cat. (Rotterdam, Museum Boymans Van Beuningen/Bruges, Groeninge-museum), Amsterdam 1965: 33-38.

Steppe 1967
Jan Karel Steppe, 'Jheronimus Bosch. Bijdrage tot de historische en de ikonografische studie van zijn werk', *Jheronimus Bosch. Bijdragen ter gelegenheid van de herdenkingstentoonstelling te 's Hertogenbosch 1967*, exh. cat. ('s Hertogenbosch, Noordbrabants Museum), Eindhoven 1967: 28-37.

Sterk 1980
Jozef Johannes Sterk, *Philips van Bourgondië (1465-1524), Bisschop van Utrecht, als protagonist van de Renaissance, zijn leven en maecenaat*, Zutphen 1980.

Stevens 1908
Henry N. Stevens, *Ptolemy's Geography. A Brief Account of all the Printed Editions Down to 1730. With Notes on some Important Variations observed in that of Ulm 1482 Including the Recent Discovery of the Earliest Printed Map of the World yet Known on Modern Geographical Conceptions in which some Attempt was made to Depart from Ancient Traditions*, London 1908.

Stevenson 1921
Edward Luther Stevenson, *Terrestrial and Celestial Globes: their History and Construction, including a Consideration of their Value as Aids in the Study of Geography and Astronomy*, vol. 1, New Haven 1921.

Stewart 1989
Alison G. Stewart, 'Sebald Beham's Fountain of Youth-Bathhouse Woodcut: Popular Entertainment and Large Prints by the Little Masters', *Register of the Spencer Art Museum* 6, 6 (1989): 64-88.

Stewart 2008
Alison G. Stewart, 'Woodcuts as Wallpaper: Sebald Beham and Large Prints from Nuremberg', *Grand Scale: Monumental Prints in the Age of Dürer and Titian*, Larry Silver & Elizabeth Wyckoff (eds.), exh. cat. (Wellesley, Davis Museum and Cultural Center/ New Haven, Yale University Art Gallery/Philadelphia, Philadelphia Museum of Art), New Haven/London 2008.

Steyaert 2007
Griet Steyaert, 'The Master of the Embroided Foliage and Aert van den Bossche', *Le Maître au Feuillage brodé: démarches d'artistes et méthodes d'attribution d'oeuvres à un peintre anonyme des anciens Pays-Bas du XVème siècle = The Master of the Embroidered Foliage: artist's processes and attribution methods to an anonymus flemish painter of the XVth century*, Florence Gombert & Didier Martens (eds.), Lille 2007: 173- 183.

Steyaert 2013
Griet Steyaert, 'Aert van den Bossche, Adam en Eva', *De erfenis van Rogier van der Weyden. De schilderkunst in Brussel 1450-1520*, Véronique Bücken & Griet Steyaert (eds.), exh. cat. (Brussels, Koninklijke Museum voor Schone Kunsten), Brussels 2013: 258-259.

Stielau 2011
Allison Stielau, 'The Case of the Case for Early Modern Objects and Images', *Kritische Berichte - Die Kunst und die Dinge. Perspektiven einer schwierigen Beziehung* 39, 3 (2011):

Stillwell 1970
Margaret Bingham Stillwell, *The Awakening Interest in Science during the first Century of Printing 1450-1550: An annotated Checklist of first Editions viewed from the Angle of their Subject Content: Astronomy, Mathematics, Medicine, Natural Science, Physics, Technology*, New York 1970.

Stock 2016
Lorraine Stock, *The Medieval Wild Man*, London 2016.

Stockholm 1970
Frän drottning Blanka tille drottning Astrid, En utställing om svensk-belgiska förbindelser, Stockholm 1970.

Stoffers 2001
Johannes Stoffers, 'Die Republik Utopia. Erläuterung und Gedanken zu einer alten Karte', *Auswärtiger Dienst* 62 (2001): 1-50.

Stols, Piccoli, De Peuters & Petre 2011
Eddy Stols, Valéria Piccoli, Patricia de Peuter & Anne Petre, *Terra Brasilis*. exh. cat. (Brussels, Cultuurcentrum ING), Brussels/Antwerp 2011.

Strauss 1974
Walter L. Strauss, *The Complete Drawings of Albrecht Dürer*, 6 vols., New York 1974.

Strauss 1980
Walter L. Strauss, *Albrecht Dürer. Woodcuts and Woodblocks*, New York 1980.

Strauss 1981
Walter L. Strauss (ed.), *The Illustrated Bartsch - vol. 10 (Commentary): Sixteenth Century German Artists. Albrecht Dürer*, New York 1981.

Streng 1990
Jean C. Streng, 'The Leiden Engraver Frans van Bleuswyck (1671-1646)', *Quaerendo* 10, 2 (1990): 111-136.

Strömberg 1965
Elisabeth Strömberg, '"Astronomie", en frank medeltidsgobeläng', *Röhsska Konstslöjdmuseets, Arsbok 1963-1964* (1965): 14-28, 46-47.

Strong 1969
Roy C. Strong, *National Portrait Gallery, Tudor & Jacobean Portraits*, 2 vols., London 1969.

Stroo & Goetghebeur 2003
Cyriel Stroo & Nicole Goetghebeur, 'Anonymous Master. Triptych of Saints Christopher, Jerome and Anthony', *The Mayer van den Bergh Museum Antwerp*, Hélène Mund, Cyriel Stroo, Nicole Goetghebeur, Hans Nieuwdorp & Michael Lomax (eds.), mus. cat. (Antwerp, The Mayer van den Bergh Museum), Corpus of the fifteenth-century painting in the Southern Netherlands and the Principality of Liège, Brussels 2003: 378-402.

Sturtevant 1988
William C. Sturtevant, 'La tupinambisation des Indiens d'Amérique du Nord', *Les figures de l'Indien*, Gilles Thérien (ed.), Montréal 1988.

Sumberg 1944
Samuel Sumberg, *The Nuremberg Schembart Festival*, New York 1944.

Sumira 2014
Sylvia Sumira, *The Art and History of Globes*, London 2014.

Surtz & Hexter 1965
Edward Surtz & J. H. Hexter (eds.), *The Complete Works of St. Thomas More. Volume 4 [Utopia]*, New Haven/London 1965.

Talbot 1971
Charles W. Talbot (ed.), *Dürer in America. His Graphic Work*, exh. cat. (Washington D.C., National Gallery of Art), New York/London 1971.

Tapié 2012
Alain Tapié (ed.), *Fables du paysage flamand. Bosch - Bles - Brueghel - Bril*, exh. cat. (Lille, Palais des Beaux-Arts), Paris 2012.

Taylor 1930
E. G. R. Taylor, *Tudor Geography, 1485-1583*, London 1930.

Tchalenko 1990
John Tchalenko, 'Earliest Wild-Man Sculptures in France', *Journal of Medieval History* 16, 3 (1990): 217-234.

Testa 2000
Judith Anne Testa, 'Simon Bening and the Italian High Renaissance: some unexplored sources', *Oud-Holland* 114, 2-4 (2000): 107-124.

Thiem 1963
Gunther Thiem, *Sammlung Max Kade*, exh. cat. (Stuttgart, Staatsgalerie-Graphische Sammlung/Munich, Staatliche Graphische Sammlung) Stuttgart 1963.

Thompson 1939
C. R. Thompson, 'The Translations of Lucian by Erasmus and S. Thomas More', *Revue belge de philologie et d'histoire* 18/4 (1939): 855-881.

Tiemann 1977
Hermann Tiemann, 'Der Roman von der Königin Sibille. In drei Prosafassungen des 14. und 15. Jahrhunderts', *Veröffentlichungen aus der Staats- und Universitätsbibliothek Hamburg* 10, Hamburg 1977: 117-173.

Tietze 1951
Hans Tietze, *Dürer als Zeichner und Aquarellist*, Vienna 1951.

Tietze & Tietze-Conrat 1938
Hans Tietze & Erica Tietze-Conrat, *Kritisches Verzeichnis der Werke Albrecht Dürers*, 3 vols., Basel/Leipzig 1938.

Tóth-Ubbens 1968
Magdy M. Tóth-Ubbens, *Schilderijen en beeldhouwwerken 15e en 16e eeuw - Koninklijk Kabinet van Schilderijen Mauritshuis, catalogus 1*, The Hague 1968.

Toulouse 2007
Sarah Toulouse, 'Marine Cartography and Navigation in Renaissance France', *The History of Cartography. Cartography in the European Renaissance*, vol. 3, pt. 2, David Woodward (ed.), Chicago 2007: 1550-1568.

Trapp & Schulte Herbrüggen 1977
Joseph Burney Trapp & Hubertus Schulte Herbrüggen, *"The King's Good Servant", Sir Thomas More, 1477/8-1535*, exh. cat. (London, National Portrait Gallery), London 1977.

Traub 2000
Valerie Traub, 'Mapping the Global Body', *Early Modern Visual Culture Representation, Race, and Empire in Renaissance England*, Peter Erickson & Clark Hulse (eds.), Philadelphia 2000: 44-97.

Triest 1998
Monika Triest, *Het besloten hof. Begijnen in de Zuidelijke Nederlanden*, Leuven 1998.

Troescher 1927
Georg Troescher, *Conrat Meit von Worms. Ein rheinischer Bildhauer der Renaissance*, Freiburg im Breisgau 1927.

Trousson 1999
Raymond Trousson, *Voyages aux pays de nulle part, histoire littéraire de la pensée utopique*, Brussels 1999.

Turner 1987
Anthony Turner, *Early Scientific Instruments - Europe 1400-1800*, London 1987.

Turner 2006
Nancy K. Turner, 'The Suggestive Brush: Painting Techniques in Flemish Manuscripts', *Flemish Manuscript Painting in Context: Recent Research*, Elizabeth Morrison & Thomas Kren (eds.), Los Angeles 2006: 57-86.

Tuve 1966
Rosemond Tuve, *Allegorical Imagery: Some Medieval Books and Their Posterity*, Princeton 1966.

Ullmann 1984
Ernst Ullmann (ed.), *Geschichte der deutschen Kunst 1470-1550 - Architektur und Plastik*, Leipzig 1984.

Unverfehrt 1980
Gerd Unverfehrt, *Hieronymus Bosch. Die Rezeption seiner Kunst im frühen 16. Jahrhundert*, Berlin 1980.

Unverfehrt 2007
Gerd Unverfehrt, *Da sah ich viel köstliche Dinge – Albrecht Dürers Reise durch die Niederlande*, Göttingen 2007.

Utrecht/Leuven 1959
Herdenkingstentoonstelling Paus Adrianus VI, exh. cat. (Utrecht, Centraal Museum/Leuven, Stadhuis), Utrecht 1959.

Van Acker 1960
K. G. van Acker, 'Nieuwe biografische gegevens over Raphaël de Marcatellis, abt van de Sint-Baafsabdij te Gent, bisschop van Rhosus en post-Bourgondisch mecenas, stichter van een bijzondere librije van verluchte handschriften, n. 1437, † 1508, 3 aug.', *Gentse Bijdragen tot de Kunstgeschiedenis en de Oudheidkunde* 18 (1960): 77-94.

Van Asperen 2009
Hanneke van Asperen, *Pelgrimstekens op perkament, Originele en nageschilderde bedevaartssouvenirs in religieuze boeken (ca. 1450 - ca. 1530)*, PhD dissertation, Radboud Universiteit, Nijmegen 2009.

Van Boheemen & Dirkse 1994
Petra van Boheemen & Paul Dirkse, *Duivels en demonen: De duivel in de Nederlandse beeldcultuur*, Utrecht 1994.

Van Bueren 1999
Truus van Bueren, *Leven na de dood. Gedenken in de late Middeleeuwen*, exh. cat. (Utrecht, Museum Catharijneconvent), Turnhout 1999.

Van Buren & Wieck 2011
Anne H. van Buren & Roger S. Wieck. *Illuminating Fashion: Dress in the Art of Medieval France and the Netherlands, 1325-1515*, exh. cat. (New York, The Morgan Library & Museum), London 2011.

Van Buyten 1975
Leo van Buyten (ed.), *Leuven, een stad die groeit. Een gemeenschap bouwt aan haar milieu*, exh. cat. (Leuven, Stedelijk Museum Vander Kelen-Mertens), Leuven 1975.

Van Cleempoel 2002
Koenraad Van Cleempoel, *A Catalogue Raisonné of Scientific Instruments from The Louvain School, between 1530 and 1600*. De diversis artibus. Collection of Studies from the International Academy of the History of Science, 65, Turnhout 2002.

Van Cleempoel 2006
Koenraad Van Cleempoel, 'Catalogue of European Astrolabes', *Astrolabes at Greenwich: a Catalogue of the Astrolabes in the National Maritime Museum, Greenwich*, Kristen Lippincott & Koenraad Van Cleempoel (eds.), Oxford 2006.

Van Cleempoel 2009
Koenraad Van Cleempoel, 'Philip II's Escorial and its Collection of Scientific Instruments', *European Collections of Scientific Instruments, 1550-1750*, Giorgio Strano, Stephen Johnston, Mara Miniati & Alison Morrison-Low (eds.), Leiden 2009: 101-127.

Van Cleempoel, Van Damme, L'Estrange Turner et al. 1997
Koenraad Van Cleempoel, Jacques van Damme, Gérard L'Estrange Turner et al. (eds.), *Instrumentos científicos del siglo XVI: La Corte española y la Escuela de Lovaina*, exh. cat. (Madrid, Fundación Carlos de Amberes), Madrid 1997.

Vandamme 1981
Erik Vandamme, 'Het "besloten hofje" in het Koninklijk Museum voor Schone Kunsten te Antwerpen. Bijdrage tot de studie van de kunstnijverheid in de provinciale Zuidnederlandse centra omstreeks 1500', *Archivum Artis Lovaniense. Bijdragen tot de geschiedenis van de kunst der Nederlanden, opgedragen aan Prof. Em. Dr. J.K. Steppe*, M. Smeyers et al. (ed.), Leuven 1981: 143-149.

Vandamme et al. 1988
Erik Vandamme et al., *Catalogus schilderkunst oude meesters*, mus. cat. (Antwerp, Koninklijk museum voor schone kunsten), Antwerp 1988.

Vandamme 2003
Idalie Vandamme, *Een bijzondere Aanbidding. Het vederdrieluik met de Aanbidding der Koningen uit het Museo de America te Madrid*, 2 vols., Master dissertation, KU Leuven, Leuven 2003.

Vandamme 2006
Idalie Vandamme, 'A Particular Adoration. The 16th-century Feather Triptych Depicting the Adoration of the Magi in the Museo de America in Madrid', *Jaarboek Koninklijk Museum voor Schone Kunsten Antwerpen* 2003 (2006): 93-125.

Van den Bergen-Pantens 1989
Christiane van den Bergen-Pantens (ed.), *Manuscrits à peintures 1460-1486*, exh. cat. (Brussels, Bibliothèque royale Albert Ier), Brussels 1989.

Vandenberghe 1983
S. Vandenberghe, 'Het marianum van het voormalig Onze-Lieve-Vrouwgasthuis te Mechelen', *Handelingen van de Koninklijke Kring voor Oudheidkunde, Letteren en Kunst van Mechelen* 87 (1983): 137-149.

Van den Boogert 1993
Bob van den Boogert, 'Portretbuste van Karel V', *Maria van Hongarije koningin tussen keizers en kunstenaars, 1505-1558*, Jacqueline Kerkhoff & Bob van den Boogert (eds), exh. cat. (Utrecht, Rijksmuseum Het Catharijneconvent/'s-Hertogenbosch, Noordbrabants Museum), Zwolle 1993: 237.

Van den Brink, Allart & Van der Glas 2001
Peter van den Brink, Dominique Allart & Marijke van der Glas, *De Firma Brueghel*, Ghent 2001.

Vandenbroeck 1981
Paul Vandenbroeck, 'Problèmes concernant l'oeuvre de Jheronimus Bosch: le dessin sous-jacent en relation avec l'authenticité et la chronologie', *Le dessin sous-jacent dans la peinture. Colloque 4*, Roger Van Schoute & Dominique Hollanders-Favart (eds.), Louvain-la-Neuve 1981: 107-119.

Vandenbroeck 1985
Paul Vandenbroeck, 'Bubo significans. Die Eule als Sinnbild der Schlechtigkeit und Torheit, vor allem in der niederländischen und deutschen Bilddarstellung und bei Jheronimus Bosch. I.', *Jaarboek Koninklijk Museum voor Schone Kunsten Antwerpen* (1985): 19-135.

Vandenbroeck 1987
Paul Vandenbroeck, *Over wilden en narren, boeren en bedelaars. Beeld van de andere, vertoog over het zelf*, exh. cat. (Antwerp, Koninklijk Museum voor Schone Kunsten), Antwerp 1987.

Vandenbroeck 1989
Paul Vandenbroeck, 'Jheronimus Bosch' zgn. 'Tuin der Lusten'', *Jaarboek Koninklijk Museum voor Schone Kunsten Antwerpen* (1989): 9-210.

Vandenbroeck 1990
Paul Vandenbroeck, 'Jheronimus Bosch' zgn. 'Tuin der Lusten'', *Jaarboek Koninklijk Museum voor Schone Kunsten Antwerpen* (1990): 9-192.

Vandenbroeck 1991a
Paul Vandenbroeck, 'Amerindian Art and Ornamental Objects in Royal Collections. Brussels, Mechelen, Duurstede, 1520-1530', *America, Bride of the Sun. 500 Years Latin America and the Low Countries*, Ine Pisters (ed.), exh. cat. (Antwerp, Koninklijk Museum voor Schone Kunsten), Ghent 1991: 99-119.

Vandenbroeck 1991b
Paul Vandenbroeck, 'Adoration of the Magi', *America, Bride of the Sun. 500 ears Latin America and the Low Countries*, Ine Pisters (ed.), exh. cat. (Antwerp, Koninklijk Museum voor Schone Kunsten), Ghent 1991: 417-418.

Vandenbroeck 1994
Paul Vandenbroeck et al., *Hooglied: De Beeldwereld van religieuze vrouwen in de Zuidelijke Nederlanden, vanaf de 13de eeuw*, exh. cat. (Brussels Paleis voor Schone Kunsten), Ghent/Brussels 1994.

Vandenbroeck 1997
Paul Vandenbroeck, *De kleuren van de geest: dans en trance in Afro-Europese tradities*, exh. cat. (Antwerp, Koninklijk Museum voor Schone Kunsten), Ghent 1997.

Vandenbroeck 2001
Paul Vandenbroeck, 'High stakes in Brussels, 1567. The 'Garden of Earthly Delights as the Crux of the conflict between William the Silent and the Duke of Alva', *Hieronymus Bosch. New insights into his Life and Work*, Jos Koldewij, Bernard Vermet & Barbara van Kooij (eds.), Rotterdam/Ghent 2001: 87-92.

Vandenbroeck 2002
Paul Vandenbroeck, *Jheronimus Bosch. De verlossing van de wereld*, Ghent/Amsterdam 2002.

Vandenbroeck 2004
Paul Vandenbroeck, 'Joachim Patinir en het ontstaan van de Vlaamse landschapskunst', *De uitvinding van het landschap. Van Patinir tot Rubens 1520-1650*, exh. cat. (Antwerp, Koninklijk Museum voor Schone Kunsten), Antwerp 2004: 33-51.

Vandenbroeck 2007
Paul Vandenbroeck, 'Sharing a Table with Strangers. The Conception of Man, behavioural Codes and Alterity in the new Iconography in Tapestries from the Southern Netherlands around 1500', *Jaarboek Koninklijk Museum voor Schone Kunsten Antwerpen* (2007): 121-153.

Vandenbroeck 2009
Paul Vandenbroeck, 'Meaningful Caprices. Folk Culture, Middle Class Ideology (ca. 1480-1510) and Aristocratic Recuperation (ca. 1530-1570): A Series of Brussels Tapestries after Hieronymus Bosch', *Jaarboek Koninklijk Museum voor Schone Kunsten Antwerpen* (2009): 212-269.

Vandenbroeck 2012
Paul Vandenbroeck, 'La relation entre ordre naturel et ordre moral dans le paysage flamand vers 1500/1520', *Fables du paysage flamand. Bosch Bles Brueghel Bril*, exh. cat. (Lille, Palais des Beaux-Arts), Paris 2012: 105-109.

Vandenbroeck 2014
Paul Vandenbroeck (ed.), *De eeuw der Vlaamse Primitieven: Laatmiddeleeuwse kunst in het Koninklijk Museum voor Schone Kunsten Antwerpen*, mus. cat. (Antwerp, Koninklijk Museum voor Schone Kunsten), Kontich 2014.

Vandenbroeck 2016
Paul Vandenbroeck, *Utopia's Droom: the Graal of Lust, the Sect of the Free Spirit and Hieronymus Bosch's Garden of Earthly Delights*, Arts and Religion 6, Leuven/Walpole 2016.

Van den Broecke 2004
Marcel van den Broecke, 'De Utopia Kaart van Ortelius', *Caert-Thresoor* 23, 4 (2004): 89-94.

Van den Broecke 2011
Marcel van den Broecke, *Ortelius Atlas Maps. An Illustrated Guide*, Utrecht 2011.

Van den Broecke & Ormeling 2005
Marcel van den Broecke & Ferjan Ormeling, 'What's in a Name?', *The Map Book*, Peter Barber (ed.), London 2005: 132-133.

Vandendriessche 1993
Gaston Vandendriessche, 'The Children of the Planets in the Late Medieval Calender-Clock in the City Museum of Leuven', *Arca Lovaniensis artes atque historiae reserans documenta* 22 (1993): 43-80.

Vandendriessche & Smeyers 1998
Gaston Vandendriessche & Maurits Smeyers, 'Uur- en kalenderwijzerplaat - Leuven, ca. 1500', *Leven te Leuven in de late Middeleeuwen*, Lutgarde Bessemans, Maurits Smeyers, Veronique Vandekerchove et al. (eds.), exh. cat. (Leuven, Stedelijk Museum Vanderkelen-Mertens), Leuven 1998: 191-192.

Van den Eynde 2015
Stéphanie van den Eynde, *Het jongste volk ter wereld: Een kritische status quaestionis en iconografisch onderzoek van de fries L'Ile du Brésil*, Unpublished Bachelor Paper, KU Leuven 2015.

Van den Gheyn 1927
Joseph van den Gheyn, *Catalogue des manuscrits de la Bibliothèque Royale de Belgique*, vol. 11, Ronse 1927.

Van der Coelen, Rümelin, Trapman et al. 2008
Peter van der Coelen, Christian Rümelin, Hans Trapman et al., *Images of Erasmus*, exh. cat. (Rotterdam, Museum Boymans-van Beuningen), Rotterdam 2008.

Van der Haagen 1837
Everhardus Stephanus van der Haagen, 'De H. Hieronymus', *Levens der heiligen, kerkvaders en martelaren - vol. 1*, The Hague 1837: 819-874.

Van der Heijden & Roegiers 1994
Henk A. M. van der Heijden & Jan Roegiers (eds.), *Eenheid op papier. De Nederlanden in kaart van Keizer Karel tot Willem I*, exh. cat. (Leuven, Onze-Lieve-Vrouw-ter-Predikheren), Leuven 1994.

Van der Kemp 1964
Gérald van der Kemp (ed.), *Vienne à Versailles*, exh. cat. (Versailles, château de Versailles), Paris 1964.

Van der Krogt 1984
Peter van der Krogt, *Old Globes in the Netherlands - A Catalogue of Terrestrial and Celestial Globes made prior to 1850 and preserved in Dutch Collections*, Utrecht 1984.

Van der Krogt 1993
Peter van der Krogt, *Globi Neerlandici: The Production of Globes in the Low Countries*, Utrecht 1993.

Van der Krogt 1995
Peter van der Krogt, 'Erdgloben, Wandkarten, Atlanten: Gerhard Mercator kartiert die Erde', *Gerhard Mercator: Europa und die Welt*, Ruth Löffler & Gernot Tromnau (eds.), exh. cat. (Duisburg, Kultur- und Stadthistorischen Museum), Duisburg 1995: 81-130.

Van der Stock 1991
Jan Van der Stock (ed.), *La ville en Flandre. Culture et Société 1477-1787*, exh. cat. (Brussels, Galerie du Crédit Communal/Schallaburg, Renaissanceschloß) Brussels 1991.

Van der Stock 1993
Jan Van der Stock (ed.), *Antwerp, story of a metropolis*, exh. cat. (Antwerp, Hessenhuis), Ghent 1993.

Vanderwiele 1981
Leo Jules Vanderwiele, *Geschiedenis van de farmacie in België met een inleiding tot de algemene geschiedenis van de farmacie*, Beveren 1981.

Van de Velde & Van Hout 2013
Hildegard van de Velde & Nico van Hout, *Het Gulden Cabinet Koninklijk Museum bij Rockox te gast*, exh. cat. (Antwerp, Museum Rockox House), Antwerp 2013.

Van Doorslaer 1912
Georges van Doorslaer, *L'enseignement de l'exposition d'art ancien de Malines en 1911: Par le Dr. G. Van Doorslaer*, Antwerp 1912.

Van Duzer 2013
Chet van Duzer, 'Hic sunt dracones: the Geography and Cartography of Monsters', *The Ashgate Research Companion to Monsters and the Monstrous*, Asa Simon Mittman & Peter Dendle (eds.), Farnham 2013: 387-436.

Van Duzer 2015
Chet van Duzer, *The World for a King: Pierre Desceliers' World Map of 1550*, London 2015.

Van Duzer & Dines 2016
Chet Van Duzer & Ilya Dines, *Apocalyptic Cartography. Thematic Maps and the End of the World in a Fifteenth-Century Manuscript*, Leiden/Boston 2016.

Van Even 1895
Edward van Even, *Louvain dans le passé et dans le présent*, Leuven 1895.

Van Gijsen 1994
Annelies van Gijsen, 'De Astrologie', *Gerardus Mercator Rupelmundanus*, Marcel Watelet (ed.), Antwerp 1994: 221-233.

Van Grieken 2013
Joris van Grieken, 'Triptych "The End of Time, Heaven and Hell"', *Hieronymus Cock. The Renaissance in Print*, Joris van Grieken, Ger Luijten & Jan Van der Stock (eds.), exh. cat. (Leuven, M - Museum Leuven/Paris, Fondation Custodia), New Haven/London 2013: 248-249.

Van Hasselt 1961
R. van Hasselt, 'De stadsverdedigingswerken van Leuven. De oudste stadsmuur', *Mededelingen van de Geschied- en Oudheidkundige Kring voor Leuven en omgeving* 1 (1961): 132-143.

Van Heesch 2014
Daan van Heesch, 'New Evidence on Pieter Huys as a Draughtsman and Designer of Prints', *Delineavit et Sculpsit. Journal for Dutch and Flemish Prints and Drawings* 37 (2014): 2-11.

Van Hoorebeeck 2007
Céline van Hoorebeeck, *Livres et lectures des fonctionnaires des ducs de Bourgogne (ca 1420-1520)*, vol. 2, PhD dissertation, l'Université de Namur, Namur 2007.

Van Luttervelt 1955
Remmet van Luttervelt (ed.), *De triomf van het maniërisme: de Europese stijl van Michelangelo tot El Greco*, exh. cat. (Amsterdam, Rijksmuseum), Amsterdam 1955.

Van Mander 1604
Karel van Mander, *Schilder-boeck*, Haarlem 1604.

Van Mander 1994
Karel van Mander, *The Lives of the Illustrious Netherlandish and German Painters from the first Edition of the* Schilder-boeck *(1603-1604) … - vol. 1: The Text*, Hessel Miedema (ed.), Doornspijk 1994.

Van Marle 1932
Raimond van Marle, *Iconographie de l'art profane au Moyen-Âge et à la Renaissance, et la décoration des demeures: Allégories et symboles*, vol. 2, The Hague 1932.

Van Ortroy 1902
Fernand van Ortroy, *Bibliographie de l'oeuvre de Pierre Apian*, Besançon 1902.

Vanpaemel & Padmos 2000
Geert Vanpaemel & Tineke Padmos (eds.), *Wereldwijs. Wetenschappers rond Keizer Karel*, Leuven 2000.

Van Raemdonck 1873-1875
Jan van Raemdonck, 'Les sphères terrestre et céleste de Gérard Mercator', *Annalen Oudheidkundige Kring Land van Waas* 5 (1873-1875): 259-324.

Van Schendel 1948
Arthur van Schendel, *De Hollandsche schilderkunst van Jeroen Bosch tot Rembrandt. Keuze van meesterwerken uit de Nederlandsche musea*, exh. cat. (Antwerp, Koninklijk Museum voor Schone Kunsten/Brussels, Paleis van Schone Kunsten), Brussels 1948.

Van Sommeren 1890
Jan Frederik van Sommeren, *Catalogus van gegraveerde portretten van Nederlanders*, pt. 2, Amsterdam 1890.

Van 't Hoff 1955
B. van 't Hoff, 'Grote stadspanorama's gegraveerd in Amsterdam sinds 1609', *Jaarboek van het genootschap Amstelodamum* 47 (1955): 81-131.

Van Uytven 1964
Raymond van Uytven (ed.), *Leuvens verleden*, exh. cat. (Leuven, Stadhuis), Leuven 1964.

Van Wachem 1954
Louis J. van Wachem, 'Zu unseren Kunstdrucktafeln, Quentin Massys, Männerbildnis', *Westermanns Monatshefte* 4, 95 (1954): 71.

Van Wegen 2005
Derk Hendrik van Wegen, *Het Vlaamse schilderkunstboek*, Zwolle 2005.

Van Werveke 1955
Hans van Werveke (ed.), *Charles Quint et son temps*, exh. cat. (Ghent, Museum voor Schone Kunsten), Brussels 1955.

Vasari 1981
Giorgio Vasari, *Les vies des meilleurs peintres, sculpteurs et architectes*, André Chastel (transl. & ed.), Paris 1981.

Verdi 2007
Richard Verdi, *The Parrot in Art from Dürer to Elizabeth Butterworth*, exh. cat. (Birmingham, Barber Institute of Fine Arts), London 2007.

Vergara 2007
Alejandro Vergara (ed.), *Patinir. Essays and Critical Catalogue*, exh. cat. (Madrid, Museo del Prado), Madrid 2007.

Vermeylen 2014
Filip Vermeylen, *Painting for the Market: Commercialization of Art in Antwerp's Golden Age*, (Studies in European Urban History (1100-1800), 2, Turnhout 2003.

Versyp 1950
Jacqueline Versyp, 'Twee Gentse bestiaria uit de XVde eeuw', *Handelingen der Maatschappij voor geschiedenis en oudheidkunde te Gent* 4, 2 (1950): 32-40.

Veth & Muller 1918
J. Veth & S. Muller, *Albrecht Dürers Niederländische Reise*, 2 vols., Berlin/Utrecht 1918.

Vey & Von der Osten 1969
Horst Vey & Gert Von der Osten, *Painting and Sculpture in Germany and the Netherlands 1500 to 1600*, Harmondsworth 1969.

Vielle 2009
Christophe Vielle, *D'Anvers à Calicut: Pieter Gillis, Utopia, sanskrit et malayāḷam*, Brussels/Louvain-la-Neuve 2009.

Vielle 2013
Christophe Vielle, 'La langue de l'île d'Utopie. Les Indes orientales vues des Flandres à la Renaissance', *Acta Orientalia Belgica* 26 (2013): 203-222.

Vincent & Chandler 1969
Clare Vincent & Bruce Chandler, 'Nighttime and Easter Time', *The Metropolitan Museum of Art Bulletin* 27, 8 (1969): 372-384.

Vöge 1908
Wilhelm Vöge, 'Konrad Meit und die Grabdenkmäler in Brou', *Jahrbuch der Preußischen Kunstsammlungen* 39 (1908): 77-118.

Vogel 1994
Klaus Anselm Vogel, 'Hartmann Schedel als Kompilator: Notizen zu einem derzeit kaum bestellten Forschungsfeld', *500 Jahre Schedelsche Weltchronik. Akten des interdisziplinären Symposions vom 23./24. April 1993 in Nürnberg*, Stephan Füssel (ed.), Nuremberg 1994: 73-97.

Vogelaar, Filedt Kok, Leeflang et al. 2011
Christiaan Vogelaar, Jan Piet Filedt Kok, Huigen Leeflang et al., *Lucas van Leyden en de Renaissance*, exh. cat. (Leiden, Museum De Lakenhal), Ghent/Amsterdam 2011.

Volckaert & Delmarcel 1993
An Volckaert & Guy Delmarcel, 'Wandtapijten in de trant van Jheronimus Bosch', *Koninklijke pracht in goud en zijde. Vlaamse wandtapijten van de Spaanse Kroon*, exh. cat. (Amsterdam, Nieuwe Kerk/Rijksmuseum), Amsterdam 1993: 92-112.

Von Bassermann-Jordan 1961
Ernst von Bassermann-Jordan, *Uhren*, Braunschweig 1961.

Von Bertele 1961
Hans von Bertele, *Globes and Spheres*, Lausanne 1961.

Von der Osten 1973
Gert von der Osten, *Deutsche und Niederländische Kunst der Reformationszeit*, Cologne 1973.

Voragine 1969
Jacobus de Voragine, *The golden legend of Jacobus de Voragine*, Granger Ryan & Helmut Ripperger (transl. & eds.), New York 1969.

Voss 1943
Wilhelm Voss, 'Eine Himmelskarte vom Jahre 1503 mit den Wahrzeichen des Wiener Poetenkollegiums als Vorlage Albrecht Dürers', *Jahrbuch der Königlich Preussischen Kunstsammlungen* 64 (1943): 89-150.

Waagen 1838
Gustav Friedrich Waagen, *Kunstwerke und Künstler in England*, vol. 2, Berlin 1838.

Waagen 1854
Gustav Friedrich Waagen, *Treasures of Art in Great Britain*, vol. 3, London 1854.

Waagen 1857
Gustav Friedrich Waagen, *Galleries and Cabinets of Art in Great Britain*, London 1857.

Warner 1979
Deborah J. Warner, *The Sky Explored: Celestial Cartography 1500-1800*, New York/Amsterdam 1979.

Watelet 1994
Marcel Watelet, *Gerardus Mercator Rupelmundanus*, Ghent 1994.

Watteeuw 2006
Lieve Watteeuw, 'Flemish Manuscript Production, Care, and Repair: Fifteenth Century Sources', *Flemish manuscript painting in context: recent research*, Elizabeth Morrison & Thomas Kren (eds.), Los Angeles 2006: 75-86.

Watteeuw 2009
Lieve Watteeuw, 'Het 'Hennessy getijdenboek', een middeleeuws manuscript in de negentiende eeuw: de boekarcheologische focus', *Miscellanea in memoriam Pierre Cockshaw (1938-2008). Aspecten van het culturele leven in de Zuidelijke Nederlanden (14de-18de eeuw)*, pt. 2, Brussels 2009: 599-611.

Weale 1864
W. H. J. Weale, *Catalogue des objets d'art religieux du Moyen Age, de la Renaissance et des temps modernes*, exh. cat. (Mechelen, Hôtel Liedekerke), Mechelen 1864.

Webb 2005
Diana Webb, 'The Domestic and Sacred Space in the Fifteenth-Century Netherlands', *Defining the Holy. Sacred Space in Medieval and Early Modern Europe*, Andrew Spicer & Sarah Hamilton (eds.), Ashgate 2005: 27-49.

Weddigen 2008
Tristan Weddigen, 'Resteverwertung', *Kritische Berichte - Rest, Fetisch, Reliquie* 36, 3 (2008): 3-6.

Weemans 2012
Michel Weemans, '"The Earthly Paradise": Herri met de Bles's Visual Exegesis of Genesis 1-3', *The Authority of the Word. Reflecting on Image and Text in Northern Europe, 1400-1700*, Celeste Brusati, Karl A. E. Enenkel & Walter S. Melion (eds.), Leiden 2012: 263-312.

Weemans 2013
Michel Weemans, *Herri met de Bles. Les ruses du paysage au temps de Bruegel et d'Érasme*, Paris 2013.

Wegemer 1990
Gerard Wegemer, '*The City of God* in Thomas More's *Utopia*', *Renascense* 44 (1992): 115-135.

Wehle & Salinger 1947
Harry B. Wehle & Margaretta Salinger, *The Metropolitan Museum of Art. A Catalogue of Early Flemish, Dutch and German Paintings*, New York 1947.

Wells 1981
Kathleen Wells, 'The Iconography of Thomas More', *Studies: An Irish Quarterly Review* 70 (1981): 55-71.

Wells 1991
William Wells, 'An 'Astronomy' Tapestry reinterpreted: François I and Giovanni da Verrazzano', *Apollo* 133 (1991): 33-37.

Weiss 1888
Edmund Weiss, 'Albrecht Dürers Geographische, Astronomische und Astrologische Tafeln', *Jahrbuch der kunsthistorischen Sammlungen des allerhöchsten Kaiserhauses*, vol. 7, Vienna 1888: 209-213.

Werfel 2001
Silvia Werfel, 'Die "Schedelsche Weltchronik" von 1493', *Journal für Druckgeschichte* 7, 4 (2001): 3.

Wetherbee 1973
The Cosmographia of Bernardus Silvestris, Winthrop Wetherbee (ed.), New York 1973.

Weyer 1859
Johann Peter Weyer, *Beschreibung des Inhalts der Sammlung von Gemälden älterer Meister des Herrn Johann Peter Weyer*, Cologne 1859.

White 1978
Lynn J. R. White, *Medieval Religion and Technology. Collected Essays*, Berkleley/Los Angeles/London 1978.

White 1986
Hayden White, *Tropics of Discourse: Essays in Cultural Criticism*, London/Baltimore 1986.

Whitfield 2010
Peter Whitfield, *The Image of the World. 20 Centuries of World Maps*, London 2010.

Wied 2003
Alexander Wied, 'Herri met de Bles. "Das irdische Paradies"', *Die Flämische Landschaft 1520-1700*, Alexander Wied, Klaus Ertz & Karl Schütz (eds.), exh. cat. (Essen, Villa Hügel/Vienna, Kunsthistorisches Museum) 2003: 286-287.

Wieser & Giorgi 2000
Marie Therese Wieser & Barbara Giorgi, *Die handschriftliche Überlieferung der Werke des heiligen Augustinus - 8: Belgien, Luxemburg und Niederlände*, 2 vols., Vienna 2000.

Wijsman 2006
Hanno Wijsman, *Handschriften voor het hertogdom: De mooiste verluchte manuscripten van Brabantse hertogen, edellieden, kloosterlingen en stedelingen*, Alphen aan de Maas 2006.

Wijsman 2010a
Hanno Wijsman, *Luxury Bound. Illustrated Manuscript Production and Noble and Princely Book Ownership in the Burgundian Netherlands (1400-1550)*, Turnhout 2010.

Wijsman 2010b
Hanno Wijsman (ed.), *Books in Transition at the Time of Philip the Fair. Manuscripts and Printed Books in the Late Fifteenth and Early Sixteenth Century Low Countries*, Turnhout 2010.

Wilhelm-Hack-Museum 1979
Katalog der Kunst des Mittelalters im Wilhelm-Hack-Museum, Ludwigshafen am Rhein, mus. cat. (Ludwigshafen am Rhein, Wilhelm-Hack-Museum), Neustadt/Landau 1979.

Wilkins 1999
Nigel E. Wilkins, *La musique au diable*, Sprimont 1999.

Williamsen 1994
Amy R. Williamsen, *Co(s)mic chaos: Exploring "Los trabajos de Persiles y Sigismunda"*, vol. 14, Newark 1994.

Wilson 1976
Adrian Wilson, *The Making of the Nuremberg Chronicle*, Amsterdam 1976.

Winkler 1915
Friedrich Winkler, 'Studien zur Geschichte der niederländischen Miniaturmalerei des XV. und XVI. Jahrhunderts', *Jahrbuch der Kunsthistorischen Sammlungen des allerhöchsten Kaiserhauses* 32, 3 (1915): 279-342.

Winkler 1924a
Friedrich Winkler, *Die altniederländische Malerei. Die Malerei in Belgien und Holland von 1400–1600*, Berlin 1924.

Winkler 1924b
Friedrich Winkler, 'Konrad Meits Tätigkeit in Deutschland', *Jahrbuch der Preußischen Kunstsammlungen* 45 (1924): 43-61.

Winkler 1925
Friedrich Winkler, *Die flämische Buchmalerei des XV. und XVI. Jahrhunderts: Künstler und Werke von den Brüdern van Eyck bis zu Simon Bening*, Leipzig 1925.

Winkler 1936-1939
Friedrich Winkler, *Die Zeichnungen Albrecht Dürers*, 4 vols., Berlin 1936-1939.

Winkler 1964a
Friedrich Winkler, 'Die deutschen Werke Nicolaus Hogenbergs von München', *Zeitschrift des deutschen Vereins für Kunstwissenschaft* 18 (1964).

Winkler 1964b
Friedrich Winkler, *Das Werk des Hugo van der Goes*, Berlin 1964.

Wittkower 1942
Rudolf Wittkower, 'Marvels of the East: A Study in the History of Monsters', *Journal of the Courtauld and Warburg Institutes* 5 (1942): 159-197.

Winkler 1957
Friedrich Winkler, *Albrecht Dürer – Leben und Werk*, Berlin 1957.

Winter 1952
Heinrich Winter, 'Notes on the World Map in "Rudimentum Novitiorum"', *Imago Mundi* 9 (1952): 102.

Wittkower 1977
Rudolf Wittkower, *Allegory and the Migration of Symbols*, London 1977.

Wolfschmidt 1995
Gudrun Wolfschmidt, 'Planeten, Kometen, Finsternisse. Peter Apian als Astronom und Instrumentenbauer', *Peter Apian: Astronomic, Kosmographie und Mathematik am Beginn der Neuzeit*, Karl Röttel (ed.), Buxheim 1995: 93-106.

Wolfson 2001
Michael Wolfson, *Ein Rundgang durch Kloster Ebstorf*, Königstein im Taunus 2001.

Wolfthal 2010
Diane Wolfthal, *In and Out of the Marital Bed: Seeing Sex in Renaissance Europe*, New Haven/London 2010.

Woltmann 1866
Alfred Woltmann, 'Holbein und Quentin Massys in Longford Castle', *Zeitschrift für Bildende Kunst* (1866): 198-205.

Woltmann 1874
Alfred Woltmann, *Hans Holbein und seine Zeit*, 2nd rev. ed., vol. 1, Leipzig 1874.

Woodall 2014
Joanna Woodall, '*De wisselaar*, Quentin Matsys's Man weighing Gold Coins and his Wife, 1514', *Netherlands Yearbook for History of Art* 64 (2014): 38-75.

Woods, Richardson & Lymberopoulou 2007
Kim W. Woods, Carol M. Richardson & Angeliki Lymberopoulou (eds.), *Viewing Renaissance Art. Renaissance Art Reconsidered*, vol. 3, New Haven/London 2007.

Woodward 1987
David Woodward (ed.), *Art and Cartography: Six Historical Essays*, Chicago/London 1987.

Woodward 1991
David Woodward, 'Maps and the Rationalization of Geographic Space', *Circa 1492: Art in the Age of Exploration*, Jay A. Levenson (ed.), exh. cat. (Washington, National Gallery of Art), New Haven/London 1991: 83-87.

Woodward 2007
David Woodward (ed.), 'Appendix 30.1: Pre-1530 manuscript maps showing the relationship between the Old and the New Worlds', *The History of Cartography - Volume 3: Cartography in the European Renaissance*, vol. 1, Chicago/London 2007: 759-795.

Wornum 1867
Ralph Nicholson Wornum, *Some Account of the Life and Works of Hans Holbein, Painter, of Augsburg*, London 1867.

Wright 1972
Cyril Ernest Wright, *Fontes Harleiani: A Study of the Sources of the Harleian Collection of Manuscripts in the British Museum*, London 1972.

Wright 2011
Christopher Wright, *A Collector's Eye. Cranach to Pissarro*, exh. cat. (Liverpool, Walker Art Gallery), Liverpool 2011.

Wright 2014
Christopher Wright (ed.), *The Schorr Collection. Catalogue of Old Master and Nineteenth-Century Paintings*, 2 vols., London 2014.

Yasui 2016
Ayumi Yasui, 'Where has Paris Gone? Sebald Beham's " Fountain of Youth" Reconsidered', *Sacred and Profane in Early Modern Art*, Kayo Hirakawa (ed.) (*Kyoto Studies in Art History*, vol. 1, 2016), Kyoto 2016: 19-36.

Zahlten 1995
Johannes Zahlten, 'Eine nackte Prinzessin reitet Moses', *Flanders in a European Perspective. Manuscript Illumination Around 1400 in Flanders and Abroad* conference proceedings, Leuven, 7-10 September 1993), Maurits Smeyers & Bert Cardon (eds.), (Corpus of Illuminated Manuscripts, 8, Leuven 1995: 659-672.

Zahn 1996
Peter Zahn, 'Hartmann Schedels Weltchronik. Bilanz der jüngeren Forschung', *Bibliotheksforum Bayern II* 24 (1996): 230-248.

Zanetti 1771
Antonio Maria Zanetti, *Della pittura Veneziana e delle opere pubbliche de Veneziani maestri libri V*, Venice 1771.

Zehnpfennig 1979
Marianne Zehnpfennig, *"Traum" und "Vision" in Darstellungen des 16. und 17. Jahrhunderts*, PhD dissertation, Eberhard-Karls-Universität, Tübingen 1979.

Zimerman 1885
Heinrich Zimerman, 'Urkunden und Regesten aus dem K.u.K. Haus-, Hof- und Staats-Archiv in Wien', *Jahrbuch der Kunsthistorischen Sammlungen des Allerhöchsten Kaiserhauses* 3 (1885): CXIX-CXX.

Zink 1968
Fritz Zink, *Die Handzeichnungen bis zur Mitte des 16. Jahrhunderts. Die deutschen Handzeichnungen*, vol. 1, mus. cat. (Nuremberg, Germanischen Nationalmuseums), Nuremberg 1968.

Zinner 1956
Ernst Zinner, *Deutsche und Niederländische astronomische Instrumente des 11.-18. Jahrhunderts*, Munich 1956.

Zinner 1964
Ernst Zinner, *Geschichte und Bibliographie der astronomischen Literatur in Deutschland zur Zeit der Renaissance - 2. unveränderte Auflage der Erstauflage von 1941 mit einem Nachtrag von 622 Nummern*, Stuttgart 1964.

Index

Lenders to the Exhibition

AUSTRIA

Kunsthistorisches Museum Wien, Kunstkammer, Vienna

BELGIUM

Koninklijk Museum voor Schone Kunsten, Antwerp
M – Museum Leuven
MAS | Museum aan de Stroom, Antwerp
Musea en Erfgoed, Mechelen – Collection Gasthuiszusters
Museum Plantin-Moretus/Prentenkabinet, Antwerp – UNESCO Werelderfgoed
Museum voor Schone Kunsten, Gent
Royal Library of Belgium, Brussels
Royal Museums of Fine Arts of Belgium, Brussels
Saint Bavo's Chapter, Ghent
University Library, KU Leuven, Special Collections
University Library, Ghent University

CZECH REPUBLIC

The Moravian Gallery, Brno

FRANCE

Bibliothèque nationale de France, Paris
De Jonckheere Gallery, Paris
Musée Unterlinden, Colmar
Musée de la Chartreuse, Douai
Musée départemental des Antiquités, Rouen
Musée nationale de la Renaissance, Ecouen

GERMANY

Bayerisches Nationalmuseum, Munich
Landesmuseum für Kunst und Kulturgeschichte, Oldenburg
Museum Lüneburg
Staatliche Kunstsammlungen Dresden, Kupferstichkabinett
Staatliche Museen zu Berlin, Kupferstichkabinett
Städel Museum, Frankfurt am Main
Wilhelm-Hack-Museum, Ludwigshafen

ITALY

Galleria dell'Accademia di Venezia, Venice
Gallerie Nazionali di Arte Antica di Roma, Rome
Museo Poldi Pezzoli, Milan

NETHERLANDS

Koninklijke Bibliotheek, National Library of the Netherlands, The Hague
Museum Boijmans Van Beuningen, Rotterdam
Museum Meermanno – House of the Book, The Hague
Rijksmuseum, Amsterdam

PORTUGAL

Caixa Geral de Depósitos | em depósito no Museu Nacional de Arte Antiga, Lisbon

SPAIN

Cabildo Catedral Primada, Toledo
Fundación Lazaro Galdiano, Madrid
Museo Arqueológico Nacional, Madrid
Museo de América, Madrid
Patrimonio Nacional, Madrid (San Lorenzo de El Escorial)

SWEDEN

The Röhsska Museum, Göteborg

UNITED KINGDOM

National Portrait Gallery, London
Royal Collection Trust / Her Majesty Queen Elizabeth II, London
The Ashmolean Museum, Oxford
The British Library, London
The National Gallery, London
The Schorr Collection, London
The Weiss Gallery, London

UNITED STATES

Adler Planetarium & Astronomy Museum, Chicago, Illinois
The J. Paul Getty Museum, Los Angeles
The Metropolitan Museum of Art, New York
University of Arizona, Museum of Art, Tucson

PRIVATE COLLECTIONS IN BELGIUM, FRANCE AND THE NETHERLANDS

Photo Credits

Unless otherwise specified, illustrations have been provided by the owners of the works of art or by the lending institutions. Every effort has been made to contact copyright-holders. Any copyright-holders we have been unable to reach or who have been inaccurately acknowledged are invited to contact the publisher.

Adler Planetarium, Chicago: cat. 65
Albertina, Vienna: fig. 3, fig. 30.1
Ashmolean Museum, Oxford: cat. 21
Bayerische Staatsbibliothek, Munich: fig. 56.1
Bibliothèque nationale de France: fig. 31, cat. 53
Bpk – Klassik Stiftung Weimar: fig. 25.1, fig. 29.1
Bpk – Staatliche Kunstsammlungen Dresden – Herbert Boswank: cat. 10
Bpk – Kupferstichkabinett - Staatliche Museen zu Berlin – Dietmar Katz: fig. 15, cat. 17, fig. 17.1, fig. 56.2, cat. 58, cat. 60B
British Museum Images: fig. 47.1, fig. 56.2
Bruno Vandermeulen – KU Leuven: fig. 2, fig. 13, cat. 7.1, cat. 7.2, cat. 8, cat. 12
Cabildo Primado, Toledo: fig. 42, cat. 69
Caixa Geral de Dépositos: cat. 46
Chicago Art Institute: fig. 54.1
Collection Galerie De Jonckheere, Paris: cat. 56
Collection of the University of Arizona Museum of Art, Tucson; Gift of Samuel H. Kress Foundation: cat. 70
Denver Art Museum Collection: Funds from Helen Dill bequest: fig. 36.1
Erich Lessing – lessingimages.com: fig. 19, fig. 40, fig. 28.1
Fundación Lázaro Galdiano, Madrid: cat. 36
Galleria Nazionale d'Arte Antica, Palazzo Barberini, Rome – Alinari Archives, Florence: cat. 14A
Germanisches Nationalmuseum: fig. 60A.1, fig. 60B.1
Johan Geleyns, Roscan: fig. 1
KBR, Brussels: fig. 8, fig. 14, fig. 38, fig. 43, cat. 1, cat. 2, cat. 27, fig. 39.1, cat. 42, cat. 44, fig. 59.1, cat. 60A, cat. 61, cat. 73, cat. 76
KHM, Vienna: cat. 25, cat. 31
KIK-IRPA, Brussels: fig. 29, fig. 30, cat. 32, cat. 33, cat. 34
National Library of the Netherlands, The Hague: fig. 34, fig. 37, cat. 52, fig. 58.1
Royal Museums of Fine Arts of Belgium, Brussels – Johan Geleyns, Roscan: cat. 40
Kunstmuseum Basel, Kupferstichkabinett – Martin P. Bühler: fig. 15, fig. 5.3, fig. 10.1
Longford Castle Collection: cat. 13B
Lukas - Art in Flanders vzw: fig. 5, fig. 25, fig. 27, cat. 14B, cat. 24, cat. 26, cat. 49, cat. 74
Lukas – Art in Flanders vzw – Hugo Maertens: cat. 19
Michael Lammgard: cat. 77
Moravská Galerie, Brno: cat. 71
Musée Unterlinden, Dist. RMN-Grand Palais: cat. 20
Musée départemental des Antiquités, Rouen: cat. 54
Musée du Quai Branly, Dist. RMN-Grand Palais – Michel Urtado – Thierry Ollivier: fig. 35
Museo Arqueológico Nacional, Madrid – Raul Fernandez Ruiz: cat. 72
Museo Nacional del Prado: fig. 20-24
Museo Poldi Pezzoli: cat. 68
Museo de América, Madrid – Ministry of Education, Culture and Sport: cat. 48
Museum of Fine Arts, Budapest: fig. 12
Museum Lüneburg: cat. 63
Museum Meermanno, House of the Book: cat. 35
Museum Plantin Moretus, Antwerp: cat. 9, cat. 45
Museu Nacional de Arte Antiga: fig. 18.1
National Trust Images: fig. 5.1
Patrimonio Nacional: cat. 22
Philadelphia Museum of Art, John G. Johnson Collection: fig. 11
Photo Archive – Fondazione Musei Civici di Venezia: cat. 39
Rijksmuseum, Amsterdam: fig. 18.2, cat. 29, fig. 29.2, cat. 30, fig. 57.1
RMN-Grand Palais (musée de la Renaissance, château d'Ecouen) – René-Gabriel Ojéda: cat. 62
RMN-Grand Palais (musée du Louvre) – Michèle Bellot: fig. 17
RMN-Grand Palais (musée du Louvre) – Gérard Blot: fig. 16.1
Rotterdam, Museum Boijmans Van Beuningen: cat. 38, cat. 59
Royal Collection Trust – Her Majesty Queen Elizabeth II 2016: fig. 9, fig. 10, fig. 3.1, fig. 6.1, cat. 13A
The Frick Collection, New York: fig. 4
The J. Paul Getty Museum, Los Angeles: cat. 37
University Library Ghent: fig. 28, cat. 6
Wilhelm-Hack-Museum, Ludwigshafen: cat. 28

Colophon

THIS PUBLICATION ACCOMPANIES THE EXHIBITION

IN SEARCH OF UTOPIA
M – MUSEUM LEUVEN
20 October 2016 – 17 January 2017

IN SEARCH OF UTOPIA is part of the citywide festival 500 YEARS OF UTOPIA (Leuven, 2016–2017)

500 YEARS OF UTOPIA is an initiative of KU[N]ST Leuven, a partnership between the city of Leuven and the KU Leuven.

Citywide Festival 500 Years of Utopia

Lien De Keukelaere (KU[N]ST Leuven), general coordination
Hanna van Zutphen (KU[N]ST Leuven), head of communications
Jessica Peetermans (KU[N]ST Leuven), communications
Silke Vandaele (KU[N]ST Leuven), communications

Exhibition In Search of Utopia

CURATORIAL COMMITTEE

Jan Van der Stock (Illuminare – KU Leuven) curator
Annelies Vogels (Illuminare – KU Leuven) co-curator

SCIENTIFIC AND ADVISORY COMMITTEE (2009-2016)

Barbara Baert, Maarten Bassens, Julie Beckers, Floortje Clerix, Hannah De Moor, Rozewinde Gerits, Hannah Iterbeke, Robrecht Janssen, Jeroen Luyckx, Diana Stoilova, Jochen Suy, Daan van Heesch, Lieve Watteeuw, Wendy Wauters (Illuminare – KU Leuven)

Lorne Campbell (The National Gallery, London), Hans Cools (KU Leuven – Fryske Akademy, Leeuwarden), Marcus de Schepper (Royal Library of Belgium), Tim Heysse (KU Leuven), Jos Koldeweij (Radboud Universiteit Nijmegen), Jan Papy (KU Leuven), Jan Roegiers (KU Leuven), Dirk Sacré (KU Leuven), Koenraad Van Cleempoel (Universiteit Hasselt), Veronique Vandekerchove (M–Museum Leuven), Paul Vandenbroeck (KU Leuven/Koninklijk Museum voor Schone Kunsten Antwerp), Joris Van Grieken (Royal Library of Belgium), Geert Vanpaemel (KU Leuven)

Organisation

ORGANIZING COMMITTEE

Peter Carpreau, Lien De Keukelaere, Aldwin Dekkers, Luc Delrue, Isabel Lowyck, Jan Van der Stock, Hanna van Zutphen, Hélène Verreyke, Annelies Vogels, Eva Wittocx

PRODUCTION

Bram Denkens, Goedele Pulinx, Tom Van Camp, Jan Van der Stock, Hélène Verreyke, Annelies Vogels, Lieve Watteeuw

COMMUNICATIONS

Veerle Ausloos, Lien De Keukelaere, Annelies Evens, Lucie Van Gerven, Hanna van Zutphen, Annelies Vogels

PUBLIC SERVICES

Lien De Keukelaere, Marthy Locht, Jasper Stevens, Hanna van Zutphen, Sofie Vermeiren, Hélène Verreyke, Annelies Vogels

SCENOGRAPHY

FELT architecture & design
Jasper Stevens, Karel Verstraeten, Bram Denkens, Bartolomeo Zanotti

The teams of KU[N]ST Leuven, M – Museum Leuven, Illuminare (KU Leuven) & Tourism Leuven

Catalogue

COORDINATION

Annelies Vogels
Katrien De Vreese
Peter Frison

EDITING

Hans Devisscher
Wendy Wauters

DESIGN

Jirka De Preter
Els Ruppol (typesetting)

PHOTOGRAVURE AND ICONOGRAPHY

Johan Geleyns, Roscan

TRANSLATION

Wouter Meeus
Lee Preedy
Irene Schaudies
Patrick Lennon

BOARD AND GENERAL ASSEMBLY OF KU[N]ST LEUVEN

Denise Vandevoort (co-chair), Katlijn Malfliet (co-chair), Geert Bouckaert, Els Buelens, Bert Demarsin, Piet Forger, Kristien Jacobs, Sabine Vandenplas, Mieke Vandermotte, Jan Van der Stock, Geert Vanpaemel, Dirk Vansina, Mieke Wellens

PUBLISHERS

Van der Stock, Jan (ed.)
In Search of Utopia,
Art and Science in the Era of Thomas More

Produced under the license of Davidsfonds.
'Davidsfonds' is a registered trademark of Davidsfonds vzw with headquarters at 3000 Leuven, Quinten Metsysplein 12.
www.davidsfonds.be
www.twitter.com/davidsfonds
www.facebook.com/davidsfondsuitgeverij
D/2016/0034/208
ISBN 978 90 5908 811 5
NUR 680

www.aup.nl
ISBN 978 94 6298 407 3

THE EXHIBITION IN SEARCH OF UTOPIA IS AN INITIATIVE OF:

THE EXHIBITION IN SEARCH OF UTOPIA WAS MADE POSSIBLE WITH THE SUPPORT OF:

THE EXHIBITION IN SEARCH OF UTOPIA IS SUPPORTED BY VISIT FLANDERS

In Memory of Melvin R. Seiden

With the support of the Van der Weyden Chair – Paul & Dora Janssen (KU Leuven)
With the support of the Veronique Vandekerchove Chair of the City of Leuven (KU Leuven)
With the support of the Chair for Medieval Sculpture in the Netherlands (Illuminare – KU Leuven)
With the support of the Illuminare Fund (KU Leuven)

THE ORGANIZERS ARE GRATEFUL FOR THE ENTHUSIASM AND SUPPORT OF:

Jos Daniëls, Bart Fransen, André Hebbelinck, Leen Huet, Nadine Malfait, Hans Nieuwdorp and Griet Steyaert